כתבי האקדמיה הלאומית הישראלית למדעים

PUBLICATIONS OF THE ISRAEL ACADEMY OF SCIENCES AND HUMANITIES

SECTION OF SCIENCES

—

FLORA PALAESTINA

EQUISETACEAE TO UMBELLIFERAE

by

MICHAEL ZOHARY

ERICACEAE TO ORCHIDACEAE

by

NAOMI FEINBRUN-DOTHAN

FLORA PALAESTINA

PART FOUR • TEXT

ALISMATACEAE TO ORCHIDACEAE

BY

NAOMI FEINBRUN-DOTHAN

JERUSALEM 1986

THE ISRAEL ACADEMY OF SCIENCES AND HUMANITIES

ISBN 965-208-000-4
ISBN 965-208-004-7

Printed in Israel
Type set at Monoline Press, Benei Beraq

ACKNOWLEDGEMENTS

The four parts of *Flora Palaestina* treat some 2470 species known at present to occur within the boundaries of British Mandatory Palestine, now the States of Israel and Jordan. An additional volume now in preparation will contain Addenda and Corrigenda to Parts One and Two.

On approaching the completion of Part Four, I can literally repeat all that was said in 1978 before the appearance of Part Three. Again I sincerely thank everyone who co-operated with me then and continued in the same spirit up to now. In particular, I feel the need of expressing my gratitude and great appreciation to Dr Irene Gruenberg-Fertig, Mr A. Grizi and Mrs Stephanie Grizi who accompanied me in the day-to-day work and enabled me to bring the *Flora* to its completion. The expertise of Dr Irene Gruenberg-Fertig in checking literature and unravelling complicated cases of nomenclature was invaluable. Prof. Chaia C. Heyn helped as before in solving problems of nomenclature. I am grateful to Dr Fania Kollmann, the expert on the genus *Allium* of the Middle East, who prepared the account on that genus for *Flora Palaestina*.

To the list of young botanists who continued to co-operate with me throughout the years and were always helpful, I wish to add the name of a young colleague, Mr A. Liston.

Dr Auguste Horovitz edited the English text of Part Four and was always helpful in answering questions concerning cultivated plants. Her co-operation is much appreciated and gratefully acknowledged.

My thanks and appreciation are expressed to Mr R. Amoils who copy-edited the text and corrected the proofs expertly and intelligently. The co-operation of the Publications Department of the Israel Academy of Sciences and Humanities, especially that of Mr S. Reem, is acknowledged thankfully.

N.F.-D.

We mourn the sudden death of Mr A. Grizi on June 3, 1985.

CONTENTS

GLOSSARY

See also *Glossary* in Parts One and Three

Arillate Provided with an aril.

Aristula A short weak awn.

Bulbils Small bulbs formed in the inflorescence.

Bulblets Small increase bulbs formed by mother bulb.

Callus An extension of the lemma below its point of insertion, fused with the rachilla of the spikelet.

Cataphyll Usually membranous leaf lacking chlorophyll, borne on shoot below the green leaves.

Claviform Club-shaped.

Concolorous Of the same colour.

Cucullate Hooded or hood-shaped.

Cupuliform Cup- or cupule-shaped.

Effuse Loosely spreading, expanded.

Equitant Leaves in two ranks, folded together lengthwise and overlapping, as in *Iris*.

Erecto-patent Spreading from an axis at an angle of about 45°.

Evanescent Lasting only a short time, vanishing.

Excurrent (vein) Extending beyond tip or margin, as mucro or awn.

Fastigiate (umbel) With parallel and erect pedicels.

Fistulous Hollow.

Fruitlet A single fruiting carpel of an apocarpous fruit.

Fugacious Transitory, ephemeral.

Hypanthial tube Tube formed by the receptacle of the flower below the perianth.

Hypocrateriform Salver-shaped, with a narrow tube abruptly expanded into a spreading limb.

Hysteranthous (leaves) Appearing after the flowers.

Intorto-pedatisect (leaf) Palmately lobed or divided, with side-lobes in turn lobed and twisted.

Intortus Twisted or bent upon itself.

Lingulate Tongue-shaped.

Lorate Strap-shaped.

Medifix Attached at the middle.

Napiform Turnip-shaped.

Nari Indurated pedogenic calcareous crust (calcrete), up to a few metres thick, formed by infilling and alteration of porous chalk and marl (D.H. Yaalon & S. Singer, *Jour. Sedimentary Petrology* 44:1016–1023, 1974).

Nectar guide Lines, spots, etc., directing to the nectary.

Panduriform Fiddle-shaped.

Patent Spreading, diverging from the axis at almost 90°.

Penicillate Shaped like a camel-hair brush.

Scarious (leaves) Of thin membranous texture and not green.

Synanthous Appearing with the flowers.

Trichotomous Having divisions or branches in threes.

Tunicate With concentric layers or coats.

Turion A generally short and fleshy shoot produced in aquatic plants on branches or rootstocks.

SPERMATOPHYTA

ANGIOSPERMAE

MONOCOTYLEDONEAE

SYNOPSIS OF FAMILIES IN PART FOUR

Helobiae (Alismatales)

110 Alismataceae
111 Butomaceae
112 Hydrocharitaceae
113 Potamogetonaceae
114 Ruppiaceae
115 Zannichelliaceae
116 Cymodoceaceae
117 Najadaceae

Liliiflorae (Liliales)

118 Liliaceae
119 Amaryllidaceae
120 Dioscoreaceae
121 Iridaceae

Juncales

122 Juncaceae

Graminales (Poales)

123 Gramineae (*Poaceae*)

Principes (Arecales)

124 Palmae

Spathiflorae (Arales)

125 Araceae
126 Lemnaceae

Pandanales

127 Sparganiaceae
128 Typhaceae

Cyperales

129 Cyperaceae

Microspermae (Orchidales)

130 Orchidaceae

MONOCOTYLEDONEAE

Stems with closed vascular bundles; cambium absent. Leaves mostly with parallel venation. Flowers predominantly 3-merous. Seeds with 1 cotyledon.

110. ALISMATACEAE

Aquatic perennials, rarely annuals. Stems scapose. Leaves basal or alternate, simple, entire, petiolate, sheathing at base, emergent, floating or submerged, with small scales in axils. Inflorescence umbel-, raceme- or panicle-like, with branches bearing cymes in whorls and with 2–3 bracts at base of each whorl. Flowers hermaphrodite or unisexual, actinomorphic, 3-merous. Perianth in 2 whorls; sepals 3, free; petals 3, free, caducous. Stamens hypogynous, 3, 6 or more, free; anthers with 2 thecae. Gynoecium apocarpous; carpels 3 to numerous, in a whorl or spiral, free or connate at base, 1- to several-ovulate; style apical or subventral. Fruit usually a whorl of fruitlets (1-seeded nutlets or druplets, or 2- to several-seeded follicles). Seeds devoid of endosperm; embryo horseshoe-shaped.

Thirteen genera and about 90 species in all parts of the globe.

Literature: F. Buchenau, Alismataceae, in: Engler, *Pflanzenreich* 16 (IV. 15): 1–66, 1903. J.E. Dandy, Alismataceae, in: K.H. Rechinger (ed.), *Flora iranica* 78:1–5, 1971. P.E. Boissier, *Flora orientalis* 5:9–11, 1882. G.E. Post, *Flora of Syria, Palestine and Sinai* ed. 2, 2:537–538, 1933.

1. Carpels numerous, free; style lateral. Fruitlets about 1–3 mm, short-beaked.
   1. Alisma
 – Carpels 6–10, connate at base; style apical. Fruitlets 6–13 mm, ending in a long beak and spreading stellately. **2. Damasonium**

1. ALISMA L.

Glabrous aquatic perennials. Leaves emergent or submerged, sometimes floating. Inflorescence emergent, pyramidal, paniculate, of cymes arranged in whorls at different levels on main axis. Flowers hermaphrodite. Sepals herbaceous, persistent. Petals larger, deciduous. Stamens 6. Carpels numerous, in 1 whorl, free, 1-ovulate; styles lateral. Fruit of a whorl of numerous nutlets; nutlets coriaceous, compressed, with a short lateral beak.

Nine species, N. temperate and Australian.

Literature: G. Samuelsson, Die Arten der Gattung *Alisma* L., *Ark. Bot.* 24A, no. 7:1–46 (1932). I. Björkquist, Studies in *Alisma* L., *Op. Bot. (Lund)* 17:1–128 (1967); 19:1–138 (1968).

1. Leaf-blades subcordate or rounded at base. Stigmas minutely papillose.
   1. A. plantago-aquatica
 – Leaf-blades cuneate at base. Stigmas coarsely papillose. **2. A. lanceolatum**

1. Alisma plantago-aquatica L., Sp. Pl. 342 (1753); Post, Fl. 2:538; Boiss., Fl. 5:9 (*"A. plantago"*) excl. syn. & var. ß. [Plate 1]

Perennial, 40–80 cm. Leaves long-petiolate, mostly emergent; blades 6–14 × 3–8 cm, ovate or elliptic-ovate, usually subcordate or rounded at base. Petals white or purplish-white. Styles equalling ovaries or longer, filiform, stigmatose on ⅛–⅕ of their length. Fl. March–September.

Hab.: In water; marshes. Rare. Gilead (Post); Hula Plain.

Area: Borealo-Trop.

2. Alisma lanceolatum With., Arr. Brit. Pl. ed. 3, 2:362 (1796); Boiss., Fl. 5:9 *pro syn. A. plantago-aquatica* L. var. *lanceolatum* (With.) Koch, Syn. Fl. Germ. 669 (1837); Post, Fl. 2:538 as var. *lanceolata* (With.) Aarons. [Plate 2]

Perennial, 40–80 cm. Leaves long-petiolate, emergent; blades up to 25(–60) × 5 cm, lanceolate to elliptic, cuneate at base, acuminate. Petals mostly purplish-pink. Styles equalling ovaries or longer, filiform, stigmatose on ½–⅔ of their length. Fl. March–September.

Hab.: In water; marshes, water-courses. Acco Plain, Coast of Carmel, Sharon Plain, Philistean Plain; Esdraelon Plain, Samaria, Shefela; Hula Plain, Upper Jordan Valley; Golan, Gilead.

Area: Euro-Siberian, Mediterranean and Irano-Turanian.

2. DAMASONIUM Mill.

Glabrous aquatic annuals or perennials. Leaves in a basal rosette, partly submerged, partly emergent or floating. Inflorescence panicle- or umbel-like with whorled branches of cymes. Flowers hermaphrodite. Sepals herbaceous, persistent. Petals larger, deciduous. Stamens 6. Carpels 6–10, in 1 whorl, more or less connate at base, 2- to several-ovulate; style apical. Fruit stellate, of a whorl of follicles; follicles more or less connate at base, laterally compressed, tapering to a straight apical beak, indehiscent or tardily dehiscent at base. Seeds (1–)2 to several.

Five to six species in the Mediterranean and Irano-Turanian regions in Europe, Australia and W. N. America.

1. Damasonium alisma Mill., Gard. Dict. ed. 8, no. 1 (1768). *Alisma damasonium* L., Sp. Pl. 343 (1753). *D. bourgaei* Coss., Not. Pl. Crit. 47 (1849); Boiss., Fl. 5:10; Post, Fl. 2:538. *D. polyspermum* Coss., *loc. cit.* (1849). [Plate 3]

Annuals or perennials, 10–40(–60) cm. Leaves long-petiolate; blades 3–6 cm, oblong to ovate-oblong, cuneate, subcordate to truncate at base, obtuse or rounded at apex, 3–5-veined. Scapes thickish, somewhat longer than leaves; whorls few, 5–7-flowered, more or less remote. Petals white, yellow at base. Follicles 6–13 mm, veined, 1–12-seeded. Fl. April–June.

Hab.: Ditches, pools. Acco Plain, Coast of Carmel, Sharon Plain, Philistean Plain; Upper and Lower Galilee, Esdraelon Plain, Samaria, Judean Mts.; Hula Plain, Beit Shean Valley; Golan.

Area: Mediterranean and Irano-Turanian.

111. BUTOMACEAE

Aquatic perennials. Stems scapose. Leaves linear, erect. Inflorescence cymose, umbel-like, with an involucre of bracts. Flowers hermaphrodite, actinomorphic, 3-merous. Perianth in 2 whorls, persistent; sepals 3, petaloid; petals 3. Stamens 9, free; anthers with 2 thecae, introrse. Gynoecium apocarpous; carpels 6(-9), 1-locular, connate at base, many-ovulate; style persistent; the stigmatic area somewhat decurrent along style. Fruit a whorl of follicles dehiscing on ventral side. Seeds minute, devoid of endosperm; embryo straight.

One temperate genus.

Literature: F. Buchenau, Butomaceae, in: Engler, *Pflanzenreich* 16 (IV. 16):1-12, 1903. C. D. K. Cook et al., *Water Plants of the World*, The Hague, 1974. A. L. Takhtajan, Outline of the classification of flowering plants (Magnoliophyta), *Bot. Rev.* 46:301 (1980). P. Uotila, Butomaceae, in: P. H. Davis (ed.), *Flora of Turkey* 8:3-4, 1984. P. E. Boissier, *Flora orientalis* 5:11-12, 1882. G. E. Post, *Flora of Syria, Palestine and Sinai* ed. 2, 2:539, 1933.

1. BUTOMUS L.

Monotypic. Characters as given for family.

1. Butomus umbellatus L., Sp. Pl. 372 (1753); Boiss., Fl. 5:12; Post, Fl. 2:539. [Plate 4]

Aquatic perennial, 40-150 cm, with thick creeping rhizome. Scape terete, mostly longer than leaves. Leaves in a basal rosette, erect, emergent, linear, sheathing at base, 6-10 mm broad, 3-quetrous below, flat and acute in upper part. Inflorescence terminal, umbel-like, many-flowered; pedicels 5-10 cm; involucral bracts 3, acuminate. Flowers showy, 2-2.5 cm in diameter, insect-pollinated; sepals and petals ovate to obovate, pinkish-white, with darker pink veins; sepals somewhat smaller than petals, greenish along midvein at back. Follicles about 1 cm, beaked. Fl. April-June.

Hab.: Marshes and slowly flowing water. Sharon Plain; Dan Valley, Hula Plain; Golan, Gilead.

Area: Euro-Siberian, Mediterranean and Irano-Turanian; naturalized in N. America.

112. HYDROCHARITACEAE

Aquatic perennials, sometimes marine, submerged or floating, monoecious or dioecious. Leaves alternate, opposite or in whorls, simple, sheathing at base and often with small axillary scales within sheath (*intravaginal squamules*). Flowers mostly unisexual, rarely hermaphrodite, actinomorphic or slightly zygomorphic, 3-merous, solitary or in cymose inflorescences subtended by a spathe; spathe of 1-2 bracts, free or more or less connate. Sepals 3(-2); petals 3(-2) or rudimentary. Stamens (1-)2-15 in 1 or more whorls, the inner sometimes staminodal; anthers with 2 thecae, opening by longitudinal slits. Ovary inferior, of 2–15 connate carpels, 1-locular, often beaked; styles 2–15, simple or 2-lobed. Fruit dry or berry-like, mostly irregularly torn. Seeds several to numerous; endosperm absent; embryo straight.

Fifteen cosmopolitan genera and 80 trop. and temperate species.

Literature: T. Eckardt, Hydrocharitaceae, in: A. Engler's *Syllabus der Pflanzenfamilien* ed. 12, 2:503–505, Berlin, 1964. H.E. Hess, E. Landolt & R. Hirzel, *Flora der Schweiz* 1:220–225, Basel & Stuttgart, 1967. C. den Hartog, Hydrocharitaceae, in: The sea-grasses of the world, *Verh. Konink. Nederl. Akad. Wet. Afd. Nat.* Ser. 2, 59(1):213–268 (1970) J. F. Dandy, Hydrocharitaceae, in: K. H. Rechinger (ed.), *Flora iranica* 80:1–4, 1971. P.E. Boissier, *Flora orientalis* 5:1–8, 1882. G.E. Post, *Flora of Syria, Palestine and Sinai* ed. 2, 2:536–537, 1933.

1. Submerged marine herbs with creeping rhizome. Leaves in pairs at nodes of rhizome.   **3. Halophila**
– Submerged or floating freshwater herbs. Leaves in basal rosette   2
2. Leaves petiolate, orbicular-reniform. Petals about 1 cm, much larger than sepals. Floating herbs (rarely rooting).   **1. Hydrocharis**
– Leaves sessile, strap-shaped. Petals minute or absent. Submerged rooting herbs.
                                                                          2. Vallisneria

1. HYDROCHARIS L.

Freshwater perennials, mostly floating, stoloniferous, with shoots at nodes, dioecious or rarely monoecious. Large buds borne on thin stolons drop to bottom in autumn and sprout in following year. Leaves in a rosette, petiolate, floating or emergent. Flowers unisexual, entomophilous, pedicellate, subtended by 1–2-valved spathe; staminate flowers 1–4 in a bud subtended by a long-pedunculate spathe; pistillate flowers single, subtended by a sessile spathe and borne on a stout pedicel. Sepals 3; petals 3, much larger than sepals. Staminate flower with 9–12 stamens in 3–4 whorls. Pistillate flower with 3–6 staminodes and pistil; styles 6, 2-lobed. Fruit berry-like, ellipsoid to globose, irregularly bursting at apex.

Six species in trop. and temperate regions of the Old World and Australia.

1. Hydrocharis morsus-ranae L., Sp. Pl. 1036 (1753); Boiss., Fl. 5:5. [Plate 5]
Plant free-floating or rooting; roots in clusters at nodes of stolons. Leaves orbicular-reniform, cordate at base, entire, long-petiolate; stipules large, scarious. Flowers about 2 cm in diameter. Petals broadly obovate, white, yellow-spotted at base. Fl. summer.
Hab.: Collected in the Hula Lake. Disappeared since the drying of the Lake.
Area: Euro-Siberian and Mediterranean.

2. VALLISNERIA L.

Submerged freshwater perennials, dioecious, stoloniferous. Leaves strap-shaped, finely to grossly dentate at margin, obtuse, many-veined. Flowers small, unisexual, pollinated on water surface. Staminate flower-buds very numerous, enclosed in a spathe borne on a long peduncle (up to 10 cm); after dehiscence of spathe, buds break off singly, rise to water surface, open and float freely; sepals 3, white, about 1 mm, reflexed at anthesis; stamens (1–)2–3. Pistillate flowers solitary, enclosed in a tubular spathe borne on a long peduncle, reaching water surface; perianth persistent, 2–3 mm in diameter; sepals white, connate into a tube not longer than spathe; petals minute; staminodes 3; ovary cylindrical, enclosed in spathe; stigmas 3, 2-lobed, exposed to contact with stamens of floating staminate flowers. After fertilization, peduncle coils spirally and draws fruit into water; fruit up to 10 cm, cylindrical, many-seeded.

Six to ten trop. and subtrop. species in both hemispheres.

1. Vallisneria spiralis L., Sp. Pl. 1015 (1753); Boiss., Fl. 5:3. [Plate 6]
Leaves all basal, up to 1 cm broad, 40 cm long, obtuse, finely denticulate at margin in upper half. Fl. summer.
Hab.: Lake Hula. Disappeared since the drying of the Lake.
Area: Trop. and Subtrop. of both hemispheres.

3. HALOPHILA Thouars

Submerged marine perennials, monoecious or dioecious. Rhizome creeping, rooting at nodes, with 2 scales at each node; one clasping rhizome, the other a lateral often undeveloped shoot. Leaves in pairs, sessile or petiolate, with a midvein and 2 intramarginal veins, usually connected by cross-veins. Flowers unisexual, subtended by a sessile 2-valved, usually 1-flowered spathe. Petals absent. Staminate flower pedicellate within spathe; sepals 3; stamens 3; anthers sessile; pollen-grains united into moniliform chains; pollination under water. Pistillate flower sessile or subsessile; ovary ellipsoid or ovoid, 1-locular, with a long hypanthium at apex, crowned by 3 reduced sepals; styles 3–5, filiform, free. Fruit ovoid or globose, beaked, with a membranous pericarp, opened by decay of pericarp.

Ten trop. species.

Literature: Y. Lipkin, On the male flower of *Halophila stipulacea*, *Israel Jour. Bot.* 24:198–200 (1975); Seagrass vegetation in Sinai and Israel, in: C.P. McRoy & C. Helfferich, *Seagrass Ecosystems* 263–293, New York & Basel, 1977.

1. Leaves linear to oblong; petiole much shorter than blade; scales on rhizome about 1 cm.   **2. H. stipulacea**
– Leaves elliptic to obovate; petiole longer than blade; scales on rhizome 3–4 mm.
                                                                                        1. H. ovalis

1. Halophila ovalis (R. Br.) Hook. fil., Fl. Tasman. 2:45 (1858); den Hartog, *op. cit.* 240; Aschers. in Boiss., Fl. 5:2, excl. syn. (*H. ovata* Gaudin); Post, Fl. 2:537. *Caulinia ovalis* R. Br., Prodr. Fl. Nov. Holl. 339 (1810). [Plate 7]

Dioecious perennial. Internodes of rhizome 1–5 cm. Scales suborbicular or obovate, transparent. Leaves on 1–10 mm long shoots, long-petiolate; blades usually 1–1.5 × 0.4–0.5 cm, elliptic to obovate, with ascending cross-veins joining intramarginal vein; margin smooth. Sepals of staminate flower 4 mm, those of pistillate flower minute. Ovary 1–1.5 mm, globose. Fl. & fr. June–August (Lipkin, 1977:282).

Hab.: In sea-water at temperatures above 10°C, from mid-tide level to depth of 10–12 m. Gulf of Elat.

Area: Trop. (Red Sea, Indian Ocean, S.W. Pacific).

H. ovalis is one of the least common seagrasses. It accompanies *H. stipulacea* and *Halodule uninervis* in several seagrass communities or forms monotypic communities in small areas.

2. Halophila stipulacea (Forssk.) Aschers., Sitz.-Ber. Ges. Naturf. Freunde Berlin 1867:3 (1867); den Hartog, *op. cit.* 258; Aschers. in Boiss., Fl. 5:3; Post, Fl. 2:537. *Zostera stipulacea* Forssk., Fl. Aeg.-Arab. 158 (1775). [Plate 8]

Dioecious perennial. Internodes of rhizome 1–4 cm. Scales elliptic or obovate, white or transparent, folded. Leaves on 1–15 mm long shoots, short-petiolate; blades 2.5–6 × 0.4–0.8 cm, linear to oblong, obtuse, green, 3-veined, with distinct ascending cross-veins joining intramarginal vein; margin serrulate. Staminate flowers rare (Lipkin, *op. cit.*, 1975), borne on a 1.5 cm pedicel; petals 4.5 × 1.5 mm, elliptic; stamens 3 mm. Ovary 3–4 mm, ovoid to ellipsoid. Fl. May–October.

Hab.: In shallow water on sandy or muddy bottom, down to 7 m in depth. Very common. Gulf of Elat.

Area: Trop., W. Indian Ocean, Red Sea, penetrating through the Suez Canal into some parts of the E. Mediterranean (Malta, Samos, Rhodes, Cyprus).

Literature: C. den Hartog, Range extension of *Halophila stipulacea* (Hydrocharitaceae) in the Mediterranean, *Blumea* 20:154 (1972).

According to Lipkin (1977), *H. stipulacea* is the most common seagrass in the Gulf of Elat, growing in a wide range of habitats. It forms pure communities or grows together with *Halodule uninervis*. It has penetrated into the Mediterranean through the Suez Canal, but has not been found on the coast of Israel or Lebanon.

113. POTAMOGETONACEAE

Aquatic perennials, rarely annuals, mainly of fresh water. Rhizomes present or absent. Stems elongate, mostly branched. Leaves submerged or floating, alternate or rarely opposite, 2-stichous, simple, with a membranous sheathing scale in leaf-axil; the scale is either free *(intrapetiolar stipule)* or adnate to leaf-base *(stipular sheath)* but with the apex free *(ligule)*; in young leaves stipule and stipular sheath surround stem and may be convolute or tubular. Inflorescence a pedunculate spike, emergent and wind-pollinated or submerged and water-pollinated. Flowers hermaphrodite, small, actinomorphic, 4-merous, ebracteate. Perianth of 4 segments, green, often interpreted as dorsal outgrowths of connectives. Stamens 4, free; anthers with 2 thecae. Gynoecium superior, 4(-1)-carpellate; carpels free or connate at base, 1-ovulate; styles short or stigma sessile. Fruit apocarpous; fruitlets nutlet- or druplet-like, indehiscent. Seeds devoid of endosperm; embryo hook-shaped or spiral.

Two genera and about 100 mostly temperate species in both hemispheres.

Literature: K.O.P.P. Graebner, Trib. Potamogetoneae, in Ascherson & Graebner, Potamogetonaceae, in: Engler, *Pflanzenreich* 31 (IV. 11):1-145, 1907. A.R. Clapham, T.G. Tutin & E.F. Warburg, *Flora of the British Isles* ed. 2, 942-959, Cambridge, 1962. E.W. Clason, Potamogetonaceae, in: *Flora neerlandica* 1(6):37-79, 1964. T. Eckardt, Trib. Potamogetoneae, in: Engler's *Syllabus der Pflanzenfamilien* ed. 12, 2:508-509, Berlin, 1964. H.E. Hess, E. Landolt & R. Hirzel, *Flora der Schweiz* 1:192-205, Basel & Stuttgart, 1967. J.E. Dandy, Potamogetonaceae, in: K.H. Rechinger (ed.), *Flora iranica* 83:1-9, 1971. Y. Waisel & Nili Liphschitz, *Water Plants of Israel*, Authority for Nature Reserves of Israel, Tel-Aviv, 1971 (in Hebrew). P.E. Boissier, *Flora orientalis* 5:14-19, 1882. G.E. Post, *Flora of Syria, Palestine and Sinai* ed. 2, 2:540-543, 1933.

1. POTAMOGETON L.

Mostly perennial herbs with elongate stems sprouting from rhizomes or stolons. Leaves all submerged or some floating, alternate; involucral leaves subopposite. Spikes ovoid to cylindrical, dense, lax or interrupted, either submerged and pollinated under water, or emergent and wind-pollinated. Flowers protogynous. Fruitlets drupe-like, usually short-beaked and generally with fleshy exocarp and bony endocarp; owing to air-containing spaces in pericarp, fruitlets dispersed by floating. Vegetative reproduction by subterranean cormlets, or by thickened tips of branches or of rhizomes that become detached from mother-plant as dormant winter-buds *(turions)*. The turions germinate after a resting period, usually in spring.

About 100 nearly cosmopolitan species.

1. Leaves linear or filiform, not exceeding 3 mm in width   2
- Leaves ovate, elliptic, lanceolate or oblong, at least 5 mm broad   5
2. Stipule free from leaf-base   3
- Stipule adnate to leaf-base, forming white-margined stipular sheath, its apex free as short ligule   4
3. Leaves not exceeding 1.5 mm in width (usually less than 1 mm). Carpels 1-3, mostly 1. Fruitlets about 3 mm, usually 1 per flower.   **6. P. trichoides**
- At least part of leaves 2 mm broad. Carpels 4. Fruitlets 2-2.5 mm, usually 4 per flower.   **5. P. berchtoldii**
4(2). Leaves 0.5-1 mm broad. Fruitlets 2-2.75 mm, with a short subapical beak.
                                                                        7. P. filiformis
- Leaves 1-3 mm broad. Fruitlets 3-5 mm, with a short subventral beak.
                                                                        8. P. pectinatus

5(1). At least part of leaves floating, with coriaceous opaque elliptic blades.
                                                                1. P. nodosus
– All leaves submerged; blades translucent, not coriaceous   6
6. Leaves short-petiolate, cuneate at base, acute, mucronate or cuspidate at apex; stipules 3–8 cm. Peduncles of spikes thicker than stems.   **2. P. lucens**
– Leaves sessile, not as above; stipules not exceeding 1 cm. Peduncles of spikes not thicker than stems   7
7. Leaves cordate at base, as long to twice as long as broad.   **3. P. perfoliatus**
– Leaves oblong, about 5 times as long as broad.   **4. P. crispus**

Subgen. 1. POTAMOGETON. Rhizome present or absent. Leaves submerged or floating, sessile or petiolate; intrapetiolar stipule free from leaf-base.

1. Potamogeton nodosus Poir. in Lam., Encycl. Méth. Bot. Suppl. 4:535 (1817); Dandy, Fl. Iran. 83:3. *P. fluitans* Roth, Tent. Fl. Germ. 1:72 (1788) p.p. & auctt.; Boiss., Fl. 5:16. Eig, Bull. Inst. Agr. Nat. Hist. Tel-Aviv 4:36 (1926) & Post, Fl. 2:541 as *P. natans* L. [Plate 9]

Rhizome creeping, stout. Leaves floating and submerged; floating leaves long-petiolate; blades 6–15 × 2.5–4 cm, coriaceous, opaque, elliptic, cuneate to obtuse at base; submerged leaves petiolate, translucent, lanceolate, veined, sometimes reduced to a petiole; stipules up to 8 cm, conspicuous. Peduncles long; spikes 3–6 cm. Fruitlets 3–4 mm, with a short beak on ventral side of apex. Fl. May–August.

Hab.: In slowly flowing water. Acco Plain, Sharon Plain, Philistean Plain; Upper Galilee; Hula Plain, Upper Jordan Valley.

Area: Mediterranean and Irano-Turanian, warmer parts of Europe and America.

2. Potamogeton lucens L., Sp. Pl. 126 (1753); Boiss., Fl. 5:16; Post, Fl. 2:542. [Plate 10]

Rhizome stout. Leaves all submerged, short-petiolate, (3–5–)8–20 × (1.5–)2–3 cm, translucent, glossy, elliptic to lanceolate, cuneate at base, acute, mucronate or cuspidate, minutely serrulate, often undulate; stipules 3–8 cm, conspicuous, membranous to herbaceous; upper leaves opposite; older leaves chalk-encrusted. Peduncles thicker than stems. Spikes dense, up to 6 cm, cylindrical, emergent at anthesis. Fruitlets 2.5–4 mm, obliquely ellipsoid, subcompressed. Fl. May–September.

Hab.: Rivers, lakes. Sharon Plain (Birkat Ata, with leaves 3–5 × 1.5–2 cm); Hula Plain.

Area: Euro-Siberian, Boreo-American, Mediterranean and Irano-Turanian.

In 1934 it could be stated that "*P. lucens* is common in the Lake of Hula, where it forms rather large pure stands at a considerable distance from the shore" (Eig, Zohary & Feinbrun, in Schedae Fl. Exsicc. Pal., Cent. III:1, no. 203, 1934). However, since Lake Hula has been dried, stands of *P. lucens* are only preserved in the Hula National Reserve.

3. Potamogeton perfoliatus L., Sp. Pl. 126 (1753); Boiss., Fl. 5:17; Post, Fl. 2:542. [Plate 11]

Rhizome creeping. Stems long, branched above. Leaves all submerged, sessile, 2–3 × 1.3–2 cm, translucent, ovate to broadly ovate, cordate at base and amplexicaul, obtuse or rounded at apex, minutely denticulate; stipules up to 1 cm, membranous, whitish, evanescent. Peduncle longer than spikes, about as thick as stem. Spikes 1–4.5 cm, short-cylindrical. Fruitlets 2.5–4 mm, ovoid, short-beaked. Fl. March–September.

Hab.: Rivers, lakes, fish ponds, water reservoirs. Philistean Plain (Revadim reservoir); Esdraelon Plain (Kfar Nahum); Judean Mts. (Solomon's Pools); Jordan River, Sea of Galilee (Lake Kinnereth); Golan (Birkat Ram).

Area: Borealo-Trop.

4. Potamogeton crispus L., Sp. Pl. 126 (1753); Boiss., Fl. 5:17; Post, Fl. 2:542, incl. var. *phialensis* Post in Post & Autran, Bull. Herb. Boiss. 1:409 (1893). [Plate 12]

Rhizome creeping, slender. Stems compressed. Leaves all submerged, sessile, 2.5–6 × 0.5–1.5 cm, oblong, semi-amplexicaul, obtuse or acute, 3(–5)-veined, undulate or crispate; margin serrulate; stipules up to 1 cm, membranous, evanescent. Spikes short, up to 1.5 cm, becoming interrupted. Fruitlets long-beaked, 4–5 mm (including beak), ovoid, compressed. Fl. June–September.

Hab.: Rivers and lakes. Rare. Sharon Plain (Birkat Battich near Hadera); Golan (Birkat Ram).

Area: Borealo-Trop.

According to Waisel & Liphschitz (1971), turions are formed on branch-tips during summer; in winter, after being exposed to low temperatures, the turions germinate and produce a rhizome and leafy branches.

5. Potamogeton berchtoldii Fieber in Berchtold & Opiz, Ökon.-Techn. Fl. Böhm. 2(1):277 (1838). *P. obtusifolius* sensu Boiss., Fl. 5:18 (1882) non Mert. & Koch (1823). [Plate 13]

Rhizome absent; winter-buds terminating lateral branches. Stems with 2 conspicuous glands at nodes. Leaves all submerged, sessile, 2–2.75 mm broad, linear, more or less obtuse and short-mucronate, 3–5-veined; stipules 3–10 mm, free, open, convolute, evanescent (persistent in axils of uppermost leaves). Peduncles thickish. Spikes dense, oblong. Fruitlets usually 4 per flower, 2–2.5 mm, obliquely obovoid, short-beaked. Fl. summer.

Hab.: Lakes, streams. Dan Valley, Hula Plain.

Area: Temperate and arctic regions of Europe, Asia and N. America.

6. Potamogeton trichoides Cham. & Schlecht., Linnaea 2:175, t.4 f.6 (1827). *P. phialae* Post in Post & Autran, Bull. Herb. Boiss. 1:409 (1893); Post, Fl. 2:542. [Plate 14]

Rhizome absent. Stems filiform, divaricately branched, leafy, with small glands at nodes; turions formed at tips of lateral branches. Leaves all submerged, sessile, 0.1–1 mm broad, filiform, gradually tapering to a long fine point, 1-veined; stipules 8–15(–20) mm, free, open, convolute, evanescent. Peduncles long, filiform. Spikes interrupted, about 1.5 cm, 4(–8)-flowered. Fruitlets usually 1 per flower, rarely up to 3, 2–2.5 mm, ovoid, somewhat compressed, short-beaked, often tuberculate at dorsal margin. Fl. April–September.

Hab.: Rivers and ponds. Sharon Plain (Birkat Ata), Philistean Plain (Revadim reservoir); W. Negev; Golan (Birkat Ram).

Area: Euro-Siberian and Mediterranean; also S. Africa.

Subgen. 2. COLEOGETON Reichenb. (1845). Rhizome creeping. Leaves submerged, sessile, narrowly linear; stipule adnate to leaf-base.

7. Potamogeton filiformis Pers., Syn. Pl. 1:152 (1805); Dandy, Fl. Iran. 83:7, t.2. [Plate 15]

Rhizome creeping. Stems slender, terete. Leaves submerged, setaceous, mostly less than 1 mm broad, obtuse or rounded at apex; stipule adnate to leaf-base forming white-margined stipular sheath; sheaths closed at base when young, open later on; ligule membranous. Spikes interrupted even before anthesis, borne on long peduncles, submerged at anthesis. Fruitlets 2–2.75 mm, obliquely ovoid, more or less compressed, with a short subapical beak. Fl. April–October.

Hab.: Ponds and pools, stagnant or slowly running water, often brackish to a certain degree. Philistean Plain; Shefela, Judean Mts. (Solomon's Pools).

Area: Temperate and arctic regions of Europe, Asia, N. America.

Related to *P. pectinatus* from which it differs in much finer leaves and smaller nutlets with a centrally placed beak, as well as in the more stagnant water occupied.

8. Potamogeton pectinatus L., Sp. Pl. 127 (1753); Eig, Bull. Inst. Agr. Nat. Hist. Tel-Aviv 4:36 (1926); Boiss., Fl. 5:18; Post, Fl. 2:542. [Plate 16]

Rhizome creeping. Stem leafy, filiform, much branched above; branches spreading in a fan-like manner. Leaves all submerged, sessile, 1–3 mm broad, narrowly linear to filiform, acute to apiculate or mucronate, the lower broader, the upper narrower; stipule adnate to the leaf-base forming stipular sheath; sheaths long (8–15 mm), open, convolute, white-membranous at margin, ending in a membranous ligule. Peduncles slender, up to 5 cm. Spikes 2.5–4(–6) cm, cylindrical, interrupted, submerged at anthesis. Fruitlets 3–5 mm, obovoid or semi-obovoid, more or less compressed, with a short subventral beak; ventral margin straight. Fl. May–October. Flowering rare and sporadic (Eig, *loc. cit.*).

Hab.: Rivers and lakes. Dan Valley (Tel el Qadi), Hula Plain, Upper Jordan Valley.

Area: Borealo-Trop.

114. RUPPIACEAE

Submerged perennials of saline or brackish water. Stems elongate, slender. Leaves 2-stichous, alternate, simple, linear to filiform, sheathing at base, entire except at minutely denticulate apex; sheaths shortly auriculate. Flowers hermaphrodite, small, ebracteate, in 2-flowered spikes; spike borne on a peduncle and subtended by 2 opposite leaves; peduncles filiform, more or less elongated after flowering, sometimes spirally coiled. Perianth lacking or vestigial. Stamens 2, free; anthers sessile, with 2 thecae; thecae opening longitudinally. Gynoecium superior; carpels 4(-10), free, 1-ovulate, sessile, becoming long-stipitate in fruit; stigmas sessile, peltate. Fruitlets of apocarpous fruit indehiscent, ripening under water. Seeds devoid of endosperm.

A single genus with several species of temperate and subtrop. regions.

Literature: G. Reese, Zur intragenerischen Taxonomie der Gattung *Ruppia* L. Ein cytosystematischer Beitrag, *Zeitschr. Bot. (Stuttgart)* 50:237-264 (1962). S.J. v. Ooststroom & T.J. Reichgelt, Ruppiaceae, in: *Flora neerlandica* 1(6):80-83, 1964. J.C. Gamerro, Observaciones sobre la biologia floral y morphologia de la Potamogetonacea *Ruppia cirrhosa* (Petagna) Grande (=*R. spiralis* L. ex Dumort.), *Darwiniana* 14:575-608, Pl. 1-4 (1968). C. den Hartog, De Nederlandse *Ruppia*-soorten, *Gorteria* 5(7/10):148-153 (1971). J.E. Dandy, Ruppiaceae, in: *Flora iranica* 84:1, 1971. P.E. Boissier, *Flora orientalis* 5:19-20, 1882. G.E. Post, *Flora of Syria, Palestine and Sinai* ed. 2, 2:543-544, 1933.

1. RUPPIA L.

Characters as given for family.

1. Peduncle much elongated after anthesis, 8 cm or longer, raising inflorescence to water surface, spirally coiled in fruit; pollination at water surface.   **1. R. cirrhosa**
- Peduncle shorter than 6 cm, slightly elongating and recurved after anthesis; pollination under water.   **2. R. maritima**

1. Ruppia cirrhosa (Petagna) Grande, Bull. Orto Bot. Napoli 5:58 (1918). *Buccaferrea cirrhosa* Petagna, Inst. Bot. 5:1826 (1787). *R. spiralis* L. ex Dumort., Fl. Belg. 164 (1827); Boiss., Fl. 5:19. Post, Fl. 2:543 as *R. maritima* L. var. *spiralis* (L.) Boiss. [Plate 17]

Perennial herb. Stems filiform, much branched. Leaves filiform. Peduncles very long after flowering (8 cm or more according to depth of water) and spirally coiled in fruit. Pollination at water surface by floating pollen; flowers protandrous. Fruitlets erect, ovoid, gradually attenuate at apex, borne on stalks 4-10 times as long as fruitlets. According to Reese (1962), 2n=40. Fl. & fr. (April-)June-July (Lipkin, 1977:287).

Hab.: Saline marshes mostly near the sea. Acco Plain, Coast of Carmel, Sharon Plain; N. Negev; Dead Sea area.

Area: Borealo-Trop.

2. Ruppia maritima L., Sp. Pl. 127 (1753). *R. rostellata* Koch in Reichenb., Pl. Crit. 2:66, t.174 f.306 (1824); Boiss., Fl. 5:20. Post, Fl. 2:543 as *R. maritima* L. var. *rostrata* Agardh (1823). [Plate 18]

Peduncles, in contrast to those in *R. cirrhosa,* are recurved after anthesis, do not exceed 6 cm and are not spirally coiled in fruit. Pollination under water; flowers protogynous. According to Reese (1962), 2n=20.

Hab.: Saline marshes, saline springs. Acco Plain, Sharon Plain, Philistean Plain (Gaza, Post); N. Negev; Lower Jordan Valley, Dead Sea area, Arava Valley.

Area: Borealo-Trop.

115. ZANNICHELLIACEAE sensu Dumort.

Submerged perennial or ephemeral herbs of fresh and brackish water, monoecious, more rarely dioecious. Roots unbranched. Stems filiform. Leaves 2-stichous, alternate, subopposite or pseudo-whorled, simple, narrowly linear, each with a stipule that is free or adnate to leaf-base. Flowers unisexual, small, solitary or in monochasial cymes, enclosed in a membranous spathe. Staminate flowers usually devoid of perianth, consisting of 1 stamen with 1 or more pairs of thecae; pollen globose. Pistillate flowers with a small cup or a few small scales as perianth; gynoecium apocarpous, of 1–9 free 1-ovulate carpels; styles simple, with a funnel-shaped stigma. Pollination under water while flowers enclosed in spathe. Fruit of several indehiscent fruitlets, sometimes viviparous. Seeds devoid of endosperm; embryo coiled.

Three genera and 6 cosmopolitan species.

Literature: K.O.P.P. Graebner, Trib. Zannichellieae, in Ascherson & Graebner, Potamogetonaceae, in: Engler, *Pflanzenreich* 31 (IV. 11):153–162, 1907. T. Eckardt, Trib. Zannichellieae, in: Engler's *Syllabus der Pflanzenfamilien* ed. 12, 2:509–510, Berlin, 1964. S. J. v. Ooststroom & T. J. Reichgelt, Zannichelliaceae, in: *Flora neerlandica* 1(6):84–87, 1964. J. E. Dandy, Zannichelliaceae, in: *Flora iranica* 85:1–4, 1971. P. B. Tomlinson & U. Posluszny, Generic limits in the Zannichelliaceae (sensu Dumortier), *Taxon* 25:273–279 (1976). A. L. Takhtajan, *A System and Phylogeny of the Flowering Plants,* Moscow & Leningrad, 1966; Outline of the classification of flowering plants (Magnoliophyta), *Bot. Rev.* 46:303 (1980). P. E. Boissier, *Flora orientalis* 5:14–15, 1882. G. E. Post, *Flora of Syria, Palestine and Sinai* ed. 2, 2:540–541, 1933.

Contrary to Graebner in Ascherson & Graebner (1907) and Eckardt (1964), we accept the delimitation of the family Zannichelliaceae proposed by Takhtajan (1966) and Tomlinson & Posluszny (1976). According to these authors, fam. Zannichelliaceae comprises plants of fresh and brackish water only, as distinct from the marine Cymodoceaceae.

1. ZANNICHELLIA L.

Monoecious perennials or ephemerals. Roots single or in pairs on lower nodes. Stems sparsely branched. Leaves linear; their stipule free or adnate to leaf-base, enclosing stem. Flowers 1-staminate and 1-pistillate, usually together at each pseudo-whorl. Staminate flower devoid of perianth; stamen 1, with 2 connate anthers. Pistillate flower surrounded by a membranous cupular deciduous perianth; carpels (1-)4-5(-8), compressed, each with a short persistent style expanded into an asymmetrical funnel-shaped stigma. Fruitlets somewhat compressed, coriaceous when ripe.

One cosmopolitan species.

1. Zannichellia palustris L., Sp. Pl. 969 (1753); Boiss., Fl. 5:14; Post, Fl. 2:541. [Plate 19]

Submerged perennial with a creeping rhizome. Stems filiform, rooting. Leaves mostly opposite (on sterile branches alternate), 0.5–2 mm broad, narrow-linear, tapering to a fine point; stipule free, tubular below, membranous, caducous. Carpels mostly 4. Fruitlets sessile or stipitate, oblique, often curved, beaked, 2–3 mm (excl. beak), mostly finely muricate at back. Fl. March–May.

Hab.: Lakes, ponds, springs, streams. Coast of Carmel, Sharon Plain, Philistean Plain; N. Negev; Hula Plain; Golan (Birkat Ram), Edom.

Area: Borealo-Trop.

116. CYMODOCEACEAE

Submerged marine perennials, dioecious, with creeping rhizome. Leaves alternate, 2-stichous, linear or subulate, sheathing at base; leaf-sheath 2-auriculate and ligulate, nearly or completely amplexicaul, leaving open or closed circular scars when shed. Flowers unisexual, devoid of perianth, solitary or in a cymose inflorescence. Staminate flower consisting of 2 dorsally connate anthers; each anther with 2 thecae; pollen thread-like. Pistillate flower consisting of 2 free ovaries each with a long undivided style or a short style divided into 2–3 strap-shaped stigmas. Fruitlet 1-seeded, indehiscent.

Differs from Zannichelliaceae in marine habitat, absence of perianth, thread-like pollen, and structure of pistillate flower with 2 free ovaries.

Five genera, mostly trop.

Den Hartog (1970, p. 145) remarks on the nature of the staminate flower: "The male 'flower' of the *Cymodoceoideae* consists of 2 dorsally connate anthers. It is distinctly stalked in the genera *Halodule, Cymodocea* and *Syringodium*. As two anthers are involved, the stalk cannot be regarded as a filament but must be a pedicel. However, according to Ascherson (1907) and Markgraf (1936), each anther represents a separate flower, and the so-called male 'flower' is, in their opinion, an inflorescence and their common stalk, therefore, a peduncle".

Literature: N. Taylor, *North American Flora* 17(1):31–32, 1909. T. Eckardt, Trib. Cymodoceeae (Cymodoceaceae), in: Engler's *Syllabus der Pflanzenfamilien* ed. 12, 2:510, Berlin, 1964. C. den Hartog, Subfam. Cymodoceoideae, in: The sea-grasses of the world, *Verh. Konink. Nederl. Akad. Wet. Afd. Nat.* Ser. 2, 59(1):144–212 (1970). P. B. Tomlinson & U. Posluszny, Generic limits in the Zannichelliaceae (sensu Dumortier), *Taxon* 25:273–279 (1976). Y. Lipkin, Seagrass vegetation in Sinai and Israel, in: C.P. McRoy & C. Helfferich, *Seagrass Ecosystems* 263–293, New York & Basel, 1977. A.L. Takhtajan, *Bot. Rev.* 46:303 (1980). P. E. Boissier, *Flora orientalis* 5:20–24, 1882. G. E. Post, *Flora of Syria, Palestine and Sinai* ed. 2, 2:544–545, 1933.

1. Rhizome sympodial, woody. Leaf-blade shed with its sheath. Anthers subsessile.
4. Thalassodendron
- Rhizome monopodial, herbaceous. Leaf-sheath persisting longer than leaf-blade. Anthers stalked 2
2. Leaves subulate, round in cross-section. Flowers in a cymose inflorescence.
3. Syringodium
- Leaves flat. Flowers solitary 3
3. Leaf-tip 1–3-veined. Anthers not attached at the same height. Style long, undivided. **1. Halodule**
- Leaf-tip 7–17-veined. Anthers attached at the same height. Style short, divided into 2 stigmas. **2. Cymodocea**

1. HALODULE Endl.

Dioecious marine herbs with herbaceous creeping rhizome. Erect stems short, each bearing 1–4 leaves. Leaf-sheath persisting longer than blade and leaving a circular scar on stem when shed; blade linear, entire, 2–3-cuspidate or variously dentate at tip. Flowers solitary and terminal. Staminate flower stalked, consisting of 2 anthers attached to a common stalk at different levels and dorsally connate by their lower parts. Pistillate flower subsessile, of 2 free ovaries; each ovary with a long undivided style. Fruitlets subglobose-ovoid, more or less compressed, with a stony pericarp and a short beak.

Six species of trop. Atlantic and Indo-Pacific coasts.

1. Halodule uninervis (Forssk.) Aschers. in Boiss., Fl. 5:24 (1882); den Hartog, *op. cit.* 147. *Zostera uninervis* Forssk., Fl. Aeg.-Arab. CXX, 157 (1775). *Diplanthera uninervis* (Forssk.) Aschers. in Engler & Prantl, Pflanzenfam. Nachtr. 1:37 (1897); Post, Fl. 2:545. [Plate 20]

Perennial. Leaf-blade 6–15 cm × 0.25–3.5 mm, narrowed at base, sometimes falcate; midrib conspicuous; leaf-tip truncate, with 1 median and 2 lateral teeth. Staminate flower on a 6–20 mm stalk; anthers 2–3 mm, red. Pistillate flower short-stalked; ovaries ovoid, 1 mm; style 3–4 cm. Fruitlets 2–2.5 mm, beaked. Fl. June–July; fr. August (Lipkin, 1977:276).

Hab.: Gulf of Elat.

Area: Coasts of the Indian Ocean and the W. Pacific.

"*Halodule uninervis* is... a sublittoral species, but penetrates into the intertidal belt.... It is a typical pioneer species dominant in all kinds of places which are less suitable for other sea-grasses. Thus *H. uninervis* can occur on firm sand and on soft mud.... Where the vegetation has been destroyed by digging, boating or storms *H. uninervis* is usually the first species to appear.... After some time other sea-grasses appear, e.g., *Cymodocea rotundata*..." (den Hartog, *op. cit.* 150).

According to Lipkin (1977), *Halodule uninervis* is the second most common sea-grass in the Gulf of Elat, after *Halophila stipulacea*. It grows in mixed populations and has not penetrated into the Mediterranean. It varies in the width of leaves, the broader leaves having been found in deeper water.

2. CYMODOCEA C. König

Dioecious marine herbs, with herbaceous creeping rhizome. Stems short, each bearing 2–7 leaves. Leaf-sheath open, persisting longer than blade, leaving an open or closed circular scar when shed; blade linear, entire, but often spinulose near tip. Flowers solitary and terminal. Staminate flower stalked, consisting of 2 anthers dorsally connate and attached at same height to a common stalk; stalk simulating a filament. Pistillate flower sessile or shortly stalked, of 2 free ovaries; each ovary with a short style divided into 2 strap-shaped stigmas. Fruitlet semicircular to semi-ovate or elliptic in outline, laterally compressed, with dorsal ridges and a beak; pericarp stony.

Four species widely distributed in trop. and subtrop. seas of the Old World.

1. Leaf-tip 9–15-veined. Fruitlet with coarsely dentate dorsal ridges.   **2. C. rotundata**
 – Leaf-tip 7–9-veined. Dorsal ridges on fruitlet not dentate.   **1. C. nodosa**

1. Cymodocea nodosa (Ucria) Aschers., Sitz.-Ber. Ges. Naturf. Freunde Berlin 1869:4 (1869); den Hartog, *op. cit.* 161; Boiss., Fl. 5:21; Post, Fl. 2:544. *Zostera nodosa* Ucria, Nuov. Racc. Opusc. Aut. Sic. 6:256 (1793). [Plate 21]

Perennial. Rhizome robust, with 1 strongly branched root. Stems short, each with 2–5 leaves. Leaf-sheath 2.5–7 cm, light purple, linear to obconical, still entire when shed, leaving a circular scar on stem; blade 10–30 × 0.2–0.4 cm, entire, more or less spinulose near tip; leaf-tip 7–9-veined, obtusely rounded. Staminate flower on a 7–10 cm long stalk; anthers 11–15 mm. Pistillate flower sessile; ovary 3 mm; style 2–3 mm; stigma 22–25 mm. Fruitlets sessile, 8 × 6 × 1.5 mm, with 3 entire to slightly crenulate dorsal ridges. Fl. June–July; fr. July–August (Lipkin, 1977:286).

Hab.: Sheltered habitats, between sea level and 5 m depth. Sandy depressions in the sublittoral, in sand enriched by organic matter. Acco Plain (Acco Harbour), Sharon Plain (Michmoret), Philistean Plain (Tel-Aviv).

Area: Widely distributed in the Mediterranean Sea and along the Atlantic coast of Africa southwards to the Tropic of Cancer.

2. Cymodocea rotundata Ehrenb. & Hemprich ex Aschers., Sitz.-Ber. Ges. Naturf. Freunde Berlin 1870:84 (1870); den Hartog, *op. cit.* 166; Boiss., Fl. 5:21; Post, Fl. 2:544. *Phucagrostis rotundata* Hemprich & Ehrenb. ex Aschers., *op. cit.* 83 *pro syn.* [Plate 22]

Differs from *C. nodosa* mainly in 9–15-veined leaves with lacerate sheaths when shed, in spirally coiled stigmas, in the somewhat longer fruitlets, the dorsal ridges of which are coarsely and acutely dentate, and in geographic distribution. Fl. April–June (Waisel & Liphschitz). Flowering and fruiting rarely observed.

Hab.: Shallow terraces covered with coral sand. Growing together with *Halodule uninervis*. Rare. Gulf of Elat.

Area: Red Sea, Indian and Pacific Oceans.

3. SYRINGODIUM Kützing

Dioecious marine herbs, with herbaceous creeping rhizome. Stems erect, short, arising from each node of rhizome and bearing 2–3 leaves. Leaf-sheath broad with 2 obtuse auricles, persisting longer than blade, leaving an open circular scar when shed; blade subulate, round in cross-section. Inflorescence cymose, umbel-like. Staminate flower stalked, consisting of 2 anthers dorsally connate by their lower parts and attached to a common stalk at same height. Pistillate flower sessile, of 2 free ovaries, each with a very short style and 2 short stigmas. Fruitlets with a stony pericarp, obliquely obovoid or ellipsoid, 4-angular in cross-section.

Two species, one Indo-Pacific, the other confined to the Caribbean.

1. Syringodium isoëtifolium (Aschers.) Dandy, Jour. Bot. 77:116 (1939); den Hartog, *op. cit.* 177. *Cymodocea isoëtifolia* Aschers., Sitz.-Ber. Ges. Naturf. Freunde Berlin 1867:3 (1867); Boiss., Fl. 5:22; Post, Fl. 2:544. [Plate 23]

Perennial. Leaf-sheath 1.5–4 cm, often red; blade 7–30 × 0.1–0.2 cm. Fruitlets 3.5–4 × 1.75–2 × 1.5 mm, obliquely ellipsoid. Fl. June–September (Lipkin, 1977:281).

Hab.: Muddy bottom in sublittoral belt together with *Cymodocea rotundata, Halodule uninervis*. Gulf of Elat.

Area: Indo-Pacific coasts.

4. THALASSODENDRON den Hartog

Dioecious marine herbs, with robust woody creeping rhizome. Stem and stem-branches bearing terminally a cluster of leaves; stems erect, borne on every fourth rhizome-internode, elongate. Leaf-sheath compressed, amplexicaul, leaving a closed circular scar when shed; sheath open, auriculate; leaf-blade denticulate at apex, shed together with sheath. Flowers solitary and terminal on short lateral shoots, subsessile, 4-bracteate. Staminate flower subsessile, consisting of 2 connate anthers attached at same height. Pistillate flower sub-sessile, of 2 free ovaries; ovary with a short style divided into 2 strap-shaped stigmas. False fruit consisting of 2 free ovaries (or 1 by abortion) and a fleshy innermost bract; often viviparous.

Two species, one paleotrop., the other of extratrop. W. Australia.

Literature: F. M. Isaac, *Jour. E. Afr. Nat. Hist. Soc. Nat. Mus.* 27:29–47 (1968). C. den Hartog, The sea-grasses of the world, *Verh. Konink. Nederl. Akad. Wet. Afd. Nat. Ser.* 2, 59(1):186–198 (1970).

1. Thalassodendron ciliatum (Forssk.) den Hartog, *op. cit.* 188, f.52 (1970). *Zostera ciliata* Forssk., Fl. Aeg.-Arab. 157 (1775). *Cymodocea ciliata* (Forssk.) Ehrenb. ex Aschers., Sitz.-Ber. Ges. Naturf. Freunde Berlin 1867:3 (1867) pro majore parte; Boiss., Fl. 5:23; Post, Fl. 2:545. [Plate 24]

Perennial. Stems 10–65 cm, 1–2 at every fourth node of robust rhizome. Leaf-sheath 1.3–2.5 cm, cream-coloured to pink; blade 10–15 × 0.6–1.3 cm, green, rounded, denticulate and often emarginate at tip. The 2 inner of the 4 bracts differing in shape and venation in staminate and pistillate plants. Anthers 6–7 mm or more. Ovary ellipsoid, 2 mm. False fruit 3.5–5 cm, mostly consisting of 2 fertilized carpels completely surrounded by a fleshy innermost bract, free floating when ripe, sometimes viviparous. Fl. June–August; fr. July–August (Lipkin, 1977:279).

Hab.: In the upper part of the sublittoral belt down to a depth of at least 10 m, rarely uncovered by tide; growing at its upper limit in circular patches, on sandy bottom, coral reefs and sand-covered rocks, forming extensive submarine meadows in deeper water. Colonies usually unisexual. Gulf of Elat.

Area: Red Sea and W. part of the Indian Ocean, along coasts of Trop. Africa and islands, E. Malaysia, Australia.

117. NAJADACEAE

Submerged annuals or perennials of fresh or brackish water, monoecious or dioecious. Stems slender, much branched, rigid, often brittle, rooting at base and at lowermost nodes. Leaves opposite, sometimes in pseudo-whorls of 3, simple, linear, usually spinose-dentate, sheathing at base, often auriculate. Flowers unisexual, small, sessile, mostly solitary in leaf-axils. Staminate flower enclosed in a spathe and a 2-lipped perianth closely applied to anther; stamen 1; anther subsessile, with 1–4 thecae; pollen globose. Pistillate flower naked or enclosed in a membranous spathe; ovary 1-carpellate, 1-locular, 1-ovulate; style divided into 2–4 linear stigmas. Fruit an indehiscent nutlet. Seed devoid of endosperm; embryo straight.

A single genus.

1. NAJAS L.

Characters as given for family.

About 50 species, mostly cosmopolitan.

Literature: A. B. Rendle, Naiadaceae, in: Engler, *Pflanzenreich* 7 (IV. 12):1–21, 1901. T. Eckardt, Najadaceae, in: Engler's *Syllabus der Pflanzenfamilien* ed. 12, 2:510–512, Berlin, 1964. W. J. J. O. de Wilde, Najadaceae, in: *Flora neerlandica*

1(6):93–96, 1964. J. E. Dandy, Najadaceae, in: *Flora iranica* 86:1–2, 1971. P. E. Boissier, *Flora orientalis* 5:27–28, 1882. G. E. Post, *Flora of Syria, Palestine and Sinai* ed. 2, 2:546–547, 1933.

1. Plants dioecious. Leaves coarsely spiny-dentate, 2–4 mm broad (incl. teeth). Stems with spiny internodes. **1. N. delilei**
- Plants monoecious. Leaves minutely spiny-dentate, less than 1 mm in width (incl. teeth). Stems with smooth internodes. **2. N. minor**

1. Najas delilei Rouy, Fl. Fr. 13:294 (1912). *N. armata* Lindb. fil., Itin. Medit., Acta Soc. Sci. Fenn. nov. ser. B. 1, 2:8 (1932). *N. muricata* Del., Fl. Eg. 281 (1813-1814), t.50 f.1 (1826) non Thuill. (1800), Boiss., Fl. 5.27. *N. marina* L. var. *muricata* (Del.) A. Br. ex K. Schumann in Mart., Fl. Brasil. 3(3):725 (1894) non Hartm.; Rendle, Trans. Linn. Soc. London (Bot.) ser. 2, 5:397 (1899); Eig, Bull. Inst. Agr. Nat. Hist. Tel-Aviv 4:37 (1926); Post, Fl. 2:547. [Plate 25]

Dioecious annual, rooting on substrate. Stem-internodes aculeate. Leaves 1–1.5 cm, 2–4 mm broad (incl. teeth), linear, coarsely and deeply serrate at margins and spiny-dentate at back; leaf-sheath entire. Anther-thecae 4. Pistillate flower devoid of spathe; stigmas 3. Fl. August.

Hab.: In water. Acco Plain (Na'aman River), Sharon Plain (Yarkon River), Philistean Plain (Nahal Sukreir); Hula Plain, Upper Jordan Valley (Jordan River), Dead Sea area (springs of Feshkha), Arava Valley.

Area: Mediterranean (Palestine, Egypt, Sicily, Algeria, Morocco), Trop. E. Africa, S. Asia (Kashmir, Ceylon), Australia.

Eig (1926), who collected this species for the first time in the region of Post's Flora in the years 1923–1925, writes: "Only the specimens of the Yarkon of 21.8.1924 were in blossom; we did not find any flowering specimens among the scores of plants collected in other periods".

2. Najas minor All., Auct. Syn. Stirp. Horti Taur. 3 (1773); Boiss., Fl. 5:28; Post, Fl. 2:547. [Plate 26]

Monoecious annual. Stems filiform, smooth. Leaves 1–2(–3) cm, less than 1 mm broad (incl. teeth), with 5–17 minute teeth at margin and sometimes at back; leaf-sheath denticulate, with short auricles. Anther with 1 theca. Pistillate flower devoid of spathe; stigmas 3. Fl. August–September.

Hab: In water. Acco Plain (Kishon River); Golan (Birkat Ram).

Area: Euro-Siberian, Mediterranean and Irano-Turnian.

118. LILIACEAE

Mostly perennial herbs with rhizome, bulb, corm or tuberous roots, scapose or caulescent, rarely annuals or small trees, sometimes climbers. Leaves alternate or whorled, often all basal. Inflorescence mostly racemose, rarely cymose *(Allium)*, sometimes flowers solitary. Flowers mostly hermaphrodite, rarely unisexual, actinomorphic, rarely somewhat zygomorphic, generally 3-merous (rarely 2-, 4- or 5-merous), often showy, mostly entomophilous. Perianth mostly petaloid, rarely sepaloid, generally of 6 (rarely 4 or up to 15) free or more or less connate segments, generally in 2 similar whorls. Stamens generally 6 (rarely fewer) in 2 whorls; filaments mostly free. Pistil syncarpous; ovary generally superior, 3-(rarely 2-4-5-)carpellate, generally 3-locular; ovules usually numerous; styles 1 or 3(-5); stigmas 1 or 3. Fruit a loculicidal or septicidal capsule, sometimes a berry, rarely a nut. Seeds numerous, rarely few by abortion; embryo small, straight or curved, surrounded by a fleshy or cartilaginous endosperm; aril sometimes present.

About 250 genera and 3,700 species in both hemispheres.

Literature: J.G. Baker, Revision of the genera of Scilleae and Chlorogaleae, *Jour. Linn. Soc. London (Bot.)* 13:209-292 (1873). K. Krause, Liliaceae, in: Engler & Prantl, *Pflanzenfam.* ed. 2, 15a:227-386, 1930. V. L. Komarov (ed.), Liliaceae, in: *Flora URSS* 4:1-475, Leningrad, 1935. F. Buxbaum, Die Entwicklungslinien der Lilioideae, *Bot. Arch.* 38:305-398 (1937). V. Täckholm & M. Drar, *Flora of Egypt* 3:1-339, Cairo, 1954. R. Maire, M. Guinochet & P. Quézel, *Flore de l'Afrique du Nord* 5:1-307, Paris, 1958. P. Mouterde, *Nouvelle Flore du Liban et de la Syrie* 1:203-288, tt. 42-94, Beyrouth, 1966. V. Täckholm, *Students' Flora of Egypt* ed. 2, 628-654, Beirut, 1974. An. A. Fedorov (ed.), *Flora partis europeae URSS* 4:202-290, Leningrad, 1979. T.G. Tutin et al. (eds.), *Flora europaea* 5:14-74, Cambridge, 1980. P.E. Boissier, *Flora orientalis* 5:155-343, 1882. G.E. Post, *Flora of Syria, Palestine and Sinai* ed. 2, 2:611-663, 1933.

1. True leaves reduced to small scarious scales; green branches *(phylloclades)* in axils of true leaves ovate, lanceolate, oblong, linear, filiform or needle-like. Flowers axillary   2
- Plants with green leaves and devoid of phylloclades; in some genera plants leafless at flowering   3
2. Phylloclades ovate to lanceolate, ending in a spiny tip. Flowers in a cluster borne on the midvein of phylloclade in the axil of a bract.   **21. Ruscus**
- Phylloclades oblong, linear, filiform or needle-like, often spinescent. Flowers 1-2 or several in axils of scale-like true leaves.   **20. Asparagus**
3 (1). Leaves petiolate; blade coriaceous, cordate or hastate at base; petiole with a pair of tendrils above base. Climbing dioecious plants, usually aculeate.
                                                                                 22. Smilax
- Leaves not as above. Plants without tendrils   4
4. Styles 3, free. Flowers arising from a corm   5
- Style 1; stigma sometimes shortly 3-lobed. Plants without corm   6
5. Perianth lacking tube; segments free, with a claw at base, white to pale lilac, with purple veins.   **5. Androcymbium**
- Perianth with a long tube.   **4. Colchicum**
6(4). Perianth-segments free or connate at base only   7
- Perianth-segments connate, forming a tube ⅓ or more of their length   18
7. Flowers strongly fragrant, white; perianth-segments 7-10 cm; anthers dorsifixed, versatile; style long; flowering stem leafy, 60-150 cm. Rare plants of maquis or cultivated ornamentals.   **9. Lilium**

 - Plants with different set of characters   8
8. Flowers solitary on scape, red, showy, rarely 2–3 and then white on inside with a
   yellow basal blotch.   **7. Tulipa**
 - Flowers more numerous, in umbels, racemes, spikes, panicles or cymes   9
9. Flowers in umbels or heads, enveloped in bud by a spathe of 1–3 bracts; spathe
   sometimes deciduous at beginning of flowering. Plants usually with an onion- or
   garlic-like smell.   **19. Allium**
 - Flowers not in umbels or heads, not enveloped by a spathe in bud   10
10. Perianth-segments rolled-in transversely during flowering; filaments and anthers
    orange-coloured, exserted.   **3. Eremurus**
 - Perianth-segments and filaments not as above   11
11. Flowers nodding; perianth campanulate; segments obovate or oblong-spathulate,
    greenish-yellow or plum-purple, with darker fine veins.   **8. Fritillaria**
 - Flowers not as above   12
12. Plants lacking bulb. Filaments dilated at base, investing ovary   13
 - Plants with a bulb. Filaments not as above   14
13. Flowers yellow or orange, somewhat zygomorphic; outer filaments about ½
    length of the inner.   **2. Asphodeline**
 - Flowers white or pale pink, actinomorphic; filaments more or less equal.
                                                                          1. Asphodelus
14(12). All leaves basal. Anthers dorsifixed   15
 - Some cauline leaves or green leaf-like bracts present. Perianth-segments yellow,
   with a green stripe along middle on outside; rarely perianth-segments white, with
   3–5 purplish-green veins on outside; anthers basifixed.   **6. Gagea**
15. Perianth-segments white (rarely yellowish), usually with a green stripe along
    middle on outside; bracts conspicuous.   **12. Ornithogalum**
 - Perianth-segments blue, lilac, violet or white with a dark midvein; bracts small or
   absent   16
16. Plants with synanthous leaves, flowering in February–April. Perianth-segments
    blue, lilac or violet, without a darker midvein.   **10. Scilla**
 - Plants with hysteranthous to subsynanthous leaves, flowering July–October or
   October–December. Perianth-segments variously coloured, with a dark midvein 17
17. Seeds flat. Leaves 3–8 cm broad or, if narrower, strongly undulate.   **11. Urginea**
 - Seeds ovoid or globose. Leaves 1.5–3 mm broad, not undulate.   **10. Scilla**
18(6). Bracts longer than pedicels; outer lobes of perianth falcate, 8 mm, much longer
    than connivent inner ones.   **13. Dipcadi**
 - Bracts minute   19
19. Perianth 17–25 mm; lobes oblong, recurved, slightly shorter than tube; filaments
    attached below middle of perianth-tube; style very short; seeds arillate.
                                                                         14. Hyacinthus
 - Perianth shorter; lobes or teeth not as above; filaments attached at or above middle of tube; seeds not arillate   20
20. Perianth ending in small reflexed teeth and more or less constricted at mouth of
    tube below teeth; teeth ovate-deltoid, less than ¼ length of tube   21
 - Perianth lobed, not constricted at mouth of tube; lobes ⅓–½ length of tube   22
21. Fertile flowers dark violet, dark or pale blue or lilac, not changing in colour
    during flowering; stamens in 1 row or obscurely 2-seriate.   **18. Muscari**
 - Colour of fertile flowers violet or purplish in bud, changing to ivory, greenish-

yellow or olive at anthesis and finally dirty brown; perianth with protuberances below mouth of tube; stamens distinctly 2-seriate.   **17. Leopoldia**

22(20). Filaments triangular; capsule not broader than long, 3-quetrous, with acute ribs or wings; seeds smooth, bluish-black owing to a waxy bloom.   **16. Bellevalia**

- Filaments linear; capsule depressed-globose, 3-lobed, somewhat broader than long, with lobes rounded at back; seeds wrinkled, black.   **15. Hyacinthella**

1. ASPHODELUS L.

Annuals or perennials. Roots fibrous or fleshy, fusiform or napiform. Leaves basal, linear. Inflorescence a terminal many-flowered raceme or panicle; bracts membranous, persistent; pedicels jointed near base or towards middle; fruiting pedicels often thickened above. Flowers actinomorphic, protogynous. Perianth-segments 6, free or somewhat connate at base, white or pale pink, with a purplish or greenish midvein. Filaments dilated at base, concave, investing ovary, ascending, filiform or fusiform above; anthers equal, dorsifixed, introrse. Ovary 3-locular; style 1, filiform; stigma capitate to sub-3-lobed. Capsule coriaceous, loculicidal. Seeds 6, acutely 3-quetrous, transversely sulcate at back.

Twelve Mediterranean and Irano-Turanian species.

1. Leaves flat, keeled, 6–20 mm broad. Perennial, provided with a cluster of fleshy fusiform roots.   **1. A. aestivus**
- Leaves semiterete, 1–4 mm broad   2
2. Pedicels jointed at middle or somewhat below middle. Leaves not sticky, glabrous or scabridulous   3
- Pedicels jointed near base. Leaves sticky, at least in the lower part   4
3. Perianth-segments 10–12 mm. Capsule 4–5 mm. Short-lived perennial.
                                                                  2. A. fistulosus
- Perianth-segments 5–7 mm. Capsule 3(–4) mm. Annual.   **3. A. tenuifolius**
4 (2). Fruiting pedicels reflexed; bracts abruptly acuminate. Capsule pendulous.
                                                                  5. A. refractus
- Fruiting pedicels erecto-patent; bracts apiculate. Capsule not pendulous.
                                                                  4. A. viscidulus

1. Asphodelus aestivus Brot., Fl. Lusit. 1:525 (1804); Post, Fl. 2:655. *A. microcarpus* Salzm. & Viv. in Viv., Fl. Cors. 5 (1824); Maire et al., Fl. Afr. Nord 5:28, f.753; Boiss., Fl. 5:313. [Plate 27]

Perennial, 60–150 cm, with short vertical rhizome surrounded by rigid fibres and bearing a cluster of fleshy fusiform 1–1.5 cm thick roots. Leaves flat, broadly linear, acuminate, acutely keeled, (6–)10–20 mm broad. Scape erect, terete, solid. Inflorescence paniculate, with numerous dense racemes; bracts ovate to lanceolate, cuspidate. Pedicels erecto-patent, shorter than flowers, jointed somewhat below middle, curved upward in fruit. Perianth-segments 12–16 mm, oblong, obtuse. Filaments papillose in lower part. Capsule 5–10 mm, obovoid, 6-gonous, truncate, with transverse elevated veins. Seeds transversely sulcate at back. Fl. January–April.

Hab.: Batha and fallow fields. Very common and forming large stands. Acco Plain, Sharon Plain, Philistean Plain; Upper and Lower Galilee, Mt. Carmel, Esdraelon Plain, Samaria, Shefela, Judean Mts., Judean Desert, N. and C. Negev; Upper Jordan Valley, Beit Shean Valley, Lower Jordan Valley, Dead Sea area; Golan, Gilead, Ammon, Moav, Edom.

Area: Mediterranean.

2. **Asphodelus fistulosus** L., Sp. Pl. 309 (1753); Maire et al., Fl. Afr. Nord 5:37, f.759; Boiss., Fl. 5:314; Post, Fl. 2:665. [Plate 28]

Short-lived perennial, 30–100 cm. Roots fibrous, somewhat fleshy. Leaves numerous, narrowly linear, up to 3(–4) mm broad, semiterete, fistulose, subulate-acuminate, keeled, usually scabridulous at margin, erect, shorter than scape. Scape hollow, simple or branched. Inflorescence paniculate, with lax racemes; bracts lanceolate, acuminate. Pedicels straight, much shorter than flowers, generally jointed at middle. Perianth-segments 10–12 mm, elliptic, obtuse. Capsule 4–5 mm, globose; valves transversely 2–3-sulcate; fruiting pedicels 2–3 times as long as capsule. Seeds transversely 2–3-sulcate at back and 2–3-pitted at sides. Fl. January–May.

Hab.: Batha and fallow fields. Sharon Plain, Philistean Plain; Lower Galilee, Shefela, Judean Mts., C. Negev; Upper and Lower Jordan Valley; Golan, Gilead, Moav, Edom.

Area: Mediterranean.

3. **Asphodelus tenuifolius** Cav., Anal. Ci. Nat. 3:46, t.27 (1801); Maire et al., Fl. Afr. Nord 5:39, f.760; Boiss., Fl. 5:314, incl. var. *micranthus* Boiss., *op. cit.* 315 (1882); Post, Fl. 2:655. *A. fistulosus* L. var. *tenuifolius* (Cav.) Baker, Jour. Linn. Soc. London (Bot.) 15:272 (1876). [Plate 29]

Annual, 20–30(–50) cm. Roots fibrous. Leaves up to 2(–3) mm broad, semiterete, fistulous. Scape hollow. Inflorescence paniculate; bracts triangular, acuminate. Pedicels jointed usually somewhat below middle. Perianth-segments 5–7 mm. Capsule 3(–4) mm, ovoid-globose. Seeds transversely 2–3-sulcate at back, 3–4-pitted at sides. Fl. February–April.

Hab.: Batha and steppes. Fairly common. Sharon Plain, Philistean Plain; Judean Mts., Judean Desert, W., N., C. and S. Negev; Upper and Lower Jordan Valley, Dead Sea area, Arava Valley; Golan, Gilead, Ammon, Moav, Edom.

Area: Saharo-Arabian and Sudanian, extending into the Mediterranean region.

Differs from *A. fistulosus* mainly in annual habit, smaller flowers and capsules, and in ecological requirements.

Eig writes in Schedae Fl. Exsicc. Pal., Cent. II:3, no. 116 (1932): "In Palestine we found several forms of transition between *A. tenuifolius* and *A. fistulosus*... *A. tenuifolius* is a common plant in the Saharo-Sindian territory of Palestine and less common in the light soils belt of the Mediterranean maritime plain and in some parts of the Irano-Turanian territories".

4. Asphodelus viscidulus Boiss., Diagn. ser. 1, 7:118 (1846); Maire et al., Fl. Afr. Nord 5:43, f.762; Boiss., Fl. 5:315; Post, Fl. 2:656. Incl. var. *biflorus* Aaronsohn & Opphr. in Opphr., Florula Transiord. 156, f.1, Bull. Soc. Bot. Genève ser. 2, 22:275, f.1 (1931). [Plate 30]

Annual, 20–25 cm. Roots fibrous, slender. Leaves 1–2 mm broad, subulate, semiterete, fistulous, glabrous and sticky (usually with adhering grains of sand). Scapes 1 to several, ascending, curved, the naked portion shorter than leaves. Flowers in lax racemes; bracts deltoid, apiculate. Pedicels jointed near base; fruiting pedicels erecto-patent. Perianth-segments 4–5 mm, oblong. Capsule 2–3 mm in diameter, depressed-globose; valves transversely sulcate. Seeds bluish, transversely 3-sulcate at back, smooth at sides. Fl. March–April.

Hab.: Deserts. C. Negev; Arava Valley; Moav, Edom.

Area: Saharo-Arabian.

5. Asphodelus refractus Boiss., Diagn. ser. 1, 13:23 (1854); Maire et al., Fl. Afr. Nord 5:41, f.761. *A. pendulinus* Coss. & Durieu, Bull. Soc. Bot. Fr. 4:399, 497 (1857); Boiss., Fl. 5:315; Post, Fl. 2:831, Add. [Plate 31]

Annual, 30–45 cm. Roots fibrous, slender. Leaves 7.5–10 cm, 1–2 mm broad, linear-subulate, acuminate, semiterete, fistulous, glabrous, sticky in lower part. Scapes 1 to several, ascending, branched once or twice below middle. Racemes long, lax; bracts deltoid, abruptly acuminate. Pedicels jointed near base; flowering pedicels erect; fruiting pedicels elongated, reflexed. Perianth-segments about 4 mm, oblong. Capsule 3–4 mm, pendulous, obovoid; valves often obscurely sulcate. Seeds grey marbled with black, transversely 3-sulcate at back, smooth at sides. Fl. March–April.

Hab.: Sandy desert. Arava Valley; Edom.

Area: Saharo-Arabian.

2. ASPHODELINE Reichenb.

Perennials with a cluster of cylindrical fleshy roots. Leaves linear with a dilated membranous sheathing base. Stems more or less leafy or leafless. Inflorescence a terminal many-flowered raceme or panicle; bracts membranous, persistent; pedicels jointed. Flowers somewhat zygomorphic, yellow or orange, more rarely white. Perianth-segments 6, narrow, 1-veined, connate at base forming very short urceolate tube; anterior segment narrower than others, deflexed, long-attenuate at base. Filaments dilated and concave at base, investing ovary, filiform above base and curved downwards, then ascending; inner filaments much longer than the outer; anthers dorsifixed, introrse, those of outer stamens minute and frequently sterile. Ovary 3-locular; style 1, filiform; stigma capitate to sub-3-lobed. Capsule coriaceous, loculicidal. Seeds 6, 3-quetrous, transversely sulcate at back.

Fifteen Mediterranean and Irano-Turanian species.

1. Stem leafy up to inflorescence.   **1. A. lutea**

- Stem leafy at most up to ⅔ of its length 2
2. Inflorescence paniculate, of many racemes. Fruiting pedicels upright, jointed below middle. **2. A. brevicaulis**
- Raceme simple or slightly branched. Fruiting pedicels recurved, jointed at middle.
3. A. recurva

1. Asphodeline lutea (L.) Reichenb., Fl. Germ. Excurs. 116 (1830); Boiss., Fl. 5:316; Post, Fl. 2:656. *Asphodelus luteus* L., Sp. Pl. 309 (1753). [Plate 32]

Plant about 1 m or more. Roots fleshy. Stem stout, leafy up to inflorescence. Leaves stiff, subulate, 3-quetrous, broadly sheathing at base. Raceme long, dense. Bracts large, membranous, with a green midvein, lanceolate or ovate cuspidate, longer than pedicels. Pedicels erect, jointed below middle. Perianth 2–3 cm, yellow or orange; segments oblong-lanceolate, with a green midvein. Capsule 10–15 mm, subglobose to ellipsoid, erect; valves transversely wrinkled. Seeds rugulose at sides. Fl. February–May.

Hab.: Rocky and damp places, growing in large groups. Acco Plain, Sharon Plain, Philistean Plain; Upper and Lower Galilee, Esdraelon Plain, Mt. Carmel, Samaria, Judean Mts., Judean Desert, C. Negev; Upper Jordan Valley; Golan, Gilead, Ammon, Moav, Edom.

Area: Mediterranean, extending into the W. Irano-Turanian.

2. Asphodeline brevicaulis (Bertol.) J. Gay ex Baker, Jour. Linn. Soc. London (Bot.) 15:276 (1876); Boiss., Fl. 5:317; Post, Fl. 2:657. *Asphodelus brevicaulis* Bertol., Misc. Bot. 1:20, Nov. Comm. Acad. Bonon. 5:430 (1842). [Plate 33]

Plant 40–50 cm; roots fleshy. Stem densely leafy up to ½–⅔ of its length. Leaves subulate, 3-quetrous, sheathing at base. Inflorescence a panicle of lax racemes. Bracts small, deltoid-cuspidate, membranous, with a greenish midvein. Pedicels solitary or 2 in axils of bracts, stiff, upright , jointed below middle. Perianth 1.5–2.5 cm, yellow; segments lanceolate, with a green midvein. Capsule 10–12 mm, globose to ellipsoid; valves wrinkled. Seeds smooth at sides. Fl. April–May.

Hab.: Batha, among rocks. Acco Plain; Upper Galilee, Samaria, Judean Mts.; Ammon.

Area: E. Mediterranean, extending into the W. Irano-Turanian.

3. Asphodeline recurva Post in Post & Autran, Bull. Herb. Boiss. 3:166 (1895); Post, Fl. 2:658; Zohary, Palest. Jour. Bot. Jerusalem ser., 2:184 (1941), incl. var. *edumea* Zohary, *loc. cit. A. edumea* (Zohary) Mout., Nouv. Fl. Liban Syrie 1:214, t.66 f.3 (1966). [Plate 34]

Plant 20–30(–40) cm; roots fleshy. Stem erect, sometimes flexuous, leafy up to ⅓–½ of its length. Leaves linear, subulate, 3-quetrous, somewhat flattened, sheathing at base. Raceme usually simple, rarely slightly branched, 7–10 cm, fairly dense. Bracts membranous with a greenish midvein, deltoid at base, long-acuminate to -cuspidate, shorter than pedicel to about as long. Pedicels solitary or 2 in axils of bracts, more or less arcuate and recurved after flowering, jointed towards middle. Perianth yellow, about 2 cm; segments linear-lanceolate, with a green midvein. Capsule 6–8 mm, subglobose; valves wrinkled. Seeds more or less sulcate at sides. Fl. April–May.

Hab.: Arid slopes. Golan, E. Gilead, Ammon, Moav, Edom.

Area: W. Irano-Turanian (Palestine, Lebanon, Syria, N. W. Iraq).

The doubts expressed by Mouterde *(loc. cit.)* concerning *A. recurva* Post are unjustified. The dozen or so specimens in our Herbarium which were collected by Eig & Zohary as well as by Naftolsky in Transjordan, N. E. Syria and N. W. Iraq, are uniform and fit the description by Post based on his collection from "agris inter Irbid ('Isbid') et Bosram (Auranitis)" in 1885. To his description Post added: "specimen unum in Herbario Boissiero apud Genevam et alterum in Herbario Postiano apud Berythum conservanda". Comparison of our plants with a photograph (from the Edinburgh Herbarium) of the Post Herbarium specimen and of a drawing of that specimen (from the Geneva Herbarium) confirm our conclusion that the name *A. edumea* (Zohary) Mout. is superfluous.

A discrepancy remains as to the flower colour of *A. recurva*. Post described the flowers as white, while the perianth of var. *edumea* Zohary is described as yellow. Zohary *(loc. cit.)* observed the yellow colour in living specimens and found it still recognizable in dried material. He suggested that in case there is an error in the description by Post concerning the colour, var. *edumea* should be included in the type and the diagnosis corrected. We follow this suggestion and include var. *edumea* in the typical *A. recurva*. Bract length is an unreliable character since the trait is variable even within the same inflorescence.

3. Eremurus Bieb.

Perennials similar in habit to *Asphodelus* and *Asphodeline*, with short rhizome and cluster of fleshy fusiform roots. Leaves basal, linear to lanceolate, tapering above. Stem leafless. Inflorescence a simple terminal raceme; bracts lanceolate. Flowers protogynous. Perianth tubular-campanulate or rotate, sometimes slightly zygomorphic; segments free or shortly connate at base. Stamens 6, equal, exserted during anthesis; filaments long; anthers attached dorsally above their base. Ovary 3-locular; locules 3–4-ovulate; style 1, filiform; stigma punctiform. Capsule membranous-coriaceous, loculicidal. Seeds 3-quetrous, sometimes winged at angles.

About 45 species, mostly Irano-Turanian, often montane.

Literature: O. Fedtschenko, *Eremurus*. Kritische Übersicht der Gattung, *Mém. Acad. Sci. Pétersb.* Ser. 8, 23(8):1–210 (1909).

1. Eremurus libanoticus Boiss. & Bl. in Boiss., Diagn. ser. 2, 4:97 (1859); Boiss., Fl. 5:322 & Post, Fl. 2:659 *pro syn. E. spectabilis* Bieb. var. *libanoticus* (Boiss. & Bl.) O. Fedtschenko (err. "Boiss. & Buhse"), Mém. Acad. Sci. Pétersb. ser. 8, 23(8):39 (1909). *E. wallii* Rech. fil., Ark. Bot. ser. 2, 1(5):302 (1949). [Plate 35]

Plant 50–100 cm, with cluster of fusiform roots, fibrous at neck and with membranous cataphylls. Leaves linear-lanceolate, acuminate, shorter than stem, 1–2 cm broad. Raceme 15–30 cm, many-flowered. Bracts membranous, lanceolate, long-acuminate, ciliate, about as long as to slightly longer than pedicels. Pedicels not jointed, erecto-patent in flower, thickened, stiff and curved upwards in fruit. Perianth-segments 10–12 mm, ovate-lanceolate to linear-oblong, white with a green midvein suffused with red-orange, rolled-in transversely from the top during anthesis. Filaments filiform, much elongating, exserted; filaments and anthers orange-coloured. Capsule transversely rugose, globose; fruiting pedicels upright, appressed to raceme-rachis. Seeds acutely tetraedrous, scarcely winged, blackish. Fl. March–May.

Hab.: Hills, mountains. Golan, Gilead, Ammon, Moav, Edom; Mt. Hermon. Area: E. Mediterranean and W. Irano-Turanian.

E. wallii Rech. fil., described from a specimen by Wall collected in El Bukea (betw. Gilead and Ammon in Transjordan), does not differ from E. libanoticus Boiss. & Bl. The diagnostic characters of E. wallii as given by Rechinger are the 2.5 cm long attenuate bracts forming a prominent tuft in a young raceme, as well as the white-green perianth of young flowers. Among specimens of E. libanoticus in our Herbarium, collected in Edom, some with young racemes display a conspicuous tuft of bracts. In flower colour too E. wallii does not differ from E. libanoticus. Data on flower colour of E. libanoticus in the original diagnosis by Boissier are "in sicco flavidi", of E. spectabilis in Flora orientalis "flavicantis". A colour photo of Eremurus from Mt. Hermon in our Herbarium clearly shows that perianth-segments in young flowers are white with a green midvein, soon becoming flushed with red-orange. The red-orange colour becomes more conspicuous in the raceme with the elongation of red-orange filaments and anthers, while the perianth-segments roll in.

4. COLCHICUM L.

Perennials, with a corm or rarely with a creeping stolon. Corm ovoid, enclosed by brown tunics, convex on one side, flattened and somewhat prolonged downwards on the other, replaced every year by a renewal corm which develops at side of previous year's corm at base of flowering shoot. Leaves and flowering shoot enclosed at base by a tubular, often white, cataphyll (spathe), which turns into a brown tunic in following year. Leaves 2–9, basal, linear to broadly lanceolate, synanthous or hysteranthous; outermost leaf tubular at base. Scape very short, subterranean during flowering. Flowers 2–5(–20) in a short raceme or solitary, hermaphrodite, showy. Perianth petaloid, pink, purple or white, rarely yellow, infundibular, with a long tube and a 6-partite limb. Stamens 6, shorter than lobes of perianth, inserted at throat of perianth; the 3 inner somewhat higher than the 3 outer; anthers versatile. Ovary subterranean, 3-locular; each locule with several ovules; styles 3, free, filiform, long, exserted from perianth-tube; stigmas punctiform or oblique. Capsule septicidal, ovoid to ellipsoid, 3-gonous, tapering or beaked at tip, borne on a lengthened stalk, maturing near ground level. Seeds globose or angular.

About 65 species, mainly in the Mediterranean, Irano-Turanian and Saharo-Arabian regions.

A taxonomically difficult genus. Both leaves and flowers are necessary for species identification. In hysteranthous species material should therefore be collected twice, during flowering and after the emergence of leaves.

Seeds and corms of some species contain the alkaloid colchicine and other toxic substances. Some species are grown for ornament in gardens.

Literature: B. Stefanoff, Monographie der Gattung *Colchicum* L., *Sborn. Bălg. Akad. Nauk. Sofia* 22:1–100 (1926). E.A. Bowles, *A Handbook of Crocus and Colchicum for Gardeners* ed. 2, 153–208, London, 1952. N. Feinbrun, The genus *Colchicum* in Palestine and neighbouring countries, *Palest. Jour. Bot. Jerusalem Ser.*, 6:71–95 (1953); Chromosome numbers and evolution in the genus *Colchicum, Evolution* 12:173–188 (1958). B.L. Burtt, R.D. Meikle & J.P.W. Furse, *Colchicum* and *Merendera, R.H.S. Lily Year Book* 31:90–105 (1968). B. Mathew, *Dwarf Bulbs* 53–67, London, 1973.

1. Parallel lamellae at base of inner or of inner and outer perianth-lobes, on either side of filament, fringed with 1 or more teeth 2
- Base of perianth-lobes without teeth 3
2. Both inner and outer perianth-lobes densely fringed on either side of base of filament. Leaves 4–5, sometimes up to 9, 3–5(–8) mm broad, sometimes hairy.
2. C. tuviae
- Inner perianth-lobes with 1 or more teeth on either side of base of filament. Leaves (2–)3(–4), 2–3(–4) cm broad, not hairy. **1. C. ritchii**
3(1). Leaves 1–2(–3) mm broad, synanthous or emerging a few days after flowers. Anthers yellow. Plants of batha in Mediterranean districts; flowering in October–December. **6. C. stevenii**
- Leaves broader. Plants with different set of characters 4
4. Plants flowering in September–December without leaves; leaves hysteranthous, emerging after end of flowering, usually in February (so that specimens collected contain either flowers or leaves). Capsule covered with brown or reddish dots, except in plants of Mt. Hermon 5
- Leaves synanthous or emerging shortly after flowers 9
5. Perianth-lobes 8–15(–20) mm broad 6
- Perianth-lobes 3–6 mm broad 8
6. Perianth-lobes beautifully purple-tessellated. Anthers dirty purple before opening; pollen pale yellow. Plants of N. and E. Golan. **9. C. bowlesianum**
- Perianth-lobes lilac-pink, not or faintly tessellated. Anthers yellow 7
7. Leaves 3–4, rarely 5–6; outer leaf 2.5–7 cm broad. Styles more or less equalling stamens, slender, not clavate. Plants of maquis. **7. C. decaisnei**
- Leaves (4–)5–9; outer leaf 1–2 cm broad. Styles distinctly overtopping stamens, thickened and clavate above, curved at apex. Plants of cultivated ground.
8. C. hierosolymitanum
8(5). Corm 1–1.5 cm in diameter; tunics extending short distance along cataphyll. Anthers purplish-brown. Plants of Golan and Mt. Hermon. **4. C. tauri**
- Corm 3–3.5 cm in diameter; tunics in many layers, extending to near top of cataphyll. Anthers yellow. Plants of hammada in the Negev, Moav and Edom.
10. C. tunicatum
9(4). Outer leaf (1.2–)1.5–3 cm broad (when well developed). Styles exceeded by stamens. Plants of Upper Galilee, Golan and slopes of Mt. Hermon.
5. C. brachyphyllum

- Outer leaf hardly reaching 1 cm in width. Styles exceeding stamens or equalling
  them   10
10. Plants of rocky or sandy habitats in deserts. Corm broadly ovoid, 1.5-4 cm in
    diameter.   **3. C. schimperi**
- Plants of Golan and Mt. Hermon. Corm oblong, 1-1.5 cm in diameter. **4. C. tauri**

1. Colchicum ritchii R. Br. in Denham & Clapp., Trav. Afr., App. 241 (1826);
Feinbr., Palest. Jour. Bot. Jerusalem ser., 6:78; Boiss., Fl. 5:163; Post, Fl.
2:614. *C. aegyptiacum* Boiss., Diagn. ser. 1, 5:66 (1844). [Plate 36]

Corm 2.5-3.5 × 2.5-3.5 cm, ovoid; tunics membranous, brown, extending
along lower ½ of cataphyll. Cataphyll 7-10 cm. Leaves (2-)3(-4), synanthous,
broadly lanceolate to linear, acute, recurved, glabrous; outer leaf 1-3 cm
broad, sometimes up to 4 cm. Flowers 2-8. Perianth pink or pale pink to
white; lobes elliptic, about 1½-2 times as long as stamens; outer lobes 8-10
mm broad; parallel lamellae at base of inner lobes with 1-4 fine teeth. Anthers
mostly dark purple, sometimes yellow. Styles slightly or not exceeding stamens; stigmas punctiform. Capsule 2-3.5 × 1-1.5 cm, ellipsoid. 2n=14. Fl.
December-February.

Hab.: Various soils and plant associations in deserts. Common. Judean
Desert, W., N. and C. Negev; Edom.

Area: Saharo-Arabian.

2. Colchicum tuviae Feinbr., Palest. Jour. Bot. Jerusalem ser., 6:79, f.1 & t.1
f.3 (1953). [Plate 37]

Corm 1.5-2.5(-3.5) × 1-2 cm, ovoid; tunics coriaceous, dark brown, extending along lower ½ of cataphyll or further. Cataphyll 5-10 cm, membranous,
often green above. Leaves 4-5(-9), synanthous (sometimes subsynanthous),
linear-lanceolate, acute, glabrous or sometimes densely to sparsely short-hairy
on upper or both faces; outer leaf 3-5(-8) mm broad after anthesis. Flowers
few, small. Perianth pale pink or white; lobes 15-22 × 3-5 mm, oblong-
elliptic to elliptic, twice as long as stamens or longer; parallel lamellae at base
of inner and outer lobes densely fringed. Stamens about 1 cm; anthers
brownish-purple, 2 mm after opening. Styles exceeding stamens; stigmas
punctiform. Capsule 2.5 × 1-1.5 cm, oblong. 2n=14. Fl. November-
January.

Hab.: Calcareous steppe soils; Artemisietum herbae-albae and Phlomidetum brachyodontis. Judean Mts., Judean Desert, N. and C. Negev.

Area: W. Irano-Turanian (Palestine, Syria).

3. Colchicum schimperi Janka ex Stefanoff, Sborn. Bălg. Akad. Nauk. Sofia
22:31 (1926); Janka, Magyar Növ. Lapok 6:177 (1882) in obs., *nom. nud.*;
Feinbr., Palest. Jour. Bot. Jerusalem ser., 6:80, t.1 ff.4, 5. [Plate 38]

Corm 2-3.5 × 1.5-4 cm, broadly ovoid; tunics coriaceous, dark brown, in
many layers, extending along lower ½ of cataphyll. Cataphyll 5-8 cm. Leaves
3(-4-5), synanthous, lanceolate, canaliculate, recurved; outer leaf (4-)6-9 mm

broad. Flowers few. Perianth nearly white to pale pink; lobes 18–30 × 3–4 mm, narrowly elliptic, 1½ times as long as stamens. Anthers purplish-brown. Styles exceeding or equalling stamens; stigmas punctiform. Capsule 1.5–2 × 0.5–1 cm, oblong or ovoid, often stipitate. 2n=14. Fl. December–February.

Hab.: Rocky habitats in deserts. C. Negev; E. Ammon, Edom.

Area: E. Saharo-Arabian (S. Palestine, Sinai, Egypt, Yemen).

4. Colchicum tauri Siehe ex Stefanoff, Sborn. Bălg. Akad. Nauk. Sofia 22:32 (1926); Feinbr., Palest. Jour. Bot. Jerusalem ser., 6:82; Mout., Nouv. Fl. Liban Syrie 1:207, t.63 f.3. *C. antilibanoticum* Gombault, Bull. Soc. Bot. Fr. 104:286 (1957). [Plate 39]

Corm 2–3 × 1–1.5 cm, oblong; tunics membranous to subcoriaceous, maroon-coloured, extending short distance along cataphyll. Cataphyll 7–10 cm. Leaves (2–)3–5, synanthous or (on Mt. Hermon) hysteranthous, linear, canaliculate; outer leaf 5–10 mm broad. Flowers up to 10. Perianth pink to nearly white; lobes oblong to narrowly elliptic, twice as long as stamens; outer lobes 3–4 mm broad. Anthers 4–4.5 mm, purplish-brown. Styles exceeding stamens; stigmas punctiform. Capsule 1.2–2.5 × 1–1.5 cm, ovoid. Fl. October–December, before the snowfall; leaves emerge shortly after flowering or after melting of snow, from February to April.

Hab.: Tragacanthic batha, 1,500–2,800 m, together with *Crocus hermoneus* subsp. *hermoneus*. Golan, Mt. Hermon.

Area: E. Mediterranean.

5. Colchicum brachyphyllum Boiss. & Hausskn. in Boiss., Fl. 5:164 (1882); Feinbr., Palest. Jour. Bot. Jerusalem ser., 6:76, t.1 f. 2; Post, Fl. 2:614. *C. fasciculare* (L.) R. Br. var. *brachyphyllum* (Boiss. & Hausskn.) Stefanoff, Sborn. Bălg. Akad. Nauk. Sofia 22:29 (1926). ?*C. libanoticum* Ehrenb. ex Boiss., Fl. 5:166 (1882). [Plate 40]

Corm 1–3 × 1–2 cm, ovoid; tunics subcoriaceous, brown, mostly extending along greater part of cataphyll. Cataphyll 7–12 cm. Leaves 3(–4), synanthous or subsynanthous (appearing with flowers or even shortly after them; therefore they are short in flowering specimens), up to 15 cm long; outer leaf 1.2–3 cm broad, erect, flat, oblong to lanceolate. Flowers numerous. Perianth white or pale pink; lobes elliptic-lanceolate, about twice as long as stamens; outer lobes 4–6 mm broad. Anthers brownish-purple. Styles exceeded by stamens; stigmas punctiform. Capsule up to 2 cm in diameter, ellipsoid to ovoid. Fl. December–February; fr. March–April.

Hab.: Damp basalt soil and stony habitats. Upper Galilee; Golan.

Area: E. Mediterranean (N. Palestine, Lebanon, Syria, S. E. Turkey).

6. Colchicum stevenii Kunth, Enum. Pl. 4:144 (1843); Feinbr., Palest. Jour. Bot. Jerusalem ser., 6:82; Boiss., Fl. 5:165; Post, Fl. 2:615. ?*C. parlatoris* Orph., Atti Congr. Bot. Firenze 1874:32 (1876); Boiss., *op. cit.* 163. [Plate 41]

Corm 2–3 × 1.5–2 cm, ovoid; tunics brown, extending along greater part of cataphyll. Cataphyll 5–10 cm. Leaves (4–)5–7, synanthous or subsynanthous, linear-filiform, 1–2(–3) mm broad. Flowers 3–8. Perianth purplish-pink; lobes narrowly elliptic to lanceolate, about twice as long as stamens; outer lobes 4–5 mm broad. Anthers yellow. Styles exceeding stamens, slightly curved at apex; stigmas oblique. Capsule 14–18 × 4–8 mm, ellipsoid. 2n=54. Fl. October–December.

Hab.: Batha in the Mediterranean territories of the country. Common. Coastal Galilee, Coast of Carmel, Sharon Plain, Philistean Plain; Upper and Lower Galilee, Mt. Carmel, Mt. Gilboa, Samaria, Shefela, Judean Mts.; Golan.

Area: E. Mediterranean (Palestine, Lebanon, Syria, S. Turkey, Cyprus, S. Greece).

Flowers often appear before the leaves, especially when rains are late. Leaves then emerge shortly after the flowers. Propagation bulblets are produced every year on the side opposite the renewal bud and opposite the furrow along which the flowering shoot is situated (cf. M. Zohary, On the vegetative reproduction of some oriental geophytes, Palest. Jour. Bot. Jerusalem ser., 1:35–38, 1938).

7. Colchicum decaisnei Boiss., Fl. 5:157 (1882); Feinbr., Palest. Jour. Bot. Jerusalem ser., 6:83, f.2 & t.2 f.7; Post, Fl. 2:612. *C. brevistylum* Feinbr. in Eig, Zohary & Feinbr., Anal. Fl. Palest. 355, 411 (1948) *nom. inval.* [Plate 42]

Corm 2–3 × 2.5–3 cm, ovoid; tunics membranous, maroon-coloured, extending along greater part of cataphyll. Cataphyll 8–12 cm. Leaves 3–4(–6), hysteranthous, 15–30 cm, broadly lanceolate, tapering above, subacute; outer leaf (2.5–)3–3.5(–7) cm broad. Flowers (2–)3–6, autumnal. Perianth purplish-pink; lobes elliptic, about twice as long as stamens; outer lobes 8–12(–15) mm broad. Longer stamens about 20 mm; anthers yellow. Styles not exceeding stamens as a rule, slender, not clavate above; stigmas slightly oblique. Capsule 2–4 × 1–1.5 cm, ellipsoid or ovoid, covered all over with minute golden-brown dots. 2n=54. Fl. October–December; leaves February–May.

Hab.: Maquis, Quercetum calliprini infectorietosum, also batha. Upper Galilee, Mt. Carmel; Golan.

Area: E. Mediterranean (N. Palestine, Lebanon, Syria, S. Turkey).

8. Colchicum hierosolymitanum Feinbr., Palest. Jour. Bot. Jerusalem ser., 6:84, f.3 & t.3 f.1 (1953). *C. decaisnei* auctt. fl. palaest. p.p. [Plate 43]

Corm 4–6 × 2.5–5 cm, ovoid; tunics membranous, maroon-coloured, extending along lower ½ of cataphyll. Cataphyll 12–25 cm. Leaves (4–)5–9, hysteranthous, about 15–25 cm, oblong to lanceolate, canaliculate, obtuse; outer leaf 1–2 cm broad. Flowers numerous (up to 20), autumnal. Perianth intensely to pale purplish-pink, sometimes obscurely tessellated; tube pale, up to 20 cm; lobes oblong-elliptic to oblong-obovate, 2½–3 times as long as stamens; outer lobes 8–12 mm broad. Anthers yellow. Styles overtopping stamens, thickened and clavate above and curved at apex; stigma oblique. Capsule 1.5–3 × 1.25–1.5 cm, ellipsoid, covered all over with minute golden-brown dots. Seeds

3.5-5 mm in diameter. 2n=18. Fl. October–January; leaves February–April.

Hab.: Cultivated fields, vineyards, olive groves, mainly on deep terra rossa. Upper and Lower Galilee, Samaria, Judean Mts.; Hula Plain, Upper Jordan Valley; Golan, ?Gilead, ?Ammon, Moav.

Area: E. Mediterranean (Palestine, Lebanon, Syria, S. E. Turkey).

9. Colchicum bowlesianum B.L. Burtt, Kew Bull. 1950:433 (1951); Bowles, Handb. Crocus & Colchicum ed. 2:180, t.31 (1952); Goulandris, Goulimis & Stearn, Wild Flowers of Greece 140 (1968).*

Corm 3.5 × 2 cm, ovoid; tunics maroon-coloured. Leaves 6–9, hysteranthous, about 15–20 cm, oblong to lanceolate, 1.5–2 cm broad. Flowers up to 15, autumnal. Perianth-lobes purple, beautifully and distinctly tessellated, oblong-obovate, 4.5–5 × 1.5–2 cm, obtuse, about twice as long as stamens; perianth-tube white. Stamens slightly unequal; filaments white, at base yellow and slightly thickened; anthers about 1 cm, dirty purple before opening; pollen pale yellow. Styles as long as or slightly exceeding longer stamens, thickened and hooked at apex, unilaterally stigmatose. 2n=22 (Bathia Pazi, unpubl.). Fl. October–November.

Hab.: Grassy batha; heavy and stony basalt soil. N. and E. Golan.

Area: E. Mediterranean.

This species recently found in the Golan Heights is undoubtedly related to *C. bowlesianum* described from the neighbourhood of Salonica in Greece. It is, however, still unknown whether the Golan plant is identical with *C. bowlesianum* and may later be found differing from it specifically.

10. Colchicum tunicatum Feinbr., Palest. Jour. Bot. Jerusalem ser., 6:87, t.2 f.11, t.3 ff.2,3 (1953). [Plate 44]

Corm (1.5–)2–4 × 3–3.5 cm, ovoid; tunics coriaceous, dark brown, in many layers, extending nearly up to top end of cataphyll (under dry conditions of habitat tunics of several years are preserved). Cataphyll 3–8 cm, with its apex just above ground level. Leaves (5–)6–8(–9), hysteranthous, 7–15 cm, lanceolate to linear, canaliculate, somewhat undulate, obtuse; outer leaf 3–12 mm broad, the others narrower. Flowers 2–6, autumnal. Perianth pink; lobes oblong, obtuse, 2–3 times as long as stamens; outer lobes 3–6 mm broad. Longer stamens 12–15 mm; anthers yellow, 3–4 mm; pollen yellow. Styles often considerably overtopping stamens, thickened and curved towards apex; stigmas short, oblique. Capsule 2.5–3.5 × 1.7–2.5 cm, oblong, covered all over with minute brown dots. 2n=54. Fl. September–October; leaves not appearing until at least one month after end of flowering.

Hab.: Gravelly and loessy hammada, Artemisietum herbae-albae, Hammadetum scopariae. N. and C. Negev; Moav, Edom.

Area: W. Irano-Turanian (Palestine, Syria).

* Not illustrated.

5. ANDROCYMBIUM Willd.

Perennials with corm enveloped in tunics. Stem short, mainly subterranean. Leaves basal, synanthous, crowded into a rosette forming an involucre to inflorescence. Leaves and flowers enclosed in a tubular membranous cataphyll in their lower part. Inflorescence dense, umbellate. Perianth petaloid, white, pink or lilac, marcescent; segments 6, free, subequal; each segment with a claw and an elliptic to lanceolate limb provided with 1-2 nectariferous glands at filament-insertion. Stamens 6, each inserted at base of limb; anthers versatile. Ovary 3-locular, many-ovulate; styles 3, free; stigmas punctiform. Capsule septicidal. Seeds globose, finely punctulate.

Thirty-five species in the Mediterranean region, Ethiopia and S. Africa.

1. **Androcymbium palaestinum** Baker, Jour. Linn. Soc. London (Bot.) 17:445 (1879); Post, Fl. 2:616. *Erythrostictus palaestinus* Boiss. ex Baker, *loc. cit.* (1879) *pro syn.*; Boiss., Fl. 5:170 (1882). [Plate 45]

Plant 5-10 cm. Corm 1.5-2 × 1-1.5 cm, ovoid; tunics dark brown, extending along subterranean stem. Leaves 6-10, overtopping inflorescence, linear-lanceolate, the upper broadened at base, 6-18 mm broad, long-acuminate. Perianth-segments 16-18 × 4 mm, white to pale lilac, with purple veins and sometimes spots; limb longer than claw, oblong-lanceolate, obtuse or subacute. Stamens ½ length of perianth; anthers twice as long as broad. Fl. December–February.

Hab.: Steppes and arid places; calcareous soil. S. Philistean Plain (Gaza); E. Samaria, Judean Mts., Judean Desert; Upper Jordan Valley, Beit Shean Valley, Lower Jordan Valley, Dead Sea area, Arava Valley; Gilead, Moav, Edom.

Area: E. Saharo-Arabian (Palestine, Egypt).

Reproduction by seeds and cormlets.

6. GAGEA Salisb.*

Bulbous perennials. Bulb 1 or 2-3, globose or ovoid, with 1 fleshy scale renewed every year, enveloped in tunics . Stem erect, solitary, unbranched. Basal leaves 1-2, rarely 3-4, canaliculate or semicylindrical; cauline leaves 2 or more, mostly subtending inflorescence. Inflorescence an umbel-like or corymb-like cyme, rarely raceme-like, or flowers single or twin. Lower flowers largest; the others gradually diminishing in size (measurements given refer to lower flowers). Perianth stellately expanded, rarely funnel-shaped, enlarging and persistent in fruit; perianth-segments 6, free, nearly equal, 3-7-veined, yellow inside, usually with a broad green median stripe on outside, rarely white with a greenish-purple stripe outside. Stamens 6, attached at base of perianth-segments, shorter than perianth; filaments subulate; anthers basi-

---

* Thanks are due to Prof. C.C. Heyn for help during the preparation of the account of *Gagea*, as well as for permission to use the drawings of *Gagea* species prepared under her guidance.

fixed. Ovary 3-locular; ovules numerous; style 1, subulate; stigma capitate or more or less 3-lobed. Capsule 3-gonous or 3-quetrous, sessile or substipitate, loculicidal. Seeds numerous, obovoid, globose to pyriform or flat.

Ninety species in the Mediterranean and in temperate Europe and Asia.

Literature: A.A. Pascher, Übersicht über die Arten der Gattung *Gagea*, *Lotos* 52 (nov. ser. 14):109–131 (1904). A. Terracciano, Gagearum species florae orientalis, *Bull. Herb. Boiss.* Ser. 2, 5:1061 (1905). G. Stroh, Die Gattung *Gagea* Salisb., *Beih. Bot. Centr.* 57 Abt. II:485–520 (1937). J.C.T. Uphof, A review of the genus *Gagea* Salisb., *Plant Life* 14:124–132 (1958); *ibid.* 15:151–161 (1959); *ibid.* 16:163–176 (1960). C.C. Heyn & A. Dafni, Studies in the genus *Gagea* (Liliaceae) I., *Israel Jour. Bot.* 20: 214–233 (1971); II., *ibid.* 26:11–22 (1977).

1. Perianth-segments white with a greenish-purple stripe or veins on outside; perianth funnel-shaped.   **10. G. libanotica**
- Perianth-segments yellow with a green stripe on outside; perianth stellately expanded   2
2. Bulbs 2, enclosed in a common tunic; stem emerging between bulbs. Seeds subglobose or pyriform; capsule 3-quetrous with 3 prominent ribs or wings   3
- Bulb single. Seeds flat, more or less 3-angular; capsule 3-gonous, devoid of prominent ribs   6
3. Basal leaves filiform. Plants usually less than 8 cm tall.   **3. G. bohemica**
- Basal leaves 2–5 mm broad, not filiform. Plants usually at least 8 cm tall   4
4. Cauline leaves opposite or subopposite   5
- Cauline leaves alternate; flower or branch of inflorescence in axil of lower cauline leaf. Perianth-segments pubescent on outside.   **2. G. micrantha**
5. Basal leaves 3-gonous, hollow, mostly single. Perianth-segments glabrous.
                                                                                      4. G. fistulosa
- Basal leaves 2, not hollow. Perianth-segments pubescent on outside.   **1. G. villosa**
6(2). In addition to normal roots, a group of much thicker cord-like roots emerge from one side of bulb and envelop it tightly.   **5. G. dayana**
- All roots more or less uniform and, if thickened, forming a dense net and not emerging from one side of bulb   7
7. Perianth-segments obtuse to subacute. At least 1 cauline leaf remote from inflorescence.   **6. G. chlorantha**
- Perianth-segments distinctly acute to acuminate. No cauline leaf remote from inflorescence   8
8. Plants growing in tufts in desert and steppe habitats. Bulb-tunics extend above ground, forming long collar-like sheath around base of stem. Basal leaves filiform, 1–2 mm broad.   **7. G. reticulata**
- Plants usually growing singly. Bulb-tunics mostly not as above. Basal leaves 2–6 mm broad   9
9. Bulb without thickened roots, with slender stolons ending in bulblets at least in some plants in a population. Stem between basal leaves and inflorescence fairly long.   **9. G. commutata**
- Bulb enveloped by a net of thickened roots, without stolons. Stem between basal leaves and inflorescence very short.   **8. G. fibrosa**

Subgen. 1. GAGEA. Subgen. *Eugagea* Pascher (1904).

Sect. DIDYMOBULBOS C. Koch, Linnaea 22:229 (1849). Bulbs 2. Seeds globose to pyriform. Capsule 3-quetrous, 3-ribbed or -winged.

1. Gagea villosa (Bieb.) Duby, Bot. Gall. 1:467 (1828); Heyn & Dafni, Israel Jour. Bot. 26:13 (1977). [Plates 46, 47]

Plant 8–15 cm. Bulbs 2, enclosed in a common tunic; tunic coriaceous, split. Basal leaves 2, 2–4 mm broad, linear, canaliculate, overtopping inflorescence; cauline leaves 2, opposite or subopposite, pubescent or villose, sometimes with bulblets in axils; lower leaf lanceolate, broader than the upper. Stem more or less hairy above. Inflorescence 2–15-flowered; bracts linear, pubescent; pedicels longer than flowers, more or less hairy. Perianth yellow, stellate-spreading; perianth-segments lanceolate, acute, pubescent outside, especially at base and apex. Capsule oblong-obovate in outline, obtuse, tapering at base, 3-quetrous, 3-ribbed, somewhat shorter than perianth. Seeds pyriform. 2n=60, 72.

Var. **villosa**; Heyn & Dafni, *loc. cit.* & f.3. *Ornithogalum villosum* Bieb., Fl. Taur.-Cauc. 1:274 (1808). *G. arvensis* (Pers.) Dumort., Fl. Belg. 140 (1827) *nom. illeg.*; Boiss., Fl. 5:205; Post, Fl. 2:624. *O. arvense* Pers. in Usteri, Ann. Bot. 11:8, t.1 f.2 (1794) *nom. superfl.* [Plate 46]. Plant 10–15 cm. Inflorescence (5–)10–15-flowered. Bracts nearly as many as flowers. Perianth-segments 15–20 mm. 2n=60, 72. Fl. February–March (rarely April).

Hab.: Fallow fields. Rare. Upper Galilee, Judean Mts.; Golan, Edom.

Var. **hermonis** Dafni & Heyn in Heyn & Dafni, *op. cit.* 16 (1977). [Plate 47]. Plant 8–10 cm. Inflorescence 2–5(–7)-flowered. Bracts few, thin and short or absent. Perianth-segments (10–)12–16 mm. Flowering later in season than var. *villosa*. 2n=60. Fl. April–May (rarely up to July).

Hab.: Rock crevices. Mt. Hermon, 1,500–2,300 m.

Area of species: Euro-Siberian and Mediterranean.

2. Gagea micrantha (Boiss.) Pascher, Lotos nov. ser., 14:122 (1904); Post, Fl. 2:624; Heyn & Dafni, Israel Jour. Bot. 26:16. *G. foliosa* (J. & C. Presl) Schult. & Schult. fil. var. *micrantha* Boiss., Fl. 5:205 (1882). [Plate 48]

Plant 4–7(–15) cm. Bulbs 2, enclosed in a common tunic, globose, 8–10 mm; tunic coriaceous, split. Stem sparsely pilose. Basal leaves 2, 3 mm broad and canaliculate in their upper ½, tapering to 1 mm at base, overtopping inflorescence; cauline leaves 2, more or less hairy; the lower 4 mm broad, lanceolate, shorter than inflorescence; the upper narrower and shorter. Inflorescence 2–7-flowered, more or less hairy; bracts linear to lanceolate; pedicels short. Perianth yellow, stellate-spreading; perianth-segments 7–10 mm, oblong-lanceolate, obtuse to subacute, pubescent outside mainly near base. Stigma clavate. Capsule elliptic in outline, 3-ribbed, ⅔–¾ length of perianth. 2n=24. Fl. April–June.

Hab.: Mt. Hermon, 2,000–2,200 m.

Area: E. Mediterranean. Endemic in Lebanon and Antilebanon.

3. Gagea bohemica (Zauschn.) Schult. & Schult. fil., Syst. Veg. 7:549 (1829); Heyn & Dafni, Israel Jour. Bot. 26:17; Boiss., Fl. 5:206; Post, Fl. 2:625. *Ornithogalum bohemicum* Zauschn., Abh. Privatges. Prag 2:120–121 (1776). [Plate 49]

Plant 5–7 cm, hairy or glabrous. Bulbs 2, small, enclosed in a common tunic; tunic split. Basal leaves 2, filiform, sulcate, exceeding inflorescence; lower cauline leaves alternate, lanceolate, acuminate. Inflorescence 1–3-flowered; bracts lanceolate; pedicels and bracts pubescent to villose. Perianth yellow, stellate-spreading; perianth-segments (10–)12–15 mm, oblong-spathulate, very obtuse, hairy at base. Seeds subglobose. Capsule oblong-obovate in outline, 3-ribbed. Seeds subglobose. 2n=24. Fl. March–April.

Hab.: Rare. Golan; Mt. Hermon.

Area: Mediterranean and W. Irano-Turanian, extending into the Euro-Siberian.

The records on *G. bohemica* from Jerusalem are apparently based on a collection of *G. chlorantha* by Roth, erroneously identified by Boissier as *G. bohemica* (Heyn & Dafni, *loc. cit.*, 1977).

4. Gagea fistulosa Ker-Gawler, Jour. Sci. Arts (London) 1:180 (1816); Heyn & Dafni, Israel Jour. Bot. 26:19; Post, Fl. 2:624. *Ornithogalum fistulosum* Ramond ex DC. in Lam. & DC., Fl. Fr. ed. 3, 3:215 (1805) *nom. superfl.* *G. liotardii* (Sternb.) Schult. & Schult. fil., Syst. Veg. ed. 15, 7:545 (1829); Boiss., Fl. 5:204. [Plate 50]

Plant 10–25 cm. Bulbs 2, enclosed in a common tunic. Basal leaf usually 1, sometimes 2, up to 3 mm broad, linear, semicylindrical to 3-gonous, hollow, exceeding inflorescence; cauline leaves 2, opposite or subopposite, broadly lanceolate, glabrous, not exceeding inflorescence. Inflorescence 1–5-flowered; bracts linear; pedicels pubescent or rarely glabrous. Perianth-segments yellow, 12–15 mm, lanceolate-elliptic, obtuse, glabrous. Capsule obovoid, 3-quetrous, retuse. Seeds subglobose. Fl. March–May(–July).

Hab.: Rare. Golan; Mt. Hermon.

Area: Irano-Turanian, extending into the alpine zone of mountains in Europe.

Subgen. 2. HORNUNGIA (Bernh.) Pascher (1904).

Sect. PLATYSPERMUM Boiss., Fl. 5:204 (1882). Seeds flat, thin. Capsule 3-gonous, not winged.

5. Gagea dayana Chodat & Beauverd in Dinsmore, Pl. Post. Dinsm., Publ. Amer. Univ. Beirut, Nat. Sci. Ser. no. 2, fasc. 1:8 (1932); Heyn & Dafni, Israel Jour. Bot. 20:220; Post, Fl. 2:831. [Plates 51, 52]

Plant 4–15(–20) cm, glabrous to partly pubescent. Bulb single, globose, small, with fibrous tunics. Roots of 2 kinds: numerous thin roots and a group of thick, cord-like, bent roots emerging from one side of bulb and enveloping it. Basal leaf 1(-2), 1-4 mm broad, canaliculate, several times as long as inflorescence; cauline leaves linear to narrowly lanceolate, 1-4 mm broad, often much exceeding inflorescence; no cauline leaves remote from inflorescence. Inflorescence 1-8-flowered; bracts often filiform. Perianth yellow, stellate-spreading; segments acute, glabrous or sparsely hairy. Capsule globose or ellipsoid. Seeds flat. Fl. January–February.

Var. **dayana**. [Plate 51]. Basal leaf 1-2 mm broad. Inflorescence 1-5-flowered, apricot-scented. Perianth-segments 10-15 × 2 mm. Stigma obscurely 3-lobed. Capsule 5-7 × 4-5 mm, more or less globose. 2n=24.
   Hab.: Sandy loam. Sharon Plain, Philistean Plain.
   Area of variety: E. Mediterranean (Israel, Lebanon).

Var. **conjungens** (Pascher) Heyn & Dafni, Israel Jour. Bot. 20:222, f.5B (1971). *G. taurica* Stev. var. *conjungens* Pascher, Bull. Soc. Nat. Moscou nov. ser., 19:368 (1907). *G. conjungens* (Pascher) Wendelbo, Symb. Afgh. 4, Biol. Skr. 10:158 (1958). [Plate 52]. Basal leaf 2-4 mm broad. Inflorescence 2-7-flowered. Perianth-segments (15-)20-30 × 3-4 mm. Stigma distinctly 3-lobed. Capsule 10-12 × 6-7 mm, ellipsoid. 2n=24.
   Hab.: Sands. W. and N. Negev.
   Area of variety: W. Irano-Turanian (Israel, Egypt, Turkey, Afghanistan).
   Area of species: E. Mediterranean and W. Irano-Turanian.

6. **Gagea chlorantha** (Bieb.) Schult. & Schult. fil., Syst. Veg. ed. 15, 7:551 (1829); Heyn & Dafni, Israel Jour. Bot. 20:222. *Ornithogalum chloranthum* Bieb., Fl. Taur.-Cauc. 3:264 (1819). *G. damascena* Boiss. & Gaill. in Boiss., Diagn. ser. 2, 4:105 (1859); Boiss., Fl. 5:209; Post, Fl. 2:625. *G. monticola* Paine, Palest. Explor. Soc. Third Statement 124 (1875). [Plate 53]
   Plant (4-)7-15(-30) cm. Stem more or less pubescent. Bulb single, globose to ovoid, small, with fibrous tunics, sometimes slightly extended along base of stem above ground. Roots numerous, thin, with a few thicker roots sometimes intermingled. Stem always developed. Basal leaves 1-2, filiform to linear, 1-3 mm broad, 1-2 times as long as inflorescence, striate; cauline leaves acuminate; 1-2 cauline leaves remote from inflorescence. Inflorescence (1-)2-5(-7)-flowered; bracts lanceolate to linear, short; pedicels usually pubescent; fruiting pedicels elongated, rigid, erect. Flowers with a faint vanilla smell; perianth yellow, stellate-spreading, growing in size during flowering; perianth-segments 10-15(-22) × 2-3(-4) mm, obtuse to subacute, glabrous. Stigma obscurely 3-lobed. Capsule 9-11 × 6-8 mm, ellipsoid. Seeds flat. 2n=24. Fl. December–April.
   Hab.: Batha and rocky places. Common. Sharon Plain; Upper and Lower Galilee, Esdraelon Plain, Mt. Gilboa, Samaria, Shefela, Judean Mts., Judean Desert, N. and C. Negev; Upper Jordan Valley; Golan, Ammon (Jebel 'Ausha, type of *G. monticola* Paine), Edom.
   Area: E. Mediterranean and Irano-Turanian.

7. Gagea reticulata (Pall.) Schult. & Schult. fil., Syst. Veg. ed. 15, 7:542 (1829); Heyn & Dafni, Israel Jour. Bot. 20:225; Boiss., Fl. 5:208; Post, Fl. 2:625. *Ornithogalum reticulatum* Pall., Reise 3, App. 727, t.D f.2 (1776). *O. circinatum* L. fil., Suppl. 199 (1781). *G. reticulata* var. *tenuifolia* Boiss., Fl. 5:208 (1882). *G. tenuifolia* (Boiss.) Fomin in Fomin & Woronov, Opred. Rast. Kryma Kavkasa 1:233 (1909). *G. reticulata* subsp. *eu-reticulata* Terracc. in Pamp., Pl. Trip. 53 (1914). [Plate 54]

Plant (5-)10-15(-30) cm, growing in tufts. Bulb single, globose to ovoid; tunics fibrous, long-extending along lower part of stem; sometimes bulblets are found in a group at ground level above bulbs in stemless plants. Roots thin. Stem usually developed, puberulent, ribbed. Basal leaf 1, (0.5-)1-2(-2.5) mm broad, filiform, glabrous to pubescent, up to twice as long as inflorescence, sometimes circinate; cauline leaves 3 or more, whorled, filiform to linear, sheathing at base. Inflorescence (1-)4-8(-16)-flowered; bracts few, filiform; pedicels appressed-pubescent; fruiting pedicels erect, rigid. Flowers faintly fragrant; perianth yellow, stellate-spreading; perianth-segments 12-18 × 2-3 mm, distinctly acute to acuminate, pubescent at base and apex. Stigma obscurely 3-lobed. Capsule 13-15 × 7-8 mm, elongate-ellipsoid. Seeds flat. 2n=24. Fl. March–April.

Hab.: Mainly light soils; desert and steppe associations. Common. E. Samaria, Judean Mts., Judean Desert, N. and C. Negev; Beit Shean Valley, Dead Sea area; Golan, Ammon, Moav, Edom.

Area: W. Irano-Turanian.

8. Gagea fibrosa (Desf.) Schult. & Schult. fil., Syst. Veg. ed. 15, 7:552 (1829); Heyn & Dafni, Israel Jour. Bot. 20:227. *Ornithogalum fibrosum* Desf., Fl. Atl. 1:294 (1798). *G. rigida* Boiss. & Sprun. in Boiss., Diagn. ser. 1, 7:108 (1846) p.p. *G. reticulata* (Pall.) Schult. & Schult. fil. var. *fibrosa* (Desf.) Boiss., Fl. 5:208 (1882) p.p. *G. reticulata* (Pall.) Schult & Schult. fil. subsp. *fibrosa* (Desf.) Maire & Weiller in Maire et al., Fl. Afr. Nord 5:126 (1958). [Plate 55]

Plant 2-9(-15) cm. Bulb single, ellipsoid, 5-10 mm; fibrous tunics usually not protruding above ground, rarely extending as collar around base of stem. Roots thickened, numerous, irregularly enveloping bulb in a dense net. Stem usually very short, glabrous. Basal leaf 1, 2-4 mm broad, linear, triangular in cross-section, glabrous or slightly ciliate, usually longer than inflorescence; cauline leaves 2-5, subopposite, linear, long-tapering above, ciliate to more or less pubescent. Inflorescence 1-3(-4)-flowered; bracts few, small, linear, slightly pubescent; pedicels densely pubescent; fruiting pedicels erect, rigid. Flowers with faint apricot smell; perianth yellow, stellate-spreading; perianth-segments 12-20 × 2-3 mm, distinctly acute to acuminate, broadly membranous-margined, pubescent. Stigma usually obscurely 3-lobed. Capsule 8-12 × 6-8 mm, globose to ellipsoid. Seeds flat. 2n=24. Fl. March–April.

Hab.: Arid places. E. Samaria, Judean Mts., Judean Desert; Lower Jordan Valley; Ammon.

Area: Mediterranean, extending into the Saharo-Arabian.

9. Gagea commutata C. Koch, Linnaea 22:227 (1849); Heyn & Dafni, Israel Jour. Bot. 20:229. [Plate 56]

Plant 7–30(–50) cm, more or less pubescent. Bulb single, globose, with fibrous tunics, in some habitats prolonged as a collar around base of stem; at least some plants in each population with stolons; each stolon ending in a small bulb. Roots thin, usually not enveloping bulb. Stem between bulb and inflorescence usually well developed, glabrous or pubescent. Basal leaf 1(–2), (2–)3–6 mm broad, flat or canaliculate, sometimes ciliate, 1–1½ times as long as inflorescence; cauline leaves 3 or more, whorled, linear to narrow-lanceolate, varying in length. Inflorescence (3–)7–15(–25)-flowered; bracts few, linear; pedicels pubescent. Flowers with a faint apricot or smoke smell; peri anth yellow, stellate-spreading; perianth-segments 13–33 × 2–4 mm, distinctly acute to acuminate. Stigma obscurely 3-lobed. Capsule 5–15 × 5–7 mm, oblong-ellipsoid. Seeds flat. 2n=24, 36. Fl. February–March.

Hab.: Batha and fallow fields.

Var. **commutata**; Heyn & Dafni, *op. cit.* 231, f.10A. *G. triphylla* C. Koch, Linnaea 22:229 (1849). *G. sarmentosa* C. Koch, *op. cit.* 230. "*G. reticulata* (Pall.) Schult." sensu Boiss., Fl. 5:208 (1882) p. p. [Plate 56]. Plant 7–15(–20) cm. Basal leaf 3–4 mm broad. Inflorescence 4–10(–20)-flowered. Perianth-segments 2–4 mm broad. Capsule 7–15 × 5–7 mm. 2n=24.

Hab.: Coast of Carmel, Sharon Plain, Philistean Plain; Upper Galilee, Shefela, Judean Mts., N. Negev; Golan.

Var. **procera** (Mout.) Heyn & Dafni, Israel Jour. Bot. 20:232, f.10B (1971). *G. procera* Mout., Fl. Djebel Druze 71 (1953).* Plant (10–)25–50 cm. Basal leaf 4–6 mm broad. Inflorescence (3–)10–25-flowered. Perianth-segments 3–4 mm broad. Capsules do not develop. 2n=36.

Hab.: Judean Mts.; Golan, Ammon.

Area of species: W. Irano-Turanian.

10. Gagea libanotica (Hochst.) Greuter, Israel Jour. Bot. 19:158 (1970). *Lloydia libanotica* Hochst. in Lorent, Wand. Morgenl. 326 (1845). *G. rubroviridis* Boiss. & Ky. in Boiss., Diagn. ser. 2, 4:106 (1859); Boiss., Fl. 5:202; Post, Fl. 2:623. *L. rubroviridis* (Boiss. & Ky.) Baker, Jour. Linn. Soc. London (Bot.) 14:301 (1874); incl. forma *multiflora* Aaronsohn & Opphr. in Opphr. & Evenari, Florula Cisiord., Bull. Soc. Bot. Genève ser. 2, 31:182 (1941). [Plate 57]

Plant 10–30 cm, tufted. Bulbs 2, with membranous tunics and bulblets at base. Basal leaves 1–2, often as long as scape; cauline leaves 2–4. Scape ending in a 3–5-flowered corymb (rarely 7-flowered). Bracts lanceolate. Perianth funnel-shaped; segments 12–15 mm, elliptic-spathulate, white, with a greenish-purple stripe and veins on outside. Capsule enclosed in persistent perianth. Seeds 3-angular, flat. Fl. April.

*Not illustrated.

Hab.: Rocky ground. ? Around Jerusalem; Golan, Edom (1,550 m); Mt. Hermon.

Area: E. Mediterranean (Palestine, Lebanon, Syria).

The record from Jerusalem is based on a specimen in Post's Herbarium (under *Lloydia rubroviridis*). We never found the plant in the area of Jerusalem.

7. TULIPA L.

Bulbous perennials. Bulb tunicate; outer tunics brown, with a dense felty lining of hairs on inside, or glabrous, or with a tuft of wool protruding from bulb near apex. Stem erect. Leaves cauline, alternate, somewhat fleshy, decreasing in size upwards. Flowers mostly solitary, terminal, rarely 2–4. Perianth campanulate or cup-shaped, of 6 free segments; segments deciduous, devoid of a nectariferous pore. Stamens 6, subequal, shorter than perianth; filaments dilated towards base; anthers basifixed. Ovary 3-locular; stigmas 3, sessile, reflexed. Capsule erect, 3-gonous, oblong to subglobose, septicidal. Seeds numerous, discoid.

About 100 temperate Eurasian species.

Literature: E. Regel, Enumeratio specierum hucusque cognitarum generis *Tulipae*, *Acta Horti Petrop.* 2:437–457 (1873). E. Levier, Les Tulipes de l'Europe, *Bull. Soc. Sci. Nat. Neuchâtel* 14:201–312 (1884). W.R. Dykes, *Notes on Tulip Species*, London, 1930. A. D. Hall, *The Genus Tulipa*, Roy. Hort. Soc. London, 1940. A. Horovitz, J. Galil, J. Nevet & J. Jofe, Bulb habit and reproduction in different ploidy forms of *Tulipa oculus-solis* in Israel, *Israel Jour. Bot.* 21:185–196 (1972). W. Marais, Notes on *Tulipa* (Liliaceae), *Kew Bull.* 35:257–259 (1980). P. Wendelbo, *Tulipa*, in: C.C. Townsend et al. (eds.), *Flora of Iraq* 8:82–87 (1984).

1. Filaments hairy at base; perianth-segments white on inside with a yellow basal blotch; outer segments with a greenish median stripe flushed with purple on outside; scape 1–2(–3)-flowered. **1. T. polychroma**
- Filaments glabrous; perianth-segments scarlet, usually with a black blotch at base, often edged with yellow; scape 1-flowered 2
2. Plants of districts receiving 100–350 mm annual rain (Negev, Judean Desert, E. Samaria, E. part of Mt. Gilboa, E. slopes of Lower Galilee facing Upper Jordan Valley, Moav, Edom). Inner perianth-segments nearly as long as to slightly shorter than the outer, obtuse to truncate, often with an abrupt short cusp or apiculum; outer segments acute to cuspidate; dark blotch at base of segments often without yellow margin. **3. T. systola**
- Plants of Mediterranean districts receiving 400–800 mm annual rain. Inner perianth-segments distinctly shorter (by 5–15 mm) than acuminate to long-cuspidate outer ones which project from rest of flower 3
3. Plants of Mediterranean hill districts, mostly on terra rossa, in maquis and garigue areas. Lowermost leaf 2.5–5 cm broad. **2. T. agenensis** subsp. **agenensis**
- Plants of sandy soils of coastal plain. Lowermost leaf 1–2 cm broad. Perianth yellowish on outside; segments generally smaller than in subsp. *agenensis*.
2. T. agenensis subsp. **sharonensis**

Sect. 1. ERIOSTEMONES Boiss., Fl. 5:191 (1882). Filaments with a tuft of hairs at base.

1. Tulipa polychroma Stapf, Denkschr. Akad. Wiss. Math.-Nat. Kl. (Wien) 50:18 (1885); Vved., in Fl. URSS 4:359; Hall, *op. cit.* 76, t.12. [Plate 58]

Plant 10–20 cm. Bulb 2–3 cm in diameter; tunics brown, leathery, with a dense felt-like lining. Leaves 2(–3), glaucous, linear-lanceolate, often falcate, canaliculate, the lower 1–2 cm broad. Flowers 1–2(–3), nodding in bud, saucer- to cup-shaped when open; perianth-segments 25–45 mm, subacute, white on inside, with a yellow basal blotch; inner segments ovate, broader than the outer; outer segments with a greenish midstripe flushed with purple on outside. Stamens ⅓–½ length of perianth; filaments yellow, longer than anthers, glabrous, bearded at base; anthers yellow, dark-tipped before dehiscence; pollen yellow. Capsule short (1.5–2.5 cm), 3-gonous, obtuse, abruptly apiculate. Fl. March–April.

Hab.: Steppes; rocky and sunny places. C. Negev; Moav, Edom.
Area: W. Irano-Turanian (Palestine, E. Syria, Iran, E. Caucasus).
Protected by law.

Contrary to Marais (*loc. cit.*, 1980), who cites *T. polychroma* Stapf as synonym of *T. biflora* Pall. (1776), *T. polychroma* significantly differs from *T. biflora* in leathery bulb-tunics densely lanate inside, and in larger perianth-segments (25–45 mm, not up to 25 mm).

T. lownei Baker (1874), also of Sect. *Eriostemones,* is endemic to the alpine zone of Mt. Hermon, Antilebanon and Lebanon. Apart from its ecology, it differs from *T. polychroma* in the colour of its perianth which is pink with a yellow centre on the inside and somewhat darker pink on the outside, in a usually single flower, and often in more numerous leaves.

Sect. 2 TULIPA. Sect. *Leiostemones* Boiss., Fl. 5:191 (1882). Filaments glabrous at base.

2. Tulipa agenensis DC. in Redouté, Liliacées 1:t.60, add. (1804); Marais, Kew Bull. 35:257 (1980). *T. oculus-solis* St-Amans, Rec. Soc. Agr. Sci. Agen. 1:75 (1804) *nom. nud.*; DC. in Lam. & DC., Fl. Fr. ed. 3, 3:200 (1805); Hall, *op. cit.* 108, t.23 & f.16; Boiss., Fl. 5:192; Post, Fl. 2:620. *T. montana* auctt. fl. pal. non Lindley (1827). [Plates 59, 60]

Plant 15–30(–40) cm. Bulb-tunics maroon-brown, papery, with a dense and thick felt-like lining. Leaves usually 4, glaucous, lanceolate, acuminate, erect or falcate; margins often undulate, sometimes ciliolate. Flower solitary, erect; perianth scarlet, narrowly campanulate; perianth-segments tapering towards base, with a large black basal blotch bordered with yellow on inside; outer segments ovate, acuminate to cuspidate, broadest below or at middle; the inner distinctly shorter, obovate, acute to cuspidate, broadest above middle. Filaments glabrous, blackish-purple; anthers purplish, exceeding the pistil; pollen yellow to olive-green. Capsule oblong, much varying in size. Fl. March-April.

Subsp. **agenensis**. [Plate 59]. Plant up to 40 cm, usually 30 cm. Bulb 2-3 cm in diameter. Lowermost leaf 2.5-5 cm broad. Outer perianth-segments (35-)40-62(-80) × (12-)15-30 mm, acuminate to cuspidate; the inner (30-)33-60 × 13-25 mm. 2n=24, 36.

Hab.: Maquis, garigue, batha. Fairly common. Upper and Lower Galilee, Mt. Carmel, Samaria, Judean Mts.; Golan, Gilead, Ammon.

Area of subsp.: as of species.

Protected by law.

Subsp. **sharonensis** (Dinsmore) Feinbr. comb. nov. *T. sharonensis* Dinsmore, Pl. Post. Dinsm., Publ. Amer. Univ. Beirut, Nat. Sci. Ser. no. 3, fasc. 2:12 (1934); Post, Fl. 2:621; Hall, *op. cit.* 110. *T. boissieri* Regel, Acta Horti Petrop. 2:448 (1873)*; Gartenflora 22:296 (1873). [Plate 60]. Differing from subsp. *agenensis* in size of plant and flower. Plant up to 30 cm, usually 20-25 cm. Bulb 1.5-2.5 cm in diameter. Lowermost leaf 1-2 cm broad. Perianth paler and suffused with yellow on outside; outer segments 38-45(-60) × 11-23 mm, acuminate to cuspidate; the inner (28-)35-40(-45) × 8-20 mm.

Hab.: Confined to sandy soils, calcareous sandstone hills and sandy loam of coastal plain. Coastal Galilee, Coast of Carmel, Sharon Plain, Philistean Plain.

Area of subsp.: E. Mediterranean. Endemic (Coastal Plain of Palestine).

Protected by law.

Often in large stands, but in danger of extinction, owing to urbanization of the coastal plain.

Area of species: E. Mediterranean (Palestine, Lebanon, Syria, Cyprus, S. Turkey).

Though this is doubted by some authors, *T. agenensis* subsp. *agenensis* is indigenous to above E. Mediterranean countries; there it is confined to primary plant associations and fairly common in forests, maquis and garigue, often on terra rossa among rocks. In France and Italy, where it was apparently introduced in the 16th century, it has become naturalized and is common on cultivated ground. Not recorded from European countries east of Italy (Jugoslavia, Greece).

Horovitz et al. (*op. cit.*, 1972) studied diploid (2n=24) and triploid (2n=36) plants of *T. agenensis* subsp. *agenensis* (*T. oculus-solis*) in natural stands in the Judean Mts. The diploid plants reproduce either entirely by seeds or partly vegetatively by lateral stolons ("droppers") bearing increase bulbs. The triploids are nearly entirely pollen- and seed-sterile and reproduce vegetatively forming clones. Diploid plants reproduce mainly by cross-pollination, insuring variation. The stoloniferous habit has a survival value, since the plants bearing droppers are well adapted to entering rock crevices and soil pockets where they are protected from the mole-rat (*Spalax ehrenbergii*) which endangers the tulip bulbs.

The triploid *T. aleppensis* reported by Hall (*op. cit.* 115-116) from near Beirut is apparently a triploid form of *T. agenensis*.

Dr. A. Horovitz (personal communication) has examined in Lower Galilee a population of *T. agenensis* with yellow flowers. Marais (*op. cit.* 257, 1980) is right in

*Type: coll. by Roth in Ramle (Palestine).

remarking: "In those species of tulips which have a yellow mesophyll, plants with yellow flowers are equivalent to albinos".

3. Tulipa systola Stapf, Denkschr. Akad. Wiss. Math.-Nat. Kl. (Wien) 50:17 (1885); Hall, *op. cit.* 108; Täckholm & Drar, Fl. Eg. 3:136; Wendelbo, in Rech. fil., Fl. Lowland Iraq 155 (1964); Wendelbo, Tulipa (MS.), in Fl. Iraq. *T. stapfii* Turrill, Bot. Mag. 157, t.9356 (1934); Hall, *op. cit.* 110, t.24 & f.17. *T. cuspidata* Stapf, *loc. cit.* (1885) non Regel (1884). *T. montana* sensu Boiss., Fl. 5:192 (1882) p. p. non Lindley (1827); Post, Fl. 2:621. *T. montana* var. *amblyophylla* Post, Plantae Postianae 1:13 (1890). *T. amblyophylla* (Post) Feinbr. in Eig, Zohary & Feinbr., Anal. Fl. Pal. 362 (1948) *nom. superfl. T. montana* subsp. *amblyophylla* (Post) Mout., Nouv. Fl. Liban Syrie 1:231 (1966). [Plate 61]

Plant 15–20(–35) cm. Bulb 2–3(–4) cm in diameter; tunics maroon-brown, papery, with a dense and thick felt-like lining; tunics from previous years persisting in dry climate. Leaves usually 4, glaucous, linear-lanceolate to elliptic, acuminate, erect or falcate; margin more or less ciliate, usually undulate; lowermost leaf 2–3.5 cm broad or broader. Flower solitary, erect; perianth-segments cuneate at base, warm-scarlet (as in *Ranunculus asiaticus*), slightly paler on outside, bearing an oblong basal black blotch, sometimes but not always edged with yellow; outer segments (32–)45–70(–80) × 20–30(–45) mm, ovate to rhombic, acute to cuspidate, often with a white cusp at tip; inner segments slightly shorter, obovate, very obtuse, rounded to nearly truncate, often with an abrupt ciliolate short cusp. Filaments glabrous, blackish-purple, with yellowish base and apex; anthers yellow or purplish; pollen yellow or olive-green. Ovary cylindrical, 3-gonous, generally longer than filaments. Capsule greatly varying in length, 3-quetrous, shortly beaked, straw-coloured. 2n=? Fl. end of February–April.

Hab.: Rocky calcareous hillslopes, limestone cliffs, in crevices and soil-pockets; Artemisietum herbae-albae, Zygophylletum dumosi, Ziziphetum loti. Locally common. E. slopes of Lower Galilee (W. Hammam, 130 m below sea level) and of Mt. Gilboa, E. Samaria, Judean Desert, C. and S. Negev; Upper Jordan Valley; Moav, Edom.

Area: W. Irano-Turanian (S. Palestine, Sinai, Syria, ?Cyprus, E. Turkey, Iraq, Iran).

Protected by law.

In the Negev *T. systola* grows on rocky and stony hillslopes dominated by Artemisietum herbae-albae and Zygophylletum dumosi. Further north, along the narrow strip of the Irano-Turanian territory, on the slopes of Judean Desert, E. Samaria, eastern part of Mt. Gilboa to Upper Jordan Valley, *T. systola* appears in plant associations of *Retama raetam* on rocks and, in the northern part of the Jordan Valley, in Ziziphetum loti.

T. montana Lindley has been recorded from Palestine by various authors (Boissier, 1882; Eig, 1933; Rechinger, 1952). Described from Persia by Lindley (1827), *T. montana* remains rather confused, since no type specimen has been preserved; Lindley himself later identified a tulip of the *Oculus-solis* group as *T. montana*. Hall (*op. cit.* 90) classed *T. montana* with the *Clusianae* group in which a tuft of wool protrudes from the apex of the bulb, and tunics are glabrous inside except near apex. He described the bulb-tunics of *T. montana* as "tough purplish skin with tuft of wool protruding from the apex, but only extending a short way down within".

Specimens cited under *T. montana* Lindl. by Boissier (in Flora Orientalis 1882) and described with tunics that are densely woolly inside are evidently *T. systola* Stapf, with the exception of specimen 454 collected by Roth from Ramleh (Philistean Plain), which is the type specimen of Regel's *T. boissieri*.

The most significant feature of *T. systola* as compared with *T. agenensis* is the shape of the flower, which is flat-topped because of the nearly equal length of outer and inner perianth-segments and the obtuse and often truncate apex of inner segments. In *T. agenensis* (both subsp. *agenensis* and subsp. *sharonensis*) inner segments are acute and outer segments distinctly exceed the inner ones in length and project above them.

The flowers of *T. systola* in the Negev and Edom are in general 35–45 mm, those of Mt. Gilboa and E. slopes of Lower Galilee facing Upper Jordan Valley, especially the ones from W. Hammam, north of Tiberias, are much larger and often reach 70–80 mm. They are strikingly similar to the flower depicted in the Botanical Magazine t.9356 (1934) as *T. stapfii* Turrill (synonymous with *T. systola*).

8. FRITILLARIA L.

Caulescent perennials with tunicate or squamose bulb. Leaves alternate or in whorls of 3. Flowers nodding, terminal and solitary or 2–3, or in an umbel or raceme; flower-bud erect. Perianth mostly campanulate, deciduous, of 6 segments; segments free, equal, obovate or oblong-spathulate, provided with a nectariferous pit at or above base. Stamens 6, inserted on receptacle of flower, included or subexserted; anthers basifixed. Ovary 3-locular; style 1, filiform, undivided or more or less deeply 3-fid. Capsule erect, obovoid or cylindrical, often short-stipitate, obtusely or acutely 6-gonal, loculidical. Seeds numerous, discoid.

Eighty-five N. temperate species.

Literature: E. M. Rix, *Fritillaria* (Liliaceae) in Iran, *Iran. Jour. Bot.* 1:75–95 (1977).

1. Fritillaria persica L., Sp. Pl. 304 (1753); Redouté, Liliacées 2:t.67 (1804); Ker-Gawler, Bot. Mag. 37:t.1537 (1813); Rix, Iran. Jour. Bot. 1:80 (1977); Wendelbo, Tulips and Irises of Iran 28: f.24 (1977); Boiss., Fl. 5:188; Post, Fl. 2:619. *F. libanotica* (Boiss.) Baker, Jour. Linn. Soc. London (Bot.) 14:270 (1874); Stapf, Bot. Mag. 151: t.9108 (1925); Boiss., Fl. 5:189; Post, Fl. 2:620. *Theresia libanotica* Boiss., Diagn. ser. 1, 13:20 (1854) var. *bracteata* Boiss. & var. *ebracteata* Boiss., *op. cit.* 21 (1854). *F. arabica* Gandoger, Bull. Soc. Bot. Fr. 66:291 (1920); Post, Fl. 2:619. [Plate 62]

Plant 60–75(–100) cm. Bulb large, squamose. Stem erect, leafy. Leaves all cauline, sessile, up to ⅔ of height of stem, glaucous, erect, oblong-lanceolate, subacute, 1.5–2.5 cm broad. Inflorescence a long raceme, 10–30 cm, 12–25-

flowered; flowers nodding, very often bracteate. Bracts, when present, oblong, often as long as or longer than pedicels. Pedicels erecto-patent, generally longer than flower; fruiting pedicels erect. Perianth 1.5–2 cm, mostly light greenish-yellow, finely veined with purplish-brown, sometimes plum-purple. Filaments glabrous or papillose. Capsule 6-gonal, depressed, 2–3 cm in diameter. Fl. February–April.

Hab.: Maquis, batha, rocky places. Upper and Lower Galilee, Esdraelon Plain, Mt. Carmel, Samaria, Judean Mts., N. Negev; Beit Shean Valley; Golan, Ammon, Moav, Edom.

Area: E. Mediterranean and W. Irano-Turanian (Palestine, Lebanon, Syria, Cyprus, S. and E. Turkey, Iraq, Iran)

9. Lilium L.

Caulescent bulbous perennials. Bulb of numerous overlapping scales. Stem erect, stout, leafy. Leaves numerous, alternate or in whorls, decreasing in size upwards. Inflorescence mostly a terminal raceme. Flowers large, showy. Perianth campanulate to infundibular; segments 6, free, equal, clawed, deciduous; claw with a nectariferous furrow. Stamens 6; filaments filiform, long; anthers dorsifixed, versatile. Ovary 3-locular; style 1, long; stigma 3-lobed. Capsule obovoid, 6-gonal. Seeds numerous, discoid.

About 80 temperate species.

Literature: H. B. D. Woodcock & W. T. Stearn, *Lilies of the World*, London, 1950.

1. Lilium candidum L., Sp. Pl. 302 (1753); Eig, Bull. Inst. Agr. Nat. Hist. Tel-Aviv 4:39–40 (1926); Mout., Nouv. Fl. Liban Syrie 1:236; Boiss., Fl. 5:172; Post, Fl. 2:617. [Plate 63]

Plant 60–150 cm. Bulb up to 10 cm, ovoid, yellow. Basal leaves oblanceolate, 3–5-veined, usually numerous; stem-leaves much smaller, gradually passing into bracts. Raceme lax, 2- to several-flowered, bracteate, sometimes bracteolate; pedicels erecto-patent, shorter than flower; flowers partly nodding, strongly fragrant; perianth white, broadly infundibular, 7–10 cm; segments oblanceolate, slightly recurved at apex. Filaments white; anthers and pollen yellow. 2n=24. Fl. May–June.

Hab.: Shade of maquis, rocky ground. Rare. Upper Galilee, Mt. Carmel.

Indigenous in Palestine and Lebanon. Protected by law.

Area: E. Mediterranean.

Literature: A. Eig, A contribution to the knowledge of the Flora of Palestine, *Bull. Inst. Agr. Nat. Hist. Tel-Aviv* 4:39–40 (1926). O. Warburg, Heimat und Geschichte der Lilie *(Lilium candidum)*, Feddes Repert. Beih. 56:167–204 (1929). Rivka Tadmor, *A cytological and morphological comparison between the wild and the cultivated Lily* (*L. candidum*) in *Israel*, MSc. Thesis, Hebrew Univ. Jerusalem, 1953–1954 (in Hebrew).

Wild-growing *Lilium candidum* was discovered on 25.6.1925 by N. Naftolsky in

Upper Galilee, between the villages Khurfesh and Peki'in, in the walls of a deep karst pit, among maquis shrubs. Later it was found in a similar habitat on Mt. Carmel by T. Kushnir. Post (Fl. 2:617) and Mouterde (Nouv. Fl. 1:236) record *L. candidum* growing on rocks in various localities in Lebanon.

According to Warburg (*op. cit.* 174), *L. candidum* has been found growing spontaneously in Greece including Crete, and also in W. Asia Minor.

Considering the scattered distribution of the species and the character of its habitats, Warburg assumes that its seeds are dispersed by birds.

According to Mouterde *(loc. cit.)*, the wild plants do not differ at all from cultivated White Lily. However, R. Tadmor *(op. cit.)* found significant differences in the shape and size of stem-leaves and bracts between the wild strain of Mt. Carmel and cultivated forms.

10. SCILLA L.

Bulbous and scapose perennials. Bulb ovoid to globose, renewed progressively in the course of several years; storage scales free. Leaves basal, linear to broadly strap-shaped or oblong. Scapes 1 or more. Inflorescence mostly a raceme; bracts small, linear or deltoid, sometimes obsolete; in some species bract accompanied by a bracteole; pedicels not jointed. Perianth stellate-spreading or campanulate; segments 6, petaloid, free or slightly connate at base, with a darker midvein, devoid of nectary, blue, lilac, violet, purplish-pink or greenish-purple, rarely white, deciduous or shrivelled and persistent after flowering. Stamens 6; filaments filiform, inserted at base of perianth; anthers dorsifixed. Ovary 3-locular; ovules 2–30 per locule; style 1, filiform; stigma capitate. Capsule membranous or fleshy, subglobose, 3-lobed, loculicidal. Seeds solitary or several in each locule, globose to ovoid or 3-gonous, dark brown or black, with or without aril.

About 80 species in temperate Eurasia, S. Africa, a few species in Trop. Africa.

Literature: J. G. Baker, Revision of the genera and species of Scilleae and Chlorogaleae, *Jour. Linn Soc. London (Bot.)* 13:214, 228–266 (1873). P. Chouard, Types de développement de l'appareil végétatif chez les Scillées, *Ann. Sci. Nat. Bot.* Ser. 10, 13:132–323 (1931). E. Battaglia, Filogenesi del cariotipo nel genere *Scilla*. II. Il cariotipo diploidi di *Scilla autumnalis* L., *Atti Soc. Tosc. Sci. Nat.*, Mem. Ser. B 59:146–161 (1952). R. D. Meikle, *Scilla cilicica, Bot. Mag.* 181:175–178, t.745 (1977). F. Speta, Karyological investigations in *Scilla* in regard to their importance for taxonomy, *Webbia* 34:419–431 (1979); Die Frühjahrsblühenden *Scilla*-Arten des östlichen Mittelmeerraumes, *Naturk. Jahrb. Stadt Linz* 25:19–198 (1981) & *Optima Leaflets* 113 (1981).

1. Plants flowering in autumn (October–December), just before or simultaneously with emergence of leaves. Leaves not reaching 0.5 cm in width. Perianth-segments 3–5 mm   2
-   Plants flowering in winter or spring (February–April) after emergence of leaves. Leaves synanthous, 1 cm broad or broader. Perianth-segments 6–16 mm   3

2. Lower pedicels at most twice as long as perianth-segments. Perianth blue-violet or lilac.   **3. S. autumnalis**
- Lower pedicels 6–8 times as long as perianth-segments. Perianth pale lilac with a brownish midvein.   **4. S. hanburyi**

3(1). Flowering scape 15–25 cm. Raceme few-flowered. Perianth-segments 12–16 mm. Leaves usually 3–5.   **1. S. cilicica**
- Flowering scape 30–80 cm. Raceme 40–150-flowered. Perianth-segments 6–7 mm. Leaves 8–12.   **2. S. hyacinthoides**

Subgen. 1. SCILLA. Roots annual, unbranched. Flowering in spring simultaneously with or after emergence of leaves. Inflorescence 1–15-flowered. Seeds with or without aril. Germination epigeal.

1. Scilla cilicica Siehe,* Gard. Chron. ser. 3, 44:194, f.81 (1908); Meikle, Bot. Mag. 181:175, t.745; Speta, *op. cit.* 131–134, f.25 (1981); Mout., Nouv. Fl. Liban Syrie 1:237, t.75 f.1. *S. palaestina* Feinbr. in Feinbr., Zohary & Koppel, Flora of the Land of Israel, Iconography, Plates 101–150, colour plate with text in English (p.15) (1958) *nom. invalid. sine descr. lat. S. veneris* Speta, Naturk. Jahrb. Stadt Linz 22:69 (1977). *S. cernua* auctt. fl. pal. & libano-syr. non Delaroche in Redouté (1810). [Plate 64]

Bulb ovoid, (1–)1.5–2.5 cm in diameter, with a very short neck; outer tunics papery, dark violet-purple, glossy. Roots unbranched; young bulb forming a contractile root. Leaves (2–)3–5(–6), synanthous, 1–1.2(–1.5) cm broad, shallowly canaliculate, usually longer than flowering scapes. Scapes (1–)2–4, 15–25 cm, flexuose, ascending. Raceme lax, (1–)2–6(–8)-flowered; rachis of raceme tinged with lavender-blue. Bract and bracteole small, light violet, deltoid to ovate, often spurred at base. Pedicels recurved during anthesis, not accrescent. Flowers showy, nodding. Perianth subcampanulate to stellate-spreading; perianth-segments 12–16 × 3–4 mm, lavender-blue, oblong-lanceolate, deciduous after flowering. Filaments whitish to lavender-blue, about 8 mm; anthers slaty violet-blue. Ovary violet to greenish, subglobose, about 3 mm in diameter, 3-sulcate; style about as long as filaments, whitish or tinged with violet; ovules 4–5 per locule. Fruiting scape prostrate. Capsule 6–8 mm in diameter, subglobose. Seeds 3–4 per locule, subglobose, about 2 mm, dull black, minutely rugulose, without aril. Germination epigeal. Fl. February–March.

Hab.: Shade of maquis, humus-rich soil on calcareous rocks. Rare. Western Upper Galilee, Mt. Carmel.

Area: E. Mediterranean (N. Palestine, Lebanon, Syria, Cyprus, S. E. Turkey).

Speta (*op. cit.*, 1981:133) says: "*Scilla cilicica* is one of the least known *Scilla* species". Indeed, a true-to-nature colour plate with the correct name and a detailed des-

---

*Thanks are due to H. Mordak (Leningrad) for a loan of specimens of *S. cilicica* from Turkey and for diverse information on the species.

cription of the species by Meikle *(op. cit.)* was only published as recently as 1977. From Palestine the plant has been recorded in the Flora of Post (1933) as *S. cernua* Redouté and *S. bifolia* L. In the Analytical Flora of Palestine by Eig, Zohary & Feinbrun (1948, in Hebrew, with a Catalogue of Latin names) it appeared under *S. hohenackeri* Fisch. & Mey. In the Iconography published in 1958 *(op. cit.)* the present author considered it a new species and named it *S. palaestina* Feinbr. (but not validly published). The above-cited Iconography contains a colour plate of the species by Ruth Koppel and text which reads: "This charming plant grows in shady rocks and within the maquis of Mt. Carmel and of Upper Galilee. It is also found in the Lebanon and Cyprus. Its bulb is small, the leaves linear and about 1 cm broad. The inflorescence is loose and few-flowered. The peduncle bends down to the ground after flowering. The flowering time is February to March".

Subgen. 2. PETRANTHE (Salisb.) Chouard (1931). Roots perennial, branched. Flowering in spring after emergence of leaves. Inflorescence many-flowered. Seeds without aril.

2. Scilla hyacinthoides L., Syst. Nat. ed. 12, 2:243 (1767); Boiss., Fl. 5:225; Post, Fl. 2:632. [Plate 65]

Bulb broadly ovoid, 4–6 cm in diameter; tunics greyish, forming a kind of web when torn. Roots branched. Leaves 8–12, synanthous, 30–60 cm, up to 3 cm broad, shallowly canaliculate, gradually narrowed at base and apex, ciliolate. Scape usually 1, 30–80 cm, erect, stout, longer than leaves. Raceme up to 40 cm, 40–150-flowered, more or less cylindrical; rachis of raceme amethystviolet. Bracts and bracteoles minute, ovate to deltoid, whitish, persistent. Flowering pedicels erecto-patent, amethyst-violet; the lower several times as long as flower; fruiting pedicels widely spreading, arcuate. Perianth stellatespreading; segments 6–7 × 2 mm, oblong, amethyst-violet, shrivelled and persisting after flowering. Filaments pale violet, 5 mm; anthers wine-red. Ovules 2 per locule, hanging side by side. Capsule depressed-globose, 3-sulcate. Seeds black, without aril. Germination epigeal. 2n=20. Fl. February–April.

Hab.: Batha and fallow fields. Acco Plain, Sharon Plain, Philistean Plain; Upper and Lower Galilee, Mt. Carmel, Esdraelon Plain, Samaria, Judean Mts.; Golan, Gilead.

Area: N. Mediterranean.

Subgen. 3. PROSPERO (Salisb.) Chouard (1931). Roots perennial, branched. Flowering in autumn before or with first rains. Leaves hysteranthous to subsynanthous. Inflorescence 5–60-flowered. Bracts obsolete. Seeds without aril. Germination epigeal.

3. Scilla autumnalis L., Sp. Pl. 309 (1753); Boiss., Fl. 5:224; Post, Fl. 2:631. [Plate 66]

Bulb ovoid, 1–2.5 cm in diameter; outer tunics greyish. Leaves 5–6, 1.5–2 mm broad, narrowly linear to subfiliform, subterete, sulcate on upper face, hysteranthous to subsynanthous, appearing shortly after first rains. Scapes 1-3, 5-30 cm, longer than leaves, erect, surrounded by several cataphylls. Raceme 5-25-flowered; fruiting raceme cylindrical. Bracts and bracteoles obsolete. Pedicels erecto-patent, the lower 1-1½(-2) times as long as perianth-segments. Perianth stellate-spreading; segments 3-5 mm, oblong, blue-violet or lilac, with a darker midvein. Filaments violet or lilac, flattened, somewhat shorter than perianth-segments; anthers dark purple. Capsule 2.5-3 mm in diameter, ovoid, 3-sulcate. Ovules 2 per locule, side by side. Seeds black, finely rugulose, without aril. Germination epigeal. 2n=14, 14+6 to 8B, rarely 2n=21 (3x) or 2n=28 (4x). Fl. October–December.

Hab: Batha and maquis. Common. Coast of Carmel, Sharon Plain, Philistean Plain; Upper and Lower Galilee, Mt. Carmel, Esdraelon Plain, Mt. Gilboa, Samaria, Judean Mts.; Dan Valley, Hula Plain, Upper Jordan Valley; Golan, Gilead, Ammon.

Area: Mediterranean, extending into the Euro-Siberian.

Literature: E. Battaglia, Un secondo caso di B-chromosomi (2n=14+6-8B) in *Scilla autumnalis* proveniente dalla Palestina, *Caryologia* 17:65-76 (1964).

4. Scilla hanburyi Baker, Jour. Linn. Soc. London (Bot.) 13:235 (1873); Boiss., Fl. 5:225; Post, Fl. 2:632. [Plate 67]

Bulb ovoid, 2-3(-5) cm in diameter; outer tunics greyish. Leaves 6-8, 2-3 mm broad, linear, hysteranthous. Scape 1 or more, 10-20 cm. Raceme 15-40-flowered; fruiting raceme pyramidal. Bracts obsolete. Flowering pedicels ascending; fruiting pedicels widely spreading, the lower 6-8 times as long as perianth-segments. Perianth stellate-spreading; segments 3-5 mm, oblong, pale lilac with a brownish or greyish midvein. Filaments somewhat shorter than perianth-segments. Capsule 3 mm in diameter, obovoid, 3-sulcate. Seeds black, rugulose, without aril. Fl. October–November.

Hab.: Deserts. Judean Desert, C. Negev; Dead Sea area.

Area: W. Irano-Turanian (S. Palestine, Lebanon, Syria).

11. URGINEA Steinh.

Bulbous and scapose perennials. Bulb ovoid to globose, renewed progressively in course of several years; storage scales free, fleshy. Leaves basal, often hysteranthous. Inflorescence a raceme; each flower subtended by a bract and often a bracteole; bracts deltoid or linear, often spurred at base or at middle; pedicels not jointed. Perianth persistent, stellate-spreading; segments 6, petaloid, free, subequal, devoid of nectary, often white or yellow, with a darker midvein or stripe. Stamens 6, inserted at base of perianth-segments; filaments filiform; anthers dorsifixed. Ovary 3-locular; ovules several in each locule; style 1, filiform; stigma capitate, 3-sulcate. Capsule chartaceous, 3-gonous, loculicidal. Seeds (1-)2-10 in each locule, flattened, winged, arranged one above the other.

About 10 species in the Mediterranean and Saharo-Arabian regions, in extratrop. Africa and India.

Literature: J. G. Baker, Revision of the genera and species of Scilleae and Chlorogaleae, *Jour. Linn. Soc. London (Bot.)* 13:214-224 (1873). G. Martinoli, Richerche citotassonomiche sui generi *Urginea* e *Scilla* della flora Sardoa, *Caryologia* 1:329-357 (1949).

1. Leaves 3 cm broad or broader. Perianth-segments white with a green or purplish midvein.   **2. U. maritima**
- Leaves 0.3-1 cm broad, strongly undulate. Perianth-segments not white.
                                                                              1. U. undulata

1. Urginea undulata (Desf.) Steinh., Ann. Sci. Nat. Bot. ser. 2, 1:330 (1834); Boiss., Fl. 5:223; Post, Fl. 2:630. *Scilla undulata* Desf., Fl. Atl. 1:300, t.88 (1798). *Drimia undata* Stearn, Ann. Mus. Goulandris 4:208 (1978). [Plate 68]

Plant 20-30 cm. Bulb ovoid, 2-4 cm in diameter. Leaves 5-8, hysteranthous, 3-10 mm broad, linear, grooved, strongly undulate, spread on ground, appearing with rains several weeks after end of flowering. Scape erect, slender, much longer than leaves. Raceme lax, 15-30 cm, 8-30-flowered. Bracts membranous, about 2-5 mm, whitish, spurred below base, linear or linear-lanceolate, with a small bracteole at side. Pedicels erecto-patent, about as long as perianth. Perianth stellate-spreading; segments 10-14 mm, linear to spathulate, dull purple, pink or olive-grey suffused with pink, with a purplish midvein. Filaments filiform, as long as perianth or shorter. Capsule 3-gonous, nearly orbicular in outline. Seeds 3-4 in each locule. 2n=40(4x). Fl. August-October, before rains. Leaves appearing in December, at beginning of rainy season, persisting until April.

Hab.: Steppes and deserts. S. Philistean Plain; N. and C. Negev; Arava Valley; Moav, Edom.

Area: Saharo-Arabian, extending into the Mediterranean.

2. Urginea maritima (L.) Baker, Jour. Linn. Soc. London (Bot.) 13:221 (1873); Boiss., Fl. 5:224; Post, Fl. 2:631. *Scilla maritima* L., Sp. Pl. 308 (1753). *Drimia maritima* (L.) Stearn, Ann. Mus. Goulandris 4:204 (1978). [Plate 69]

Bulb 5-10(-18) cm in diameter; tunics thick and fleshy, whitish or brown-red. Leaves 10-20, 30-60 × 3-8 cm, erect, broadly lanceolate, glaucous, hysteranthous, appearing early in winter after end of flowering. Scape 60-100 cm or taller, stout, purplish. Raceme long, dense, with 50 flowers or more. Bracts small, membranous, spurred below middle. Pedicels erecto-patent, about 2-3 times as long as perianth. Perianth stellate-spreading; segments 7-9 mm, oblong, obtuse, white with a green or purplish midvein. Filaments shorter than perianth, flattened at base; anthers greenish. Style about as long as stamens. Capsule 3-gonous, obovate in outline. Seeds 1-4 in each locule, black, glossy. 2n=40(4x). Fl. July-October, before rains. Leaves appearing in November-December, at beginning of rainy season, persisting until April or May.

Hab.: Batha, fallow fields, field borders. Very common. Sharon Plain, Philistean Plain; Upper and Lower Galilee, Mt. Carmel, Samaria, Shefela, Judean Mts., Judean Desert, C. Negev; Upper Jordan Valley; Golan, Gilead, Ammon, Moav, Edom.

Area.: Mediterranean.

Bulbs collected and used as diuretic in pharmaceutical industry.

According to Maire et al. (Flore de l'Afrique du Nord 5:165, 1958), "One finds in November–December synanthous plants with a raceme of greenish closed flowers; these are specimens parasitized by *Ustilago urgineae* Maire. This parasite, whose spores develop in the anthers, retards the development of the flowering scape till the appearance of leaves" (translated from French). The fungus has not been recorded in Israel.

12. ORNITHOGALUM L.

Bulbous and scapose perennials. Bulb tunicate. Leaves basal, linear or lorate, synanthous. Inflorescence a raceme or corymb. Bracts solitary, mostly membranous, lanceolate. Flowers protogynous and generally outcrossing. Perianth persistent, stellate-spreading; perianth-segments 6, free or almost so, subequal, white (rarely yellowish), mostly with a green median stripe on outside. Stamens 6, usually inserted on receptacle of flower; filaments petaloid, white, triangular or lanceolate to linear, sometimes 3-cuspidate; anthers dorsifixed, versatile. Ovary 3-locular, multi-ovulate; style 1, filiform; stigma capitate or obscurely 3-lobed. Capsule 3-gonous and 3-sulcate, sometimes 3-winged, loculicidal by 3 valves. Seeds obovoid or angular.

About 150 species in temperate regions of the Old World.

Literature: J. G. Baker, Revision of the genera and species of Scilleae and Chlorogaleae, *Jour. Linn. Soc. London (Bot.)* 13:257–285 (1873). I. M. Krasheninnikov, *Ornithogalum*, in: *Flora URSS* 4:379–394, 1935. N. Feinbrun, The genus *Ornithogalum* in Palestine and neighbouring countries, *Palest. Jour. Bot. Jerusalem Ser.*, 2:132–150 (1941). C. Zahariadi, Sous-genres et sections mésogéens du genre *Ornithogalum*, etc., *Rev. Biol. (Bucarest)* 10:271–291 (1965). J. Cullen & J.A. Ratter, Taxonomic and cytological notes on Turkish *Ornithogalum*, *Notes Roy. Bot. Gard. Edinb.* 27:293–339 (1967). U. Kushnir, *Reproductive Mechanisms and Cytotaxonomy in Ornithogalum L. Species of Israel*, M.Sc. Thesis, Tel-Aviv, 1971 (Hebrew with a summary in English). U. Kushnir, J. Galil & M. Feldman, Cytology and distribution of *Ornithogalum* in Israel. I. Section *Heliocharmos* Baker, *Israel Jour. Bot.* 26:63–82; II. Section *Beryllis* (Salisb.) Baker, *ibid.* 26:83–92 (1977). N.D. Agapova, Cytosystematic investigation of European representatives of the genus *Ornithogalum* L. of the USSR Flora. I. Subgen. *Beryllis* and *Myogalum*, *Bot. Zhur.* 62:970–983 (1977). C. Zahariadi, Notes on the intrageneric classification of the genus *Ornithogalum* L., *Bot. Zhur.* 62:1624–1639 (1977).

1. Perianth-segments without a green median stripe on outside, 10–15 mm broad. Ovary dark green to black.   **3. O. arabicum**
- Perianth-segments narrower than above, with a green median stripe on outside. Ovary green or yellowish   2

2. Raceme cylindrical, narrow, elongated; pedicels equal or subequal. Perianth-segments yellowish turning brownish during flowering or white with a narrow green median stripe   3

– Raceme corymbiform, ovoid, obovoid or ellipsoid to oblong; pedicels gradually shorter towards top. Green median stripe on perianth-segments broad   4

3. Perianth-segments 7–8(–12) mm, yellowish, turning brownish when dried. Anthers dirty white. Narrowed upper part of bract as long as to slightly shorter than its broader lower part.   **2. O. fuscescens**

– Perianth-segments 11–15 mm, milk-white, not brownish when dried. Anthers light yellow. Narrowed upper part of bract much longer than its broader lower part.   **1. O. narbonense**

4(2). Leaves ciliate or hairy on lower surface.   **6. O. neurostegium**

– Leaves glabrous, not ciliate   5

5. Leaves filiform, 1–1.5 mm broad. Plants of steppes.   **9. O. trichophyllum**

– Leaves 5–30 mm broad or, if narrower, with a silvery-white median stripe   6

6. Leaves with a silvery-white longitudinal median stripe, thickish, 2–8 mm broad; bulb with bulblets.   **8. O. divergens**

– Leaves without a white longitudinal median stripe; bulb without bulblets   7

7. Bracts lanceolate, long-acuminate, green-veined from near base.   **7. O. platyphyllum**

– Bracts oblong or ovate, subacute or abruptly acuminate, veined in their upper part   8

8. Bracts subacute, not acuminate. Leaves 10–30 mm broad. Raceme rounded at apex before flowering, 10–20(–30)-flowered.   **5. O. lanceolatum**

– Bracts abruptly acuminate. Leaves 2–5(–10) mm broad. Raceme acute before flowering; flowers mostly fewer.   **4. O. montanum**

Subgen. 1. BERYLLIS (Salisb.) Baker, Jour. Linn. Soc. London (Bot.) 13:273 (1873). Bulb renewed progressively through several years. Inflorescence an elongated cylindrical raceme; pedicels equal. Perianth-segments white or yellowish, with narrow green median stripe on outside.

1. Ornithogalum narbonense L., Cent. Pl. 2:15 (1756); Boiss., Fl. 5:214. *O. pyramidale* auctt. fl. palaest. non L. [Plates 70, 71]

Bulb ovoid. Leaves linear, canaliculate. Raceme cylindrical; bracts lanceolate, long-acuminate, the narrowed upper part of bract much longer than its broader lower part. Perianth-segments 11–15 mm, milk-white with a narrow green median stripe on outside. Anthers light yellow; filaments lanceolate. Flowering pedicels erecto-patent; fruiting pedicels erect, elongated, appressed to axis of raceme, longer than capsule. Capsule oblong.

1. Segetal plants, 35–70 cm, on deep alluvial soils. Flowering raceme very long (up to 30 cm), many-flowered (30–60). 2n=54.   subsp. **narbonense**

– Plants 15–30 cm, generally in habitats different from the above. Flowering raceme shorter (4–10 cm), with fewer flowers (9–20). 2n=18.   subsp. **brachystachys**

Subsp. **narbonense**. *O. narbonense* L., *loc. cit.*; Boiss., Fl. 5:214 excl. var. & syn.; Post, Fl. 2:627 excl. var. *O. narbonense* L. subsp. *typicum* Feinbr., Palest. Jour. Bot. Jerusalem ser., 2:135, f.2 (1941). [Plate 70]. Plant 35-70 cm. Bulb 2-3 cm in diameter, often 20-25 cm deep. Leaves (6-)10-20 mm broad. Flowering raceme up to 30 cm, 30-60-flowered. Germination epigeal. 2n=54. Fl. March-April.

Hab.: Segetal in cultivated fields on deep heavy alluvial soils, with *Prosopis farcta*. Acco Plain, Sharon Plain, Philistean Plain; Upper Galilee, Mt. Carmel, Esdraelon Plain, Samaria, Judean Mts.; Dan Valley, Hula Plain, Upper Jordan Valley; Golan, Gilead, Ammon.

Area of subspecies: Mediterranean.

Subsp. **brachystachys** (C. Koch) Feinbr., *op. cit.* 135, f.1 (1941). *O. brachysta-chys* C. Koch, Linnaea 22:248 (1849). [Plate 71]. Plant 15-30 cm. Bulb 1-1.5 cm in diameter. Leaves 3-7(-15) mm broad, minutely denticulate at margin. Flowering raceme 4-10 cm, 9-20-flowered. Germination epigeal. 2n=18. Fl. February-April.

Var. **brachystachys**. *O. narbonense* var. *brachystachys* (C. Koch) Bornm., Beih. Bot. Centralbl. 38 Abt. II:456 (1921); Post, Fl. 2:628; Boiss., Fl. 5:214 as "forma racemo brevi" of *O. narbonense*. Capsule 8-10(-12) mm.

Hab.: Steppes and batha. Common. Coast of Carmel, Sharon Plain, Philistean Plain; Upper and Lower Galilee, Mt. Carmel, Esdraelon Plain, Mt. Gilboa, Samaria, Judean Mts., Judean Desert, N. and C. Negev; Upper and Lower Jordan Valley, Dead Sea area; Golan, Gilead, Ammon, Moav.

Var. **macrocarpum** Feinbr., *loc. cit.* (1941). Capsule 14-15 mm.

Hab.: Less common. Sharon Plain, Philistean Plain; Judean Mts.; Ammon.

Area of subspecies: E. Mediterranean and W. Irano-Turanian.

Plants exceeding 30 cm in height or plants with longer racemes are sometimes found in less arid habitats, forming a transition between the two subspecies.

2. **Ornithogalum fuscescens** Boiss. & Gaill. in Boiss., Diagn. ser. 2, 4:107 (1859); Feinbr., *op. cit.* 136; Thiébaut, Bull. Soc. Bot. Fr. 82:194 (1935); Boiss., Fl. 5:214; Post, Fl. 2:627. [Plate 72]

Plant 40-70(-100) cm. Bulb about 3 cm in diameter. Leaves 4-6, 12-25 mm broad. Raceme cylindrical, many-flowered, dense, 10-30 cm; bracts lanceolate, long-acuminate, 3-veined, sometimes laciniate or dentate at margin before flowering; the narrowed upper part of bract as long as or slightly shorter than its broader lower part. Flowering pedicels erecto-patent, about 1 cm, as long as bracts or longer; fruiting pedicels elongated, erect. Perianth-segments 7-8(-12) mm, yellowish, turning brown when dried; green median stripe apparent on both inside and outside. Anthers dirty white. Capsule about 1 cm, ovate in outline. Germination epigeal. 2n=54+2B. Fl. March-April.

Hab.: Cultivated fields on heavy alluvial soil, with *Prosopis farcta*; less common than *O. narbonense* subsp. *narbonense*. Philistean Plain; Esdraelon Plain, Shefela; Golan, Gilead, Ammon.

Area: E. Mediterranean (Palestine, Lebanon, Syria).

Subgen. 2. CARUELIA (Parl.) Baker, Jour. Linn. Soc. London (Bot.) 13:259 (1873). Bulb renewed progressively through several years. Inflorescence a short corymb. Perianth-segments without green median stripe on outside. Style very short.

3. Ornithogalum arabicum L., Sp. Pl. 307 (1753); Feinbr., *op. cit.* 138 & 147; Boiss., Fl. 5:215; Post, Fl. 2:628. [Plate 73]

Plant 30–40 cm. Bulb large, with numerous bulblets. Leaves 5–8, broadly linear, 10–30 mm broad, about as long as scape or longer. Scape stout. Raceme corymbiform, 6–30-flowered; bracts broad, triangular-lanceolate, long-acuminate. Perianth showy, white; segments 25–30 × 7–15 mm, broadly elliptic, white on both surfaces, without a green stripe outside; outer segments ovate, mucronate. Ovary dark green to black, glossy. $2n=51(3x)$. Fl. April–May.

Hab.: Probably not indigenous. On wall in Jerusalem (Meyers & Dinsmore 5433).

Area: Mediterranean.

Reproduction to a large extent vegetative, by bulblets. Cultivated at present in Israel, also for export.

According to its general distribution, *O. arabicum* appears to be a Mediterranean species. Since it is the only Mediterranean species of the S. African section *Caruelia*, its origin is in need of a thorough study. It is recorded from the Canary Islands, Madeira, N. Africa, France, Italy and Dalmatia and in the Eastern Mediterranean from Palestine and the Lebanon. Rouy (Fl. Fr. 12:415, 1910) and Ascherson & Graebner (Syn. Mitteleur. Fl. 3:250, 1905) doubted its indigeneity in France. Mouterde (1966) cited *O. arabicum* as not indigenous in Beirut (Lebanon). As to Palestine we wrote in 1941: "We doubt the indigeneity of *O. arabicum* in Palestine. Although reported by Tristram from El-Ghor (the Lower Jordan Valley), it is known with certainty only from the walls of the Tower of David in the Old City of Jerusalem. There it was collected by Mr J.E. Dinsmore, who has since cultivated the species in his garden" (Feinbr., *op. cit.* 147, 1941).

Subgen. 3. ORNITHOGALUM. Subgen. *Heliocharmos* Baker, Jour. Linn. Soc. London (Bot.) 13:258 (1873). Bulb renewed every year. Inflorescence a usually shortened corymbiform raceme, in some species elongated in fruit; lower pedicels longest. Perianth-segments pure white, with a broad green median stripe on outside.

4. Ornithogalum montanum Cyr. in Ten., Fl. Nap. 1:Prodr. XXII & 176, t.33 (1811–1815); Feinbr., *op. cit.* 138, ff.7, 8 & t.5 f.1; Mout., Nouv. Fl. Liban Syrie 1:243 excl. syn., t.77 f.2; Boiss., Fl. 5:216 excl. var.; Post, Fl. 2:629 excl. var. [Plate 74]

Plant 5–10 cm. Bulb small, 1–1.5 cm in diameter. Leaves 3–6(-8), (2-)5–10 mm broad, linear-lanceolate, flat, glabrous, usually overtopping inflorescence. Raceme corymbiform, lax, subacute at apex before anthesis; flowers mostly not exceeding 10; lower bract 2–2.5 cm, ovate, abruptly acuminate, veined in upper part. Pedicels erecto-patent at first, becoming horizontal or reflexed later. Perianth white; outer segments (10–)15(–18) mm; green median stripe broad. Capsule about 8 mm, broadly ovoid to oblong, rounded above, with prominent ribs. Germination hypogeal. 2n=16, 16+4B. Fl. mainly December–January (–February).

Hab.: Batha, among rocks; various soils; sea level up to 1,200 m. Philistean Plain; Upper and Lower Galilee, Mt. Carmel, Esdraelon Plain, Samaria, Shefela, Judean Mts., Upper Jordan Valley; Golan, Gilead.

Area: E. Mediterranean and W. Irano-Turanian (Palestine, Lebanon, Syria, Cyprus, Turkey, Greece, Bulgaria, Italy, Iraq, Iran, Transcaucasia).

5. Ornithogalum lanceolatum Labill., Icon. Pl. Syr. 5:11, t.8 f.1 (1812); Boiss., Fl. 5:216; Post, Fl. 2:628. *O. billardieri* Mout., Nouv. Fl. Liban Syrie 1:242, t.77 f.1 (1966). [Plate 75]

Plant 5–15 cm. Bulb generally fairly large, up to 3 cm in diameter. Leaves 4–8, (10–)15(–30) mm broad, glabrous, appressed to the ground, flat, broadly lanceolate to oblong-lanceolate, overtopping flowering raceme. Raceme corymbiform, dense, many-flowered, rounded at apex before anthesis; flowers mostly 20–30; bracts oblong-elliptic, subacute, white-membranous and sparsely veined near apex; lower bract 2–3 cm. Pedicels erecto-patent in flower and in fruit. Perianth white; outer segments 17–25 mm; green median stripe very broad. Capsule 1 cm, ellipsoid, retuse. Germination hypogeal. 2n=16. Fl. mainly December–January.

Hab.: Batha, among rocks; various soils, mostly at 900–1,200 m a. s. l. Upper Galilee, Samaria; Golan.

Area: E. Mediterranean (Palestine, Syria, Lebanon, Cyprus, S. Turkey).

Sometimes cultivated (B. Mathew, Dwarf bulbs, London, 1973).

Mouterde (1960, p. 242) described *O. billardieri* Mout. from the plant generally accepted as *O. lanceolatum* Labill., and regarded the binomial *O. lanceolatum* as synonymous with *O. montanum* Cyr. He argued that the plant described by Labillardière and recorded by Boissier from the N. corner of the E. Mediterranean (Laodicea, at present Lattakia, and Sidon) could not be identical with the high elevations plant of Lebanon and Mt. Hermon. According to Mouterde, the plant in Labillardière's drawing has pedicels longer than bracts, while in the montane plants of the *Ornithogalum* concerned, bracts are longer than pedicels. From our experience with *Ornithogalum*, the length of bracts relative to that of pedicels is of minor significance, since pedicels in this genus generally lengthen in the course of flowering. On the other hand, Mouterde obviously disregarded the diagnostic value of the *shape* of bracts in *Ornithogalum* in general and in the two species in question, *O. montanum* and *O. lanceolatum*, in particular. The bracts of *O. lanceolatum* are subacute, while those of *O. montanum* are acuminate. The drawing by Labillardière clearly depicts subacute bracts. Moreover, the broad foliage leaves in Labillardière's drawing are quite characteristic of the montane *O. lanceolatum*. Mouterde's *O. billardieri* is, in our opinion, synonymous with *O. lanceolatum*, while *O. montanum* should no doubt be considered a separate species.

6. Ornithogalum neurostegium Boiss. & Bl. in Boiss., Fl. 5:222 (1882) emend. Feinbr.; Mout., Fl. Djebel Druze 75-77, t.2, ff.2-4 (1953); Contrib., Bull. Soc. Bot. Fr. 100:345-346 (1953); Nouv. Fl. Liban Syrie 1:244, t.77 f.4 (1966). [Plates 76, 77]

Plant 10-20 cm. Bulb ovoid, with or without bulblets. Leaves 4-6, 4-10 mm broad, canaliculate, linear-lanceolate, often undulate, ciliate at margin and hispid or glabrous on lower surface, often overtopping inflorescence. Raceme 5-30-flowered, corymbiform, broadly ovoid or oblong; bracts 1.5-4 cm, ovate-lanceolate, long-attenuate, acute, distinctly green-veined nearly from base. Pedicels erecto-patent, the lower more elongated. Perianth white; outer segments 12-20 mm; green median stripe broad. Capsule 7-10(-15) mm, obovoid to oblong. Germination epigeal. Fl. February– March(–April).

1. Leaves hispid on lower face; hairs often retrorsely appressed. Bulb with or without bulblets. Plants diploid (2n=18), confined to 900-1,200 m in Upper Galilee, Golan and C. Negev. subsp. **neurostegium**
- Leaves ciliate, otherwise glabrous. Bulb mostly with numerous bulblets. Plants autotetraploid (2n=36), more rarely diploid (2n=18), growing throughout the country except in deserts. subsp. **eigii**

Subsp. **neurostegium**. *O. neurostegium* Boiss. & Bl. in Boiss., *loc. cit.*; Post, Fl. 2:630. Incl. var. *crispo-undulatum* Gombault, Bull. Soc. Bot. Fr. 102:337 (1955). *O. ulophyllum* Hand.-Mazz., Ann. Naturh. Mus. (Wien) 28:19, f.3 (1914). *O. fimbriatum* Willd. var. *atrichocaulon* Gombault, Bull. Soc. Bot. Fr. 93:145 (1946). [Plate 76]

Hab.: Maquis and batha on rocky places and steppes on high plateau in Artemisietum herbae-albae. Upper Galilee; C. Negev; Golan.

Area: E. Mediterranean and W. Irano-Turanian (Palestine, Lebanon, Syria, S. Turkey, N. Iraq).

Subsp. **eigii** (Feinbr.) Feinbr. stat. nov. *O. eigii* Feinbr., Palest. Jour. Bot. Jerusalem ser., 2:139, ff.9a-e & t.5 f.2 (1941). *O. neurostegium* Boiss. & Bl. var. *ciliatum* (Boiss.) Mout., Fl. Djebel Druze 75 (1953); Nouv. Fl. Liban Syrie 1:244, t.77 f.4c. *O. fimbriatum* Willd. var. *ciliatum* Boiss., Fl. 5:221 (1882); Post, Fl. 2:630. [Plate 77]

Hab.: Batha and fallow fields. Common. Sharon Plain, Philistean Plain; Upper and Lower Galilee, Mt. Carmel, Esdraelon Plain, Mt. Gilboa, Samaria, Shefela, Judean Mts., Judean Desert, N. Negev; Lower Jordan Valley; Golan, Gilead, Ammon, Moav, Edom.

Area: E. Mediterranean (Palestine and Djebel Druze in Syria).

Area of species: E. Mediterranean and W. Irano-Turanian.

The closeness of *O. eigii* Feinbr. to *O. neurostegium* Boiss. & Bl., not recognized by Feinbrun (1941), was pointed out by Mouterde in his Contribution (Bull. Soc. Bot. Fr. 100:345-346, 1953) and in Flore du Djebel Druze (1953). Mouterde treated this taxon as *O. neurostegium* Boiss. & Bl. var. *ciliatum* (Boiss.) Mout. Specimens of typical *O. neurostegium* found by Feinbrun in Upper Galilee and subsequent studies by Kushnir (1971) and Kushnir et al. (1977) indicate that the two taxa are best to be regarded as two subspecies of *O. neurostegium*, i.e. as subsp. *neurostegium* and subsp. *eigii* (Feinbr.) Feinbr.

The cytotaxonomic studies by Kushnir (1971) and Kushnir et al. (1977) showed that subsp. *neurostegium* is diploid (2n=18) whereas subsp. *eigii* is tetraploid (2n=36), though rare diploid plants of subsp. *eigii* have also been found (near Meiron, Upper Galilee). Tetraploid plants of subsp. *eigii* are generally bulblet-forming, and vegetative propagation apparently plays an important part in the wide distribution of this taxon within its area. Geographically, subsp. *eigii* is endemic in Palestine (Cis- and Transjordan) and in Djebel Druze in Syria. Subsp. *neurostegium* has a wider distribution area. Mouterde (1966) records it from Lebanon, Syria (except in Djebel Druze), S. Turkey and N. Iraq. *O. ulophyllum* Hand.-Mazz. is synonymous with subsp. *neurostegium*, as already indicated by Mouterde (1966).

Until recently (Kushnir et al., 1977) subsp. *neurostegium* was thought to be confined to the north of the country (Upper Galilee and Golan), but it is now found to occur also in the extreme south. A reexamination of specimens in our herbarium, collected in the C. Negev and previously identified as *O. eigii*, showed that leaves are hispid on their underside as is characteristic of subsp. *neurostegium*. The strongly undulate leaves of the Negev plants resemble our specimens from N. Iraq recorded in 1941 as *O. ulophyllum*. The area of subsp. *neurostegium* in Israel is thus disjunct, and its S. part is Irano-Turanian in character. It should be noted that the hair-cover in the C. Negev specimens differs from that in typical subsp. *neurostegium*; hairs are not uniform and not retrorsely appressed but irregular in length and direction.

Significantly, a chromosome count by Bathia Pazy (Jerusalem) revealed that the C. Negev plants are diploid, with 2n=18, as are plants of subsp. *neurostegium* from Upper Galilee and Golan.

7. Ornithogalum platyphyllum Boiss., Diagn. ser. 1, 5:64 (1844); Feinbr., *loc. cit.* 145; Mout., Nouv. Fl. Liban Syrie 1:243, t.77 f.3 (1966). *O. montanum* Cyr. var. *platyphyllum* (Boiss.) Boiss., Fl. 5:217 (1882); Post, Fl. 2:629. [Plate 78]

Plant (10-)15-30 cm. Bulb ovoid. Leaves 8-12(-20) mm broad, broadly linear, glabrous, shorter than inflorescence, more rarely slightly exceeding it. Raceme many-flowered, elongated, distinctly longer than scape; bracts lanceolate, gradually long-acuminate, green-veined nearly from base; the lower 18-25 mm. Pedicels erecto-patent. Perianth white; outer segments about 15 mm; green median stripe broad. Capsule 10-15 mm, erect, ellipsoid, tapering towards base, ribbed above. Fl. April–May.

Hab.: Hills. Golan.

Area: W. Irano-Turanian (Golan, Lebanon, Syria, Turkey, N. Iraq, Iran).

8. Ornithogalum divergens Boreau, Not. Pl. Fr. No. 3 (1847) & Fl. Centre Fr. ed. 2, 2:507 (1849); Gren. in Gren. & Godr., Fl. Fr. 3:190 (1855); Feinbr., *op. cit.* 140; Boiss., Fl. 5:218. Post, Fl. 2:629 as *O. umbellatum*. [Plate 79]

Plant (15-)20-30(-40) cm. Bulb with numerous bulblets germinating after separation from mother-plant. Leaves 6-9, overtopping inflorescence, linear, 2-5(-8) mm broad, thickish, canaliculate, glabrous, with a silvery-white longitudinal median stripe. Raceme corymbiform, lax, few-flowered; lower bract

3-4.5(-6) cm, lanceolate, gradually long-acuminate, green-veined from near base. Pedicels horizontal to reflexed after flowering. Perianth white; outer segments (15-)20-25 mm; green median stripe broad. Capsule 12 mm, oblong. 2n=45(5x). Fl. February–March.

Hab.: Cultivated ground. Rare. Sharon Plain, Philistean Plain; Upper Galilee, Mt. Carmel, Esdraelon Plain, Samaria, Shefela, Judean Mts., Judean Desert; Dan Valley, Beit Shean Valley; Golan, Ammon, Edom.

Area: Mediterranean.

Reproduction to a large extent vegetative, by bulblets.

Kushnir (1971) found no seed formation after artificial cross-pollination in *O. divergens* and concluded that this was due to the hybrid nature of the pentaploid plants studied. The lack of seeds is presumably compensated by the abundant production of bulblets.

9. Ornithogalum trichophyllum Boiss. & Heldr. in Boiss., Diagn. ser. 2, 4:108 (1859); Feinbr., *op. cit.* 142, f.13a–d & t.5 f.3. *O. tenuifolium* Guss. var. *trichophyllum* (Boiss. & Heldr.) Boiss., Fl. 5:219 (1882); Post, Fl. 2:630. [Plate 80; plant enlarged by 1½]

Plant 8-10(-15) cm. Bulb ovoid, non-proliferous; outer tunics grey, the inner white, in herbarium glossy and transversely rugose. Leaves 4–6(–8), 1–1.5 mm broad, filiform, glabrous, with a thin longitudinal furrow, longer than inflorescence. Raceme 2–6 cm, corymbiform, dense, 5–10-flowered; bracts ovate-lanceolate, acuminate, as long as or longer than pedicels; lower bract 2–3.5 cm. Flowering pedicels erecto-patent, shorter than to about as long as flower; fruiting pedicels not elongating. Perianth white; segments 10–20 mm, with a broad green median stripe outside. Capsule 8–12 mm, ovoid or oblong. Germination epigeal. 2n=18. Fl. December–February.

Hab.: Steppes. Judean Desert, N. and C. Negev; Arava Valley; Moav, Edom.
Area: Saharo-Arabian and W. Irano-Turanian (S. Palestine, Sinai, Egypt).

13. DIPCADI Medik.

Bulbous and scapose glabrous perennials. Leaves basal, usually linear. Inflorescence a raceme; bracts membranous. Perianth gamophyllous, 6-lobed, tubular-infundibular or campanulate; lobes as long as tube or longer, dissimilar, outer lobes falcate, the inner shorter, connivent. Stamens 6, inserted at throat of perianth; filaments short, filiform; anthers dorsifixed. Ovary globose, 3-locular; ovules numerous; style 1, straight; stigma capitate. Capsule membranous, dehiscing loculicidally at apex. Seeds black, strongly compressed, horizontal.

Fifty-five mainly paleotrop. species.

1. Dipcadi erythraeum Webb & Berth., Phyt. Canar. 3(2):341 (1848); Muschler, Man. Fl. Eg. 220; Post, Fl. 2:647. *Uropetalum erythraeum* (Webb &

Berth.) Boiss., Fl. 5:286 (1882). *Hyacinthus serotinus* Forssk., Fl. Aeg.-Arab. Suppl. 209 (1775) non L. (1753). [Plate 81]

Plant 15–25 cm. Bulb ovoid, 1–4 cm; tunics white, papery. Leaves 2–4, fleshy, linear, canaliculate, as long as or longer than scape. Raceme lax, 6–12-flowered; bracts deltoid-lanceolate, acuminate, veined, about as long as pedicels or longer; pedicels short, nodding before anthesis, erect during anthesis. Perianth 13 mm, tube greyish-green; lobes dirty orange, dirty salmon or brownish-red; outer lobes oblong, cucculate, ending in a recurved cusp; the inner rounded, much shorter than tube. Capsule 12 mm, 3-quetrous; valves broader than long. Seeds plano-convex, ovate in outline. Fl. February–March.

Hab.: Sandy soil in deserts. N. and C. Negev, Upper Jordan Valley, Dead Sea area, Arava Valley; Edom.

Area: Saharo-Arabian.

14. Hyacinthus L. emend. Chouard

Bulbous and scapose perennials. Bulb tunicate; outer tunics membranous, not extended into neck. Leaves basal. Inflorescence a raceme; bracts minute. Perianth gamophyllous, tubular-urceolate or tubular, 6-fid, with long, spreading or recurved lobes. Stamens 6, included to subexserted, 1-seriate; filaments short, inserted below middle of perianth-tube; anthers dorsifixed near their base. Ovary 3-locular; locules 6–8-ovulate; style 1; stigma obtuse. Capsule loculicidal, depressed-globose, 3-lobed, somewhat fleshy; lobes rounded at back. Seeds black.

Three species, *H. litwinowii* E. Czerniak. and *H. transcaspicus* Litw. in mountains of N. E. Iran and Turkmenistan, and *H. orientalis* in the E. Mediterranean (B. Bentzer et al., 1974).

Literature: P. Chouard, Révision de quelques genres et sous-genres de Liliacées bulbeuses *(Scilla, Endymion, Hyacinthus)*, *Bull. Mus. Nat. Hist. Nat. (Paris)* Ser. 2, 3:176–179 (1931). A.S. Losina-Losinskaya, *Hyacinthus*, in: *Flora URSS* 4:311–312, 1935. C.D. Darlington, J.B. Hair & R. Hurcombe, The history of the garden hyacinths, *Heredity* 5:233–252 (1951). B. Bentzer, R. v. Bothmer & P. Wendelbo, Cytology and morphology of the genus *Hyacinthus* L. s. str. (Liliaceae), *Bot. Not.* 127:297–301 (1974). K. Persson & P. Wendelbo, The artificial hybrid *Hyacinthus orientalis* × *transcaspicus* (Liliaceae), *Bot. Not.* 132:207–209 (1979). P. Wendelbo, Notes on *Hyacinthus* and *Bellevalia* in Turkey and Iran, *Notes Roy. Bot. Gard. Edinb.* 38:423–434 (1980).

1. Hyacinthus orientalis L., Sp. Pl. 317 (1753); Boiss., Fl. 5:309; Post, Fl. 2:654. [Plate 82]

Plant 15–30 cm. Bulb subglobose or depressed, 1.5–3 cm in diameter; tunics whitish or purplish. Scape erect. Leaves 3–6, 6–20(–25) mm broad, canaliculate. Raceme 4–15-flowered. Bracts membranous, often 2-fid, shorter than pedicels. Pedicels short, erecto-patent, later nodding. Flowers fragrant. Perianth 17–25 mm, blue, constricted above ovary; lobes oblong, somewhat shorter than ventricose tube. Stamens included; filaments shorter than anthers. Style shorter than ovary, not reaching stamens. Capsule subglobose-depressed, 3-sulcate. Seeds black, finely granulated on surface, with a large aril, adapted to myrmecochory. 2n=16. Fl. January–March.

Hab.: Maquis and fields. Upper and Lower Galilee.

Area: Indigenous in the E. Mediterranean (Palestine, Lebanon, Syria, Cyprus, S. Turkey). Naturalized in some Mediterranean countries.

Widely grown for ornament; taken into cultivation in Turkey and introduced to Europe in 1560. The cultivars vary in flower colour, size of flowers and density of the inflorescence and are diploid, triploid or aneuploid (Darlington et al., 1951).

According to Wendelbo (*op. cit.* 1980), Israel plants should be classed with subsp. *orientalis* growing at altitudes between 400 m and 1,600 m. Subsp. *chionophilus* Wendelbo is described as confined to higher altitudes (1,800–2,500 m) in Turkey and as having perianth-lobes of similar length to the tube.

15. HYACINTHELLA Schur

Bulbous and scapose perennials. Bulb tunicate, small. Leaves basal, (1–)2(–3), linear, usually with elevated veins formed by 2 strands of thick-walled fibres accompanying the vascular bundle. Inflorescence a cylindrical raceme or spike; bracts minute. Flowering pedicels erecto-patent or horizontal, not nodding. Flowers white or light blue. Perianth gamophyllous, 6-lobed, campanulate, tubular or infundibular; lobes ¼–⅓ length of tube, somewhat spreading; perianth split or torn longitudinally after flowering and remaining attached at base through fruiting stage. Stamens included to subexserted, sub-2-seriate to 1-seriate, inserted below bases of perianth-lobes; anthers dorsifixed near their base. Ovary 3-locular; style 1, filiform. Capsule loculicidal, depressed-globose, 3-lobed, coriaceous; lobes rounded at back. Seeds 1–2 in each locule, black, wrinkled, often glossy, not arillate.

Eleven species, mainly Mediterranean and Irano-Turanian.

Literature: F. Schur, *Hyacinthella, Österr. Bot. Wochenbl.* 6:227–229 (1856). A.S. Losina-Losinskaya, *Hyacinthella,* in: *Flora URSS* 4:407–408, 1935. N. Feinbrun, Revision of the genus *Hyacinthella* Schur, *Bull. Res. Counc. Israel* 10D:324–347 (1961). K. Persson & P. Wendelbo, Taxonomy and cytology of the genus *Hyacinthella* (Liliaceae-Scilloideae) with special reference to the species of S.W. Asia. Part I, *Candollea* 36:513–541 (1981); Part II, *ibid.* 37:157–175 (1982).

1. Hyacinthella nervosa (Bertol.) Chouard, Bull. Mus. Nat. Hist. Nat. (Paris) ser. 2, 3:178 (1931); Feinbr., *op. cit.* 336. *Hyacinthus nervosus* Bertol., Miscell. Bot. 1:21 (1842). *Bellevalia nervosa* (Bertol.) Boiss., Fl. 5:306 (1882); Post, Fl. 2:653. *B. aleppica* Boiss., Diagn. ser. 2, 4:111 (1859). *B. haynei* (Baker) Boiss., Fl. 5:308 (1882). *Hyacinthus haynei* Baker, Jour. Bot. 12:7 (1874). [Plate 83]

Plant 6–10(–15) cm. Bulb ovoid, 1.2–2 cm in diameter. Leaves 2, rarely 1 or 3, oblong or linear-lanceolate, about as long as scape or longer, often canaliculate, ciliolate; outer leaf 5–15 mm broad. Scapes 1–2. Raceme dense. Flowers subsessile. Perianth 6–8 mm, tubular-campanulate, pale blue to pale violet; lobes ovate. Filaments longer than anthers, inserted on upper ¼ of perianth-tube; anthers dark violet, subexserted. Fruiting pedicels hardly elongating. Capsule 2.5–5 × 3.5–5 mm. Seeds ovoid. Fl. February–March.

Hab.: Steppes on calcareous soil. Rather rare. E. Samaria, Judean Desert; Lower Jordan Valley; Moav, Edom.

Area: W. Irano-Turanian (Palestine, Syria, E. Turkey, N. Iraq).

$2n=24$ were counted in specimens from S. E. Turkey.

16. BELLEVALIA Lapeyr.

Bulbous and scapose perennials. Bulb tunicate, renewed in course of several years. Leaves basal, 2–6(–7–10), linear, lanceolate or lorate, often ciliate or scabrous at hyaline margin. Inflorescence a raceme or spike, often with a few sterile flowers at tip. Bracts minute. Perianth gamophyllous, campanulate, tubular or infundibular, 6-lobed, not constricted under lobes, white, violet or lilac (rarely yellowish), mostly changing during flowering to greyish-brown or dirty brown; flower-bud often purplish, pink or violet; after flowering perianth abscising at base and falling off or remaining on top of capsule; lobes ¼–½ length of tube, rarely as long as tube, erect or more or less spreading, each lobe, especially outer ones, with a protuberance on outside. Stamens 6, equal, included or subexserted; filaments triangular, inserted at base of perianth-lobes; anthers purple, violet, lilac or blue, rarely yellow or greenish. Ovary 3-gonous, 3-locular; style 1, elongate; stigma obtuse. Capsule 3-quetrous, with 3 acute ribs or wings, loculicidal or in a few species indehiscent and deciduous as a whole. Seeds 2 or more, ovoid to globose, smooth, bluish-black, covered with a waxy bloom, not arillate; hilum white. $x=8$.

About 50 Mediterranean, Irano-Turanian and Saharo-Arabian species.

Literature: A. S. Losina-Losinskaya, *Bellevalia*, in: *Flora URSS* 4:395–405, 1935. N. Feinbrun, A monographic study on the genus *Bellevalia* Lapeyr. (Caryology, taxonomy, geography), *Palest. Jour. Bot. Jerusalem Ser.*, 1:42–54, 131–142 (1938), 336–409 (1940). K. Persson & P. Wendelbo, *Bellevalia hyacinthoides*, a name for *Strangweja spicata* (Liliaceae), *Bot. Not.* 132:65–70 (1979). R. v. Bothmer & P. Wendelbo, Cytological and morphological variation in *Bellevalia*, *Nord. Jour. Bot.* 1:4–11 (1981).

For the determination of *Bellevalia* species fruiting racemes are indispensable; exact notes should be made of the colour of perianth and of flower-buds, as well as of changes in coloration during anthesis.

1. Flowering and fruiting raceme broadly conical, very lax, about as long as broad. Flower-buds deep violet. Lower fruiting pedicels 9–15 cm. Capsule 17–23 mm, about twice as long as broad. **9. B. longipes**
- Flowering and fruiting raceme cylindrical or oblong, or, if conical or ovoid, fruiting pedicels and capsules shorter than above 2
2. Flowers pendulous during flowering. Flower-buds deep violet; flowering perianth

violet or lilac in its lower part, turning dirty brown or greenish-brown during
flowering   3
- Flowers not pendulous, or rarely pendulous, and then flower-buds not deep violet
  and flowering perianth not as above   4
3. Perianth 12–14(–16) mm, narrowly tubular-campanulate. Flowering pedicels as
   long as or shorter than flower.   **1. B. trifoliata**
- Perianth 8–11 mm, broadly tubular-campanulate. Flowering pedicels longer than
  flower.   **2. B. macrobotrys**
4(2). Lobes of perianth about as long as tube. Ripe capsules falling entire. Flowering and fruiting pedicels 1–3 mm.   **8. B. desertorum**
- Lobes of perianth nearly ½ length of tube or shorter. Ripe capsules persistent,
  loculicidal. Flowering and fruiting pedicels longer than in above (rarely very
  short)   5
5. Flowering and fruiting raceme conical or broadly ovoid, not or slightly longer
   than broad   6
- Flowering and fruiting raceme cylindrical or oblong, distinctly longer than broad
                                                                             7
6. Flowering raceme 3.5–4 cm long; fruiting raceme 4–5 cm long and 5–6 cm broad.
                                                                   11. B. zoharyi
- Flowering raceme longer than above (5.5–9 cm); fruiting raceme 7–16 cm long,
  5–14 cm broad.   **10. B. stepporum**
7(5). Perianth (9–)10–14 mm. Flowering raceme 10 cm or longer   8
- Perianth 6–10 mm. Flowering raceme usually shorter   9
8. Leaves prostrate, ciliate. Flowers not pendulous. Raceme nearly sessile on ground.
   Plants of steppes, mostly on loess soils in Negev.   **4. B. eigii**
- Leaves erect, ciliolate or scabrous at margin. Flowers pendulous for some time
  during flowering. Raceme borne on a long scape. Plants of fields on deep alluvial
  soils.   **3. B. warburgii**
9(7). Anthers yellow or greenish. Perianth yellowish with 5 dark veins; lobes white
   with a dark midvein in herbarium specimens; flower-bud light purple. Plants of
   Golan and of slopes of Mt. Hermon at 1,800 m or higher up.   **7. B. densiflora***
- Anthers violet or purple. Plants with different set of characters   10
10. Perianth tubular-campanulate, 7–10 mm, bluish-white before flowering, white
    during flowering, remaining whitish in herbarium. Plants of steppes (Judean
    Desert, C. Negev, Edom).   **6. B. mosheovii**
- Perianth turbinate, 6–9(–10) mm, white at beginning of flowering, then turning
  dirty greyish-brown, remaining dark in herbarium; flower-buds white, rarely
  pink or lilac. Plants mainly of Mediterranean batha.   **5. B. flexuosa**

Sect. 1. NUTANS Feinbr., *op. cit.* 337 (1940). Fruiting and flowering raceme
cylindrical; pedicels nearly equal in length. Perianth mostly green-veined,
violet or pale blue, sometimes yellowish-green or greenish-white in bud, dirty
yellowish-brown or greenish-purple during flowering. Valves of capsule

---

* Another *Bellevalia* species on Mt. Hermon is *B. douinii* Pabot & Mout. (in Mout.
1966, p.249) which is conspicuous by its large, 2–2.5 cm broad and 1.5 cm long capsule. Its flowering raceme, measuring 6–10 cm in length, is much longer than that of
B. densiflora.

rounded, often retuse at apex, about as long as broad. Leaves as long as or shorter than scape, rarely longer.

Subsect. COLORATA Feinbr., *loc. cit.* Flowers violet in bud.

1. Bellevalia trifoliata (Ten.) Kunth, Enum. Pl. 4:308 (1843); Feinbr., *op. cit.* 343; Boiss., Fl. 5:303; Post, Fl. 2:651. *Hyacinthus trifoliatus* Ten., Fl. Nap. 3:376, t.136 (1824–1829). [Plate 84]

Plant 25–60 cm. Bulb 2–4 cm in diameter, deep in the ground. Leaves 2–4(–6), erect, lanceolate, canaliculate; outer leaf 1.2–3 cm broad. Scapes 1–3. Flowering raceme cylindrical, 20–50-flowered, 2.5–4 cm broad; rachis often purplish; flowering pedicels as long as or shorter than flower, erecto-patent before flowering, nearly horizontal to recurved during flowering. Flowers pendulous. Fruiting raceme cylindrical, up to 5 cm broad; fruiting pedicels not or somewhat longer than flower, horizontal or slightly curved upwards, somewhat thickened. Perianth 12–14(–16) mm, narrowly tubular-campanulate, often slightly zygomorphic, violet in bud, lilac at base at beginning of flowering, turning greenish-brown later; lobes olive-coloured, green-veined, ovate or oblong, ¼–⅓ length of tube. Anthers violet. Capsule persistent, loculicidal; valves broadly ovate. 2n=8. Fl. February–March.

Hab.: Fields on deep heavy and humid alluvial soil. Acco Plain, Coast of Carmel, Sharon Plain, Philistean Plain; Upper Galilee, Mt. Carmel, Esdraelon Plain, Samaria, Shefela, Judean Mts.; Hula Plain.

Area: Mediterranean.

2. Bellevalia macrobotrys Boiss., Diagn. ser. 1, 13:35 (1854); Feinbr., *op. cit.* 345, t.17 f.1; Boiss., Fl. 5:303; Post, Fl. 2:651. [Plate 85]

Plant 30–60 cm. Bulb 3–5 cm in diameter, deep in the ground. Leaves 2–6, erect, lanceolate or lorate, canaliculate; outer leaf 1.5–3.5 cm broad. Scapes 1–2. Flowering raceme long, cylindrical, 20–50-flowered, usually longer than scape, 4–6 cm broad; rachis generally purplish above, with some sterile flowers at apex; flowering pedicels up to 1½ times as long as flower, erecto-patent before flowering, nearly horizontal to recurved during flowering. Flowers pendulous. Fruiting raceme cylindrical, broad; fruiting pedicels 1.5–3 cm, horizontal to slightly curved upwards. Perianth 8–11 mm, broadly tubular-campanulate, zygomorphic and somewhat gibbous at base, deep violet in bud, violet at base at beginning of flowering, turning dirty brown later; lobes green-veined, ovate, ¼–½ length of tube; anterior lobe longer than others Anthers violet. Capsule persistent, loculicidal; valves broadly ovate to suborbicular. 2n=8. Fl. March–April.

Hab.: Fields on heavy alluvial soils. Sharon Plain, Philistean Plain; Upper Galilee, Mt. Carmel, Esdraelon Plain, Shefela, Judean Mts., N. and W. Negev· Gilead, Ammon.

Area: E. Mediterranean and W. Irano-Turanian.

Subsect. ALBIFLORA Feinbr., *op. cit.* 337 (1940). Flowers white, green-veined in bud.

3. **Bellevalia warburgii** Feinbr., Palest. Jour. Bot. Jerusalem ser., 1:355, t.17 f.2 (1940). [Plate 86]

Plant 40-60 cm. Bulb 3-4 cm in diameter, deep in the ground. Leaves (2-)3-6, (25-)30-40 cm, erect to ascending, lanceolate to lorate, canaliculate, ciliolate to scabrous at margin; outer leaf 2-3.5 cm broad. Scapes 1-2(-3). Flowering raceme long, cylindrical, lax, 25-70-flowered, usually about as long as scape; flowering pedicels 1½-2(-3) times as long as flower, erecto-patent, later horizontal and arcuate-spreading. Flowers patent, sometimes pendulous. Fruiting raceme much elongated, 5-9 cm broad; fruiting pedicels 2.5-5 cm, horizontal to more or less curved upwards. Perianth 10 mm, tubular-campanulate, white before flowering, white and green-veined in bud and at beginning of flowering, turning dirty brown later; lobes greenish, broadly ovate, ¼-⅓ length of tube. Anthers violet or lilac. Capsule persistent, loculicidal; valves ovate, 1 cm or somewhat longer. 2n=16. Fl. March-April.

Hab.: Fields on heavy alluvial soils. Sharon Plain, Philistean Plain; Upper Galilee, Esdraelon Plain, Shefela, Judean Mts.

Area: E. Mediterranean.

4. **Bellevalia eigii** Feinbr., Palest. Jour. Bot. Jerusalem ser., 1:357, t.17 f.3 (1940). *B. alexandrina* Feinbr., *op. cit.* 356, t.17 f.5. [Plate 87]

Plant 20-30 cm. Bulb large, 3-4 cm in diameter. Leaves 3-7, prostrate on ground, lanceolate or oblong, somewhat undulate, ciliate; outer leaf 1.5-3(-4.5) cm broad. Scapes 1-4. Flowering raceme cylindrical or oblong, 25-50-flowered, 10-20 cm long, 5-6 cm broad in lower part, longer than scape; flowers not pendulous; flowering pedicels erecto-patent, 1½-2 times as long as flower. Fruiting raceme cylindrical, 20 × 8 cm; fruiting pedicels 2-2.5 cm, nearly horizontal. Perianth (9-)10-14 mm, tubular-campanulate, greenish-white, turning dirty brown later; lobes ovate, green-veined, ⅓-½ length of tube; veins prominent. Anthers violet. Capsule persistent, loculicidal, short-stipitate; valves ovate to elliptic. 2n=24. Fl. February-April.

Hab.: Steppes, loess and calcareous soils. Common. Judean Desert, N. and C. Negev; Lower Jordan Valley; Edom.

Area: E. Saharo-Arabian (S. and E. Palestine, Sinai, Egypt).

Since the publication of the *Bellevalia* monograph, large-scale observations of *B. eigii* have been made in the field, its chromosomes have been counted and herbarium material has been collected. As a result it is concluded that *B. alexandrina* is synonymous with *B. eigii*. The area of the latter thus extends to Lower Egypt.

Sect. 2. BELLEVALIA. Sect. *Patens* Feinbr., *op. cit.* 337 (1940). Fruiting raceme cylindrical; flowering raceme cylindrical, oblong or ovoid; flowering pedicels erecto-patent, shorter than flower or nearly 0. Perianth violet, lilac or greenish-white in bud, dirty yellowish-brown during flowering. Valves of capsule ovate, round or elliptic, rounded at base.

Subsect. BELLEVALIA. Subsect. *Romana* Feinbr., *op. cit.* 338 (1940). Flowers and capsules pedicellate. Capsules persistent, dehiscing at maturity.

5. Bellevalia flexuosa Boiss., Diagn. ser. 1, 13:36 (1854); Feinbr., *op. cit.* 379, f.34 & t.18 f.11; Boiss., Fl. 5:303; Post, Fl. 2:652. [Plate 88]

Plant 10–30 cm. Bulb 1.25–2.5 cm in diameter. Leaves (3–)4–5(–7–10), prostrate and somewhat undulate or erect and not undulate, linear-lanceolate, attenuate towards apex and base, ciliolate or scabrous at margin; outer leaf 1–3 cm broad. Scapes 1–4. Flowering raceme cylindrical, lax, 15–20(–50)-flowered, usually as long as scape; flowering pedicels generally about as long as flower, sometimes up to 1½–2 times as long, erecto-patent, sometimes changing to horizontal or even nodding. Fruiting raceme cylindrical, 3–4 cm broad; fruiting pedicels 8–15 mm, horizontal to erecto-patent. Perianth 6–9(–10) mm, turbinate, white or rarely pink or lilac in bud, white and green-veined and soon turning dirty greyish-brown from base up during flowering, becoming dark in herbarium; lobes ovate, yellowish-veined or greenish-veined, ½ length of tube. Anthers violet to purple. Capsule persistent, loculicidal; valves 8–10 mm, orbicular to ovate. Seeds 2 mm. 2n=8. Fl. February–March.

Hab.: Batha, especially Sarcopoterietum spinosi, fallow fields, rarely steppes. Common. Sharon Plain, Philistean Plain; Upper and Lower Galilee, Mt. Carmel, Esdraelon Plain, Mt. Gilboa, Samaria, Shefela, Judean Mts., Judean Desert, N. Negev (rare); Dan Valley, Upper and Lower Jordan Valley, Dead Sea area; Golan, Gilead, Ammon, Edom.

Area: E. Mediterranean (Palestine, Lebanon, Syria).

6. Bellevalia mosheovii Feinbr., Palest. Jour. Bot. Jerusalem ser., 1:380, f.35 & t.18 f.12 (1940). [Plate 89]

Plant 14–20 cm. Bulb 1.5–3 cm in diameter. Leaves 3–5(–7), lorate or lanceolate, undulate, ciliate, much longer than scape; outer leaf 0.8–2 cm broad. Scapes 1–4. Flowering raceme cylindrical, 20–30-flowered, 3–6 × 2–2.5 cm, longer than scape; flowering pedicels usually shorter than flower, 3–6 mm, erecto-patent, sometimes horizontal, lilac. Flowers erecto-patent. Fruiting raceme cylindrical, 4–10 × 2–4 cm; fruiting pedicels 7–11 mm, erecto-patent to horizontal, usually thickened. Perianth 7–10 mm, tubular-campanulate, bluish-white in bud, white during flowering, remaining whitish in herbarium; lobes green-veined, ⅓–¼ length of tube. Anthers violet. Capsule persistent, loculicidal; valves ovate. Seeds about 2 mm. Fl. March–April.

Hab.: Steppes, Artemisietum herbae-albae. Judean Desert, C. Negev; Edom.

Area: W. Irano-Turanian (Palestine, Syrian Desert, Iraq).

7. Bellevalia densiflora Boiss., Diagn. ser. 1, 7:109 (1846); Feinbr., *op. cit.*

377; Mout., Nouv. Fl. Liban Syrie 1:251, t.80 f.2; Boiss., Fl. 5:304; Post, Fl. 2:652, incl. var. *longipes* Post, Pl. Postianae 1:14 (1890). *Hyacinthus densiflorus* (Boiss.) Baker, Proc. Linn. Soc. London (Bot.) 11:432 (1871). ?*B. hermonis* Mout., Nouv. Fl. Liban Syrie 1:250, t.78 f.5 (1966). *B. subalpina* Shmida & Zohary in Zohary, New Anal. Fl. Israel 443 (1976) *nom. inval.* [Plate 90]

Plant 10–30(–40) cm. Bulb 1.5–3 cm in diameter. Leaves 3–5, linear-oblong, canaliculate; outer leaf 1–2 cm broad. Scapes 1–3. Flowering raceme oblong, 3–5(–7) cm, 15–35-flowered, dense; flowering pedicels erecto-patent to horizontal, ½–1½ times as long as flower. Fruiting pedicels hardly accrescent. Perianth 8–10 mm, tubular-campanulate, light purple in bud, yellowish with 6 dark veins during flowering, turning greyish-brown later; lobes about ⅓–½ length of tube, whitish with a dark midvein in herbarium. Anthers yellow or greenish. Capsule persistent, loculicidal. Fl. February–March.

Hab.: Wet habitats, maquis, Zizyphetum loti on basalt soil and tragacanthic batha, 1,800 m and higher up on slopes of Mt. Hermon. Upper Jordan Valley; Golan.

Area: W. Irano-Turanian (N. Palestine, Syria, S. Turkey).

Subsect. CAVAREA (Mattei) Feinbr., *op. cit.* 338 (1940). Flowers and capsules sessile or subsessile. Capsules deciduous, indehiscent.

8. Bellevalia desertorum Eig & Feinbr., Beih. Bot. Centralbl. 49 Abt. II:666, f.1 (1932); Feinbr., *op. cit.* 383. [Plate 91]

Plant 8–15 cm. Bulb 2–3(–4) cm in diameter. Leaves (2–)4–5(–7), prostrate, lanceolate to lorate, sometimes canaliculate, somewhat fleshy, glaucous, often marked with white spots, not ciliate; outer leaf 0.8–1.8 cm broad. Scapes 1–3, short. Flowering raceme dense, oblong, 10–25-flowered, 2–4 × 1.5–2 cm; bracts more or less conspicuous; flowering and fruiting pedicels very short (1–3 mm); flowers erecto-patent. Fruiting raceme spike-like. Perianth 8–13 mm, light blue or light lilac; lobes about as long as tube. Anthers purple or violet. Capsule indehiscent, falling entire, winged; valves nearly orbicular, 7–12 mm in diameter. Seeds large. 2n=8. Fl. December–March.

Hab.: Steppes. Artemisietum herbae-albae, Noaetum mucronatae, etc. S. Judean Mts., Judean Desert, N. and C. Negev; Lower Jordan Valley, Dead Sea area; Edom.

Area: W. Irano-Turanian and E. Saharo-Arabian.

Sect. 3. CONICA Feinbr., *op. cit.* 337 (1940). Fruiting raceme conical; flowering raceme conical, ovoid or oblong. Perianth violet, lilac or greenish-white in bud, and during flowering dirty yellowish-brown or dirty violet turning greenish-brown. Valves of capsule retuse at apex, cuneate at base, mostly oblong or obovate. Leaves mostly shorter than scape.

Subsect. ORIENTALIS Feinbr., *loc. cit.* Flowers violet in bud.

9. Bellevalia longipes Post in Post & Autran, Bull. Herb. Boiss. 3:165 (1895); Feinbr., *op. cit.* 359; Post, Fl. 2:651. *B. ciliata* sensu Boiss., Fl. 5:302 quoad pl. Ciliciae, Syriae & Mesopotamiae non (Cyr.) Nees (1833–1837). *B. ciliata* var. *paniculata* Post, Jour. Linn. Soc. London (Bot.) 24:440 (1888). [Plate 92]

Plant 50–60 cm, 70–80 cm in fruiting stage. Bulb 2.5–3 cm in diameter, deep in the ground. Leaves 3–5, lanceolate to lorate, not ciliate at margin, blackish when dried; outer leaf 2–3(–3.5) cm broad. Scape usually 1. Flowering raceme broadly conical, very lax, 20–45-flowered, 15–35 cm, about as long as broad; sterile flowers generally absent; flowering pedicels erecto-patent, sometimes partly recurved, often purplish; lower flowering pedicels 6–10 cm. Fruiting raceme conical, 25–35 cm broad at base; fruiting pedicels rigid, horizontal or partly recurved, the lower ones up to 15 cm. Perianth 9–11(–13) mm, campanulate, dark violet in bud, dirty violet at beginning of flowering, turning greenish-brown at last; lobes ovate-oblong, about ½ length of tube. Anthers violet or purple. Capsule oblong, 17–23 mm, retuse at apex. Seeds large, 3.25 mm in diameter. 2n=8. Fl. March. Fr. April–May.

Hab.: Fields on alluvial soils. Acco Plain; Upper and Lower Galilee; Esdraelon Plain, Samaria, Shefela, Judean Mts.; Golan, Moav.

Area: W. Irano-Turanian (Palestine, Syria, S. Turkey, N. Iraq).

Subsect. OCCIDENTALIS Feinbr., *op. cit.* 337 (1940). Flowers white, and green-veined or yellowish-veined in bud.

10. Bellevalia stepporum Feinbr., Palest. Jour. Bot. Jerusalem ser., 1:370, ff.30–32 & t.18 ff.6, 7 (1940). *B. ciliata* (Cyr.) Nees subsp. *stepporum* (Feinbr.) Mout., Nouv. Fl. Liban Syrie 1:249 (1966). [Plates 93, 94]

Plant 15–40 cm. Bulb about 3 cm in diameter. Leaves 3–6, prostrate, lanceolate or lorate, undulate, ciliate; outer leaf 2(–3) cm broad. Scapes often 2–3. Flowering raceme conical or ovoid, 22–50-flowered; flowering pedicels spreading, rarely nodding; the lower twice to several times as long as flower, later horizontal or arcuate. Fruiting raceme conical; fruiting pedicels thickened. Perianth 8–10 mm, tubular-campanulate, umbilicate at base, changing in colour from white (or rarely pale lilac) with green or yellow veins in bud to greenish-white, and turning gradually to greyish-brown later; lobes pale, ovate or oblong, about ½ length of tube. Anthers lilac. Capsule persistent, loculicidal; valves 10–17 mm, oblong or obovate, cuneate at base, retuse at apex. 2n=16. Fl. March–April.

Var. *stepporum* (var. *typica* Feinbr., *loc. cit.,* f.30, 1940) is confined to the Syrian Desert and to S. Turkey. Plants are 20–40 cm tall, with a dense conical raceme. Flowering pedicels are several times as long as flower; lower fruiting pedicels 4–7 cm. Capsule 15–17 mm.

In Palestine the species is represented only by var. *transjordanica* and var. *edumea*.

1. Raceme lax, 22–35-flowered; fruiting raceme 10–16 × 10–14 cm; lower fruiting pedicels 5–7 cm.   var. **transjordanica**

– Raceme dense, 30–50-flowered; fruiting raceme 7–10 × 5–6(–9) cm; lower fruiting pedicels 2.5–3(–4) cm. **var. edumea**

Var. **transjordanica** Feinbr., *op. cit.* 370, f.32 & t.18 f.7 (1940). [Plate 93]. Plant 20–40 cm. Raceme lax; lower flowering pedicels 3–5 times as long as flower. Perianth devoid of green veins, sometimes yellow-veined. Capsule 10–14 mm.
Hab.: Steppes. E. Gilead, E. Ammon, Moav.

Var. **edumea** Feinbr., *op. cit.* 371, f.31 & t.18 f.6 (1940). [Plate 94]. Plant 15–18 cm. Raceme dense, longer than scape; pedicels spreading or curved upwards at tip, shorter than in other two varieties; lower flowering pedicels 2–4 times as long as flower. Perianth green-veined. Capsule 10–14 mm.
Hab.: Steppes, Artemisietum herbae-albae. C. Negev; E. Ammon, Edom.
Area of species: W. Irano-Turanian (Palestine, Egypt, Syria, Iraq).

11. Bellevalia zoharyi Feinbr., Palest. Jour. Bot. Jerusalem ser., 1:372, t.18 f.10 [Plate 95]

Plant 10–15 cm. Bulb 1.2–3.5 cm in diameter. Leaves 3–5, prostrate, lanceolate or lorate, glaucous, often undulate, ciliate or scabrous at margin; outer leaf 0.5–2 cm broad. Scapes 1–2. Flowering raceme 3.5–4 cm, ovoid, nearly sessile on ground, 15–35-flowered; flowering pedicels erecto-patent, generally longer than flower. Fruiting raceme deltoid, 5–6 cm, not longer to slightly longer than broad; fruiting pedicels thick, horizontal; the lower up to 2 cm. Perianth 8–9 mm, tubular-campanulate, white without conspicuous veins before flowering, soon turning dirty white; lobes ½ length of tube, in herbarium darker in colour than tube. Capsule persistent, loculicidal, 8–13 × 6–9 mm. Seeds 2 mm, globose. Fl. March–April.
Hab.: Steppes. Rare. Judean Desert, ?Negev; Ammon, Moav, Edom.

Var. **zoharyi**. Subsp. *typica* Feinbr., *op. cit.* 373 (1940). Capsule ovoid, 8–10 mm.

Var. **pyricarpa** (Feinbr.) Feinbr. comb. nov. Subsp. *pyricarpa* Feinbr., *op. cit.* 373 (1940). Capsule pyriform, 13 mm.
Area of species: Endemic. W. Irano-Turanian.

The infraspecific taxa of *B. zoharyi*, differing in shape and size of capsules, deserve no more than a varietal rank. Contrary to earlier data, the distribution area of the two taxa is the same.

17. LEOPOLDIA Parl.

Bulbous and scapose perennials. Bulb tunicate. Leaves basal, linear to lanceolate, canaliculate, more or less flaccid. Scape erect. Inflorescence an elongate many-flowered lax raceme. Bracts minute. Flowers dimorphic; the upper sterile, generally brightly coloured; the others fertile, purple or violet in bud, their colour changing during flowering, often becoming olive-green, greyish-brown, yellow, etc. Perianth gamophyllous, 6-dentate, abscissed at base and falling off after flowering; in fertile flowers cylindrical-urceolate, somewhat zygomorphic, constricted at throat, ending in short reflexed subequal teeth; tube more or less angular towards throat, with prominent protuberances on shoulders around throat during flowering. Stamens 6, included, distinctly 2-seriate; 3 filaments inserted above middle of perianth-tube, the 3 others higher up. Ovary ovoid, nearly 3-gonous, 3-locular; style 1, subulate; stigma minutely 3-lobed. Capsule 3-quetrous, loculicidal, in some species indehiscent and deciduous. Seeds 2 in each locule, globose, ovoid or pear-shaped, black, reticulate-rugose, not arillate. n=9 (the majority of species are diploid).

About 25 species, mainly Mediterranean and Irano-Turanian

A monographic study of the genus would most probably reduce the number of species of *Leopoldia*.

The taxonomy of *Leopoldia* species is mainly based on the following characters: (1) shape of raceme (cylindrical vs. conical), (2) colour of perianth-teeth in fertile flowers lighter than tube (pale yellowish) or darker than tube (blackish, maroon, brown), (3) colour of perianth in fertile flowers, (4) shape of sterile flowers, (5) shape of capsule-valves. In contrast, width of leaves and length of perianth of fertile flowers vary considerably within species (e.g. in *L. comosa, L. longipes* and others) and thus cannot be used for delimitation of species.

Literature (cf. also literature of *Muscari*): F. Parlatore, *Flora palermitana* 1:435–438, Firenze, 1845; *Flora italiana* 2:493–498, Firenze, 1857. T. v. Heldreich, Über die Liliaceen-Gattung *Leopoldia* und ihre Arten, *Bull. Soc. Imp. Nat. Moscou* 53:56–75 (1878). A.S. Losina-Losinskaya, *Leopoldia*, in: *Flora URSS* 4:409–411, 1935. N. Feinbrun, Revision of the genus *Hyacinthella* Schur, *Bull. Res. Counc. Israel* 10D:324–347 (1961). D.C. Stuart, *Muscari* and allied genera, *R.H.S. Lily Year Book* 29:125–138 (1966); Chromosome numbers in the genus *Muscari* Mill., *Notes Roy. Bot. Gard. Edinb.* 30:189–196 (1970). F. Garbari, Sul rango tassonomico di *Leopoldia* Parl., *Muscarimia* Kostel., *Muscari* Mill., *Gior. Bot. Ital.* 101:300–301 (1968a); Il genere *Muscari* (Liliaceae): Contributo alla revisione citotassonomica, *ibid.* 102:87–105 (1968b); Nuove osservazioni citologiche sui generi *Muscari* e *Leopoldia, ibid.* 103:1–9 (1969); Le specie del genere *Leopoldia* Parl. (Liliaceae) in Italia, *Webbia* 28:57–80 (1973). F. Garbari & W. Greuter, On the taxonomy and typification of *Muscari* Mill., etc., *Taxon* 19:329–335 (1970). B. Bentzer, Taxonomy, variation and evolution in representatives of *Leopoldia* Parl. (Liliaceae) in the Southern and Central Aegean, *Bot. Not.* 126:69–132 (1973).

1. Leaf 1. Pedicels of fertile flowers at most 3 mm. Perianth-tube ivory-coloured; throat and teeth dark purple to blackish. Capsule deciduous at maturity, indehiscent.   **4. L. eburnea**
- Leaves 2 or more. Pedicels of fertile flowers longer than above; lower pedicels about as long as or longer than flower. Perianth-tube not ivory-coloured. Capsule persistent, dehiscent at maturity   2
2. Perianth-teeth of fertile flowers pale yellowish, lighter in colour than tube; sterile flowers long-pedicellate forming a conspicuous corymb-like tuft to raceme; sometimes tuft nearly suppressed.   **1. L. comosa**

- Perianth-teeth of fertile flowers blackish-purple to nearly black, much darker than tube 3
3. Flowering and fruiting raceme conical; lower flowering pedicels about 3 cm; lower fruiting pedicels 4-9 cm. Perianth of fertile flowers (7-)8-11 mm. Capsule-valves triangular, acute, longer than broad. Plants of loess and rocky ground in steppe districts. **5. L. longipes**
- Flowering and fruiting raceme cylindrical (rarely narrowly conical); flowering and fruiting pedicels shorter than above. Perianth of fertile flowers generally shorter. Capsule-valves not longer than broad 4
4. Plants of sandstone and sandy soils of coastal plain. Sterile flowers dark violet, inflated, somewhat longer than broad. Pedicels of fertile flowers hardly elongating towards end of flowering (reaching 10-12 mm). **2. L. bicolor**
- Plants of desert mountain slopes. Sterile flowers not as above, often narrowly tubular. Pedicels of lower fertile flowers elongating up to 18 mm towards end of flowering. **3. L. deserticola**

1. Leopoldia comosa (L.) Parl., Fl. Palerm. 1:438 (1845); Bentzer, Bot. Not. 126:75, f.29A (in colour). *Hyacinthus comosus* L., Sp. Pl. 318 (1753). *Muscari comosum ("comosus")* (L.) Mill., Gard. Dict. ed. 8, no. 2 (1768); Boiss., Fl. 5:291; Post, Fl. 2:648 excl. var. [Plate 96]

Plant 25-40 cm. Bulb 1.5-2.5(-3) cm in diameter, ovoid, usually at depth of 10-30 cm. Leaves (2-)3-4(-5), 5-15(-20) mm broad, linear, more or less canaliculate, ½-1½ times as long as scape. Raceme 15-25 × 3-6 cm, cylindrical, ending in a conspicuous corymb-like tuft of long-pedicellate sterile flowers; sometimes tuft of sterile flowers nearly suppressed. Pedicels of fertile flowers horizontal, about as long as flower. Fertile flowers 7-9 mm, 1¼-2 times as long as broad; perianth purple in bud, during flowering olivaceous-yellow in the lower part of the tube, dirty brown in upper; teeth pale yellowish, lighter than tube. Sterile flowers much smaller, purple, obovoid, borne on ascending arcuate pedicels 3-4 times as long as flower. Capsule-valves broadly ovate. Fl. February–April.

Hab.: Cultivated ground. Fairly common. Coastal Galilee, Acco Plain; Sharon Plain, Philistean Plain: Upper and Lower Galilee, Mt. Carmel, Esdraelon Plain, Samaria, Judean Mts., Judean Desert, W. and N. Negev; Dan Valley; Golan, Gilead, Ammon, Moav, Edom.

Area: Mediterranean and Irano-Turanian, extending into the Euro-Siberian.

2. Leopoldia bicolor (Boiss.) Eig & Feinbr., Palest. Jour. Bot. Jerusalem ser., 4:58 (1947); Bentzer, Bot. Not. 125:329 (1972). *Muscari bicolor* Boiss., Fl. 5:294 (1882); Täckholm, Stud. Fl. Eg. ed. 2, 642. *M. maritimum* auct. fl. palaest. non Desf.; Boiss., Fl. 5:293; Post, Fl. 2:649; incl. var. *dolichobotrys* Opphr. & forma *longiscapum* Opphr. in Opphr. & Evenari, Florula Cisiord., Bull. Soc. Bot. Genève ser. 2, 31:194, f.8 (1941). [Plate 97]

Plant (10-)15-50 cm. Bulb 1.5-2.5(-3.5) cm in diameter. Leaves 4-6, 2-7(-13) mm broad, linear, canaliculate, flaccid, about as long as or shorter than scape. Raceme cylindrical, 6-14 × 2-4 cm, 25-50-flowered. Flowering pedicels horizontal to reflexed, as long as to slightly longer than flower, hardly elongating towards end of flowering. Fertile flowers 6–9 mm, about twice as long as broad; perianth greenish, suffused with purple in bud, during flowering olivaceous-yellow in the lower part of tube, dirty brown in upper; teeth and throat blackish-purple. Sterile flowers few, 4–5 × 2.5–3.5 mm, dark violet, inflated, on pedicels shorter to somewhat longer than flower. Capsule-valves 5 × 4 mm, nearly orbicular. Fl. March–April.

Hab.: Sandstone and sandy loam on the coastal plain. Fairly common. Coastal Galilee, Acco Plain, Sharon Plain, Philistean Plain.

Area: E. Mediterranean (N. Egypt, Palestine, S. Lebanon).

Leopoldia bicolor is recorded by authors of the Palestinian flora under *Muscari maritimum* Desf. or *Leopoldia maritima* (Desf.) Parl. However, the authentic *M. maritimum* of Desfontaine is confined to the W. coast of N. Africa (Tunisia, Algeria, Morocco). It differs from *L. bicolor* in the fertile flowers being yellow or pale yellow with bright yellow teeth.

Bentzer (Bot. Not. 125:329, 1972) found 2n=18 in specimens of *Leopoldia bicolor* from Mersa Matruh (Lower Egypt).

3. **Leopoldia deserticola** (Rech. fil.) Feinbr. comb. nov. *Muscari deserticolum* Rech. fil., Ark. Bot. ser. 2, 5(1):87 (1959); Mout., Nouv. Fl. Liban Syrie 1:254. [Plate 98]

Plant 10–15 cm. Bulb up to 2.5 cm in diameter. Leaves 3–8, 1–2 mm broad, folded, very flaccid, longer than scape. Raceme cylindrical to narrowly conical, 5–10 × 2–3 cm, 20–30-flowered, compact, later lax. Fertile flowers 3–4 times more numerous than sterile ones. Pedicels of fertile flowers more or less horizontal, in flower 4 mm, in fruit up to 18 mm. Perianth 5–7(-9) × 2 mm; teeth and throat blackish. Sterile flowers few, dense, shortly pedicellate; perianth 1–3 mm, in some flowers narrowly tubular. Fl. March–April.

Hab.: Stony desert, mountain slopes, 350–1,200 m. N. Negev; E. Ammon, Moav, Edom (with *Retama raetam*).

Area: W. Irano-Turanian (Palestine, Syrian Desert).

4. **Leopoldia eburnea** Eig & Feinbr., Palest. Jour. Bot. Jerusalem ser., 4:58, t.1 (1947); Täckholm, Kosinová & Chrtek, Novit. Bot. Horti Bot. Univ. Carol. Pragen. 1967:55 (1968). *Muscari eburneum* (Eig & Feinbr.) Stuart, Notes Roy. Bot. Gard. Edinb. 30:190 (1970). [Plate 99]

Plant 15–30 cm. Bulb 2–2.5 cm in diameter. Leaf 1, broadly linear, 25–35 × 1–2 cm, canaliculate-plicate, narrowed towards apex and base, flaccid, ciliolate at margin, glaucous, much longer than flowering scape. Scape thickish. Raceme cylindrical, 5–8 cm in flower, much elongating at maturity. Pedicels of fertile flowers horizontal, 3 mm, somewhat elongated in fruit. Fertile flowers 6–8 mm, ivory-coloured; teeth blackish-purple, nearly black. Sterile flowers few, subsessile, dense, oblong-cylindrical, dark violet. Anthers dark violet. Capsule 3-quetrous, retuse at apex, indehiscent and deciduous when ripe;

valves suborbicular, broader than long (7–9 × 9–12 mm), subcordate at base. Seeds pyriform, 2.25 mm. Fl. February–March.

Hab.: Sands, sandy soil, sandy loess. Rare. S. Philistean Plain; W., N. and C. Negev.

Area.: E. Saharo-Arabian (S. Palestine, Egypt).

Found in Egypt between Cairo and Alexandria.

This synaptospermous desert plant, with subsessile capsules deciduous after maturation, is reminiscent of *Bellevalia desertorum* Eig & Feinbr. and *B. sessiliflora* (Viv.) Kunth.

5. Leopoldia longipes (Boiss.) Losinsk., in Fl. URSS 4:410 (1935). *Muscari longipes* Boiss., Diagn. ser. 1, 13:37 (1854); Fl. 5:290; Post, Fl. 2:648. [Plates 100, 101]

Plant 20–40(–50) cm. Bulb 2–4(–6) cm in diameter. Leaves 3–8, canaliculate-plicate, varying in width, usually shorter than scape. Flowering and fruiting raceme conical. Pedicels more or less horizontal; the lowermost at least 3 times as long as perianth in flower, 4–9 cm in fruit. Fertile flowers (7–)8–11 mm, about 2–2½ times as long as broad; perianth purplish in bud, turning greenish-yellow to olivaceous-brown during flowering; teeth blackish-purple. Sterile flowers few, violet, mostly short-pedicellate. Fruiting raceme 10–20(–30) cm; pedicels elongated, rigid; stalk of fruiting raceme breaking off at base when dry and fruiting raceme driven by wind as a tumbleweed dispersing the seeds. Capsule conical, persistent, dehiscing at maturity; valves triangular, acute, distinctly longer than broad. Fl. March–May.

Subsp. **longipes**. *Muscari longipes* Boiss., Diagn. ser. 1, 13:36 (1854); Feinbr., Zohary & Koppel, Fl. Land Israel, Iconography, t.43, Jerusalem, 1949; Boiss., Fl. 5:290; Post, Fl. 2:648. *M. aaronsohnii* Opphr. & Beauverd in Opphr., Florula Transiord. 159, Bull. Soc. Bot. Genève ser. 2, 22:278, f.3 (1931). *L. aaronsohnii* (Opphr. & Beauverd) Feinbr., in Eig, Zohary & Feinbr., Analyt. Fl. Palaest. 410, 1948. [Plate 100]. Leaves 4–8, linear-lorate, (15–)30–50 × 0.6–2 cm. Bulb 2–4 cm in diameter.

Hab.: Fertile and mostly cultivated loess and sandy-loess soils. Common, mostly in fields, often in Achileetum santolinae. Rare on stony soils. W. and N. Negev; E. Gilead, E. Ammon, Moav, Edom.

Area of subsp.: W. Irano-Turanian (Palestine, Syria, Iraq, Iran).

Within subsp. *longipes* variation in width of leaves is considerable. Narrow-leaved plants of Moav and Edom have been described as *Muscari aaronsohnii* Opphr. & Beauverd; however, we failed to find differences in other traits (shape of perianth-teeth or of ovary and stigma, as given by Oppenheimer & Beauverd) between these plants and the typical subspecies.

Subsp. **negevensis** Feinbr. & Danin.* [Plate 101]. Leaves 3–5, broadly oblong,

* See Appendix, p. 397.

generally falcate, 10–15(–20) × 2–4 cm. Bulb 3–6 cm in diameter, larger than in subsp. *longipes*.

Hab.: Stony and rocky slopes, in steppe areas dominated by the *Artemisia herba-alba–Reaumuria negevensis* association or by Zygophylletum dumosi. C. and S. Negev, N. Sinai.

Area of subsp.: E. Saharo-Arabian (Palestine, N. Sinai).

Area of species: W. Irano-Turanian and E. Saharo-Arabian.

Subsp. *longipes* is common in fields on loess and sandy-loess soils of W. and N. Negev, and grows successfully even in non-cultivated soils. It is rare on stony soils. In contrast, subsp. *negevensis* grows mainly on stony and rocky slopes in steppe areas of the highlands of C. Negev and Gebel Halal in N. Sinai at elevations of 300–1,000 m. In some parts of the C. Negev, such as Nahal Zin, subsp. *negevensis* is found adjacent to rocky slopes in shallow wide wadi-beds which benefit from a good water regime.

18. MUSCARI Mill.

Bulbous and scapose perennials. Bulb tunicate. Leaves basal, narrowly linear or filiform, sometimes canaliculate. Scape erect, simple. Inflorescence a short dense raceme or spike. Bracts minute. Pedicels usually shorter than flower. Upper flowers sterile, often differing in colour and shape from fertile flowers; fertile flowers nodding or ascending. Perianth gamophyllous, 6-dentate, in fertile flowers obovoid to subglobose or urceolate, mostly more or less constricted under throat, violet, lilac, blue or purple, not changing in colour during anthesis; teeth short, ovate-deltoid, mostly more or less reflexed. Stamens 6, included, 1-seriate or obscurely 2-seriate, inserted about half-way up perianth-tube. Ovary 3-gonous, 3-locular, with 2 ovules in each locule; style 1, filiform; stigma capitate. Capsule 3-quetrous, chartaceous to membranous, loculicidal. Seeds ovoid to globose, black, usually wrinkled, not arillate.

About 30 species, mainly Mediterranean, Irano-Turanian and Euro-Siberian.

Literature: A.S. Losina-Losinskaya, *Muscari* Mill., in: *Flora URSS* 4:412–422, 1935. N. Feinbrun, A monographic study on the genus *Bellevalia* Lapeyr., *Palest. Jour. Bot. Jerusalem Ser.*, 1:391–401 (1940); Revision of the genus *Hyacinthella* Schur, *Bull. Res. Counc. Israel* 10D:324–347 (1961). D.C. Stuart, *Muscari* and allied genera. A Lily group discussion, *R.H.S. Lily Year Book* 29:125–138 (1966); Chromosome numbers in the genus *Muscari* Mill., *Notes Roy. Bot. Gard. Edinb.* 30:189–196 (1970). F. Garbari, Il genere *Muscari* (Liliaceae): Contributo alla revisione citotassonomica, *Gior. Bot. Ital.* 102:87-105 (1968); Note sul genere *Pseudomuscari* (Liliaceae), *Webbia* 27:369–381 (1972).

1. Fertile flowers light lilac to pale blue, with darker veins. Plants flowering in autumn before or with first rains.   **3. M. parviflorum**
- Fertile flowers differing in colour from above. Plants flowering in winter or spring

2. Perianth blackish-violet with concolorous teeth. Valves of capsule broader than long, not distinctly retuse at apex. **1. M. commutatum**
- Perianth-tube deep greyish-blue to greyish-violet; teeth white. Valves of capsule not broader than long, slightly retuse at apex. **2. M. pulchellum**

1. Muscari commutatum Guss., Pl. Rar. 145 (1826); Mout., Nouv. Fl. Liban Syrie 1:256; Boiss., Fl. 5:296; Post, Fl. 2:650. *M. inconstrictum* Rech. fil., Ark. Bot. ser. 2, 2(5):314, t.3 (1952). [Plate 102]

Plant 10–30 cm. Bulb ovoid, 1–2.5 cm in diameter. Leaves synanthous, 1.5–3 mm broad, linear with a fine groove on upper face. Raceme 2–4 cm, 10–22-flowered, dense, ovoid. Pedicels of fertile flowers shorter than to as long as flowers. Flowers nodding. Sterile flowers few. Perianth of fertile flowers 5–6(–8) mm, deep blackish-violet, oblong-obovoid, not or slightly constricted at throat, with concolorous teeth; teeth triangular-ovate, not longer than broad, slightly recurved, connivent before anthesis. Valves of capsule 5–5.5 × 5–6 mm, suborbicular, broader than long, rounded to cordate at base, not distinctly retuse at apex. Fl. December–March.

Hab.: Rocky and stony ground. Sharon Plain; Upper and Lower Galilee, Esdraelon Plain, Mt. Gilboa, Samaria, Judean Mts., Judean Desert, N. and C. Negev; Lower Jordan Valley, Dead Sea area; Golan, Ammon, Edom.

Area: Mediterranean and W. Irano-Turanian.

We join Mouterde (*loc. cit.*) in doubting the taxonomic value of *M. inconstrictum* Rech. fil. and the diagnostic significance of the degree of constriction at the throat of perianth in plants from the *M. commutatum* group.

2. Muscari pulchellum Heldr. & Sart. ex Boiss., Diagn. ser. 2, 4:109 (1859); Heldr., Atti Congr. Internat. Bot. Firenze 27 (1876); Boiss., Fl. 5:295 & Post, Fl. 2:649 as *M. racemosum* (L.) Mill. *M. racemosum* auct. fl. pal. non (L.) Mill. [Plate 103]

Plant 15–25 cm. Bulb ovoid, 1–2 cm. Leaves synanthous, 4–6, 1–3 mm broad, linear with a fine groove on upper face. Raceme 2–3 cm, 10–20-flowered, shortly cylindrical, rather lax. Pedicels shorter than flowers, spreading, later deflexed. Sterile flowers light blue, ovoid to cylindrical-clavate, narrow, the upper gradually shorter. Perianth of fertile flowers 4–6 mm, deep greyish-blue to greyish-violet, ovoid-urceolate, constricted at throat; teeth very short, white, recurved, triangular-ovate, obtuse, as broad as long. Valves of capsule about 6 mm, broadly elliptic to suborbicular, cuneate to rounded at base, slightly retuse at apex. Fl. January–April.

Hab.: Batha and arid hills. Sharon Plain, Philistean Plain; Lower Galilee, Mt. Carmel, Esdraelon Plain, Mt. Gilboa, Samaria, Judean Mts., Judean Desert, C. Negev; Golan, Gilead, Ammon, Moav, Edom.

Area: Mainly E. Mediterranean and W. Irano-Turanian, extending to the S. Med. islands.

Differs from the European *M. neglectum* Guss. ex Ten. [*M. racemosum* (L.) Lam. &

DC., *M. atlanticum* Boiss. & Reut.] in longer pedicels, more lax cylindrical racemes, non-imbricate fertile flowers and light blue sterile flowers.

3. Muscari parviflorum Desf., Fl. Atl. 1:309 (1798); Boiss., Fl. 5:299; Post, Fl. 2:650. [Plate 104]

Plant 10–20 cm, autumn flowering. Bulb ovoid, 1–1.5 cm in diameter. Leaves 3–5, subequal, nearly filiform, semiterete with a fine groove on upper face, emerging shortly after flowers. Raceme 1.5–3 cm, 6–12-flowered, cylindrical, very lax. Pedicels erecto-patent to horizontal, about as long as flower, each subtended by a setaceous bract and bracteole. Sterile flowers minute or nearly absent. Perianth 3–5 mm, light lilac to pale blue, with darker veins, ovoid-urceolate, slightly constricted above; teeth short, concolorous, recurved, ovate. Anthers dark violet. Valves of capsule 3–3.5 mm, suborbicular, broader than long, with elevated veins. Fl. September–November.

Hab.: Sandy places, cultivated ground; usually in large groups. Sharon Plain, Philistean Plain; Upper and Lower Galilee, Mt. Carmel, Samaria, Shefela, Judean Mts. (rare); Dan Valley.

Area: Mediterranean.

19. ALLIUM L.*

Bulbous and scapose perennials. Bulb tunicate, solitary or in cluster attached to a rhizome. Leaves sheathing at base, filiform, linear or elliptic, flat or cylindrical to semicylindrical, often fistulous. Flowers in a terminal umbel, in bud totally enclosed within a spathe; spathe divided into 2 or more valves, or entire, often caducous; pedicels often with bracteoles at base. Perianth stellate or narrowly campanulate or ovoid-urceolate; perianth-segments persistent, free or slightly connate at base, 1-veined. Stamens 6, free or connate at base into an annulus, inserted at base of segments and sometimes adnate to their base; filaments in many species with lateral cusps; anthers ellipsoid-oblong, dorsifixed. Ovary 3-locular; ovules mostly 2 in each locule, sometimes many. Style 1, filiform; stigma mostly entire, dot-like or capitate, rarely more or less 3-lobed. Capsule loculicidal; seeds 1–2 in each locule, rarely more, 3-quetrous or compressed, rarely globose, black.

About 400 species in the N. hemisphere.

Literature: E. Regel, Alliorum adhuc cognitorum monographia, *Acta Horti Petrop.* 3:1–266 (1875). A. I. Vvedensky, *Allium,* in: *Flora URSS* 4:112–280, 1935. F. Hermann, Sectiones et subsectiones nonnullae europeae generis *Allium, Feddes Repert.* 46:57–58 (1939). N. Feinbrun, *Allium* sectio *Porrum* of Palestine and the neighbouring countries, *Palest. Jour. Bot. Jerusalem Ser.,* 3:1–21 (1943). W. T. Stearn, Notes on the genus *Allium* in the Old World, *Herbertia* 11:11–34 (1946). N. Feinbrun, Further studies on *Allium* of Palestine and the neighbouring countries, *Palest. Jour. Bot. Jerusalem Ser.,* 4:144–157 (1948); Chromosome counts in Palestinian *Allium*

* By Dr Fania Kollmann.

species, *Palest. Jour. Bot. Jerusalem Ser.*, 5:13–16 (1949); Chromosomes and taxonomic groups in *Allium, Caryologia* 6 (Suppl.):1036–1041 (1954). H.A. Jones & L.K. Mann, *Onions and their Allies*, London & New York, 1963. C. Zahariadi, *Allium*, in: Săvulescu (ed.), *Flora Republicii Socialiste România* 11:187–269, 1966. P. Wendelbo, New subgenera, sections and species of *Allium, Bot. Not.* 122:25–37 (1969); Alliaceae, in: K.H. Rechinger (ed.), *Flora iranica* 76:1–100, tt.1–28, 1971. F. Kollmann, *Allium descendens* L., a nomen ambiguum, *Taxon* 19(5):789–792 (1970); New chromosome counts in *Allium* species of Palestine and Mt. Hermon, *Israel Jour. Bot.* 19:245–248 (1970); Karyotypes of three *Allium* species of the *erdelii* group, *Caryologia* 23:647–655 (1970); Karyology of some species of *Allium* Sect. *Molium* in Israel, *Israel Jour. Bot.* 22:92–112 (1973). R. von Bothmer, Biosystematic studies in the *Allium ampeloprasum* complex, *Opera Bot.* 34:1–104 (1974). B.E.E. Wilde-Duyfjes, *A Revision of the Genus Allium (Liliaceae) in Africa*, Wageningen, 1976. W.T. Stearn, European species of *Allium* and allied genera of Alliaceae: a synonymic enumeration, *Ann. Mus. Goulandris* 4:83–198 (1978); *Allium*, in: T.G. Tutin et al. (eds.), *Flora europaea* 5:49–69, Cambridge, 1980. F. Kollmann, *Allium* L., in: P.H. Davis (ed.), *Flora of Turkey* 8:98–211, 1984. P.E. Boissier, *Flora orientalis* 5:229–285, 1882. G.E. Post, *Flora of Syria, Palestine and Sinai*, ed. 2, 2:633–647, 1933.

When collecting specimens of the genus *Allium* for herbarium studies, notes should be made on living plants of the following characters of systematic value: colour of perianth and stamens, which may change in herbarium; leaves in cross-section, whether solid or hollow, keeled or not; structure of outer bulb-tunics and shape of bulblets. Plants pressed with the bulb before flowering may develop abnormally shaped umbels due to etiolation, longer than normal pedicels, fewer flowers and non-typical colour of perianth.

1. Filaments of inner stamens 3-cuspidate, the median cusp bearing the anther   2
 - All filaments simple (not 3-cuspidate) or rarely inner filaments with a small tooth on either side of base   10
2. Leaves hollow, terete or semiterete or flattened. Spathe persistent, rarely caducous and then perianth-segments green or yellowish-green, obtuse and rounded at apex, stamens long-exserted and bulb-tunics reticulate   3
 - Leaves not hollow, flat. Spathe caducous   8
3. Outer bulb-tunics reticulate   4
 - Outer bulb-tunics membranous or coriaceous   7
4. Perianth blue or blue-violet. Stamens included.   **20. A. hierochuntinum**
 - Perianth dull green, yellowish-green or white. Stamens exserted or subexserted   5
5. Plant (60–)70–150 cm tall. Perianth-segments (2.5–)3–3.5 mm, green or yellowish-green, white-margined, obtuse and rounded at apex, the inner emarginate. Stamens long-exserted.   **23. A. dictyoprasum**
 - Plants 6–30(–40) cm. Perianth-segments at least 3.5 mm, white with a purple or green midvein, not rounded at apex. Stamens subexserted   6
6. Perianth-segments 6–7 mm. Leaves as long as stem or longer; stem (6–)10–15 cm; plants buried in sand.   **22. A. sinaiticum**
 - Perianth-segments 3.5–5 mm. Leaves shorter than stem; stem 15–30(–40) cm.   **21. A. artemisietorum**
7(3). Plants (60–)80–100 cm. Cusps of inner filaments as long as their ciliolate basal lamina (undivided part).   **18. A. phanerantherum**

- Plants (10-)15-30(-60) cm. Cusps of inner filaments distinctly shorter than basal lamina.   **19. A. curtum**

8(2). Anthers exserted   9

- Anthers included or only slightly exserted; outer perianth-segments dark purple, the inner conspicuously paler in colour, often almost white, 5-5.5 mm.
  17. A. scorodoprasum subsp. **rotundum**

9. Perianth 2.5-4 mm, dark purple, rarely green; inner segments truncate and minutely denticulate at tip, somewhat longer than the outer; segments overlapping or laterally touching one another during anthesis. Bulblets borne on stolons several cm long.   **16. A. truncatum**

- Perianth 4-5 mm; inner segments not truncate, not longer than the outer; segments not touching one another laterally during anthesis. Bulblets numerous, nearly sessile, remaining close to mother-bulb.   **15. A. ampeloprasum**

10(1). Leaves hollow, terete, semiterete or flattened. Spathe 2-valved; valves mostly unequal, each valve ending in a slender appendage. Valves as long as or longer than umbel or if shorter, outer bulb-tunics reticulate (Sect. *Codonoprasum*)   11

- Leaves not hollow, flat. Spathe shorter than umbel   16

11. Outer and inner perianth-segments ovate or lanceolate, acute; filaments included. Plants of steppe and desert   12

- Perianth-segments obtuse; filaments exserted or included   13

12. Outer bulb-tunics reticulate. Perianth campanulate, 5-6 mm.   **12. A. sindjarense**

- Outer bulb-tunics not reticulate. Perianth tubular-urceolate, 6-7(-8) mm.
  11. A. desertorum

13(11). Filaments and anthers exserted   14

- Filaments and anthers included or anthers slightly exserted   15

14. Perianth-segments connivent, light green to brownish-green; outer segments broadest near apex, apiculate or notched. Stem 40-60(-85) cm, straight; anthers pale yellow.   **14. A. albotunicatum**

- Perianth-segments not connivent, purplish-pink, sometimes yellowish-green; outer segments rounded at apex. Stem 10-25(-35) cm, mostly flexuose; anthers yellow.   **13. A. stamineum**

15(13). Perianth milky-white; segments often with a purplish or greenish midvein; anthers partly exserted. Flowering pedicels more or less equal. Spathe-valves mostly ovate at base, not exceeding umbel.   **10. A. pallens**

- Perianth not milky-white; anthers included. Flowering pedicels very unequal. At least one spathe-valve exceeding umbel.   **9. A. paniculatum**

16(10). Umbel 10-40 cm in diameter, polygamous with hermaphrodite, staminate or neuter flowers; pedicels very unequal, the longest generally not bearing fruit.
  29. A. schubertii

- Umbel not exceeding 10 cm in diameter; all flowers hermaphrodite   17

17. Leaf-sheaths reaching above ground level and sheathing at least ¼ of stem. Ovules 2 in each locule of ovary. Perianth-segments erect after flowering and enveloping capsule (Sect. *Molium*)   18

- Leaf-sheaths all subterranean, or just reaching ground level. Ovules more than 2 in each locule. Perianth-segments withered after flowering, usually reflexed and not enveloping capsule (Sect. *Melanocrommyum*)   25

18. Perianth stellate   19

- Perianth campanulate; segments erect   20

19. Umbel nodding in bud. Anthers pale green. Leaves glabrous, often scabridulous at margin. Stem 3-quetrous, rarely terete. **1. A. neapolitanum**

  - Umbel erect in bud. Anthers yellow. Leaves (at least sheath) hairy or long-ciliate at margin, rarely glabrous. Stem terete. **2. A. trifoliatum**

20(18). Stem 80–130 cm. Leaf-sheath densely and retrorsely hairy; blade usually hairy on both faces, rarely glabrous on upper face. Perianth pink-purple or cream-coloured. Filaments exserted. **8. A. carmeli**

  - Stem 10–60(–70) cm. Filaments included or slightly exserted. Plants with a different set of characters 21

21. Perianth-segments obtuse, pink with a darker midvein. Leaves glabrous. Bulb bearing several bulblets. Plants of sands in W. Negev and southwards.
7. A. roseum var. **tourneuxii**

  - Plants with a different set of characters 22

22. Leaves pilose on both faces and ciliate at margin, rarely glabrous. Perianth straw-coloured, rarely pinkish; segments 8–12 mm. **4. A. erdelii**

  - Leaves minutely and densely velutinous on lower face along veins, or leaf-sheaths covered with clusters of retrorse club-shaped hairs. Perianth white or, if pale yellow, perianth-segments 4–6 mm 23

23. Perianth-segments 10–13 mm, milky-white, with an indistinct midvein, rarely suffused with pink. **6. A. negevense**

  - Perianth-segments (4–)5–7 mm, white or pale yellow 24

24. Plants of desert sands or sandy loess. Leaf-sheaths and part of blade covered with clusters of club-shaped retrorse hairs. Perianth-segments white with a distinct purplish midvein. **3. A. papillare**

  - Plants of steppe-like habitats on calcareous and basalt hills. Leaves minutely velutinous along veins. Perianth-segments pale yellow or white. **5. A. qasyunense**

25(17). Low plants (6–20 cm) of rocky places in deserts. Perianth-segments white with a purple midvein. Filaments dark purple, white at tip. Ovary dark purple.
28. A. rothii

  - Plants with a different set of characters 26

26. Perianth, filaments and anthers deep purple. Stem 30–80 cm. Leaves 2–4(–5) cm broad, shorter than stem. Plants of steppe and desert districts.
25. A. aschersonianum

  - Perianth and filaments pale purple, pale lilac, white or greenish-white 27

27. Plants of sandy loam and kurkar of coastal plain. Stem 15–25(–40) cm. Leaves 1–2 cm broad. Perianth pale purple. Filaments about as long as perianth. Ovary green. **26. A. tel-avivense**

  - Plants of other habitats and with a different set of characters 28

28. Perianth stellate; segments white, pale lilac or mauvish-pink. Stem 60–100 cm, rarely 25–50 cm and then ovary dark purple and filaments dark purple in their lower part. Leaves 2.5–8 cm broad. **24. A. nigrum**

  - Perianth not stellate; segments more or less erect, greenish-white. Ovary green. Filaments white. Stem 15–40 cm. Leaves 1–2 cm broad (sometimes in Judean Mts. up to 4 cm). **27. A. orientale**

Sect. SCHOENOPRASUM Dumort., Fl. Belg. 140 (1827). Cultivated.

Allium schoenoprasum L., Sp. Pl. 301 (1753); Jones & Mann, Onions 43, f.3 & t.8b (1963); Stearn, in Fl. Eur. 5:55 (1980).*

Bulbs very narrow, conical, clustered in dense clumps on a short rhizome. Stem hollow, 5–50 cm. Leaves narrow, hollow. Spathe short, 2–3-lobed, persistent. Umbel hemispherical or ovoid, dense; pedicels short. Perianth campanulate; segments lilac or pale purple, rarely white, usually lanceolate, acute or acuminate. Anthers included.

Cultivated for its edible leaves *(chives)*. Little grown in Israel.

Sect. CEPA (Mill.) Prokh., Bull. Appl. Bot. Pl.-Breed. (Leningrad) 24:176 (1931). Cultivated.

Allium cepa L., Sp. Pl. 301 (1753); Sm. in Sibth. & Sm., Fl. Graeca 4:22, t.326 (1823); Regel, *op. cit.* 92; Jones & Mann, Onions 31, tt.3, 4; Stearn, Ann. Mus. Goulandris 4:133 (1978); Boiss., Fl. 5:249; Post, Fl. 2:638.*

Bulb depressed-globose. Stem hollow, tapering from inflated lower part. Leaves hollow, almost semicircular in section. Spathe persistent, often 3-valved, not exceeding pedicels. Umbel globose. Perianth stellate; segments 3–4 mm, greenish-white. Stamens exserted; inner filaments with a small tooth at base on each side.

Cultivated for its edible bulbs *(onions)*. Unknown in wild state but possibly allied to *A. oschaninii* B. Fedtsch. (*A. cepa* var. *sylvestre* Regel) of C. Asia (Stearn, *loc. cit.*, 1978), *A. pskmense* B. Fedtsch. and *A. galanthum* Kar. & Kir.

The name *A. ascalonicum* L. has long been erroneously applied to bulblet-forming variants of *A. cepa (shallot)*, but is actually based on the wild *A. hierochuntinum* described by Boissier (1882).

Sect. 1. MOLIUM G. Don ex Koch, Syn. Fl. Germ. ed. 3, 715 (1857) p.p. Sect. *Crommyum* Webb & Berth. subsect. *Haplostemon* series *Molia* Boiss., Fl. 5:231 (1882) p.p.; incl. sect. *Rhodoprason* F. Hermann, Feddes Repert. 46:57 (1939). Bulb ovoid or subglobose; outer tunics often thick, variously sculptured. Stem terete or angular. Leaves almost basal, with short above-ground sheaths, flat. Spathe mostly persistent, 1-valved, often 2- to several-lobed. Perianth stellate or campanulate. Filaments simple. Ovary with distinct nectariferous pores. Ovules 2 in each locule.

1. Allium neapolitanum Cyr., Pl. Rar. Neap. 1:13, t.4 (1788); Boiss., Fl. 5:274; Post, Fl. 2:644. [Plate 105]

Bulb 1–2 cm in diameter, subglobose; outer tunics membranous or crustaceous. Stem 20–50 cm, 3-quetrous, rarely terete. Leaves usually 2, 0.5–2(–3) cm broad, flat, broadly linear, attenuate-acuminate, keeled beneath, glabrous and often scabridulous at margin, usually shorter than stem. Spathe persistent, 1(–2)-valved, ovate, acuminate, shorter than pedicels. Umbel 5–8(–10) cm in diameter, lax, fastigiate or hemispherical, many-flowered, nodding in bud; pedicels 1.5–3.5 cm. Perianth stellate; segments 0.7–1.2 cm, white, elliptic-ovate, obtuse. Filaments white, flattened, tapering towards tip, acute, ½ length of perianth-segments; anthers pale green. Capsule 5 mm, enclosed in perianth. 2n=14, 21, 28, 28+1B. Fl. March–April.

* Not illustrated.

Hab.: Maquis, garigue, batha, rocky and shady places. Common. Coastal Galilee, Coast of Carmel, Acco Plain, Sharon Plain, Philistean Plain; Upper and Lower Galilee, Mt. Carmel, Mt. Gilboa, Samaria, Shefela, Judean Mts., Judean Desert, C. Negev; Hula Plain, Upper Jordan Valley; Golan, Gilead, Ammon, Moav, Edom.

Area: Mediterranean.

Apart from typical *A. neapolitanum* growing in the Mediterranean districts and habitats of Palestine, plants have been found in more arid parts of the country, which differ mainly in their stem being terete and not triquetrous. These plants were first found in C. Negev. In a paper by F. Kollmann on the cytology of several *Allium* species (Israel Jour. Bot. 22:92–112, 1973) these aberrant plants have been provisionally named *A. palaestinum* Kollmann (*nom. nud.*).

2. Allium trifoliatum Cyr., Pl. Rar. Neap. 2:11, t.3 (1792) subsp. **hirsutum** (Regel) Kollmann, Israel Jour. Bot. 24:204, f.1 (1975). [Plate 106]

Bulb 1–1.2 cm in diameter, subglobose; outer tunics membranous. Stem 15–40 cm, terete. Leaves 3–5, 5–7 mm broad, linear, flat, almost basal, about as long as stem; blade and sheath densely covered with white hairs, rarely glabrous. Spathe persistent, 1(–2)-valved. Umbel 2.5–4 cm in diameter, fastigiate to hemispherical; pedicels 1½–3 times as long as perianth-segments. Perianth stellate; segments 7–10 mm, lanceolate, obtuse or subacute, white, often with a purple midvein, sometimes reddening with age. Filaments ½ length of perianth, subulate; anthers yellow. Capsule 4–5 mm, enclosed in perianth. Fl. March–April.

Var. **hirsutum**; Kollmann, Israel Jour. Bot. 24:204 (1975). *A. hirsutum* Zucc., Abh. Akad. Wiss. (München) 3:232, t.2 f.2 (1843) non Lam. (1778); Boiss., Fl. 5:271; Post, Fl. 2:643. *A. subhirsutum* L. var. *hirsutum* Regel, Acta Horti Petrop. 3:221 (1875). ?*A. hierosolymorum* Regel, Acta Horti Petrop. 11:301 (1890). [Plate 106]. Sheaths and blades densely hairy. 2n=14.

Hab.: Rocky places in maquis and batha. Common. Upper and Lower Galilee, Mt. Carmel, Samaria, Judean Mts.; Golan, Gilead, Ammon, Moav.

A. trifoliatum subsp. *hirsutum* var. *hirsutum* grows in stands in batha and maquis of the hill districts in Palestine. Extensive studies of populations showed that the plant is very variable in degree of leaf-hairiness, as remarked also by Wilde-Duyfjes (1976, pp. 136–137). In some populations leaves are ciliate on margin with scattered hairs on blades; sometimes hairiness occurs on sheaths and base of blades only. The plant is also variable in the shape and width of perianth-segments and in the occurrence of a purple midvein on segments.

Var. **sterile** Kollmann, Israel Jour. Bot. 24:204 (1975). *A. dinsmorei* Rech. fil., Ark. Bot. ser. 2, 2(5):312–313 (1952). Sheaths and blades of leaves glabrous. Plants often triploid (2n=21) with very low percentage of fertile pollen and low seedset and reproducing by bulblets.

Hab.: Ploughed fields, olive groves, fallow fields. Upper Galilee; Golan.

This taxon is confined to secondary habitats in the northern part of the country.

Area of subspecies: E. Mediterranean.

Area of species: Mediterranean.

A. dinsmorei Rech. fil. does not differ from *A. trifoliatum* var. *sterile* in either habit or glabrosity of leaves.

3. Allium papillare Boiss., Diagn. ser. 1, 13:27 (1854); Wilde-Duyfjes, Revis. Allium Africa:147–150, f.27; Boiss, Fl. 5:271; Post, Fl. 2:643. [Plate 107]

Bulb 1–1.5 cm in diameter, ovoid; outer tunics coriaceous, greyish-brown, pitted under outer epidermis, inner tunics membranous. Stems 1(–2), (10–)15–30 cm, thickish. Leaves 2–6, 4–6 mm broad in lower ⅓ to ¼, flat, canaliculate, narrowly linear becoming gradually subulate, margins rolled in, shorter than to as long as stem, rarely somewhat longer; sheath and blade somewhat above sheath covered with clusters of club-shaped retrorse hairs. Spathe persistent, often purplish, 1-valved, divided into 3–4 ovate cuspidate lobes. Umbel (2–)3–5 cm in diameter, fastigiate to hemispherical, few- to many-flowered; flowering pedicels subequal, twice as long as flowers. Flowers sweet-smelling. Perianth campanulate; segments 6–7 mm, white with a purplish midvein, obtuse, the outer ovate, the inner oblong-ovate. Stamens included; filaments slightly shorter than perianth; anthers yellow. Stigma minutely 3-lobed. Capsule 5 mm, nearly enclosed in perianth, subglobose, angular, often purplish. $2n=14$. Fl. end of March – beginning of April.

Hab: Sands and sandy loess. W. and N. Negev.

Area: Saharo-Sindian (S. Palestine, Sinai, Egypt).

4. Allium erdelii Zucc., Abh. Akad. Wiss. (München) 3:236, t.5 (1843); incl. var. *roseum* Boiss., Fl. 5:270 (1882) and var. *hirtellum* Opphr., Florula Transiord. 158, Bull. Soc. Bot. Genève ser. 2, 22:277 (1931)*; Kollmann, Israel Jour. Bot. 18:65 (1969); Boiss., Fl. 5:269; Post, Fl. 2:642. *A. philistaeum* Boiss., Diagn. ser. 1, 13:26 (1854). [Plate 108]

Bulb 8–17 mm in diameter, globose to ovoid; outer tunics coriaceous, greyish-brown, the inner pale yellow. Stem 10–30(–40) cm, terete. Leaves 3–6, flat, 4–8 mm broad in lower ⅓ or ¼, tapering and much narrowed above, usually ciliate at margin and pilose on both faces (hairs sparse on upper face), rarely glabrous. Spathe persistent, sometimes pinkish or purplish, 1-valved, 3–4-lobed, somewhat shorter than pedicels. Umbel 4–5 cm in diameter, hemispherical or fastigiate, many-flowered, dense; flowering pedicels subequal, about 1½ times as long as perianth. Perianth oblong-campanulate; segments 8–12 mm, straw-coloured, rarely pale cream to pinkish, with a pinkish or greenish midvein, oblong-lanceolate, subacute to acute. Filaments shorter than or as long as perianth; anthers yellow, subexserted. Style longer than

---

* Var. *hirtellum* was discarded by Oppenheimer in Vegetatio 1:165 (1949).

stamens, exserted; stigma minutely 3-lobed. Capsule 3-3.5 mm, depressed-globose, enclosed in perianth. 2n=16. Fl. end of February–April.

Hab.: Batha and steppes. Coastal Galilee, Coast of Carmel, Acco Plain, Sharon Plain, Philistean Plain; Lower Galilee, Samaria, Shefela, Judean Mts., N. and C. Negev; Upper Jordan Valley, Beit Shean Valley, Lower Jordan Valley; S. Golan, Gilead, Ammon, Moav.

Area: E. Mediterranean and W. Irano-Turanian (Syria, Palestine, Egypt, Libya).

5. Allium qasyunense Mout. (err. *"gasyunense"*), Bull. Soc. Bot. Fr. 100:348 (1953); Nouv. Fl. Liban Syrie 1:280, t.89 f.1; Kollmann, Israel Jour. Bot. 18:66 (1969). *A. erdelii* Zucc. var. *lasiophyllum* Nábělek, Publ. Fac. Sci. Univ. Masaryk 105:35, f.5 & t.1 (1929) & var. *micranthum* Opphr., Florula Transiord. 158, Bull. Soc. Bot. Genève ser. 2, 22:277 (1931). [Plate 109]

Bulb 0.5-1 cm in diameter, ovoid; outer tunics coriaceous, greyish-brown; the inner membranous, white. Stem 40-60(-70) cm, striate. Leaves 3-5, flat, 4-6(-8) mm broad in the lower ½, gradually narrowed towards tip, linear, shorter than stem; sheaths and both faces of blade densely and minutely velutinous along veins. Spathe persistent, shorter than umbel, 1-valved, 4-lobed; lobes acute. Umbel 3-4 cm in diameter, fastigiate to hemispherical, many-flowered; flowering pedicels subequal, about 1½-2 times as long as perianth. Perianth campanulate; segments (4-)5-6 mm, pale yellow or white, ovate-oblong, acute. Filaments slightly longer than perianth; anthers yellow, exserted. Style long-exserted. Capsule 2.5-3 mm, depressed-globose, much shorter than and enclosed in perianth. 2n=14. Fl. March–April.

Hab.: Calcareous and basalt hills. E. Samaria, Judean Desert, N. Negev; Upper Jordan Valley, Beit Shean Valley, Lower Jordan Valley; Gilead, Ammon, Moav, Edom.

Area: W. Irano-Turanian (E. Palestine, Syria).

6. Allium negevense Kollmann, Israel Jour. Bot. 18:69, f.3 (1969). [Plate 110]

Bulb about 1.5 cm in diameter, globose; outer tunics coriaceous, brownish-grey, pitted; the inner white, membranous. Stem 20-40 cm, leafy in lower part. Leaves 2-3, flat, 5-10(-15) mm broad, linear, canaliculate in lower part, densely and minutely velutinous along veins on lower face and on sheaths (rarely hairs seen under lens as small tubercles); upper face of blade glabrous or subglabrous. Spathe persistent, shorter than pedicels, 1-valved, 1-3 lobed; lobes acuminate. Umbel fastigiate, becoming more or less globose after anthesis and reaching 7-10 cm in diameter; pedicels 2-3 times as long as perianth. Perianth narrowly campanulate; segments 10-13 mm, erect, milky-white, greenish at base, rarely suffused with pink, oblong, acute or obtuse, with an indistinct midvein. Filaments shorter than perianth; anthers included, greenish-yellow. Ovary depressed-globose; style somewhat exceeding stamens. Capsule about 4 mm, much shorter than and enclosed in perianth. 2n=20. Fl. March–April.

Hab.: Calcareous hills, stony ground; Zygophylletum dumosi, Varthemietum iphionoides, Suaedetum asphalticae. Judean Desert, C. Negev; Dead Sea area.

Area: W. Irano-Turanian. Endemic.

Similar to *A. qasyunense* in the velutinous leaf indumentum (Kollmann, *loc. cit.*, 1969) and to both *A. erdelii* and *A. qasyunense* in shape of the spathe, the fastigiate umbel and shape of the perianth (narrowly campanulate with erect segments). In shape of umbel and perianth *A. negevense* also resembles *A. roseum* but can generally be distinguished by its milky-white perianth. Rare white-flowered *A. roseum* can be distinguished from *A. negevense* by its glabrous non-velutinous leaves.

7. Allium roseum L., Sp. Pl. 296 (1753) var. **tourneuxii** Boiss., Fl. 5:274 (1882); Täckholm & Drar, Fl. Eg. 3:78-79. [Plate 111]

Bulb 1-1.5 cm in diameter, ovoid or subglobose with numerous ovoid bulblets; outer tunics greyish or brown, pitted after decay of outer epidermis. Stem 10-60(-70) cm. Leaves 3-4, 3-7 mm broad, linear, flat, long-acuminate, glabrous. Spathe persistent, 1-valved, deeply 3-4-lobed. Umbel 2.5-3.5 cm in diameter, fastigiate or hemispherical, without bulbils; pedicels 7-10 mm, subequal. Flowers fragrant. Perianth campanulate; segments 7-8 mm, pink, elliptic-oblong, obtuse; outer segments longer than the inner ones, often retuse or slightly eroded at apex. Filaments broadly triangular, ½ length of segments. Style not exceeding stamens. Capsule 4 mm, subglobose. Fl. February–April.

Hab.: Semi-stabilized sands; Artemisietum monospermae. W. Negev.
Area of variety: Saharo-Sindian (S. Palestine, Sinai, Egypt, Libya).

8. Allium carmeli Boiss., Diagn. ser. 1, 13:28 (1854); Nábělek, Publ. Fac. Sci. Univ. Masaryk 105:36 (1929); Feinbr., Palest. Jour. Bot. Jerusalem ser., 4:152 (1948); Boiss., Fl. 5:273; Post, Fl. 2:644. [Plate 112]

Bulb 1.5-2.5 cm in diameter, ovoid; tunics white, thickish. Stem 80-130 cm, up to 1 cm thick, bearing leaves up to ¼ of its length. Leaves (0.8-)1-1.5(-3) cm broad at base, linear, flat, canaliculate, long-tapering, shorter than stem, hirsute on both faces and ciliate, or glabrous with scattered hairs on upper face; sheath velvety with dense soft velutinous retrorse hairs. Spathe persistent, shorter than umbel, 1-valved, divided into 2-3 ovate, shortly caudate lobes. Umbel 4-6 cm in diameter, globose, many-flowered; flowering pedicels subequal, purplish, 4 times as long as perianth. Perianth broadly campanulate; segments 5-7(-8) mm, purplish-pink or cream-coloured with a purple midvein, concave, oblong-ovate, obtuse. Filaments tapering from broad lanceolate base, exceeding segments by ¼. Ovary globose; style thickish, exserted. Capsule 4-5 mm, depressed-globose. Fl. April–May.

Var. **carmeli.** Flowers cream-coloured. 2n=14.
Hab.: Rocks. Mt. Carmel, Mt. Gilboa.

Var. **roseum** Zohary, Beih. Bot. Centr. 51 Abt. II:293 (1934). *A. trichocoleum* Bornm., Feddes Repert. 17:452 (1921); Feinbr., Palest. Jour. Bot. Jerusalem

ser., 4:152 (1948); Kollmann, Israel Jour. Bot. 26:146 (1977); Post, Fl. 2:639.
A. lownei Baker in herb. unpubl. Flowers purplish-pink. 2n=14, 21.

Hab.: Rocky ground, maquis, batha, ploughed fields and vineyards. More common and widespread than var. *carmeli*. Upper and Lower Galilee, Esdrelon Plain, Mt. Gilboa, Judean Mts.; Golan.

Area of species: E. Mediterranean (Palestine, Lebanon, Syria).

Bornmüller (1921) described *A. trichocoleum* from a plant collected by Dinsmore near Jerusalem. According to Bornmüller, *A. trichocoleum* differs from *A. carmeli* in indumentum of leaves and in colour and length of perianth-segments; leaves are glabrous on their upper surface in *A trichocoleum* and hairy on both surfaces in *A. carmeli*; and perianth-segments are purplish-pink with a darker midvein and 3 mm long in *A. trichocoleum* and cream-coloured and 5-6 mm long in *A. carmeli*. Our examination of type specimens of *A. trichocoleum* showed leaves to be hairy on both surfaces and perianth-segments to be 5-6 mm and not 3 mm long. Moreover, our studies of populations of *A. carmeli* var. *roseum* in different parts of Israel showed that leaves are generally hairy on both surfaces, as they are in var. *carmeli*. Populations of plants with nearly glabrous leaves (with few hairs scattered on the upper face and at margin) were found on Mt. Gilboa only.

It is noteworthy that even in plants of var. *carmeli* with cream-coloured flowers the midvein and pedicels are purplish. On Mt. Gilboa, intermediates with pale pink flowers occur. Purple pigment appears especially in buds, tips of perianth-segments and pedicels.

Sect. 2. CODONOPRASUM Reichenb. in Mössler, Handb. ed. 2, 1:588 (1827). Sect. *Crommyum* subsect. *Haplostemon* series *Codonoprasa* Boiss., Fl. 5:254 (1882). Bulb ovoid, oblong or globose. Leaves sheathing up to ⅔ of stem. Spathe 2-valved; valves unequal in length, each with an ovate or lanceolate, usually veined base, narrowed above into a tail-like appendage; at least one valve longer than umbel. Perianth campanulate or cylindrical, never stellate. Filaments simple. Ovary with or without minute inconspicuous nectariferous pores; ovules 2 in each locule.

9. Allium paniculatum L., Syst. Nat. ed. 10, 2:978 (1759); Boiss., Fl. 5:259; Post, Fl. 2:640. [Plate 113, 114, 115]

Bulb 1.5 cm in diameter; outer tunics membranous. Stem 30-70 cm. Leaves 3-5, sheathing lower ⅓-½ of stem; blades (1-)2(-5) mm broad, hollow, flattened, prominently ribbed on lower face. Spathe persistent, 2-valved; valves unequal in length, ovate to narrowly lanceolate at base, each contracted above base into a long appendage; one or both valves exceeding umbel. Umbel 3.5-10(-14) cm in diameter, varying in shape, usually many-flowered; pedicels very unequal, the outer curving outwards, the inner erect. Perianth campanulate or cylindrical; segments 4-7 mm, narrowly obovate to oblong, usually obtuse, sometimes apiculate. Stamens included. Ovary ellipsoid, 2-3 times as long as broad, papillose, narrowed at base and apex. Capsule about 5 mm.

1. Perianth-segments 4-5 mm. Inner bulb-tunics and lower part of leaf-sheaths not purple. Valves of spathe separating early. Fl. April–July.   subsp. **paniculatum**

- Perianth-segments (5–)6–7 mm. Inner bulb-tunics and lower part of leaf-sheaths purple. Valves of spathe separating tardily. Fl. September–October.   subsp. **fuscum**

Subsp. **paniculatum**. *A. longispathum* Delaroche in Redouté, Liliacées 6:t. 316 (1811). *A. intermedium* DC. in Lam. & DC., Fl. Fr. ed. 3, 5:318 (1815). *A. paniculatum* var. *longispathum* (Delaroche) Regel, Acta Horti Petrop. 3:191 (1875). [Plate 113, 114]. Perianth-segments pink, greenish-white or greenish-brown. 2n=16, 24, 32. Fl. April–July.

Hab.: Rocks, batha and maquis. Lower Galilee, Mt. Gilboa, Samaria, Judean Mts., Judean Desert; Upper Jordan Valley, Beit Shean Valley, Lower Jordan Valley; Golan, Gilead, Ammon, Moav.

Area of subsp.: as of species.

Subsp. **fuscum** (Waldst. & Kit.) Arcangeli, Comp. Fl. Ital. ed. 2, 136 (1894); Stearn, in Fl. Eur. 5:60. *A. fuscum* Waldst. & Kit., Pl. Rar. Hung. 3:267, t.241 (1808–1809); Reichenb., Icon. Fl. Germ. 10:21, t. 485 (1848); Regel, Acta Horti Petrop. 3:90 (1875). [Plate 115]. Perianth-segments cream, whitish-green or greenish with a broad greenish or purplish-brown midvein and with purple streaks, often turning purplish-brown when dry. Fl. September–October.

Exceptional in its late flowering time.

Hab.: Dolomite rocks. Rare. Mt. Carmel.

Area of subsp.: E. Mediterranean, extending into Romania.

Area of species: Mediterranean, extending into C. and E. Europe.

Subsp. *paniculatum* is variable in shape of perianth, colour and shape of perianth-segments and length of pedicels. Three main types have been observed in subsp. *paniculatum* in Palestine. Plants from the Upper Jordan Valley have a campanulate, pinkish perianth with subacute segments (Plate 113, upper right). Plants from the Lower Jordan Valley are characterized by connivent greenish-white truncate perianth-segments (Plate 113, centre). Specimens from Judean Mts., Upper Galilee and Golan have mostly longer pedicels and a more effuse umbel with greenish, broadly campanulate flowers and subacute segments (Plate 114).

A. chloranthum Boiss. recorded by Feinbrun (1948) from various parts of Palestine has been found to be *A. paniculatum* subsp. *paniculatum*. The record of *A. chloranthum* by Tristram (1884) is apparently to be referred to that taxon.

10. **Allium pallens** L. Sp. Pl. ed. 2, 427 (1762); Wilde-Duyfjes, Taxon 22:74 (1973). *A. coppoleri* Tineo, Cat. Pl. Horti Panorm. 275 (1827); Feinbr., Palest. Jour. Bot. Jerusalem ser., 4:154 (1948). *A. pallens* L. var. *coppoleri* (Tineo) Parl., Fl. Ital. 2:550 (1857). *A. paniculatum* L. var. *pallens* (L.) Gren. & Godr., Fl. Fr. 3:209 (1855); Boiss., Fl. 5:260; Post, Fl. 2:640. *A. stamineum* Boiss. var. *nigro-pedunculatum* Opphr., Florula Transiord. 158, Bull. Soc. Bot. Genève ser. 2, 22:277 (1931). *A. amblyanthum* Zahar., Biol. Gallo-Hellen. 6:53 (1975). [Plate 116]

Bulb 1–1.5 cm in diameter; outer tunics membranous. Stem (12–)20–30 cm, more rarely up to 65–110 cm. Leaves 3–4, 1–2(–5) mm broad, hollow, filiform to linear. Spathe persistent, 2-valved; valves narrowly ovate or lanceolate at base, each contracted above base into a slender appendage. Umbel 1.5–4.5 cm in diameter, subglobose, hemispherical or fastigiate, compact; flowering pedicels more or less equal, sometimes dark purple to black. Perianth narrowly campanulate; segments 3–5 mm, milky-white, often with a purplish or greenish midvein, slightly broader above middle, truncate, rounded at apex, sometimes apiculate. Filaments included; anthers partly exserted, purple. Capsule 4 mm. 2n=16, 32. Fl. May–July.

Hab.: Batha and steppes and other arid places. Common. Acco Plain, Sharon Plain, Philistean Plain; Upper and Lower Galilee, Mt. Carmel, Esdraelon Plain, Shefela, Judean Mts., Judean Desert, W. and C. Negev; Upper and Lower Jordan Valley; Golan.

Area of subsp.: Mediterranean, extending into the Irano-Turanian territories of Palestine.

Variable in height of plant, length of spathe and of perianth and in colour of pedicels. In the desert (Negev and Judean Desert) it appears in a dwarfed form (12–15 cm) with a few-flowered umbel and abruptly narrowed, short valves of the spathe (Feinbrun, 1948). On the other hand, plants from saline marshes and heavy soils in the Sharon Plain (Mikhmoreth) are very tall tetraploids (2n=32), 110 cm high, with large umbels and 4–5 mm long perianth-segments. Pedicels vary from green to dark purple. Perianth-segments in herbarium specimens from the Negev usually show pink colour diffusing from the purplish midvein.

11. Allium desertorum Forssk., Fl. Aeg.-Arab. 72 (1775); Kollmann, Notes Roy. Bot. Gard. Edinb. 33:437–440 (1975). *A. modestum* Boiss., Diagn. ser. 1, 13:33 (1854); Feinbr., Palest. Jour. Bot. Jerusalem ser., 4:155 (1948); Boiss., Fl. 5:261; Post, Fl. 2:640. [Plate 117]

Bulb 1–2 cm in diameter, ovoid-oblong; outer tunics greyish-black, thick, bark-like, splitting into strips or fibres; the inner yellowish, striate. Stem 7–20 cm, striate, somewhat flexuose. Leaves up to 3, 1–3 mm broad, hollow, subtriquetrous, slender, canaliculate, striate, usually as long as or longer than stem. Spathe persistent, 2-valved; valves membranous, red-veined, shorter to somewhat longer than umbel. Umbel 1.5–5 cm in diameter, nodding in bud, few- to many-flowered, effuse; flowering pedicels unequal, 1½–3 times as long as perianth. Perianth narrow, urceolate; segments 6–7(–8) mm, pinkish-white or pink, with a greenish or reddish midvein, oblong-lanceolate, acute, somewhat recurved at tip. Filaments included, ⅔–¾ length of perianth, dilated at base, subulate towards tip. Style included; stigma capitate. Capsule about 3.5 mm, depressed-globose, shorter than perianth. 2n=16. Fl. March–May.

Hab.: Desert; stony ground, calcareous rocks, grey desert soil; Artemisietum herbae-albae, Zygophylletum dumosi, Salsoletum vermiculatae. Judean Desert, N. and C. Negev; Lower Jordan Valley, Dead Sea area.

Area: Saharo-Sindian (Palestine, Sinai, Egypt).

12. Allium sindjarense Boiss. & Hausskn. ex Regel, Acta Horti Petrop. 3:121

(1875); Kollmann, Notes Roy. Bot. Gard. Edinb. 33:439, f.1B (1975); Boiss., Fl. 5:266; Post, Fl. 2:641. ?*A. azaurenum* Gomb., Bull. Soc. Bot. Fr. 109:204 (1962). [Plate 118]

Bulb 1–1.5 cm in diameter, oblong-ovoid; outer tunics brownish, densely reticulate-fibrous. Stem 10–25(–30) cm, stout, flexuose. Leaves 2–3, 1–3 mm broad, hollow, semiterete, caniculate on upper face, shorter to somewhat longer than stem; sheaths ribbed, very shortly soft-hairy. Spathe persistent, 2-valved; valves much shorter than umbel. Umbel up to 8 cm in diameter, lax, hemispherical to globose; pedicels purplish, 3–8 times as long as perianth, curved upwards under flower. Perianth campanulate; segments 5–6 mm, whitish, with a greenish-brown or purple midvein, ovate lanceolate, acute. Filaments included, purple; anthers purple, included or partly exserted. Ovary with 3 purplish stripes. Capsule. 3–4 mm, subglobose, enclosed in persistent perianth. 2n=16. Fl. April–May.

Hab.: Steppes and deserts, sandy and loess soils; Anabasetum articulatae. W., N. and C. Negev; E. Gilead, Edom.

Area; W. Irano-Turanian and Saharo-Arabian (Palestine, Syrian Desert, Iraq, S. Iran, Saudi Arabia).

A. sindjarense has been described as having an intensely purple perianth but plants from Palestine have whitish perianth-segments with a green, brown or purple midvein (Kollmann, *op. cit.* 439).

The data recorded by Feinbrun (1948) under *A. desertorum* Forssk. should be referred to *A. sindjarense*.

13. Allium stamineum Boiss., Diagn. ser. 2, 4:119 (1859); Fl. 5:257; Post, Fl. 2:639 excl. vars.; Wendelbo, Fl. Iran. 76:63, t.6 f.88, t.17 f.1. [Plates 119, 120] See p. 413.

Bulb 0.75–1.5 cm in diameter, ovoid; outer tunics blackish or ash-grey, torn into strips. Stem 10–25(–35) cm, terete, flexuose, mostly stout. Leaves 3–4, about 1 mm broad, hollow, filiform, reaching and sometimes overtopping the umbel, sheathing lower 1/3–1/2 of stem. Spathe persistent, 2-valved; valves unequal, longer than umbel; each valve tapering into a slender appendage. Umbel 3–5(–7) cm in diameter, lax, usually many-flowered; pedicels unequal, several times longer than perianth. Perianth campanulate; segments oblong-elliptic or -ovate, the outer rounded at apex, the inner obtuse. Filaments 1¼–2 times as long as perianth; anthers yellow. Ovary subglobose, sessile; style filiform, exserted. Capsule 3–4 mm, depressed-globose, 3-gonous. Fl. March–July.

Subsp. **stamineum**. [Plate 119]. Spathe-valves in not fully open umbels lanceolate at base and with prominent veins; each valve tapering into a long slender appendage 2–3 times as long as lanceolate base. Perianth-segments 3.5–4.5(–5) mm, purplish-pink with a darker midvein; rarely flowers greenish. Filaments 1¼–1½ times as long as perianth. 2n=16, 16+2 to 4B.

Hab.: Batha, mainly in the Mediterranean territories of the country. common. Coast of Carmel, Sharon Plain, Philistean Plain; Upper and Lower Galilee, Mt. Carmel, Esdraelon Plain, Samaria, Shefela, Judean Mts.; Dan Valley, Upper Jordan Valley; Golan, Gilead, Ammon.

Area of subspecies: as of species.

Subsp. **decaisnei** (C. Presl) Kollmann, comb. nov. *A. decaisnei* C. Presl, Bot. Bemerk. 114 (1844). *A. pallens*? Decne., Florula Sinaica, Ann. Sci. Nat. Bot. ser. 2, 2:18, no. 69 (1834) non L. (1762). [Plate 120]. Plants somewhat taller than subsp. *stamineum*. Spathe-valves in not fully open umbels yellowish, ovate at base with weak veins; each valve abruptly narrowed into a slender appendage 1–1½ times as long as ovate base. Perianth-segments 3–4 mm, yellowish-green to greenish-grey, with a green midvein. Filaments 1½–2 times as long as perianth. $2n=16,16+1B$.

Hab.: Rocks and rock-crevices, loess, hammada; various plant associations (Artemisietum herbae-albae, Noaeetum mucronatae and others). Judean Desert, N. and C. Negev; Lower Jordan Valley, Dead Sea area.

Area of subspecies: W. Irano-Turanian. Endemic in Palestine and Sinai.

Area of species: E. Mediterranean and W. Irano-Turanian (Palestine, Lebanon, Syria, Turkey, N. Iraq, N. Iran).

Decaisne in Florula Sinaica (*loc. cit.*, 1834) described a plant (no. 69) collected by Bové near Tor (Sinai Peninsula) as *A. pallens*? and noted that it differs from typical *A. pallens* in its rounded rather than truncate apex of perianth-segments. Later, C. Presl (*loc. cit.*, 1844) named Decaisne's plant and similar plants collected by Schimper (no. 258) on Mt. Katherina (Sinai) *A. decaisnei*, in honour of Decaisne. The binomial *A. decaisnei* can be considered as validly published. Schimper's specimens and recently collected plants from Mt. Katherina and from other localities in Sinai and S. Palestine have been studied by us. They are close to *A. stamineum* but differ from typical *stamineum* in several morphological and eco-geographical traits: (a) morphology of spathe-valves in young partly open umbels, (b) colour of perianth, (c) length of filaments relative to perianth, (d) habitat and distribution area. Accordingly, we decided that *A. decaisnei* should be considered a distinct subspecies within *A. stamineum*.

A. stamineum subsp. *stamineum* is one of the most common *Allium* species of Palestine. It is confined mainly to batha associations in the Mediterranean territories and is characterized by its purplish-pink flowers. Occasionally, however, plants with a greenish perianth may occur within populations of typical *A. stamineum*.

In steppe and desert habitats of the Negev and Sinai, plants of the Irano-Turanian subsp. *decaisnei* have exclusively yellowish-green or greenish-grey flowers.

14. Allium albotunicatum O. Schwarz, Feddes Repert. 36:73 (1934) subsp. **albotunicatum**; Kollmann & Shmida, Israel Jour. Bot. 26:139–141, f.6 (1977). [Plate 121]

Bulb 1.2–1.5 cm in diameter, ovoid or subglobose, often bearing bulblets; outer tunics greyish, split from base. Stem 40–60(–85) cm. Leaves 3–5, 1–2 mm broad, hollow, linear, semiterete, canaliculate, sheathing lower ½ of stem. Spathe 2-valved, persistent; valves unequal, linear, 2–3 times as long as umbel. Umbel 3–6(–8) cm in diameter, usually many-flowered, lax; pedicels very unequal, bracteolate, 3–15 times as long as perianth; after anthesis pedicels curve towards centre, causing umbel to appear dense and globose. Perianth obconical-campanulate, narrowed at apex; segments 4–5 mm, obtuse, connivent, greenish-brown or pale green with a green or greenish-purple midvein; often straw-coloured or reddish after flowering or when dried; outer segments oblong, broadest near apex, apiculate or notched. Filaments 1½–2 times as long as perianth; anthers pale yellow. Style exserted. Capsule 3 mm, depressed-globose. 2n=16, 24, 32. Fl. May–June.

Hab.: Batha, borders of cultivated fields, calcareous, basalt or volcanic-ash soils, among rocks. Samaria, Judean Mts.; Golan.

Area of subsp.: as of species.

Area of species: E. Mediterranean (Palestine, Mt. Hermon, S. Lebanon. W. Turkey).

A second subspecies, endemic to Mt. Hermon, was described as subsp. *hermoneum* Kollmann & Shmida (1977).

Sect. 3. ALLIUM. Sect. *Alliotypus* Dumort., Fl. Belg. 140 (1827). Sect. *Porrum* (Mill.) Reichenb. in Mössler, Handb. ed. 2, 1:541 (1827). Sect. *Crommyum* Webb & Berth. subsect. *Porrum* Boiss., Fl. 5:229 (1882). Bulb ovoid or subglobose. Leaves linear, flat and solid or terete to semiterete and hollow. Spathe 1- or 2-valved, usually beaked and caducous, rarely persistent. Perianth campanulate to ovoid; segments connivent. Filaments unequal; outer usually simple, rarely dentate or 3-cuspidate; the inner with a broader flat *basal lamina*, 3-cuspidate, rarely 5–7-cuspidate; the median cusp anther-bearing, the lateral cusps sterile, usually much elongated. Ovary with distinct nectariferous pores; ovules 2 in each locule.

Allium sativum L., Sp. Pl. 296 (1753); Reichenb., Icon. Fl. Germ. 10:22, t.488 (1848); Jones & Mann, Onions 36, 210, tt.5b, 12a; Stearn, Ann. Mus. Goulandris 4:171 (1978); Post, Fl. 2:635.*

Bulb depressed-ovoid, composed of 5–15(–60) bulblets of more or less equal size. Leaves flat, keeled. Spathe 1-valved with a long beak, caducous. Umbel with bulbils only, or rarely also with a few indehiscent flowers. Flowers greenish-white or pink, rarely white.

Commonly cultivated from very ancient times for its bulbs (*garlic*) used for flavouring. Probably derived from *A. longicuspis* Regel, a wild species from Central Asia (Vvedensky, *op. cit.* 243; Jones & Mann, *op. cit.* 36; Stearn, *loc. cit.*).

15. **Allium ampeloprasum** L., Sp. Pl. 294 (1753); Feinbr., Palest. Jour. Bot. Jerusalem ser., 3:6 (1943); Kollmann, Israel Jour. Bot. 20:13–20 & 263–272 (1971); Caryologia 25, 3:295–312 (1972); Bothmer, Opera Bot. (Lund) 34:21 (1974); Boiss., Fl. 5:232; Post, Fl. 2:634. *A. halleri* G. Don, Monogr. Allium 15 (1827). [Plate 122]

* Not illustrated.

Bulb 2-6 cm in diameter, broadly ovoid or subglobose; outer tunics ash-grey, papery, easily split; bulblets 3-5 mm in diameter, numerous, yellowish, borne on short stalks or nearly sessile and remaining crowded close to mother-bulb even after disintegration of bulb-tunics. Stem 50-180 cm, stout. Leaves 4-10, 5-20 mm broad, linear, flat, keeled, sheathing lower ⅓-½ of stem. Spathe up to 8 cm, 1-valved, long-beaked in bud, caducous. Umbel 5-9 cm in diameter, usually globose, many-flowered, dense; flowering pedicels 20-36 mm, purple, reddish, lilac or green, unequal, several times as long as peri-anth. Perianth cup-shaped or broadly campanulate; segments 4-5 mm, some-what remote from one another laterally (not overlapping), white, purple, lilac or green, with or without a green midvein, more or less scabrous at back; outer segments mostly oblong-lanceolate, concave, subacute; the inner broadly ovate, obtuse and apiculate. Filaments purple, white or lilac, some-what exserted, ciliolate near base; inner filaments 3-cuspidate; median cusp of inner filaments shorter than lateral cusps; lateral cusps thin and twisted; anthers purple or yellow. Style exserted. Capsule 4 mm. 2n=32, 40, 48. Fl. April-May.

Hab.:Cultivated and disturbed ground, fields, vineyards, batha, sandy loam and heavy alluvial soil. Common. Coastal Galilee, Acco Plain, Sharon Plain, Philistean Plain; Upper and Lower Galilee, Mt. Carmel, Esdraelon Plain, Samaria, Shefela, Judean Mts., W., N. and C. Negev; Dan Valley, Upper Jordan Valley; Golan, Gilead, Ammon.

Area: Mediterranean and Irano-Turanian, extending into W. Europe.

A. ampeloprasum varies in flower-colour. Perianth-segments may be purple or lilac with or without a green midvein, or white with a green vein. A white-flowered variant of *A. ampeloprasum* is mistakenly recorded as *A. leucanthum* in Post's Flora (ed. 2, 2:634). The true *A. leucanthum* C. Koch is found in the Caucasus and in N. Iran and is apparently endemic there (Kollmann, *op. cit.* 267, 1971).

Allium porrum L., Sp. Pl. 295 (1753). *A. ampeloprasum* L. var. *porrum* (L.) J. Gay, Ann. Sci. Nat. Bot. ser. 3, 8:218 (1847); Jones & Mann, Onions 40, t.7a (1963); Stearn, Ann. Mus. Goulandris 4:174 (1978); Post, Fl. 2:635.*

Bulbs poorly or not at all developed. Spathe long-beaked. Umbel globose, many-flowered, dense. Commonly cultivated as a vegetable (*leek*). Probably derived from *A. ampeloprasum*. A cultigen not known in wild state (Jones & Mann, *loc. cit.*; Stearn, *loc. cit.*).

Allium kurrat Schweinf. ex Krause, Notizbl. Bot. Gart. Berlin 9:524 (1926); Täck-holm & Drar, Fl. Eg. 3:85-86; Jones & Mann, Onions 41, t.7b.*

Similar to *A. porrum* but of smaller stature and with narrower leaves. Like *A. porrum* it may have originated from *A. ampeloprasum*. Cultivated in S. Palestine (Hebron) (Täckholm & Drar, *loc. cit.*, 1954) and on Mt. Katherina (Sinai). Known only in cultivation. Leaves consumed raw or used for flavouring.

Both species are autotetraploid.

* Not illustrated

16. Allium truncatum (Feinbr.) Kollmann & D. Zohary comb. & stat. nov. *A. ampeloprasum* L. subsp. *truncatum* (Feinbr.) Kollmann, Israel Jour. Bot. 20:271 (1971). *A. ampeloprasum* L. var. *truncatum* Feinbr., Palest. Jour. Bot. Jerusalem ser., 3:6 (1943). *A. ampeloprasum* L. var. *portorii* Gomb., Bull. Soc. Bot. Fr. 96:10–11 (1949). [Plate 123]

Bulb 2–4 cm in diameter, subglobose or ovoid; outer tunics scarious, ash-grey; inner tunics white, membranous; renewal bulb usually single or none; bulblets few (usually not more than 15), 4–10 mm in diameter, borne on long (up to 25 cm) horizontal or oblique stolons. Stem (80–)100–150 cm, stout. Leaves sheathing lower ⅓ of stem; blades shorter than scape, 8–15 mm broad, flat, keeled, slightly scabrous or glabrous at margin, withered at time of flowering. Spathe up to 8 cm, 1-valved, long-beaked in bud, caducous. Umbel 3–8 cm in diameter, subglobose, rather dense, many-flowered; flowering pedicels 12–20 mm, unequal. Perianth cylindrical, campanulate; segments dark purple or more rarely green, 2.5–4 mm, connivent; outer segments oblong-ovate, obtuse, scabrous; inner segments oblong to obovate, truncate, often irregularly denticulate at tip, somewhat longer than the outer ones, less scabrous. Filaments exserted; median cusp of inner filaments shorter than the lateral ones; the lateral thin, twisted and as long as or shorter than ciliolate basal lamina. Style exserted. Capsule globose. 2n=16, 24, 32. Fl. April–June.

Hab.: In Irano-Turanian territories in crop fields on loess (Negev), Lissan marls and basalt soils (Jordan Valley); in Mediterranean territories on alluvial and sandy soils, on rendzina derived from Eocene and Senonian calcareous rocks. Sharon Plain, Philistean Plain; Upper Galilee, Mt. Carmel, ?Samaria, Shefela, Judean Mts., W., N. and C. Negev; Upper Jordan Valley, Beit Shean Valley, Lower Jordan Valley, Dead Sea area; Golan, Gilead, Ammon.

Area: W. Irano-Turanian (Palestine, Syria, N. Egypt).

In 1971 Kollmann emphasized morphological differences of flower parts and of organs of vegetative reproduction between *A. ampeloprasum* L. subsp. *ampeloprasum* and subsp. *truncatum* (Feinbr.) Kollmann, the latter subspecies having smaller flowers, a truncate apex of inner perianth-segments, and bulblets remote from the mother-bulb. She found that the differences between the two taxa are effectively maintained in sites where they grow side-by-side. Further studies (Kollmann, 1972) provided evidence that the two taxa differ also in meiotic behaviour and in ploidy levels, and that their ecological requirements and general geographical areas also differ. It was therefore decided to raise subsp. *truncatum* to the rank of species.

In Palestine *A. truncatum* occurs mainly in the Irano-Turanian territories but extends, to a small degree, also into Mediterranean territories, where it grows in crop fields and is often found together with *A. ampeloprasum*. The latter does not penetrate into more arid territories of the country. The general distribution area of *A. truncatum* is more W. Irano-Turanian and more restricted than that of *A. ampeloprasum*.

On the other hand, *A. truncatum* shows similarity to *A. bourgeaui* Rech. fil., an endemic species of cliffs on the Aegean islands. As pointed out by Bothmer (Opera Bot. 34:97, 1974), both taxa have short, more or less uniformly truncate inner perianth-segments, the same scabrosity of perianth-segments and exserted stamens.

The two species differ mainly in the mode of vegetative reproduction; *A. truncatum* has "mobile" bulblets, borne at the ends of long horizontal or oblique stolons, remote from the mother-bulb, whereas *A. bourgeaui* is characterized by the "stationary" type of bulblest, crowded near the mother-bulb (Galil, 1965). Furthermore, the two differ in several other morphological characters. *A. truncatum* has a taller stem (80–150 cm) which is thinner beneath the umbel, its lower leaves are narrower at base, its leaf-margin (but not keel) is slightly scabrous, and its ligule and the beak of the spathe are relatively shorter. Also the habitats of the two species differ considerably in character.

17. Allium scorodoprasum L., Sp. Pl. 297 (1753) subsp. **rotundum** (L.) Stearn, Ann. Mus. Goulandris 4:178 (1978) & Fl. Eur. 5:65. *A. rotundum* L., Sp. Pl. ed. 2, 423 (1762); Reichenb., Icon. Fl. Germ. 10:24, t.492 (1848); Regel, Acta Horti Petrop. 3:57 (1875) p.p.; Bothmer, Bot. Not. 125:63, ff.1A, 2A, 3F–G (1972); Kollmann & Shmida, Israel Jour. Bot. 26:133–135 (1977); Boiss., Fl. 5:233; Post, Fl. 2:635. *A. porphyroprasum* Heldr., Bull. Herb. Boiss. 6:394 (1898). [Plate 124]

Bulb 1–2 cm in diameter, depressed-globose to ovoid, often with stipitate bulblets; outer tunics brown, breaking up into fibrous strips; bulblets numerous, black, brown or violet. Stem 25–90 cm. Leaves 2–6, 2–8(–15) mm broad, linear, flat, often canaliculate, scabrous on margins, sheathing lower ⅓–½ of stem. Spathe 1-valved, about 1.5 cm, shortly beaked, caducous. Umbel 1–4(–6) cm in diameter, subglobose, nodding in bud, with numerous conspicuous bracteoles at base of pedicels; flowering pedicels unequal; the outer very short, the inner up to 8 times as long as perianth. Perianth ovoid; segments connivent; the outer 4.5 mm, dark purple, lanceolate or narrowly ovate, scabrous at báck or along keel; the inner 5–5.5 mm, paler, whitish, sometimes with a purple mid-stripe, oblong, obtuse, sometimes short-apiculate. Filaments included, ciliolate at base; median cusp about ¼–⅓ length of basal lamina; lateral cusps 2–3 times as long as median cusp. Capsule about 4–5 mm. 2n=32. Fl. May–July.

Hab.: Basalt and heavy soil, hard limestone and marls. Upper Galilee, Judean Mts. (sporadic); Golan, Mt. Hermon.

Area of subsp.: Euro-Siberian, Mediterranean and W. Irano-Turanian (Palestine, Lébanon, Syria, Turkey, N. Iraq, W. Iran, Talysh, Caucasus, European Russia, C. and S. Europe).

18. Allium phanerantherum Boiss. & Hausskn. in Boiss., Fl. 5:235 (1882); Feinbr., Palest. Jour. Bot. Jerusalem ser., 3:11, f.20 (1943); Mout., Nouv. Fl. Liban Syrie 1:267; Wendelbo, Fl. Iran. 76:51, t.5 f.69; Kollmann, Notes Roy. Bot. Gard. Edinb. 33:305 (1974). *A. davisianum* Feinbr., Palest. Jour. Bot. Jerusalem ser., 3:13, ff.4 & 23 (1943) emend. Kollmann, Notes Roy. Bot. Gard. Edinb. 31:121 (1971). *A. descendens* auct. non L. (1753) nec Sm. in Sibth. & Sm. (1823); Feinbr., *op. cit.* 12, f.22 (1943). [Plate 125]

Bulb 1–2 cm in diameter, subglobose; outer tunics membranous, greyish.

Stem (60–)80–100 cm. Leaves 2–5, 2–5 mm broad, hollow, terete, sheathing lower ½ of stem. Spathe 1–1.5 cm, persistent, reflexed and split into several lobes. Umbel 2–4 cm in diameter, globose to ellipsoid; flowering pedicels 2–3 times as long as perianth. Perianth subcoriaceous, umbilicate at base, tubular to oblong-ovoid, pale green or purple at tip or all over; segments 4–5 mm, the outer oblong-elliptic, obtuse, cymbiform; the inner ovate, subacute. Filaments somewhat exserted; inner filaments 3-cuspidate; median cusp of inner filaments shorter and somewhat thicker than the longer and finer lateral cusps; basal lamina ciliolate, about as long as cusps; anthers 2.2 mm, oblong, yellowish or purplish-brown, exserted. Ovary oblong; style long-exserted. Capsule 4 mm, overtopped by persistent perianth. 2n=16. Fl. June–July.

Hab.: Rocks and hard limestone cliffs, maquis of *Quercus calliprinos* and *Q. ithaburensis*, batha of hemicryptophytes. Upper and Lower Galilee, Mt. Carmel, Mt. Tabor, Samaria, Judean Mts.; Dan Valley, Hula Plain, Upper and Lower Jordan Valley; Golan, Gilead.

Area: Mediterranean and W. Irano-Turanian (Palestine, Lebanon, Syria, Turkey, Iraq, Iran).

19. Allium curtum Boiss. & Gaill. in Boiss., Diagn. ser. 2, 4:116 (1859). [Plate 126]

Bulb 1.5–2.5 cm in diameter, globose or depressed-globose, bearing bulblets; outer tunics membranous. Stem (10–)15–35(–60) cm, stout. Leaves 3, 2 mm broad, hollow, semiterete, canaliculate. Spathe persistent, 2(–3)-valved, ovate, very short, reflexed. Umbel 2–4.5 cm in diameter, globose to ovoid; outer pedicels very short, the inner 2–3 times as long as perianth. Perianth ovoid-campanulate or hemispherical; segments 3–5 mm, ovate, obtuse. Filaments exserted; median cusp of inner filaments thicker and longer to shorter than lateral cusps; basal lamina longer than cusps. Capsule about 2 mm, depressed-globose. 2n=16. Fl. (March–) April–May.

Two subspecies and two varieties in one of the subspecies are distinguished; they differ mainly in colour of perianth and of anthers, as well as in eco-geographical character.

1. Perianth ovoid, 3–3.5 mm; segments purple and white- or light purple-margined; anthers purple; antheriferous cusp of inner filaments longer than lateral cusps. Plants of sandy loam or calcareous sandstone of coastal plain.   subsp. **curtum**
 – Perianth ovoid-oblong, 4–5 mm; segments pale green, white-margined; antheriferous cusp not longer than lateral cusps   2
2. Perianth-segments with a green midvein; anthers yellowish; antheriferous cusp shorter than lateral ones; style green. Plants of calcareous hills in Mediterranean territories.   subsp. **palaestinum** var. **palaestinum**
 – Perianth-segments with a broad purple, greenish-purple or green midvein; anthers and upper part of filaments purple; antheriferous cusp about as long as lateral ones; style purple. Plants of sandy soils of Negev and of basalt soils in Upper Jordan Valley and Lower Galilee.   subsp. **palaestinum** var. **negevense**

Subsp. **curtum**; Feinbr., Palest. Jour. Bot. Jerusalem ser., 3:14–16, f.24 (1943); Mout., Nouv. Fl. Liban Syrie 1:269; Boiss., Fl. 5:245; Post, Fl. 2:638. *A. sphaerocephalum* L. subsp. *curtum* (Boiss. & Gaill.) Duyfjes, Revis. Allium Africa 52 (1976). [Plate 126]. 2n=16.

Hab.: Calcareous sandstone, sandy loam; *Desmostachya bipinnata–Centaurea procurrens* association. Coastal Galilee, Acco Plain, Coast of Carmel, Sharon Plain, Philistean Plain, W. Negev.

Subsp. **palaestinum** Feinbr., Palest. Jour. Bot. Jerusalem ser., 3:14, ff.6 & 25 (1943); Mout., *loc. cit.*

Var. **palaestinum.** 2n=16.

Hab.: Among rocks, terra rossa; batha in the hills. Upper Galilee, Mt. Tabor, Mt. Gilboa, Esdraelon Plain, E. Samaria, Judean Mts.; Golan, Edom.

Var. **negevense** Kollmann var. nov.*

Hab.: Sandy loess soil, basalt soil. Lower Galilee; Upper Jordan Valley; C. Negev.

Area of species: E. Mediterranean and W. Irano-Turanian (N. Egypt, Palestine, Lebanon, S. Syria, S. Turkey, Cyprus).

Specimens of subsp. *palaestinum* have been recorded as *A. sphaerocephalum* L. var. *viridi-album* (Tin.) Boiss. from Jerusalem and Tiberias in Post's Flora (2:636), from Motza (Judean Mts.) in Rech. fil., Ark. Bot. ser. 2, 2(5):314 (1952) and from Mt. Canaan (Upper Galilee) in Oppenheimer & Evenari, Florula Cisiord., Bull. Soc. Bot. Genève ser. 2, 31:184 (1941).

20. Allium hierochuntinum Boiss., Fl. 5:244 (1882); Feinbr., *op. cit.* 20, f.33; Post, Fl. 2:637. *A. ascalonicum* L., Amoen. Acad. 4:454 (1759) *nom. confus.*; Stearn, Bull. Brit. Mus. (Bot.) 2:181 (1960). [Plate 127]

Bulb 0.8–1.5 cm in diameter, ovoid-oblong; tunics pale brown, reticulate-fibrous. Stem (12–)15–35 cm. Leaves filiform, hollow, shorter than or as long as stem, sheathing the stem up to ½ or less. Spathe persistent, purplish, 2-valved; valves ovate-triangular, subacute, much shorter than umbel. Umbel (1.5–)2–2.5(–3) cm, subglobose; outer pedicels very short, the inner as long as or somewhat longer than perianth. Perianth cylindrical-campanulate; segments 6–7 mm, blue or blue-violet, lanceolate, subacute. Stamens included; filaments ⅔ length of perianth; inner filaments 3-cuspidate, the median cusp about ½ length of lateral cusps. Capsule 2.5–3.5 mm, globose. 2n=16. Fl. March–May.

Hab.: Steppes and deserts; gravelly Senonian hills. Judean Desert; Lower Jordan Valley, Dead Sea area; E. Gilead, E. Ammon, Moav, Edom.

Area: Irano-Turanian and Saharo-Sindian (Palestine, Syria).

The reports of *A. scabriflorum* by Paine (1875) and by Tristram (1884) from Moav apparently refer to *A. hierochuntinum. A. scabriflorum* Boiss. is an endemic of Anatolia.

* See Appendix, p. 397.

21. Allium artemisietorum Eig & Feinbr. in Feinbr., Palest. Jour. Bot. Jerusalem ser., 3:18–19, f.30 (1943). [Plate 128]

Bulb (1–)1.5–2.5(–3) cm in diameter, oblong-ovate; outer tunics brown, reticulate-fibrous, somewhat extended around base of stem; bulblets few. Stem 15–30(–40) cm, flexuose. Leaves 2–3, 1–2 mm broad, hollow, semiterete, striate, sheathing lower ¼ of stem, often withering before anthesis. Spathe persistent, many-veined, ovoid in bud, later lacerate, often longer than lower pedicels. Umbel (1–)1.5–3(–4) cm in diameter, dense, many-flowered, fastigiate; flowering pedicels often shorter than, as long as or somewhat longer than perianth, bracteolate; inner pedicels up to 2 cm; bracteoles 2–4 mm, lanceolate or linear. Perianth campanulate; segments 3.5–5 mm, white with a reddish or green midvein, ovate-oblong, subacute, scabridulous along midvein or all over. Filaments glabrous; inner filaments 3-cuspidate; median cusp slightly shorter, equal or somewhat longer than lateral cusps, shorter than basal lamina, subexserted; anthers purplish, subexserted. Style exserted. Capsule 3 mm, ovate, shorter than perianth. 2n=16. Fl. March–May.

Hab.: Steppes, rocky or chalky soil; Artemisietum herbae-albae. Judean Desert, N. and C. Negev; Lower Jordan Valley, Dead Sea area; Arava Valley; E. Ammon, Edom.

Area: W. Irano-Turanian (Palestine, Egypt).

22. Allium sinaiticum Boiss., Diagn. ser. 1, 13:31 (1854); Täckholm & Drar, Fl. Eg. 3:67; Boiss., Fl. 5:244; Post, Fl. 2:637. [Plate 129]

Bulb 2–2.5 cm in diameter, ovoid; outer tunics brown, reticulate-fibrous. Stem 6–10(–15) cm, stout. Leaves 2.2–2.5 mm broad, hollow, terete, sheathing lower ⅓ of stem, attenuate towards apex, much longer than stem. Spathe persistent, ovoid in bud, many-veined, shortly and abruptly mucronate, later nearly as long as pedicels and irregularly lacerate. Umbel 2–4 cm in diameter, hemispherical to globose, few- to many-flowered; flowering pedicels subequal, 2–2½ times as long as perianth or shorter. Perianth campanulate; segments 6–7 mm, white, with a pale purple or green midvein, subobtuse and shortly mucronate, scabrous all over. Filaments subexserted, broadened and ciliate in lower part; inner filaments 3-cuspidate; cusps about ⅓ length of basal lamina; median cusp scarcely shorter than lateral cusps. Capsule 4 mm, globose-3-quetrous. Fl. March–April.

Hab.: Hammada and sands; nearly buried in sand. Arava Valley; Edom.

Area: E. Saharo-Arabian (S. Palestine, Egypt).

23. Allium dictyoprasum *("dyctioprasum")* C.A. Mey. ex Kunth, Enum. Pl. 4:390 (1843); Vvedensky, in Fl. URSS 4:236, t.14 ff.1 & 1a; Boiss., Fl. 5:243; Post, Fl. 2:637. *A. viride* Grossh., Fl. Kavk. 1:201 (1928); Wendelbo, Fl. Iran. 76:51, t.5 f.68. ?*A. emarginatum* Rech. fil., Ark. Bot. ser. 2, 1(14):505 (1951). [Plate 130]

Bulb 2–2.5 cm in diameter, ovate-oblong; outer tunics brown, reticulately fibrous, prolonged into a long neck, 2–3 times as long as bulb itself. Stem

(60-)70-150 cm. Leaves 3-4, 3-11 mm broad, hollow, terete, sheathing lower ⅓ of stem. Spathe 2-valved. Umbel 2.5-4.5(-5) cm in diameter, globose, many-flowered, dense; flowering pedicels unequal, 2-5 times as long as perianth. Perianth subglobose; segments 2.5-3.5 mm, dull green or yellowish-green, white-margined; outer segments somewhat concave, broadly elliptic, obtuse and rounded at apex; the inner somewhat longer, elliptic-ovate, emarginate. Filaments longer than perianth-segments, glabrous or rarely ciliate at base; the outer sometimes with 1-2 small teeth on either side; the inner 3(-5-7)-cuspidate; anthers yellow or purple. Capsule about 3.5 mm. 2n=16. Fl. June-July.

Hab.: Rocky slopes. C. Negev; Upper and Lower Jordan Valley; Golan, Moav (Paine).

Area: W. Irano-Turanian (Palestine, ?Syria, Iraq, Iran, S. and E. Turkey, Caucasus).

Plants from Upper Jordan Valley and Golan have a yellowish-green perianth and yellow anthers, those from C. Negev have a green perianth and purple anthers.

Examination of *A. emarginatum* Rech. fil. (1951) from Zahle showed that the specimens do not differ from Palestinian *A. dictyoprasum* in shape of perianth and perianth-segments and probably represent rather underdeveloped plants of *A. dictyoprasum* with small umbels.

Sect. 4. MELANOCROMMYUM Webb & Berth., Phyt. Canar. 3:347 (1848); Boiss., Fl. 5:229; Wendelbo, Fl. Iran. 76:67 (1971). Bulb subglobose. Leaves basal, flat, with no above-ground sheaths. Stem mostly longer than leaves. Spathe persistent, 1-valved, becoming 2-4-fid. Perianth stellate to broadly campanulate; segments reflexed and more or less curled after flowering. Filaments simple. Ovary with 3-8(-10) ovules in each locule.

24. Allium nigrum L., Sp. Pl. ed. 2, 430 (1762); Redouté, Liliacées 2:t.102 (1804); Sm. in Sibth. & Sm., Fl. Graeca 4:20, t.323 (1823); Ker-Gawler, Bot. Mag. 29:t.1148 (1808) as *A. magicum* L.; Boiss., Fl. 5:279; Post, Fl. 2:645. *A. dumetorum* Feinbr. & Szelubsky in Feinbr., Palest. Jour. Bot. Jerusalem ser., 4:146 (1948); Szelubsky, Palest. Jour. Bot. Jerusalem ser., 5:1-12 (1949). *A. nigrum* L. var. *dumetorum* (Feinbr. & Szelubsky) Mout., Nouv. Fl. Liban Syrie 1:281-282 (1966). *A. asclepiadeum* auct. non Bornm. (1917). [Plate 131]

Bulb 2-3(-5.5) cm in diameter, ovoid; outer tunics membranous. Stem 60-100 cm, rarely 25-30 cm. Leaves 3-6, 2.5-8 cm broad, flat, broadly lanceolate, acute. Spathe persistent, 2-4-lobed; lobes up to 3 cm, narrowly triangular, acuminate. Umbel 5-10 cm in diameter, hemispherical or fastigiate, many-flowered; pedicels 2.5-4.5 cm. Perianth stellate at anthesis, later flaccid or reflexed; segments 6-9 mm, white or pale lilac to mauvish-pink, with a greenish midvein, oblong-ovate, obtuse. Filaments about ½-¾ length of perianth, white or pale lilac, sometimes purple at least in their lower part. Ovary dark green or dark purple. Capsule 6-8 mm, globose. 2n=16. Fl. March-April.

Hab.: Maquis of *Quercus calliprinos* and of *Q. ithaburensis,* fallow fields, hedges. Upper and Lower Galilee, Mt. Tabor, Mt. Carmel; Golan (common on basalt soil).

Area: Mediterranean.

In Palestine *A. nigrum* varies considerably in several characters. The height of plants varies from 25 cm to 100 cm; colour of perianth is mostly white, but in some cases pale lilac or mauvish-pink with a green midvein (in plants of maquis on Mt. Carmel, described as *A. dumetorum* Feinbr. & Szelub. in 1948); filaments are white or lilac, or in some cases purple in the lower ½; colour of ovary is dark green or dark purple. According to Smith in Sibthorp & Smith (*op. cit.*), the name *nigrum* refers to the almost black ovary found in some plants of the species.

In some parts of its distribution area *A. nigrum* is recorded as growing on cultivated ground and in waste places. In Palestine this species is found in natural plant associations, such as *Quercus calliprinos* maquis, or some garigue associations.

Plants from Mt. Meiron in Upper Galilee, conspicuous by their purple filaments and ovary and only 30 cm in height, have been recorded by Feinbrun (1948) as *A. asclepiadeum* Bornm. described from S. Turkey (formerly N. Syria). However, specimens with similar flowers but a height of 60–80 cm have been since collected in Lower Galilee. They can all be included in *A. nigrum* L.

25. Allium aschersonianum W. Barbey in C. & W. Barbey, Herb. Levant 163, t.4 (1882); Feinbr., Palest. Jour. Bot. Jerusalem ser., 4:146–148, t.34 (1948); Szelubsky, Palest. Jour. Bot. Jerusalem ser., 5:1–12 (1949); Boiss., Fl. 5:283; Post, Fl. 2:646. [Plate 132]

Bulb 2–3 cm in diameter, ovoid or subglobose; outer tunics greyish, inner tunics white. Stem 30–80 cm, stout. Leaves (2–)3–5(–6), 2–4(–5) cm broad, flat, more or less erect, lorate-lanceolate, tapering above, slightly undulate, scabridulous at margin, shorter than scape. Spathe persistent, 1-valved, 2–3-lobed. Umbel 4–8 cm, hemispherical, many-flowered, dense; flowering pedicels 3–4 times as long as perianth. Perianth deep purple, stellate, finally reflexed; segments 6–7 mm, linear, obtuse. Stamens erect, somewhat longer than perianth; filaments and anthers deep purple; pollen greenish-grey. Ovary green or purplish; style as long as stamens or distinctly shorter. Capsule 7–8 mm; valves of capsule somewhat longer than broad. 2n=16. Fl. February–March.

Hab.: Sandy and rocky ground; Artemisietum monospermae, Noaeetum mucronatae. E. Samaria, Judean Desert, S. Shefela, W., N. and C. Negev; Lower Jordan Valley, Dead Sea area; Moav, Edom.

Area: W. Irano-Turanian (Palestine, Syria, S. Turkey, Egypt, Libya).

26. Allium tel-avivense Eig in Eig, Zohary & Feinbr., The Plants of Palestine, Analytical Key, Jerusalem, 75 (1931; in Hebrew); Feinbr., Palest. Jour. Bot. Jerusalem ser., 4:148, t.3A ff.1, 3 (1948); Szelubsky, *ibid.* 5:1–12 (1949). *A. aschersonianum* W. Barbey subsp. *tel-avivense* (Eig) Opphr. in Opphr. & Evenari, Florula Cisiord., Bull. Soc. Bot. Genève ser. 2, 31:189, f.7 (1941). [Plate 133]

Bulb 2–3.5 cm in diameter, ovoid; outer tunics brownish-black; the inner thin, often purplish. Stem 15–25(–40) cm, stout. Leaves 5–9 (usually 6–7), 1–2 cm broad, flat, lorate-linear, mostly longer than stem, often undulate and sca-bridulous at margin. Spathe persistent, 2–2.5 cm, 2–3-lobed, with purplish veins, in bud conical and acuminate. Umbel 5–8 cm, hemispherical to sub-globose, many-flowered; pedicels 1.5–3 cm. Perianth stellate; segments 7–9 mm, pale purple, linear, subobtuse. Filaments pale purple, about as long as perianth; anthers purple. Ovary green, depressed-globose; style shorter than stamens. Capsule about as long as broad, slightly emarginate at tip. 2n=16. Fl. March.

Hab.: Sandy loam and kurkar (calcareous sandstone) hills. Acco Plain, Coast of Carmel, Sharon Plain, Philistean Plain.

Area: E. Mediterranean. Endemic.

The statement by Feinbrun (1948, p.149) that *A. tel-avivense* is found also on the Phoenician coast (S. of Beirut), has not been confirmed by later studies; we have not seen specimens of this species from the Lebanese coast and Mouterde (1966) did not record it from there.

The name *A. tel-avivense* was published in 1931 in Hebrew. The name is valid according to the Intern. Code of Bot. Nomenclature. The first Latin description, accompanied by a good drawing of the plant, was given by Oppenheimer (in Opphr. & Evenari, 1941) under the name *A. aschersonianum* subsp. *tel-avivense* (Eig) Opphr. *A. tel-avivense* was treated at species rank by Feinbrun (1948) after a careful study was undertaken together with R. Szelubsky who examined extensive material for her com-parative caryological and morphological study of *A. tel-avivense*, *A. aschersonianum* and *A. dumetorum*.

27. Allium orientale Boiss., Diagn. ser. 1, 13:25 (1854); Feinbr., Palest. Jour. Bot. Jerusalem ser., 4:149; Boiss., Fl. 5:282; Post, Fl. 2:645. [Plate 134]

Bulb 1.5–3 cm in diameter, ovoid to subglobose; outer tunics greyish, the inner membranous, white. Stem 15–40 cm. Leaves 2–6, 1–2 cm broad (some-times in Judean Mts. up to 4 cm), flat, linear-lorate, long-tapering above, often undulate. Spathe persistent, 1-valved, 2–4-lobed, finally reflexed; lobes ovate. Umbel 2.5–5 cm in diameter, hemispherical, many-flowered; flowering pedicels 2–3 times as long as perianth. Perianth-segments 6–7(–9) mm, yellowish-white with a pale green midvein, oblong-elliptic, obtuse, more or less erect, finally flaccid or slightly reflexed. Filaments white, ¾ length of perianth-segments; anthers pale yellow. Ovary green; style about as long as filaments. Capsule 5–6 mm, globose, slightly depressed. 2n=16. Fl. February–April.

Hab.: Calcareous hills, gravelly soil, basalt soil; batha and fallow fields. Sharon Plain, Philistean Plain; Upper and Lower Galilee, Esdraelon Plain, Mt. Gilboa, Samaria, Shefela, Judean Mts., Judean Desert, N. Negev; Upper Jordan Valley, Dead Sea area; Golan, Gilead, Ammon.

Area: E. Mediterranean (Palestine, Lebanon, Syria, Cyprus, Turkey, Libya).

Specimens cited by K.H. Rechinger in Ark. Bot. ser. 2, 2(5):313 (1952) as *A. libani* from the Judean Desert and Upper Jordan Valley were examined and found to be.

A. orientale. A. libani recorded by Tristram (1884) is probably also identical with *A. orientale. A. libani* Boiss. (1854) is an endemic of Mt. Hermon and Lebanon, confined to high altitudes (1,700–2,800 m) (Kollmann & Shmida, pp. 146–147, f.9, 1977).

28. Allium rothii Zucc., Abh. Akad. Wiss. (München) 3:235, t.4 (1843); Boiss., Fl. 5:283; Post, Fl. 2:646. [Plate 135]

Bulb 3–4 cm in diameter, ovoid; outer tunics bark-like, lacerate. Stem 6–15 cm, stout. Leaves numerous, 0.6–3 cm broad, flat, spread out on ground, as long as or longer than stem, lorate-lanceolate, wavy and twisted; margins smooth. Spathe persistent, 1-valved, 2–4-lobed, shorter than umbel. Umbel (2.5–)3–5(–6) cm in diameter, hemispherical, many-flowered, dense; pedicels 2–2½ times as long as perianth. Perianth stellate; segments 4–4.5 mm, white with a purple or greenish midvein, oblong, obtuse. Filaments dark purple, white at tip, much dilated at base, gradually attenuate, acute, as long as perianth; anthers dark purple; ovary dark purple or blackish. Style short. Capsule about 5 mm, globose. 2n=16. Fl. March–April.

Hab.: Rocky slopes; Zygophylletum dumosi. Judean Desert, N. and S. Negev; Dead Sea area; Edom.

Area: W. Irano-Turanian (Palestine, Syrian Desert).

It should be noted that the perianth-segments are white with a purple midvein, and not purple at the base as given by Boissier.

Sect. 5. KALOPRASUM C. Koch, Linnaea 22:235 (1849); Wendelbo, Fl. Iran. 76:89 (1971); Tscholokaschvili, Not. Syst. Inst. Bot. Tbilisi 31:52 (1975). Bulb subglobose. Stem stout at maturity, tapering towards base. Leaves basal, with no above-ground sheaths. Umbel more or less lax. Pedicels often markedly unequal. Spathe persistent. Perianth stellate to broadly campanulate, after flowering rigid, straight. Filaments simple. Ovary with 6–10 ovules in each locule.

29. Allium schubertii Zucc., Abh. Akad. Wiss. (München) 3:234, t.3 (1843); Hook. fil., Bot. Mag. 124, tt.7587–8 (1898); Wilde-Duyfjes, Revis. Allium Africa 189, f.36; Boiss., Fl. 5:278; Post, Fl. 2:644. [Plate 136]

Bulb 3–5 cm in diameter; outer tunics coriaceous, brown. Stem 30–60 cm, stout, hollow. Leaves 4–8, up to 5–7 cm broad, flat, lorate-lanceolate, more or less undulate and scabridulous at margin. Spathe persistent, 1-valved, 2–3-lobed. Umbel 10–40 cm in diameter, hemispherical to subglobose, very many-(50–200)-flowered, polygamous; pedicels purplish, rigid, thickened at apex, very unequal, 3–20 cm; flowers on longest pedicels generally staminate or neuter, mostly with a reduced ovary, rarely in some specimens hermaphrodite. Perianth stellate; segments 7–9 mm, mauve-purple, lanceolate, subacute. Filaments mauve, about ½ length of segments; anthers pale yellow. Ovary green; style short. Capsule 6–8 mm. 2n=16. Fl. March–May.

When in fruit, stem breaks up at base and fruiting inflorescence is driven by wind as tumbleweed.

Hab.: Fields, chiefly on heavy alluvial soil of plains and broad valleys. Sharon Plain, Philistean Plain; Upper and Lower Galilee, Mt. Carmel, Esdraelon Plain, Samaria, Shefela, Judean Mts.; Upper Jordan Valley; Golan, Ammon.

Area: E. Mediterranean (Palestine, Lebanon, Syria, S. Turkey, Libya).

Dubious Species

1. Allium sprengeri Regel (1889) from Jaffa remains doubtful. It is described as having bulbs attached to a rhizome, flat leaves and yellowish flowers. As stated by Feinbrun(1948), in Palestine it "has never been found again, neither was another species of Sect. *Rhizirideum*".

2. Allium lachnophyllum Paine (1875) from Ziza (Moav). We could not locate Paine's specimen and do not know to what species it should be referred. None of the species known to us fits Paine's description.

3. Allium hierosolymorum Regel (1890) recorded from Jerusalem is probably *A. trifoliatum* subsp. *hirsutum* (see p. 79).

4. Allium ascalonicum L. was based on a Palestinian plant later named *A. hierochuntinum* Boiss. but the name *A. ascalonicum* has long been applied to variants of *A. cepa* cultivated as *shallots* [see pp. 78, 93 and Stearn, Bull. Brit. Mus. (Bot.) 2:181, 1960 and Flora europaea 5:56, 1980].

20. Asparagus L.

Perennial herbs or half-shrubs, dioecious or polygamous, with subterranean rhizome. Rhizome producing overground much-branched stems, climbing in some species. Twigs and branchlets bearing small scale-like proper leaves and green phylloclades. Phylloclades needle-shaped or leaf-like and mostly linear, solitary or more often in clusters, in axils of scale-like proper leaves. Scale-like leaves often produced at base into a spur, and sometimes spinescent at tip. Flowers unisexual, rarely hermaphrodite, pedicellate, 1–2 or several in racemes or clusters in axils of scale-like leaves; pedicels jointed. Perianth petaloid, marcescent, campanulate or stellate-spreading; segments 6, subequal, free or connate at base. In staminate flowers stamens 6, free, equal, included, inserted at the base of perianth-segments; anthers dorsifixed; in pistillate flowers stamens reduced to staminodes. Ovary 3-locular, with 2–8 ovules in each locule; style filiform; stigma 3-lobed, recurved. Fruit a juicy berry, 3- or by abortion 1-locular, globose, 1- or few-seeded. Seeds more or less globose, mostly black.

About 300 species in the Old World.

Literature: J.G. Baker, Revision of the genera and species of Asparagaceae, *Jour. Linn. Soc. London (Bot.)* 14:594–629 (1875). P. Misczenko, Critical species of the genus *Asparagus* of the Crym-Caucasus flora and a key to their identification, *Monit. Jard. Bot. Tiflis* 12:15–52 (1916). A. Bozzini, Revisione cito-sistematica del genere *Asparagus* L.I., *Caryologia* 12:199–264 (1959).

1. Phylloclades solitary (rarely some of them 2–3 in cluster), strongly spinescent, often 1.5–2 mm thick, 1–3 cm long, sometimes reaching 4–7 cm in length.
   4. A. stipularis
 - Nearly all phylloclades in clusters of 2–12, slenderer, not exceeding 2.5 cm in length   2
2. Phylloclades filiform, soft. Plants of moist habitats. Scale-like leaves not spurred.
   1. A. palaestinus
 - Phylloclades rigid. Plants of batha and maquis   3
3. Phylloclades 0.5–1(–1.8) cm, usually thick, angular, distinctly unequal, mostly 3–6 in a cluster (rarely part of phylloclades single or twin), abruptly ending in a brown spine.
   3. A. aphyllus
 - Phylloclades 0.3–0.8(–1) cm, slender, acicular, equal or subequal, 4–12 in a cluster, tapering to a very short yellow spinule.
   2. A. acutifolius

1. **Asparagus palaestinus** Baker, Jour. Linn. Soc. London (Bot.) 14:602 (1875); Boiss., Fl. 5:335; Post, Fl. 2:661. *A. lownei* Baker, *op. cit.* 601; Boiss., *loc. cit.*; Post, *loc. cit.* [Plate 137]

Perennial, 60–100 cm, smooth, much branched; lower branches and branchlets spreading to reflexed. Scale-like leaves devoid of a spur. Phylloclades in clusters of 3–8, 0.6–2.5 cm, filiform, soft. Flowers 1–2. Pedicels jointed at about middle, 2–2½ times as long as flower. Perianth campanulate; segments connate near base, more or less equal. Berry 8 mm in diameter, dark brown. Fl. mainly February–May.

Hab.: Thickets on river banks and marshes. Acco Plain, Coast of Carmel, Sharon Plain, Philistean Plain; Upper and Lower Galilee, Mt. Carmel, Judean Mts.; Hula Plain, Upper and Lower Jordan Valley, Dead Sea area; Transjordan (Mouterde).

Area: E. Mediterranean (Palestine, ?Lebanon, ?Syria, S. and S.E. Turkey).

Related to the Irano-Turanian *A. trichophyllus* Bunge.

2. **Asparagus acutifolius** L., Sp. Pl. 314 (1753); Boiss., Fl. 5:337; Post, Fl. 2:661; incl. var. *brachyclados* Bornm., Verh. Zool.-Bot. Ges. Wien 48:641 (1898). [Plate 138]

Climbing perennial, 50–100 cm, much branched, with woody base; twigs and ultimate branches striate, velvety. Lower scale-like leaves spurred at base, becoming prickly with time. Phylloclades 4–12 in a cluster, (0.2–)0.3–0.8(–1) cm, equal or subequal, needle-like, rigid, tapering to a yellow spinule. Flowers 1–2(–4). Pedicels jointed towards middle, slightly longer than flower. Perianth campanulate, with segments subequal, connate up to ¼–½. Berry about 5 mm in diameter, green becoming black. Fl. August–September.

Hab.: Maquis, rocks. Hula Plain; Golan.

Area: Mediterranean.

3. Asparagus aphyllus L., Sp. Pl. 314 (1753); Boiss., Fl. 5:337; Post, Fl. 2:661. [Plate 139]

Climbing perennial, up to 1 m high, with woody base. Stems intricately and spreadingly branched; branches angular, scabridulous. Lower scale-like leaves deltoid, with a spinescent spur. Phylloclades 3–6(-9) in a cluster, spreading, unequal, 0.5–1(-1.8) cm, linear, angular, rigid, spinescent at tip, thick or slender; rarely some phylloclades single or twin. Flowers 1–2. Pedicels nodding, jointed at middle, nearly as long as flower. Perianth more or less stellate; inner perianth-segments shorter than outer, incurved. Berry 6–7 mm in diameter, blackish. Fl. September–November.

Hab.: Batha and maquis, among rocks. Common. Coast of Carmel, Sharon Plain, Philistean Plain; Upper and Lower Galilee, Mt. Carmel, Esdraelon Plain, Samaria, Shefela, Judean Mts.; Upper Jordan Valley; Golan, Gilead, Ammon, Edom.

Area: S. and E. Mediterranean.

4. Asparagus stipularis Forssk., Fl. Aeg.-Arab. 72 (1775); Boiss., Fl. 5:338, incl. var. *brachyclados* Boiss. (1882); Post, Fl. 2:662. [Plate 140]

Thorny shrub, 1 m or taller, dioecious, with a short branched subterranean rhizome; roots partly slender, partly thickened into fusiform tubers giving rise to above-ground stems. Stems woody, erect, intricately and spreadingly branched, now and then climbing; branches angular. Lower scale-like leaves with a spinescent spur. Phylloclades solitary (a few sometimes in groups of 2–3), linear, woody, spinescent at tip, (0.8-)1–4(-7) cm long, 1.5–2 mm thick. Flowers 1–2, sometimes 3–8. Pedicels jointed at middle, 1½ times as long as flower. Perianth more or less stellate; segments equal. Berry 6–8 mm, subglobose, blackish. Fl. March–May and September–November.

Hab.: Hedges. Fairly common. Acco Plain, Coast of Carmel, Sharon Plain, Philistean Plain; ?Judean Mts., Judean Desert, N. and C. Negev; Lower Jordan Valley, Dead Sea area, Arava Valley; Edom.

Area: S. Mediterranean and Saharo-Arabian.

21. RUSCUS L.

Dioecious, erect, branched shrubs with a subterranean rhizome. Proper leaves small, membranous, scale-like. Ultimate branches in axils of membranous leaves transformed into green leaf-like phylloclades; phylloclades persistent, coriaceous, veined, horizontal by basal torsion. Flowers unisexual, borne half-way up on midvein on upper or lower face of phylloclade in an axil of a scale-like leaf (bract). Perianth greenish, marcescent; segments 6, free, spreading, unequal, the inner narrower than the outer. Staminate flower: stamens 3; filaments connate into a tube. Pistillate flower: 3 connate filaments devoid of anthers surrounding ovary; ovary 1-locular, with 2 ovules; style very short; stigma capitate. Fruit a fleshy berry, with 1–2 subglobose seeds.

Seven species in the Euro-Siberian, Mediterranean and Irano-Turanian regions.

Literature: P.F. Yeo, A contribution to the taxonomy of the genus *Ruscus*, *Notes Roy. Bot. Gard. Edinburgh* 28:237-266 (1968).

1. Ruscus aculeatus L., Sp. Pl. 1041 (1753); Boiss., Fl. 5:340, incl. var. *angustifolius* Boiss. (1882). Post, Fl. 2:662. [Plate 141]

Shrub, 25–60 cm, dark green. Stems rigid, perennial, branched. Proper leaves lanceolate or triangular, acuminate. Phylloclades 1.5–3(–4) cm, rigid, ovate to lanceolate, tapering to a spiny tip. Flowers 1–2, short-pedicellate, subtended by a bract. Perianth up to 2.5 mm. Berry about 1 cm in diameter, globose, red. Fl. March–May.

Hab.: Maquis, among rocks. Upper and Lower Galilee, Mt. Tabor, Mt. Carmel, Samaria; Dan Valley; Golan.

Area: Mediterranean, extending into the Euro-Siberian and W. Irano-Turanian.

Yeo (*op. cit.* 261) distinguishes var. *angustifolius* Boiss. from var. *aculeatus*, the phylloclades of the former variety being 4 times as long as broad or longer. This subdivision of the species is, in his opinion, supported by the fact that "there is in *R. aculeatus* a considerable variation in cladode width in relation to length, and that somewhat narrow cladodes are rather frequent from about Austria and Italy eastwards but are absent west of this (i.e., in West Europe and North Africa)".

In Israel plants corresponding to both varieties according to the above definition, are found and transition forms also occur. However, no ecological or geographical differentiation within Israel could be detected that would justify acceptance of two separate taxa.

R. aculeatus is collected for ornament, especially for dry bouquets, and often dyed red, yellow, etc.

Protected by law.

22. SMILAX L.

Evergreen dioecious woody climbers, often prickly, with a thick rhizome. Stems branched. Leaves mostly alternate, 2-stichous, rarely opposite, petiolate, 3- to many-veined, with a pair of tendrils borne on a petiole above its base; veins reticulate. Flowers unisexual, small, in umbel-like cymes often united into a raceme. Perianth of 6(–15) petaloid deciduous segments. Staminate flower: stamens 6, equal, inserted at base of perianth-segments; anthers introrse. Pistillate flower: filiform deciduous staminodes; ovary 3-locular, with 1–2 ovules in each locule; stigmas 3, sessile, recurved, deciduous. Fruit a globose berry, with 1–2(–3) globose seeds.

Three hundred and fifty trop. and subtrop. species.

Fossil leaves of *Smilax* found already in the Upper Cretaceous.

Literature: P. Vernet, *Contribution à l'étude écologique et systématique de Smilax aspera L.*, Thèse, Montpellier, 1965; *Bull. Soc. Bot. Fr., Mém.* 1966:140–146 (1967). S. Ferri, *Ricerche sulla struttura anatomica e su alcuni principi attivi di Smilax aspera L. e Smilax mauritanica Poir. per un contributo alla loro deffinizione sistematica, Webbia* 21:475–486 (1966).

1. Smilax aspera L., Sp. Pl. 1028 (1753); Boiss., Fl. 5:343; Post, Fl. 2:663. Incl. var. *altissima* Moris & De Not., Mem. Accad. Sci. Torino ser. 2, 2:183 (1839)=var. *mauritanica* (Poir.) Gren. & Godr., Fl. Fr. 3:234 (1855; "Desf."). S. *mauritanica* Poir., Voy. Barb. 2:263 (1789). [Plate 142]

Green climber, reaching up to 10–15 m in height, glabrous, with a thick rhizome. Stems perennial, slender, zigzag, angular, climbing on trees, shrubs or twining on rocks, armed with recurved prickles, rarely without prickles. Leaves persistent, coriaceous, hastate or cordate-triangular, more or less acuminate, without prickles or sparsely prickly on margins and sometimes along veins on lower surface; petiole up to 2 cm, prickly or unarmed. Perianth cream-coloured, in staminate flowers 5 mm, in pistillate flowers 3.5 mm. Berry 8–10 mm, red. Fl. October–January.

Hab.: Batha and maquis. Common. Acco Plain, Coast of Carmel, Sharon Plain; Upper and Lower Galilee, Mt. Carmel, Samaria, Judean Mts., Dan Valley; Golan, Gilead, Ammon, Edom.

Area: Mediterranean, extending into Abyssinia and India.

Var. *altissima* is characterized by taller growth, larger leaves and the lack of prickles. In Palestine it is usually confined to maquis and other shady habitats. Plants transitional to the typical form in the above characters are also found. Whether var. *altissima* is a genetically based ecotype or only environmentally determined, remains to be tested.

Ferri (*op. cit.*, 1966) studied plants of S. *aspera* and plants classified morphologically as S. *mauritanica* (equivalent to var. *altissima* Moris & De Not.) in Italy. She found no differences between them in anatomical structure and in specific chemical compounds and came to the conclusion that it is impossible to distinguish S. *mauritanica* from S. *aspera*.

119. AMARYLLIDACEAE

Glabrous perennial herbs with a bulb, scapose, rarely caulescent. Leaves alternate, generally all basal, often 2- or 3-stichous. Inflorescence mostly cymose and corymbiform, subtended by a spathe formed by 1 or more valves; rarely flowers solitary. Flowers hermaphrodite, actinomorphic or somewhat zygomorphic, 3-merous, often showy, mostly entomophilous, sometimes ornithophilous. Perianth petaloid, of 6 segments in 2 whorls, free or arising from a short or long hypanthial tube; in some genera a corona (paracorolla) appears as an extra whorl. Stamens 6, in 2 whorls, sometimes 3 stamens staminodial; filaments free or adnate to hypanthial tube, or connate and forming corona;

anthers introrse. Pistil syncarpous; ovary inferior, 3-carpellate, 3-locular or rarely 1-locular; ovules mostly numerous (sometimes 2–1); style 1, slender; stigma 3-lobed or capitate. Fruit a capsule, dehiscing loculicidally or irregularly, sometimes transitional to a berry. Embryo small, straight, surrounded by endosperm; endosperm of starch or cellulose.

About 85 genera and 1,100 species in warm-temperate regions, some in steppe areas of Africa and W. Asia, less numerous in the Trop.

Many species contain alkaloids.

Literature: J. G. Baker, *Handbook of the Amaryllideae*, etc., London, 1888. F. Pax & K. Hoffmann, Amaryllidaceae, in: Engler & Prantl, *Pflanzenfam.* ed. 2, 15a:391–430, 1930. V. Täckholm & M. Drar, *Flora of Egypt* 3:310 485, Cairo, 1954. P. Mouterde, *Nouvelle Flore du Liban et de la Syrie* 1:289–294, Beyrouth, 1966. P. Wendelbo, Amaryllidaceae, in: K. H. Rechinger (ed.), *Flora iranica* 67:1–8, 1970. V. Täckholm, *Students' Flora of Egypt* ed. 2, 655–657, Beirut, 1974. T. G. Tutin et al. (eds.), Amaryllidaceae, in: *Flora europaea* 5:75–84, Cambridge, 1980. P. E. Boissier, *Flora orientalis* 5:143–154, 1882. G. E. Post, *Flora of Syria, Palestine and Sinai* ed. 2, 2:604–611, 1933.

1. Caulescent herbs. Stem leafy. Flowers lilac or blue, devoid of corona.   **4. Ixiolirion**
– Scapose herbs. Flowers not as above   2
2. Flowers yellow or orange, devoid of corona.   **1. Sternbergia**
– Flowers white, cream-coloured or yellow, provided with white, yellow or orange corona   3
3. Corona 12-dentate. Stamens on edge of corona in sinuses between teeth.

3. Pancratium
– Corona entire, crenulate or irregularly lobed. Stamens inserted below corona on tube of perianth.   **2. Narcissus**

1. STERNBERGIA Waldst. & Kit.

Bulbous and scapose perennials. Bulb tunicate; outer tunics membranous, brown, often extending upwards into a collar. Leaves and scape enclosed at base in 2 or more tubular membranous cataphylls. Leaves narrowly to broadly linear. Scapes 1 or more, each bearing a single flower or umbels of 2(–3), enclosed in a spathe; spathe membranous, tubular at base. Flowers crocus-like, yellow or orange. Perianth gamophyllous, campanulate to infundibular, devoid of corona; hypanthial tube present, gradually broadened; limb 6-partite. Stamens 6, inserted at throat of perianth, 3 of them alternately shorter; filaments much longer than anthers, filiform; anthers dorsifixed, usually at middle. Ovary inferior, 3-locular, with numerous ovules; style 3-gonous; stigma 3-lobed. Capsule subglobose, somewhat fleshy, irregularly dehiscent or indehiscent. Seeds subglobose, arillate.

Eight mainly Mediterranean and W. Irano-Turanian species.

Literature: N. Feinbrun & W. T. Stearn, A revision of *Sternbergia* (Amaryllidaceae) in Palestine, *Bull. Res. Counc. Israel* 6D:167–173 (1958).

1. Perianth-segments not reaching 0.5 cm in width.   **1. S. colchiciflora**

- Perianth-segments at least 1 cm broad   2
2. Perianth-segments 2–4 times as long as tube. Leaves with a whitish mid-stripe.
                                                                   2. S. lutea
- Perianth-segments nearly as long as to somewhat shorter than tube. Leaves devoid
  of a whitish longitudinal mid-stripe.   **3. S. clusiana**

1. Sternbergia colchiciflora Waldst. & Kit., Pl. Rar. Hung. 2:172, t.159
(1803–1804); Feinbr. & Stearn, *op. cit.* 171; Boiss., Fl. 5:147; Post, Fl. 2:606. *S. aetnensis* (Rafin.) Guss., Fl. Sic. Prodr. 1:395 (1827). *S. colchiciflora* Waldst. & Kit. var. *aetnensis* (Rafin.) Guss., Fl. Sic. Syn. 2:811 (1844–1845); Rouy, Bull. Soc. Bot. Fr. 31:182 (1884); Eig, Bull. Inst. Agr. Nat. Hist. Tel-Aviv 6:52 (1927). [Plate 143]

Bulb 1.2–2 cm in diameter. Leaves hysteranthous, 2–4 mm broad, narrowly linear, longer than scape. Scape short, elongating in fruit. Flowers autumnal, small, as a rule rising slightly above ground. Perianth pale yellow; segments oblong, 2.5–3.5 × 0.3–0.4 cm. Fl. September–November.

Hab.: Batha on calcareous rocks. Rare. Upper and Lower Galilee, Mt. Gilboa, Samaria, Judean Mts.; Golan.

Area: Mediterranean, extending into the Euro-Siberian and W. Irano-Turanian.

Recorded in Fl. URSS (4:489, 1935) as occasionally cleistogamous, the flowers developing to maturity within the bulb and the capsule maturing above ground.

2. Sternbergia lutea (L.) Ker-Gawler ex Spreng.; Syst. Veg. 2:57 (1825); Feinbr. & Stearn, *op. cit.* 172; Boiss., Fl. 5:147. *Amaryllis lutea* L., Sp. Pl. 292 (1753). *S. aurantiaca* Dinsmore in Post, Fl. ed. 2, 2:607 (1933). [Plate 144]

Bulb 1.5–2.5 cm in diameter. Leaves appearing soon after the flowers, 0.5–2 cm broad, linear, dark green, marked with a whitish mid-stripe. Scape 5–12 cm. Flowers autumnal. Spathe ovate-lanceolate. Perianth yolk-yellow; tube 0.5–1 cm; segments 3–4.5 × 1.5–2 cm, obovate-elliptic, obtuse. Fl. October.

Hab.: Judean Mts. (Jerusalem).
Area: Mediterranean and Irano-Turanian.

This species is most probably not indigenous in Palestine. It was found in abandoned gardens in Jerusalem.

S. lutea is usually recorded as synanthous, but in Jerusalem its flowers appear as a rule before the leaves, in October. Amico (Nuovo Gior. Bot. Ital. n.s. 54:748–771, 1947) experimented with bulbs of *S. lutea* collected from natural habitats in Italy where its flowers appear together with leaves. He found that in vapour-saturated air at 30°C leaves appeared in June instead of September and flowering was suppressed. Amico concluded that a resting period in dry atmosphere is necessary for a full vegetative and reproductive growth cycle in *S. lutea*.

3. Sternbergia clusiana (Ker-Gawler) Ker-Gawler ex Spreng., Syst. Veg. 2:57 (1825); Feinbr. & Stearn, *op. cit.* 168; Boiss., Fl. 5:148. *Amaryllis clusiana* Ker-Gawler, Bot. Mag. 27:text below t.1089 (1808). *S. stipitata* Boiss. &

Hausskn. in Boiss., Fl. 5:148 (1882). *S. macrantha* J. Gay ex Baker, Handb. Amaryll. 28 (1888). *S. spaffordiana* Dinsmore, Feddes Repert. 24:302 (1928); Post, Fl. 2:607; Feinbr., Zohary & Koppel, Fl. Land of Israel, Iconography, Plates 51–100 (1952). [Plate 145]

Bulb 5–7.5 cm in diameter. Leaves hysteranthous, 2–3 cm broad, somewhat twisted, glaucescent. Scape 5–13 cm, its greater part often subterranean, clothed below in brown tunics. Flowers autumnal, showy. Spathe tubular in its lower part, somewhat longer than tube of perianth. Perianth yolk-yellow, large; segments broadly oblong-elliptic, about as long as or shorter than slender tube; outer segments up to 7 × 3.5 cm, often cucculate at apex or with inflexed apical appendage, 4 × 2 mm. Fl. September–November.

Hab.: Batha and maquis on various rocks and soils, in the hills, mainly eastwards of the watershed; rock crevices and wadis in desert districts. Sharon Plain; Upper Galilee, Mt. Gilboa, Samaria, Judean Mts., C. Negev; Golan, Moav, Edom.

Area: E. Mediterranean and W. Irano-Turanian (Palestine, Lebanon, Syria, Turkey, Iran).

Protected by law. Its distribution in this country has been followed up and mapped by the Authority for Nature Reserves.

2. NARCISSUS L.

Bulbous and scapose perennials. Bulb tunicate; outer tunics membranous, brown. Leaves and scape enclosed at base in tubular membranous cataphylls. Leaves linear. Flowers pedicellate, several on a scape, rarely solitary, subtended by a spathe, sometimes fragrant. Perianth gamophyllous, white or yellow, hypocrateriform, provided with a corona at throat; limb 6-partite; hypanthial tube present; corona campanulate or cup-shaped, entire, lobed or crenulate. Stamens 6, inserted on perianth-tube, below corona; anthers dorsifixed, usually more or less at middle. Ovary inferior, ovoid or oblong, 3-locular, with numerous ovules; style filiform; stigma truncate or sub-3-lobed. Capsule loculicidal, 3-valved. Seeds subglobose or angular, crustaceous, black.

Sixty species, mainly in the W. Mediterranean and W. Asia.

Literature: E.A. Bowles, *A Handbook of Narcissus*, Martin Hopkinson, London, 1934. R. Dulberger, Flower dimorphism and self-incompatibility in *Narcissus tazetta* L., *Evolution* 18:361–363 (1964). A. Fernandes, Le problème du *Narcissus tazetta* L. II., *Bol. Soc. Brot.* Ser. 2a, 40:277–319 (1966); Improvements in the classification of the genus *Narcissus* L., *Plant Life* 24:51–57 (1968); Contribution to the knowledge of the biosystematics of some species of genus *Narcissus* L., *Publ. Univ. Sevilla*, 245–284 (1969). Shulamith Weitz & Naomi Feinbrun, Cytology and systematics of *Narcissus tazetta* L. in Israel, *Israel Jour. Bot.* 21:9–20 (1972). D.A. Webb, Taxonomic notes on *Narcissus* L., *Jour. Linn. Soc. London (Bot.)* 76:298–307 (1978).

1. Scape 3–15-flowered. Leaves several, appearing with or before flowers. **1. N. tazetta**
– Scape 1(–2)-flowered. Leaves 1–2, appearing after end of flowering. **2. N. serotinus**

Sect. 1. TAZETTAE DC. in Lam. & DC., Fl. Fr. ed. 3, 3:322 (1805). Sect. *Hermione* Salisb. Leaves synanthous, flat or canaliculate. Flowers in umbels, concolorous or bicoloured; flowering in early winter or spring, rarely in autumn.

1. Narcissus tazetta L., Sp. Pl. 290 (1753); Boiss., Fl. 5:150; Post, Fl. 2:608. [Plate 146]

Plant 20–50 cm. Bulb 3.5–4 cm in diameter. Leaves (2–)3–6, 0.5–1 cm broad, lorate, glaucescent, appearing with or before flowers. Scape 3–15-flowered. Pedicels unequal. Flowers fragrant, horizontal or nodding, dimorphic. Perianth white or cream-coloured; segments ovate-elliptic, spreading; outer segments apiculate; corona yellow or orange, cup-shaped, ⅕–½ length of perianth-segments. Stamens on very short filaments, in 2 series; both series arranged in upper part of perianth-tube; upper anthers subexserted or exserted. Style varying in length, i.e. either short (ending much below anthers) or long (reaching lower or upper anthers or in between). $2n=20$. Fl. November–February.

Hab.: Wet fields on alluvial soil or rocky batha and maquis. Abundant. Coast of Carmel, Sharon Plain, Philistean Plain; Upper and Lower Galilee, Mt. Carmel, Esdraelon Plain, Samaria, Shefela, Judean Mts., C. Negev; Dan Valley, Upper and Lower Jordan Valley; Golan.

Area: Mediterranean, extending into the W. Irano-Turanian.

R. Dulberger (*op. cit.*) studied the breeding system in *N. tazetta*: "Israeli populations of *Narcissus tazetta* have invariably been found to be dimorphic, consisting of both short-styled and long-styled types. In short-styled plants the stigma level is below the lower tier of anthers; in the long-styled plants it is usually at the height of the upper anthers, but may also be situated at any height between the two tiers of anthers or even above the upper one. While each of the two floral types displays variability in style length, there is an apparent discontinuity between them. The height of the anthers also fluctuates widely; however, there is no difference in anther level between the two types... distyly is apparently not linked genetically with incompatibility. *N. tazetta* thus differs from classical cases of dimorphism".

Boissier (*loc. cit.*) described several varieties from the E. Mediterranean and some of them are cited by Post (*loc. cit.*). These varieties can be disregarded, as they are based on unstable characters of perianth-segments and corona.

A garden cultivar of *N. tazetta* with double flowers and sterile anthers (var. *plenus*) is commonly grown in hill districts of Israel. It has been found (Weitz & Feinbrun, *op. cit.*) to be triploid ($2n=30$), and reproduces by bulb division. It may be identical with the triploid cultivar which Webb (1978) regarded as having been known as *N. cypri* Sweet. Indeed we found that the same triploid cultivar is grown also in Cyprus.

Specimens with normal flowers somewhat larger than in wild plants sometimes occur among plants of this garden variety.

N. papyraceus Kew-Gawler ($2n=22$) is cultivated commercially in Israel.

Sect. 2. SEROTINI Parl., Fl. Ital. 3:157 (1858). Leaves hysteranthous, appearing after end of flowering, filiform. Flowers usually solitary, bicoloured; flowering in autumn.

2. Narcissus serotinus L., Sp. Pl. 290 (1753); Boiss., Fl. 5:151; Post, Fl. 2:608. [Plate 147]

Plant 10–25 cm. Bulb 1.2–2 cm in diameter. Leaves of young bulbs 2, of flowering bulbs 1, filiform, appearing after end of flowering time. Scape slender, 1(–2)-flowered. Flowers faintly fragrant. Perianth white; segments oblong to oblong-lanceolate, acute to mucronate, spreading; corona very short, orange, 3-fid. Stamens in 2 series; lower stamens on long filaments with anthers at about middle of perianth-tube; the upper nearly sessile, with anthers near throat of perianth, exserted or subexserted. 2n=30. Fl. November–December.

Ripe seeds found at end of December.

Hab.: Calcareous sandstone. Rare. Coast of Carmel, Sharon Plain.

Area: Mediterranean.

3. PANCRATIUM L.

Bulbous and scapose perennials. Bulb tunicate; outer tunics membranous, brown. Leaves and scape enclosed at base in tubular membranous cataphylls. Leaves linear. Flowers white, large, fragrant, in many-flowered umbels, rarely solitary, subtended by a spathe, pedicellate to subsessile; pedicels unequal; spathe 2-valved, membranous, shorter than flowers. Perianth gamophyllous, infundibular, with 6 equal segments and provided with a corona; hypanthial tube present, long. Stamens 6; filaments dilated and united into dentate corona borne on lower part of perianth; each filament flanked by a tooth on either side; anthers dorsifixed at middle. Ovary inferior, 3-locular, with numerous ovules; style filiform; stigma sub-3-lobed. Capsule subglobose, obscurely 3-gonous. Seeds angular or winged, crustaceous, black.

Fifteen species in the Mediterranean, in Trop. Asia and Trop. Africa.

Literature: Aliza Keren & M. Evenari, Ecological aspects concerning distribution and germination of *Pancratium maritimum* L., *Israel Jour. Bot.* 23:202–215 (1974). Ella Werker & A. Fahn, Seed anatomy of *Pancratium* species from three different habitats, *Bot. Gaz.* 136:396–403 (1975).

1. Perianth 2.5–4 cm. Leaves green, with a white mid-stripe. Plants of rocky Mediterranean batha.   **2. P. parviflorum**
– Perianth longer. Leaves glaucous, devoid of mid-stripe. Plants of sandy soils   2
2. Leaves subsynanthous, 1–2.5(–3) cm broad, slightly twisted. Plants of maritime sands.   **1. P. maritimum**
– Leaves hysteranthous, 4–9 mm broad, curled up into a spiral. Plants of desert sands.   **3. P. sickenbergeri**

1. Pancratium maritimum L., Sp. Pl. 291 (1753); Boiss., Fl. 5:152; Post, Fl. 2:609. [Plate 148]

Plant 50–70 cm. Bulb subglobose, usually 5–7 cm in diameter, deeply sunk, with a long neck. Leaves subsynanthous, 1–2.5(–3) cm broad, erect, broadly lorate, slightly twisted, glaucous, about as long as scape. Scape thick, compressed. Flowers fragrant, 2–8 to a spathe. Perianth 8–15 cm; tube longer than limb; segments oblong-lanceolate, with a narrow green mid-stripe on outside; teeth of corona 12, triangular, acute. Seeds hydrochorous. Fl. August–October. Leaves appearing shortly after flowers.

Hab.: Sands in the spray belt along the seashore. Common. Coastal Galilee, Acco Plain, Coast of Carmel, Sharon Plain, Philistean Plain.

Area: Mediterranean, extending along sandy shores of the Atlantic, Black and Caspian Seas.

Protected by law.

2. Pancratium parviflorum Desf. ex Del. in Redouté, Liliacées 8:t.471 (1815); Desf., Tabl. École Bot. ed. 2, 272 (1815) *nom. nud.*; Decne., Ann. Sci. Nat. Bot. ser. 2, 4:346 (1835); Boiss., Fl. 5:153; Post, Fl. 2:609. *Vagaria parviflora* (Desf. ex Del. in Redouté) Herbert, Amaryll. 61, 76, 226 (1837). [Plate 149]

Plant 25–35 cm. Bulb ovoid, 2.5–5 cm in diameter, with long neck. Leaves hysteranthous, about 1–1.5 cm broad, lorate, green, with a white mid-stripe, about as long as scape. Scape slender. Flowers not fragrant, 4–10 to a spathe. Perianth 2.5–4 cm; tube nearly as long as limb; segments linear-lanceolate, with a broad green mid-stripe on outside; teeth of corona 12, triangular, acute. Seeds arillate, myrmecochorous. Fl. September–November. Leaves appearing in December.

Hab.: Batha on calcareous rocks. Fairly rare. Coastal Galilee, Coast of Carmel, Sharon Plain; Upper and Lower Galilee, Mt. Carmel, Samaria, Shefela, Judean Mts.; Golan.

Area: E. Mediterranean (Palestine, Lebanon, Syria).

Protected by law.

3. Pancratium sickenbergeri Aschers. & Schweinf. ex C. & W. Barbey, Herb. Levant 158 (May 1882); Aschers. & Schweinf. in Boiss., Fl. 5:153 (July 1882); Aschers. & Schweinf., Garten-Zeit. (Wittmack) 2:345 *cum icone* (1883); Post, Fl. 2:609. [Plate 150]

Plant 30–40 cm. Bulb ovoid, 3–4 cm in diameter, with a long neck. Leaves hysteranthous, 4–9 mm broad, lorate, glaucous, curled up into a spiral, somewhat longer than scape. Scape 3–6 mm in diameter. Flowers 4–8 to a spathe. Perianth 6–7 cm; tube shorter than limb; segments oblong-lanceolate, with a broad green mid-stripe on outside; teeth of corona 12, triangular, acute. Seeds anemochorous. Fl. September–October. Leaves appearing in March.

Hab.: Sands in deserts. W., N. and C. Negev; Dead Sea area, Arava Valley; Edom.

Area: Saharo-Arabian.

Protected by law.

4. Ixiolirion Fisch. ex Herbert

Bulbous caulescent perennials. Bulb tunicate. Stem erect, leafy, surrounded at base with membranous sheaths. Leaves linear, grass-like, mainly in lower part of stem. Inflorescence a lax raceme or panicle, sometimes corymbiform. Flowers showy, actinomorphic. Perianth lilac or blue, infundibular, 6-partite nearly to base, devoid of corona; perianth-segments 3-veined. Stamens 6, included, 3 of them alternately longer, attached at base of perianth-segments; filaments subulate; anthers basifixed. Ovary inferior, 3-locular, with numerous ovules; style filiform; stigma 3-lobed, with lobes filiform, spreading and revolute. Capsule coriaceous, oblong-clavate, narrowed at base, dehiscing loculicidally by 3 valves. Seeds ovoid-oblong.

Three species in W. and C. Asia.

Literature: S. G. Gorshkova, in: V. L. Komarov (ed.), *Flora URSS* 4:489-491, 1935.

1. Ixiolirion tataricum (Pall.) Herbert, App. 37, t.2 (1821); Schult. & Schult. fil. in Roem. & Schult., Syst. Veg. 7:752 (1829); Gorshkova, in Fl. URSS 4:490; Wendelbo, Fl. Iran. 67:2. *Amaryllis? tatarica* Pall., Reise 3:727, t.D f.1 (1776); Herbert, Amaryllid. t.19 (1837). *I. montanum* (Labill.) Herbert, App. 37 (1821) & Amaryllid. t.20 f.3 (1837); Boiss., Fl. 5:154; Post, Fl. 2:610. *A. montana* Labill., Icon. Pl. Syr. 2:5, t.1 (1791). *I. pallasii* Fisch. & Mey. ex Ledeb., Fl. Ross. 4:116 (1852) *nom. superfl.*; Boiss., *loc. cit.*; Post, *op. cit.* 611. [Plate 151]

Plant 30–50 cm. Bulb ovoid, 1–1.5 cm in diameter, clothed in brownish tunics, usually deep in the ground. Leaves long, narrowly linear and long-attenuate. Flowers 3–17, lilac or blue, showy, pedicellate. Pedicels subtended by membranous, lanceolate-subulate bracts. Perianth 3–4 cm; segments oblong-linear, the outer long-mucronate, the inner broader, obtuse. Fl. March–May.

Hab.: Batha and steppes, loess and sandy soil, often in Artemisietum herbae-albae. Upper Galilee, E. Samaria, Judean Mts., Judean Desert, W., N. and C. Negev; ?Dan Valley, Lower Jordan Valley, Arava Valley; Golan, Gilead, Ammon, Moav, Edom.

Area: Irano-Turanian.

120. DIOSCOREACEAE

Climbing perennial herbs or shrubs, with mostly tuber-like rhizomes, mostly dioecious, rarely monoecious. Leaves alternate or opposite, petiolate, reticulate-veined, often cordate. Inflorescence a raceme, spike or panicle. Flowers small, actinomorphic, unisexual, rarely hermaphrodite, mostly bracteate. Perianth persistent, 6-lobed or 6-partite. Stamens 6, or 3 stamens fertile and 3 staminodes. Ovary inferior, 3-carpellate, usually 3-locular, rarely 1-locular; ovules usually 2 in each locule. Fruit mostly a capsule, rarely a berry.

Seeds often winged, rarely nearly globose; endosperm horny, surrounding a small embryo; second cotyledon reduced.

Ten genera and about 650 trop. and warm-temperate species.

Literature: R. Knuth, Dioscoreaceae, in: Engler & Prantl, *Pflanzenfam.* ed. 2, 15a:438–462, 1930. P. E. Boissier, *Flora orientalis* 5:343–344, 1882. G. E. Post, *Flora of Syria, Palestine and Sinai* ed. 2, 2:664, 1933.

1. TAMUS L.

Climbing dioecious perennial herbs with a fleshy underground tuber. Leaves alternate, cordate. Flowers unisexual, with a bract and bracteole at base, in axillary racemes or spikes; staminate racemes and spikes many-flowered, usually longer than pistillate ones; sometimes pistillate flowers solitary. Staminate flower: perianth campanulate, with 6 subequal lobes; stamens 6, inserted at base of perianth; rudiment of pistil short. Pistillate flower: perianth shorter than ovary, with 6 minute narrow segments; staminodes small or absent; ovary 3-locular, locules 2-ovulate; style 3-fid; stigmas dilated, 2-fid, recurved. Fruit a fleshy berry, mostly 1-locular by abortion. Seeds few, ovoid or subglobose, not winged.

Five mainly Mediterranean species.

1. Petiole short, not exceeding 3 cm. Flowers sessile. Bract and bracteole ovate.
   2. T. orientalis
 - Petiole long, exceeding 3 cm, at least in lower leaves. Flowers pedicellate. Bract and bracteole linear, pointed at tip.   **1. T. communis**

1. Tamus communis L., Sp. Pl. 1028 (1753); Boiss., Fl. 5:344; Post, Fl. 2:664. [Plate 152]

Glabrous perennial. Tuber cylindrical. Stems up to 4 m, twining, sometimes branched. Leaves 3–10 cm, broadly ovate, deeply cordate at base, acute to acuminate, ending in a more or less long point or mucro; petiole generally longer than or as long as blade. Flowers pedicellate, in racemes, with linear bract and bracteole. Staminate racemes many-flowered, longer than leaf; perianth greenish-yellow. Pistillate racemes several-flowered, shorter than leaf; perianth greenish. Berry 8–12 mm, subglobose, red, rarely yellowish. Fl. February–April.

Hab.: Maquis. Upper and Lower Galilee, Mt. Tabor, Mt. Carmel, Samaria, Judean Mts.; Dan Valley, Hula Plain; Golan, Gilead, Ammon.

Area: Mediterranean, extending into the W. Irano-Turanian and into W. Europe.

Var. *cretica* (L.) Boiss., *loc. cit.*, recorded by Post from the Judean Mts., does not seem to deserve separate status.

Berry quoted as being poisonous especially for children.

2. Tamus orientalis Thiébaut, Bull. Soc. Bot. Fr. 81:119 (1934); Mout., Nouv. Fl. Liban Syrie 1:294. [Plate 153]

Glabrous perennial. Tuber turnip-shaped, 2–3 cm. Stem rarely exceeding 1 m, twining, simple or slightly branched. Leaves not exceeding 4 cm, ovate, cordate at base, acute to acuminate, ending in a mucro; petiole usually shorter than blade. Flowers sessile, with ovate bract and bracteole, in nodding axillary spikes; axis of spike flexuous. Staminate spikes often twin, 10–20-flowered; perianth white to lilac, fragrant, campanulate, later stellate. Pistillate spikes 1–7-flowered, shorter than leaf; perianth smaller than in staminate flower. Berry oblong, 5–10 mm, red at maturity. Fl. December–January.

Hab.: Batha among rocks. Upper Galilee, Mt. Gilboa, Samaria, Judean Mts.; Golan.

Area: E. Mediterranean (Palestine, Lebanon, Syria).

121. IRIDACEAE

Perennials with bulb, corm or rhizome, caulescent or scapose. Leaves alternate, usually 2-stichous, mostly equitant. Inflorescence cymose, subtended by a spathe formed by a bract and 1 or more bracteoles; sometimes inflorescence reduced to 1–3 flowers. Flowers hermaphrodite, actinomorphic or zygomorphic, 3-merous, often showy, entomophilous or ornithophilous. Perianth petaloid, of 6 segments, mostly arising from a hypanthial tube; segments more or less equal or dissimilar. Stamens 3, opposite outer perianth-segments; anthers extrorse or opening laterally by slits. Pistil syncarpous; ovary nearly always inferior, 3-carpellate, usually 3-locular; ovules usually numerous; style divided into 3 entire or dissected branches; stigmas sometimes petaloid. Fruit a loculicidal capsule dehiscing by 3 valves. Seeds usually numerous, globose, angular or flat; embryo small, surrounded by a copious hard endosperm.

Sixty-five genera and about 1,800 temperate and tropical species, with the greatest concentration in Africa south of the equator.

Literature: J.G. Baker, *Handbook of the Irideae,* London, 1892. F.L.E. Diels, Iridaceae, in: Engler & Prantl, *Pflanzenfam.* ed. 2, 15a:463–505, 1930. V. Täckholm & M. Drar, *Flora of Egypt* 3:446–523, Cairo, 1954. R. Maire, M. Guinochet & P. Quézel, *Flore de l'Afrique du Nord* 6:104–209, 1959. P. Mouterde, *Nouvelle Flore du Liban et de la Syrie* 1:295–322, 1966. P. Goldblatt, Cytological and morphological studies in the southern African Iridaceae, *Jour. S. African Bot.* 37:317–460 (1971). V. Täckholm, *Students' Flora of Egypt* ed. 2, 659–662, Beirut, 1974. P. Wendelbo & B. Mathew, Iridaceae, in: K.H. Rechinger (ed.), *Flora iranica* 112:1–79, 1975. T.G. Tutin et al. (eds.), Iridaceae, in: *Flora europaea* 5:86–102, Cambridge, 1980. P.E. Boissier, *Flora orientalis* 5:94–143, 1882. G.E. Post, *Flora of Syria, Palestine and Sinai* ed. 2, 2:583–604, 1933.

1. Perianth zygomorphic, with a curved hypanthial tube.   **5. Gladiolus**
- Perianth actinomorphic   2
2. Style-branches filiform, not petaloid. Perianth-segments more or less equal   3
- Style-branches petaloid. Outer perianth-segments distinctly dissimilar from inner ones   4

3. Perianth-tube longer than perianth-segments. **4. Crocus**
- Perianth-tube very short, much shorter than perianth-segments. **3. Romulea**
4(2). Plants with bulb or rhizome. Hypanthial tube present. **2. Iris**
- Plants with corm. Hypanthial tube absent. Slender sterile beak of ovary bearing the other flower parts. **⁴1. Gynandriris**

1. GYNANDRIRIS Parl.

Perennials with a tunicate corm. Corm of 1 enlarged internode, with a terminal growing apex; tunics reticulate-fibrous. Shoot enclosed in 3–4 membranous cataphylls. Leaves 1–2, linear, long and slender, usually canaliculate. Stem branched; each branch bearing a few-flowered inflorescence. Inflorescence substended by a 2–valved membranous spathe. Flowers lasting several hours only, opening at about noon and fading towards evening. Perianth similar to that in *Iris* but devoid of a hypanthial tube. Outer segments, *falls*, with an ascending claw and a spreading or reflexed limb; inner segments, *standards*, narrower than falls, erect or spreading. Stamens 3, appressed to style-branches and slightly shorter; filaments partly united; anthers extrorse. Ovary inferior, sessile with a long sterile terminal beak bearing the other flower parts; style divided into 3 petaloid branches. Capsule membranous, transparent. Seeds numerous, large, dark, visible through spathes when ripe. x=6.

Nine species, seven of them S. African, two Mediterranean, Saharo-Arabian and Irano-Turanian.

Literature: P. Goldblatt, Systematics of *Gynandriris* (Iridaceae), a Mediterranean-southern African disjunct, *Bot. Not.* 133:239–260 (1980). J. Galil, Morpho-ecological studies on geophilic plants. Vegetative dispersal of *Gynandriris sisyrinchium* L., *Israel Jour. Bot.* 30:165–172 (1981).

1. Flowers small, pale blue-grey; falls 10–15(–20) mm, 4–5 mm broad; spathe 2–3 cm. Leaf often solitary. **2. G. monophylla**
- Flowers medium to large, pale to dark blue, violet, lilac or purple; falls 20–40 mm, 8–12 mm broad; spathe 3.5–6.5 cm. Leaves usually 2. **1. G. sisyrinchium**

1. Gynandriris sisyrinchium (L.) Parl., Nuovi Gen. Sp. Monocot. 49 (1854); Wendelbo & Mathew, Fl. Iran. 112:12 (1975); Goldblatt, *op. cit.* 254, ff.1, 6 (1980). *Iris sisyrinchium* L., Sp. Pl. 40 (1753); Boiss., Fl. 5:120; Post, Fl. 2:589. [Plate 154]

Plant 10–30 cm. Corm 1.2–2 cm in diameter; tunics dark-coloured, extended above into a neck. Leaves 1–2, usually exceeding inflorescence, strongly veined, often recurved, with a tubular sheath at base. Stem few-branched, occasionally simple. Spathe usually dry at flowering time; inner valve 3.5–6.5 cm, the outer shorter. Perianth blue, violet, lilac or purple; falls with a white or yellow patch and spotted, 20–40 × 8–12 mm; limb usually shorter than claw, spreading; standards 20–30 × 4–6 mm, lanceolate, erect, much narrower than falls. Filaments 7–15 mm, united for about ⅓ their length; anthers 4.5–10 mm. Style-branches 8–20 mm, petaloid, erect, 2-fid. Capsule (excl. beak) cylindrical to ellipsoid. 2n=24. Fl. January–April.

Hab.: Batha, steppes and fallow fields. Common. Acco Plain, Sharon Plain, Philistean Plain; Upper and Lower Galilee, Mt. Carmel, Esdraelon Plain, Samaria, Shefela, Judean Mts., Judean Desert, W., N. and C. Negev; Hula Plain, Upper and Lower Jordan Valley; Golan, Gilead, Ammon, Moav.

Area: Mediterranean–W. Irano-Turanian.

2. Gynandriris monophylla Boiss. & Heldr. ex Klatt, Linnaea 34:578 (1865–1866); Goldblatt, *op. cit.* 257, f.63 (1980). *G. sisyrinchium* var. *monophylla* (Heldr.) Halácsy, Consp. Fl. Graec. 3:191 (1904). *Iris sisyrinchium* L. var. *monophylla* Heldr., Atti Congr. Intern. Bot. Firenze 234 (1876); Boiss., Fl. 5:120; Post, Fl. 2:588. *I. sisyrinchium* var. *minor* sensu Täckholm & Drar, Fl. Eg. 3:465 (1954). [Plate 155]

Plant rarely exceeding 4–6 cm above ground. Corm 0.7–1.5 cm in diameter; tunics dark-coloured. Leaves 1(–2), 10–15 cm, trailing. Stem usually simple, sometimes bearing 1–2 lateral inflorescences. Spathe 2–3 cm, outer valve slightly shorter than the inner. Flowers very pale blue-grey, with nectar guide orange ringed with white; falls 10–20 × 4–5 mm; claw pale, densely spotted; limb horizontal, slightly shorter than claw; standards 9–15 × 2 mm, lanceolate, erect. Filaments 4–6 mm, united for about 1 mm; anthers 2–3(–4) mm. Style-branches 5–6 mm. Fl. March–April.

Hab.: Rare. C. Negev.

Area: E. Mediterranean–Saharo-Arabian (S. Greece, Crete, S. Palestine, Sinai, Egypt, Libya).

In S. Greece (Attica) *G. monophylla* grows side-by-side with *G. sisyrinchium,* but no intermediates between them have been found.

2. Iris L.*

Perennials with rhizome or bulb. Leaves equitant, ensiform to linear, or canaliculate, or angular. Flowers hermaphrodite, actinomorphic, in a cymose inflorescence or solitary, subtended by a spathe of 2 or more valves. Perianth with a hypanthial tube, of 6 dimorphic perianth-segments; outer segments *(falls)* spreading or recurved at outer end, the inner *(standards)* erect or spreading to reflexed. Stamens 3, free, opposite the falls; anthers oblong, basifixed. Ovary inferior, (1–)3-locular, many-ovulate, tipped by a more or less long hypanthial tube enclosing style; style divided above into 3 petaloid branches more or less connate at base, each branch convex on outer side and concave on inner, overlying stamen and claw of fall, 2-lobed above and bearing stigma in

---

* The author's gratitude is expressed to Dr M. Avishai for co-operation, especially for the loan of rare reprints and his rich collection of colour slides of *Oncocyclus* species.

form of membranous flap at base of lobes. Capsule loculicidal, 3-valved. Seeds numerous, sometimes arillate.

About 300 N. temperate species.

Literature: W. R. Dykes, *The Genus Iris,* Cambridge, 1913. J. E. Dinsmore, Plantae Postianae et Dinsmoreanae, Fasc. 2, *Publ. Amer. Univ. Beirut Nat. Sci. Ser.,* no. 3:8-11 (1934). P. H. Davis, *Oncocyclus* Irises in the Levant, *Jour. Roy. Hort. Soc.* 71(4):93-97 (1946). B. Fedtschenko, *Iris,* in: V. L. Komarov (ed.), *Flora URSS* 4:511-516, 1935. R. W. Highwood, Further notes of the *Juno* Irises of the Levant, *Brit. Iris Soc. Yearbook* 1953:149-150 (1953). G. H. M. Lawrence, A reclassification of the genus *Iris, Gentes Herb.* 7:346-371 (1953). M. Simonet, Recherches cytologiques et génétiques chez les *Iris, Bull. Biol. France Belg.* 106:255-444 (1932); Nouveaux dénombrements chromosomiques chez les *Iris, Compt. Rend. Acad. (Paris)* 235, 21:1244-1246 (1952). Z. Ginsburg, The natural growing-conditions of *Oncocyclus* Irises in Israel, *Brit. Iris Soc. Yearbook* 1956:51-63 (1956). G. I. Rodionenko, *Genus Iris,* USSR Acad. Sci., Moscow Leningrad, 1961 (in Russian). J. Galil, Development cycle and ecology of *Iris palaestina* (Baker) Boiss., *Bull. Res. Counc. Israel* 11D:17-24 (1962). K. K. Kidd, An analysis of beard hair morphology in the Sect. *Iris, Yearbook Aril Soc. Intern.* 1963:15-18 (1963). G. Schutz, Irises in Israel, *Atti del 1° Simposio Intern. dell'Iris*:225-241, Firenze, 1963. P. Werkmeister, Catalogus Iridis 1967, Namen und Synonyme des Genus *Iris, Deutsche Iris- und Liliengesellschaft* 2, 1967. P. Wendelbo, *Tulips and Irises of Iran and their Relatives,* Bot. Inst. Iran, Teheran, 1977. M. Avishai, *Species Relationships and Cytogenetic Affinities in Section Oncocyclus in the Genus Iris,* Ph. D. Thesis, Hebrew Univ. Jerusalem, 1977. M. Avishai & D. Zohary, Chromosomes in the *Oncocyclus* Irises, *Bot. Gaz.* 138:502-511 (1977); Genetic affinities among *Oncocyclus* Irises, *Bot. Gaz.* 141:107-115 (1980). N. Feinbrun, A new *Iris* from Israel, *Notes Roy. Bot. Gard. Edinb.* 37:75-78 (1979). M. Avishai, *Irises in Eretz Israel,* Society for Plant Protection, 1979 (in Hebrew). D. Shimshi, Two ecotypes of *Iris atrofusca* Bak. and their relations to man-modified habitats, *Israel Jour. Bot.* 28:80-86 (1979-80).

1. Leaves 1-2(-3), 4-gonal, 3-6 mm thick, erect, straight. Plants with a bulb; outer bulb-tunics reticulate-fibrous   2
- Leaves 4 or more, not 4-gonal. Plants with a rhizome or, if with a bulb, outer tunics membranous   3
2. Falls (outer perianth-segments) (3.5-)4-4.5(-5) cm; claw green-veined on outside; limb veined, but not dark-spotted.   **15. I. vartanii**
- Falls (4.5-)5-6 cm; claw not green-veined on outside; limb veined and dark-spotted.   **14. I. histrio**
3(1). Plants with a bulb. Standards (inner perianth-segments) spreading or reflexed, smaller than falls   4
- Plants with a rhizome. Standards erect to erect-incurved, larger than falls or, if smaller than falls, plants growing in water or on marshes   7
4. Lobes of style-branches acute to acuminate, distinctly longer than broad. Falls marked with broad veins and/or oblong spots or dots on a lighter ground; median yellow ridge along falls not crested   5
- Lobes of style-branches obtuse, rounded, about as long as broad. Falls veined but not spotted; median ridge along falls with a short yellow crest   6
5. Markings on falls yellowish-white, greenish or pale sky-blue. Leaves 5-6, about 2

cm broad; bulb-tunics only slightly prolonged above bulb.   **16. I. palaestina**
- Markings on falls purple. Leaves 4–5, 0.5–1.6 cm broad, strongly undulate; bulb-tunics much prolonged above bulb.   **17. I. edomensis**
6(4). Scape 15–40 cm. Leaves in flowering specimens 8–12. Crest on falls pale yellow, not surrounded by a yellow area.   **18. I. aucheri**
- Scape 7–15 cm. Leaves in flowering specimens 5–7. Crest on falls yolk-yellow and rising from a yolk-yellow area which is surrounded by a white margin.
                                                                    19. I. regis-uzziae
7(3). Hydrophytes. Perianth yellow; falls without a beard   8
- Plants of dry habitats or cultivated. Perianth rarely yellow; falls bearded   9
8. Plant 70–150 cm or more. Stem branched, several-flowered; base of stem not bearing prickly fibres.   **1. I. pseudacorus**
- Plant up to 60 cm. Stem unbranched, 1-flowered, with prickly fibres at base, formed from old basal leaf-sheaths.   **2. I. grant-duffii**
9(7). Scape unbranched, 1-flowered. Hairs of beard diffuse on lower part of fall. Seeds arillate   10
- Scape branched, bearing several flowers. Beard linear, narrow, along middle line of fall. Seeds without aril   18
10. Plants of sands of Negev. Flowers lilac to violet; leaves falcate, 4–6 mm broad, $\frac{1}{3}$–$\frac{1}{2}$ length of scape. Standards nearly uniformly lilac; falls darker lilac to violet, devoid of dots or of conspicuous veins; hairs of beard nearly black. Rhizome stoloniferous.   **13. I. mariae**
- Plants not as above   11
11. Perianth purple, brown-purple, chestnut-brown or nearly black (rarely yellow variants lacking anthocyanin occur); colour of standards same or nearly same as that of falls but often somewhat lighter in tint or, if considerably lighter, standards marked with dark veins and minute dense dots   12
- Perianth conspicuously bicoloured; standards very pale, with fine pale lilac, pale violet or pale purple veins (and sometimes also scattered dots) on a nearly white ground; falls distinctly darker with a characteristic pattern of veins and dots and/or spots on a white or cream-coloured ground   16
12. Leaves reaching base of spathe or overtopping it, 1–2 cm broad, mostly erect   13
- Leaves not reaching base of spathe, mostly less than 1 cm broad and mostly falcate   14
13. Perianth 10–12 cm (excl. tube and ovary), longer than broad; both standards and falls veined and densely and minutely dotted on a paler ground; standards purple; falls darker purple or chestnut-coloured. Plants 40–50 cm, mainly on Mt. Gilboa.
                                                                    8. I. haynei
- Perianth smaller, about as long as broad, more or less uniform in colour, brown to nearly black or dark purple, sometimes with very dense but distinct veins and dots. Plants 20–30 cm, of N. Negev, Judean Desert, E. Samaria, Beit Shean Valley, S. Golan and Gilead.   **9. I. atrofusca**
14(12). Plants of sandy soils of coastal plain [kurkar (calcareous sandstone), sand and sandy loam]. Rhizome thin, stoloniferous. Falls with a beard of yellow dark-tipped hairs on a yellow ground; claw of falls yellow on outside; perianth mostly chestnut-brown.   **12. I. atropurpurea**
- Plants of C. Negev, Ammon, Moav and Edom. Beard on a cream-coloured or dark ground, with hairs varying in colour; claw of falls not yellow on outside   15

15. Plants of fallow fields and steppe habitats of Ammon and Moav, usually 30–40 cm. Leaves 8–9 mm broad. Rhizome fairly thick, compact, not stoloniferous. Perianth dark claret-brown to nearly black.   **10. I. nigricans**
  - Desert plants of Edom and C. Negev, usually 20–25 cm. Leaves 3–6 mm broad, strongly falcate. Rhizome thin, stoloniferous. Perianth smaller than in *I. nigricans*, chestnut-brown or dark purple, often varying in colour.   **11. I. petrana**
16(11). Standards with small purple dots (especially near margin) in addition to fine lilac or purple veins. Falls with brown to red-brown oblong spots reaching 2–3 mm; margin of falls turned upwards. Rhizome provided with long slender stolons.   **6. I. bismarckiana**
  - Standards veined but devoid or nearly devoid of dots. Margin of falls not turned upwards. Rhizome not stoloniferous   17
17. Scape much longer than leaves. Falls with purple-brown veins and oblong purple-brown spots.   **5. I. hermona**
  - Scape about as long as leaves. Falls with purple or reddish veins and minute dots.
                                                                                      7. I. lortetii
18(9). Flowers white. Scape 40–80 cm.   **4. I. albicans**
  - Flowers pale violet. Scape 100–120 cm, much longer than leaves.
                                                                                  3. I. mesopotamica

Subgen. 1. IRIS. Plants with rhizome. Leaves equitant, flat, ensiform or linear.

Sect. 1. LIMNIRIS Tausch, Hort. Canalius 1 (1823). Stem often branched, with branches 2- to several-flowered, or unbranched, 1-flowered. Falls not bearded. Seeds without aril.

1. Iris pseudacorus L., Sp. Pl. 38 (1753); Wendelbo & Mathew, Fl. Iran. 112:18; Boiss., Fl. 5:127; Post, Fl. 2:594. [Plate 156]

Plant 70–150 cm or more. Rhizome very thick, branched. Leaves several, ensiform, acute, the basal nearly as long as stem, 2–4 cm broad; cauline leaves gradually passing into bracts. Stem erect, branched in its upper half; branches 2- to several-flowered. Spathe 6–7.5 cm. Flowers yellow, pedicellate. Tube short. Falls 5–7.5 cm, somewhat reflexed, with a large orange spot in middle and with purple veins, without a beard or crest, abruptly narrowed into a claw which is shorter than limb. Standards erect, linear-lanceolate, 1.5–3 × 0.2–0.3 cm, smaller than falls and narrower than style-branches. Fl. April–May.

  Hab.: In water, ditches, marshes. Hula Plain.
  Area: Euro-Siberian, Mediterranean and Irano-Turanian.

  Protected by law.

2. Iris grant-duffii Baker, Handb. Irid. 7 (1892); Bot. Mag. 124:t.7604 (1898); Wendelbo & Mathew, Fl. Iran. 112:21; Mout., Nouv. Fl. Liban Syrie 1:312; Post, Fl. 2:593. [Plate 157]

Plant (incl. leaves) up to 60 cm, growing in tufts. Rhizome compact, creeping; base of each shoot with very prickly fibres formed from old leaf-sheaths. Leaves 40–50(–60) × 0.3–0.5 cm, mostly longer than scape, about 6 to a shoot, flat, linear, erect, straight, strongly veined. Scape 25–40 cm, 1-flowered, unbranched. Spathe 7.5–9 cm, greenish with a papery margin. Flower greenish-yellow. Tube about 7 mm. Falls 6–7 cm, nearly panduriform; claw veined with purple or lilac on a yellowish ground; limb shorter than claw, obovate, marked with an orange-yellow stripe. Standards 5 cm, erect, oblanceolate, narrower than falls. Style-branches 4.5 × 0.8 cm, narrowly obovate, 2-lobed; lobes 1.5 cm. Fl. January–February(–March).

Hab.: Marshes. Acco Plain, Sharon Plain; Lower Galilee, Esdraelon Plain, Samaria; Upper Jordan Valley; Golan.

Area: E. Mediterranean (Palestine, Syria).

Protected by law.

Sect. 2. IRIS. Sect. *Pogoniris* (Spach) Baker, Gard. Chron. ser. 2, 1876:647 (1876). Stem often branched; branches usually with more than one flower. Falls with a beard of multicellular hairs along middle line. Seeds without an aril.

3. Iris mesopotamica Dykes, Genus Iris 176 (1913); Post, Fl. 2:601; Mout., Nouv. Fl. Liban Syrie 1:314, t.105 f.2; G. Schutz, Irises in Israel, Atti 1° Simpos. Intern. Iris, Firenze 231 (1963).*

Plant 100–120 cm, dormant in summer. Rhizome very large, creeping. Leaves 45–60 × 3–5 cm, ensiform. Inflorescence branched; branches 2–3-flowered. Spathe 5–6 cm, green at base, scarious in the upper half; pedicels very short. Flowers fragrant. Falls 9 × 4 cm, recurved, with a narrow beard of light yellow hairs; limb obovate, violet; base of limb and claw with bronze veins on a white ground. Standards erect-incurved, lilac, paler than falls, obovate, 8.5 × 4.5 cm. 2n=48. Fl. May.

Hab.: Cultivated, especially in cemeteries, and possibly wild or escaped from cultivation in several places in N. Palestine (Upper Galilee, Golan) on fallow fields among hills, sometimes in rocky places.

Area: Probably E. Mediterranean.

Boissier (Fl. 5:138) records *I. pallida* Lam. from Crete (Smith), Rhodes (Bourgeau), Syria near Tripoli (Blanche) and from N. Palestine (Hayne ex Baker). Dinsmore in Post (Fl. 2:602) cites Boissier's data for Tripoli and adds a record by Nábělek (1926) from above Banias (Golan). In addition, Dinsmore in Post (*op. cit.* 601) records *I. mesopotamica* Dykes (1913) as "cultivated everywhere in Palestine and Syria, especially in cemeteries, and apparently wild in many places".

According to Mouterde (*loc. cit.*), *I. pallida* Lam., very common in gardens in Europe and characterized by its spathe which is entirely scarious even before flowering, is absent from our region. Instead, he records *I. mesopotamica* for Syria (includ-

* Not illustrated.

ing Banias) "probablement hors culture", as well as frequently cultivated in cemeteries and gardens, but "sûrement originaire des régions côtières asiatiques de la
Méditerranée".

Recently, after a very rainy season, large stands of *I. mesopotamica* in flower were
found and photographed by Hadass and Yair Parag in fallow fields near Gush Halav
in Upper Galilee (near Mt. Meiron).

4. Iris albicans Lange, Vid. Meddel. Dansk Naturh. Foren. Kjøbenhavn
1860:76 (1861); Webb & Chater, in Fl. Eur. 5:91; G. Schutz, Irises in Israel,
Atti 1° Simpos. Intern. Iris, Firenze 230 (1963); Mout., Nouv. Fl. Intern. Iris,
Firenze 230 (1963); Mout., Nouv. Fl. Liban Syrie 1:314; Post, Fl. 2:602.*

Plant 40–80 cm, not dormant in summer. Rhizome thick, creeping, branched. Leaves 30–45 × 3–5 cm, ensiform, persisting the year round. Inflorescence
shortly branched; branches 1–2(–3)-flowered. Spathe-valves 3–5 cm. Flowers
nearly sessile, white, translucent. Tube 2.5–3 cm. Falls 6–8 × 3–4 cm, with a
narrow beard of white yellow-tipped hairs; limb obovate, white; claw shorter
than limb, pale green at back and with pale greenish-yellow veins at sides.
Standards erect-incurved, oblong-obovate, nearly as long as falls, pure white,
with a short claw pale green at back. Lobes of style-branches longer than
broad, acute to acuminate. Capsules generally empty of seeds. 2n=44. Fl.
March–April.

Cultivated in Moslem cemeteries in the Mediterranean, especially in the E.
Mediterranean (Palestine, Lebanon, Syria, Cyprus).

"It is doubtful whether their native land is Yemen, as reported in literature" (G.
Schutz, *loc. cit.*).

Sect. 3. ONCOCYCLUS (Siemss.) Baker, Gard. Chron. ser. 2, 5:787 (1876).
Plants with a rhizome, usually growing in clumps. Stem 1-flowered. Falls
with a dark patch *(signal)* at centre and a broad beard of multicellular hairs
on claw. Standards glabrous (rarely with a few hairs on claw). Seeds black
with a white aril. Leaves ensiform.

According to Avishai (Ph.D. Thesis, 1977), *Oncocyclus* irises can be divided into
several aggregates; the Palestinian species belong to three aggregates:

(1) *I. hermona* and *I. bismarckiana* belong to the *Iberica* aggregate. In these species
hairs of beard are covered all over with rather long papillae.

(2) *I. lortetii* belongs to the *Lortetii* aggregate in which papillae are short and borne
on the upper part of hairs only.

(3) *I. haynei, I. atrofusca, I. nigricans, I. petrana, I. atropurpurea* and *I. mariae*
belong to the *Haynei* aggregate. Hairs of beard in species of this aggregate are clavate
and devoid of papillae.

Colour variants deficient in anthocyanin, mostly with yellow (more rarely with
nearly white) flowers, are found in nearly all local *Oncocyclus* species, especially in
those of the *Haynei* aggregate.

* Not illustrated.

5. Iris hermona Dinsmore in Post, Fl. 2:596 (1933); Dinsmore, Pl. Post. Dinsm., Publ. Amer. Univ. Beirut, Nat. Sci. Ser., no. 3, fasc. 2:8 (1934). [Plate 158]

Plant 30–50 cm. Rhizome rather stout, without stolons. Stem 1-flowered. Leaves 6–8, 1–1.5 cm broad, ensiform, erect or somewhat recurved, partly ever-green (on slopes of Mt. Hermon), appearing very early, much shorter than scape and not reaching flower, narrower than in *I. bismarckiana*, glaucous. Spathe 10 cm, often purplish. Flower bicoloured. Falls 6.5–8.5 × 4.5–6.5 cm, obovate, recurved, with purple veins and oblong purple-brown spots on a cream-coloured ground; signal small, orbicular, almost black; claw short and broad; hairs of beard dark purple on a light ground. Standards 6.5–8.5 × 5.5–7.5 cm, suborbicular, erect-incurved, with fine purple veins on a white to pale lilac ground and nearly without dots. Lobes of style-branches brown-spotted like falls; style-branches purplish-brown. Seeds numerous. 2n=20. Fl. April–May.

Hab.: Heavy basalt soil and dark brown soil derived from hard limestone; edges of *Quercus ithaburensis* and *Q. calliprinos* maquis, batha. C. and N. Golan, Mt. Hermon. Often abundant locally.

Area: E. Mediterranean. Endemic in N. Palestine and S. Syria.

Protected by law.

6. Iris bismarckiana Regel, Wiener Ill. Gartenzeit. 1892:855, f.72 (1892); Baker, Gard. Chron. ser. 3, 1893:506 (1893); Wright, Bot. Mag. 130:t.7986 (1904); Post, Fl. 2:592. *I. nazarena* (Foster) Dinsmore in Post, Fl. 2:596 (1933); Dinsmore, Pl. Post. Dinsm., Publ. Amer. Univ. Beirut, Nat. Sci. Ser., no. 3, fasc. 2:8 (1934). *I. sari* Schott var. *nazarena* Foster, The Garden 43:133(1893). [Plate 159]

Plant 30–40 cm. Rhizome short, oblique, with long slender stolons. Stem 1-flowered. Leaves 6–7, 2–3 cm broad, broadly ensiform, more or less erect, reaching flower, glaucous. Spathe 7.5 cm. Flower about 9 cm (excl. tube and ovary), bicoloured. Falls 6 × 4 cm, orbicular-cuneate to obovate-cuneate, re-curved, marked with oblong reddish-brown spots and veins on a cream-coloured ground; margin curved upwards in front; signal large, more or less circular, blackish-purple; hairs of beard dark purple. Standards 7–9 cm, nearly orbicular to obovate-unguiculate, erect-incurved, slightly serrate-crenate, cream, bluish or pale lilac with fine lilac to pale purple veins and scattered dots or spots; claw dark red-brown. Style-branches marked with red-brown on a straw-coloured ground. Seeds often few. 2n=20. Fl. March–April.

Hab.: Batha, stony ground, terra rossa, basalt soil, greyish-white rendzina on soft calcareous Senonian and Eocene rocks. Upper and Lower Galilee, Mt. Tabor, Giv'at Hamore; W. Hula Plain; slopes of Mt. Hermon (up to 1,300 m).

Area: E. Mediterranean. Endemic in N. Palestine and S. Lebanon.

Protected by law.

In Upper Galilee *I. bismarckiana* is sometimes found in close vicinity to *I. lortetii* on terra rossa.

As noted by Mouterde *(loc. cit.)*, Charles Sprenger (Gard. Chron. ser. 3, 1904, 2:222, 1904) revealed that *I. bismarckiana* Regel and *I. sari* var. *nazarena* Foster originated from the same Palestinian source and can be regarded as synonymous, the binomial *I. bismarckiana* having priority.

7. Iris lortetii W. Barbey in C. & W. Barbey, Herb. Levant 178, t.7 (1882); Baker, Bot. Mag. 118:t.7251 (1892); Mout., Nouv. Fl. Liban Syrie 1:319; Boiss., Fl. 5:131; Post, Fl. 2:597. *I. samariae* Dinsmore in Post, Fl. 2:598 (1933). [Plate 160]

Plant 30–50 cm. Rhizome stout, short, without stolons. Stem 1-flowered. Leaves 8, 1–2 cm broad, ensiform, erect, nearly reaching base of spathe or of flower, glaucous. Spathe longer than perianth-tube, 7.5–10 cm. Flower bicoloured. Falls 5–8 × 4.5 cm, oblong-obovate, recurved, with numerous lilac to pink dots and veins on a pale cream or yellowish ground; signal rather small, brown-purple or red; beard with sparse brown hairs on a pale ground. Standards 9–11 × 7–8 cm, orbicular with a distinct claw, erect-incurved, light-coloured, with fine lilac veins and devoid of dots; margin irregularly crenate-dentate. 2n=20. Fl. April–May.

Hab.: Batha on stony terra rossa. Upper Galilee, Giv'at Hamore, Mt. Gilboa, Samaria; Dan Valley, Hula Plain.

Area: E. Mediterranean. Endemic in N. Palestine and S. Lebanon.

Protected by law.

Var. **lortetii.** Standards with pale lilac veins on a nearly white ground.
Var. **samariae** (Dinsmore) Feinbr. comb. & stat. nov. Standards with purple to brownish veins on a cream-coloured ground.

I. samariae Dinsmore has been collected in Samaria near the town of Schekhem (Nablus) by Dinsmore and later by T. Kushnir. According to Dinsmore in Post, Fl. 2:598, its flower resembles the flower of *I. lortetii*, but is larger, 13 cm in diameter, and the falls are wider, not strongly reflexed, 4–5 × 6–7 cm, with purple lines and larger purple dots on a cream-coloured ground. It flowers in April, earlier than *I. lortetii* which flowers in May.

In our view, *I. samariae* cannot be maintained as a species. It can at most be distinguished as a variety of *I. lortetii*, mainly by the cream-coloured (not white) ground colour of standards and the purple to brownish (not lilac) colour of their veins. Some weight can be given to the somewhat earlier flowering time of the Samaria plants.

8. Iris haynei Baker, Gard. Chron. ser. 2, 1876:710 (1876); Boiss., Fl. 5:138; Post, Fl. 2:601. *I. biggeri* Dinsmore in Post, Fl. 2:599 (1933). [Plate 161]

Plant 40–50 cm, growing in clumps. Rhizome stout. Stem 1-flowered. Leaves 5–8, 1–1.5 cm broad, ensiform, erect, reaching spathe. Spathe 7–8.5 cm. Flowers fragrant, 10–12 cm (excl. tube and ovary), distinctly longer than broad, mostly purple. Falls 7–8 × 4–6 cm, oblong-ovate, recurved from about middle, dark purple to brownish-purple, very densely dotted and veined; signal blackish dark-purple; beard on a yellowish or white ground, of hairs vary-

ing in colour (dark purple, white or dark-tipped yellow). Standards 9–10 ×
6–7 cm, erect-incurved, obovate, purple-veined and densely purple-dotted on a
pale ground. 2n=20. Fl. March–April.

Hab.: Terra rossa; field-borders. Mt. Gilboa, N. E. Samaria.
Area: Endemic.

Protected by law.

A variant deficient in anthocyanin, with a golden-yellow flower and a wine-red
signal on the falls, has been found growing in populations of typical plants (Tuvia
Kushnir and others).

9. Iris atrofusca Baker, Gard. Chron. ser. 3, 1893:384 (1893); Post, Fl. 2:599. *I.
loessicola* T. Kushnir, Palest. Jour. Bot. Jerusalem ser., 4:231 (1949) *nom.
nud. I. jordana* Dinsmore in Post, Fl. 2:598 (1933); Dinsmore, Pl. Post.
Dinsm., Publ. Amer. Univ. Beirut, Nat. Sci. Ser., no. 3, fasc. 2:9 (1934). *I.
atropurpurea* Baker var. *gileadensis* Dinsmore in Post, Fl. 2:601 (1933);
Dinsmore, *op. cit.*:11 (1934). *I. hauranensis* Dinsmore in Post, Fl. 2:598
(1933); Dinsmore, *op. cit.*:10 (1934); Mout., Nouv. Fl. Liban Syrie 1:316.
[Plate 162]

Plant 20–30 cm. Rhizome stout, compact. Stem 1-flowered. Leaves 5–8, 1 cm
broad, ensiform, glaucous, erect or falcate, reaching and somewhat exceeding
base of spathe. Spathe 7–8 cm. Flowers fragrant, purplish-brown to dark purple. Falls 6–7.5 × 3–4.5 cm, obovate-cuneate, recurved from about middle, dark
brown-purple to nearly black; signal broad, brown-black; hairs of beard yellow, brown-tipped on a cream-coloured ground. Standards 7–9 × 4.5–7 cm,
erect-incurved, obovate-cuneate to suborbicular-cuneate, with dark purplish-
brown to dark purple radiating veins and dense dots. 2n=20 (T. Kushnir). Fl.
April.

Hab.: Loess and calcareous hills; batha with *Phlomis brachyodon, Noaea
mucronata, Echinops polyceras, Eremostachys laciniata*. N. Negev, Judean
Desert, E. Samaria, Beit Shean Valley; S. Golan, Gilead.

Area: W. Irano-Turanian. Endemic.

Protected by law.

Yellow-flowered plants appear fairly often, single or in small groups, in natural
populations of typical *I. atrofusca*.

The area of *I. atrofusca* extends from the N. Negev eastwards into the Judean Desert
and then northwards along E. Samaria, the Beit Shean Valley, the Valley of the river
Yarmook and further north into S. Golan. In the northern part of its distribution area
I. atrofusca becomes more variable in colour of perianth, its flower becoming more
purple.

Plants with more purple and less brown flower colour, especially on the standards,
appear in the Beit Shean Valley and even in the northern part of E. Samaria. They
approach what used to be called *I. jordana*, and is regarded here as a colour variant of
I. atrofusca. Standards are dark purple-veined on a paler ground, a feature found in *I.
haynei*. This feature may have introgressed into *I. atrofusca* from *I. haynei*. However,
the characteristic shape of the flower of *I. atrofusca* is maintained throughout its

whole area, i. e., flowers are about as long as broad, not longer than broad, as in *I. haynei*.

One may assume that *I. atrofusca* has hybridized with *I. haynei* in the part of its area lying S. and E. of Mt. Gilboa where *I. haynei* predominates.

10. Iris nigricans Dinsmore in Post, Fl. 2:600 (1933); Dinsmore, Pl. Post. Dinsm., Publ. Amer. Univ. Beirut, Nat. Sci. Ser., no. 3, fasc. 2:11 (1934) Latin descr. [Plate 163]

Plant 30–40 cm. Rhizome stout, compact. Stem 1-flowered. Leaves 6–13, 8–9 mm broad, much shorter than scape and not reaching spathe, more or less falcate. Spathe about 7 cm. Flower 9–10 cm (excl. tube and ovary), dark brown-purple. Falls 6–7 × 3–4.5 cm, strongly recurved from middle, oblong-ovate, dark brown-purple; signal black; beard of dark purple hairs on a cream-coloured to pale yellow ground. Standards 9–10 × 6–7 cm, obovate-unguiculate, erect-incurved, slightly lighter in colour than falls, with dark purple to dark violet veins and very numerous and mostly coalescing dots on a paler ground. Style-branches light chocolate-coloured. 2n=20. Fl. April.

Hab.: Fallow fields and steppe habitats. Ammon (es-Salt), Moav (Madaba, Kerak, Um-el-Ammud).

Area: W. Irano-Turanian. Endemic.

11. Iris petrana Dinsmore in Post, Fl. 2:599 (1933); Dinsmore, Pl. Post. Dinsm., Publ. Amer. Univ. Beirut, Nat. Sci. Ser., no. 3, fasc. 2:10 (1934) Latin descr.; Mountfort, Portrait of a Desert t.59, London, 1965. [Plate 164]

Plant 20–25 cm. Rhizome thin, stoloniferous. Stem 1-flowered. Leaves 6–10, 3–6 mm broad, ensiform, glaucous, strongly falcate, hardly ½ length of scape. Spathe 7.5 cm. Flower 5–6 cm (excl. tube and ovary), mostly dark brown, sometimes varying in colour (dark purple, violet, rarely standards much lighter than falls). Falls 3.5–5.5 × 2.5–3 cm, recurved from middle, oblong-ovate, generally uniform in colour; signal nearly black; hairs of beard mostly yellow, dark-tipped on a cream-coloured ground. Standards 5.5–6 × 3.5–4 cm, obovate-unguiculate, erect-incurved, nearly uniform in colour or with darker veins. 2n=20. Fl. April.

Hab.: Desert plains, gravelly Neogene sand with *Anabasis articulata* and *Retama raetam*, and low calcareous hills with *Artemisia herba-alba* and *Anabasis syriaca*. C. Negev (locally abundant, mainly between Yeroham and Dimona) and Edom (from Shawbak southwards).

Area: W. Irano-Turanian. Endemic.

Protected by law.

12. Iris atropurpurea Baker, Gard. Chron. ser. 3, 1889:330 (1889); Post, Fl. 2:600. [Plate 165]

Plant 25–35 cm, growing in clumps. Rhizome stoloniferous; stolons less than 1 cm in thickness. Stem 1-flowered. Leaves 7–11, 5–8 mm broad, ensiform, glaucous, falcate, not reaching base of spathe and often only about ½ as long as scape. Spathe up to 7 cm. Tube somewhat longer than spathe. Perianth generally chestnut-brown to dark purple, varying from reddish to nearly black-purple. Falls 3.5–6 × 2.5–4 cm, oblong-ovate, recurved, uniform in colour; signal semi-circular, black, velvety; hairs of beard yellow, dark-tipped on a yellow ground. Standards 5.5–8 × 4.5–6 cm, suborbicular-unguiculate, erect-incurved, similar in colour to falls, often slightly lighter, with inconspicuous dark veins and devoid of dots. 2n=20. Fl. January–April.

Hab.: Kurkar (calcareous sandstone) hills and sandy loam. Often in large stands. Sharon Plain, Philistean Plain.

Area: Mediterranean. Endemic.

Protected by law. In danger of extinction, owing to urbanization of the coastal plain. Maintained in several small nature reserves established in the seventies.

A yellow-flowered variant which lacks anthocyanin is met with occasionally in populations of typical plants.

P.H. Davis (*op. cit.* 94, 1946) remarks about *I. atropurpurea*: "...there can be little doubt that this microspecies originated quite recently from related Irano-Turanian species".

13. Iris mariae W. Barbey, Österr. Bot. Zeitschr. 41:207 (1891) & Gartenfl. 42:488, f.106 (1893). *I.·helenae* W. Barbey* in C. & W. Barbey, Herb. Levant 159 (1882); Baker, Handb. Irid. 19; Boiss., Fl. 5:132; Post, Fl. 2:601; Täckholm & Drar, Fl. Eg. 3:461. [Plate 166]

Plant 20–30 cm, growing in clumps. Rhizome stoloniferous; stolons 3–4 cm deep; roots reddish. Stem 1-flowered. Leaves 7–8, 4–6 mm broad, ensiform, strongly falcate, glaucous, ⅓–½ length of scape. Spathe longer than tube of perianth. Flower lilac to violet, about 7 cm (excl. tube and ovary). Falls 5 × 2.5–3 cm, oblong-ovate, recurved, darker than standards, violet, with a dark velvety signal; hairs of beard nearly black. Standards 6–6.5 × 4 cm, lilac, erect-incurved, obovate with a short claw. 2n=20. Fl. February–March.

Hab.: Loose sand, usually Artemisietum monospermae; also Noaeetum mucronatae arenarium. Fairly common locally. W. and N. Negev.

Area: Saharo-Sindian (S. Palestine, Sinai).

One of the most beautiful *Oncocyclus* Irises. Protected by law.

Subgen. 2. HERMADACTYLOIDES Spach, Ann. Sci. Nat. Bot. ser. 3, 5:91 (1846). Plants with bulb consisting of a single fleshy scale; outer tunics reticulate-fibrous. Leaves basal, linear, 4-gonous apiculate. Stem short, 1- to few-flowered.

* *I. helenae* was discovered and described by W. Barbey who dedicated it to his mother, Hélène-Marie. Later, when it was pointed out to him that C. Koch previously described another species under the name *I. helena*, Barbey changed the name of his species into *I. mariae*.

14. Iris histrio Reichenb. fil., Bot. Zeit. 30:488 (1872); Hook. fil., Bot. Mag. 99:t.6033 (1873); Boiss., Fl. 5:121; Post, Fl. 2:589. [Plate 167]

Plant 15–20 cm. Bulb oblong-ovoid, 1.5 cm in diameter, with small bulblets; tunics reticulate, with longitudinal fibres connected by oblique ones; roots fibrous. Leaves 1–2(–3), 20–40 cm, appearing some time before flowers, erect, linear, 4-gonal, grooved along sides, acute, about twice as long as scape. Scape 1-flowered. Flowers shortly pedicellate. Spathe narrow, acuminate. Hypanthial tube exserted from spathe at apex, blue. Perianth blue. Falls (4.5–)5–6 cm; limb reflexed, oblong, with a golden-yellow median area, with bluish veins radiating towards margin and with scattered deep-violet spots; claw somewhat longer than limb. Standards erect, oblanceolate, smaller than falls. Lobes of style-branches lanceolate, often crenate on outer margin. 2n=20. Fl. January–February.

Hab.: Rocky places in maquis. Fairly rare. Upper Galilee; Golan.

Area: E. Mediterranean (Palestine, Lebanon, Syria, S. Turkey).

15. Iris vartanii Foster, Gard. Chron. ser. 2, 23:438 (1885); Baker, Bot. Mag. 113:t.6942 (1887); Post, Fl. 2:590. [Plate 168]

Plant 15–20 cm, tufted. Bulb ovoid to oblong-ovoid, 1–2 cm in diameter, with numerous small bulblets; tunics pale brown, reticulate, with longitudinal fibres connected by oblique ones; roots fibrous. Leaves 1–2, 20–30 cm, appearing with or sometimes before flowers, linear, 4-gonal, acute, longer than scape. Scape 1-flowered. Flower shortly pedicellate. Spathe 3.5–5 cm, with lanceolate valves. Hypanthial tube much exserted from spathe. Perianth pale slaty-blue to -lilac, varying in intensity of colour, rarely white (var. *alba* Dykes, Gard. Chron. 1911). Falls (3.5–)4–4.5(–5) cm, with a median longitudinal somewhat elevated yellow stripe; limb reflexed, oblong, 1.5 cm broad, marked with radiating veins darker than ground-colour, much shorter and broader than claw; claw green-veined on the outside. Standards erect, somewhat shorter than falls, oblanceolate. Lobes of style-branches 2 cm, oblong, acuminate, veined. 2n=20. Fl. December–February.

Hab.: Rocky batha among stones; *Sarcopoterium spinosum–Phlomis brachyodon* association. Fairly rare. Upper and Lower Galilee, Mt. Carmel, E. Samaria, Judean Mts.; Golan.

Area: E. Mediterranean. Endemic in Palestine and S. Syria.

Subgen. 3. SCORPIRIS Spach, Ann. Sci. Nat. Bot. ser. 3, 5:91 (1846). Subgen. *Juno* (Tratt. ex Roem. & Schult.) Baker, Handb. Irid. 2 (1892). Plants with a bulb consisting of several fleshy more or less entire scales; outer tunics membranous; roots more or less fleshy to swollen. Leaves distichous, canaliculate, mostly falcate-recurved. Falls with a median longitudinal somewhat elevated yellow stripe, sometimes developed as a crest, never bearded; standards reduced in size, mostly spreading or reflexed.

16. Iris palaestina (Baker) Boiss., Fl. 5:122 (1882); Mout., Nouv. Fl. Liban Syrie 1:310, t.101 f.4; Post, Fl. 2:591. *Xiphion palaestinum* Baker, Jour. Bot. (London) 9:108 (1871). [Plate 169]

Plant 10–20 cm. Bulb ovoid, 2–3 cm in diameter; outer tunics membranous, light brown, prolonged beyond neck; inner tunics fleshy, annual; storage roots fleshy, about 8 mm thick. Stem very short, 1–3-flowered. Leaves 5–6 in flower-bearing plants, up to 7 in young plants, 8–10(–15) × 1–2 cm, light green, glossy on upper face, lanceolate, acuminate, strongly falcate-recurved, plicate-canaliculate, many-veined, often undulate; margin white, ciliolate. Spathe 1-flowered, membranous, yellowish, cuspidate. Flowers yellowish-white, greenish or pale sky-blue, sweet-smelling. Hypanthial tube subex serted. Falls 4.5–5.5 cm; claw about 3–4 cm, erecto-patent, obovate, winged, with wings turned upwards and usually bluish-veined on outside; limb 1–1.5 cm, recurved, ovate, marked with radiating greyish-blue oblong and broad veins or spots with a median yellow, usually purple-dotted stripe, slightly elevated, not crested. Standards 1.5–2 × 0.3 cm, spreading or reflexed, linear-canaliculate, broadened in upper part. Style-branches as long as falls, 2-lobed; lobes oblong, acuminate, longer than broad, entire. Capsules usually buried in soil. 2n=24. Fl. January–February.

Hab.: Batha. Coastal Galilee, Acco Plain, Coast of Carmel, Sharon Plain, Philistean Plain; Upper and Lower Galilee, Mt. Carmel, Esdraelon Plain, Samaria, Shefela, Judean Mts.; Dan Valley; Golan, Gilead.

Area: E. Mediterranean (Palestine, Lebanon, S. Turkey).

17. Iris edomensis Sealy, Kew Bull. 1949:561 (1950). [Plate 170]

Plant 13–14 cm. Bulb ovoid, 2–3.5 cm in diameter; outer tunics greyish-brown, membranous. Stem very short. Leaves 4–5, 10–15.5 × 0.5–1.5 cm, falcate, linear-lanceolate, canaliculate, undulate, green on upper, glaucous on lower face, veined; margin white, conspicuously thickened. Flowers 1–2, sessile. Spathe-valves oblong, acuminate, 5–6 cm, green during anthesis. Hypanthial tube 5–6.5 cm. Falls marked with a median yellow purple-dotted stripe and with numerous oblong purple spots and dots on a white ground; claw 3–3.5 cm, broadly cuneate; limb 3 cm, ovate, recurved. Standards 2–2.5 cm, whitish with greenish veins, linear-canaliculate, becoming broadened in upper part. Style-branches 2-lobed; lobes oblong, acuminate. Fl. January–February.

Hab.: Limestone and sandstone hills; Artemisietum herbae-albae. Edom (around Petra).

Area: W. Irano-Turanian. Endemic in S. Transjordan.

18. Iris aucheri (Baker) Sealy, Kew Bull. 1949:562 (1950); Wendelbo & Mathew, Fl. Iran. 112:50, t.6 f.4 & t.18 f.2 (1975). *Xiphion aucheri* Baker, Jour. Bot. (London) 9:110 (1871). *Juno aucheri* (Baker) Klatt, Bot. Zeit. 30:498 (1872). *I. fumosa* Boiss. & Hausskn. ex Boiss., Fl. 5:123 (1882). *I. sindjarensis*

Boiss. & Hausskn. ex Boiss., Fl. 5:122 (1882); Baker, Bot. Mag. 116:t.7145 (1890). *I. moabitica* Dinsmore in herb.*

Plant 15–30 cm. Bulb oblong-ovoid; tunics membranous, long-extended above neck; storage roots long, fleshy but not swollen. Stem completely concealed by leaves at flowering stage, 3–6-flowered. Leaves in flower-bearing plants 6–10, up to 25 cm, 2.5–4.5 cm broad at base, falcate, lanceolate, acute, plicate-canaliculate, prominently many-veined, glossy green above, without white margin; upper leaves spathe-like; in plants not flowering in current season leaves 3–5, 3–5 mm broad. Spathe-valves 7–9 cm, long-acuminate. Flowers slaty-lilac, -blue or nearly white. Tube 6 cm. Falls 4.5–5 cm; claw 3–3.5 cm, obovate, with wings turned sharply upwards; limb orbicular, 1.5 cm, with a well-raised 1 cm erose-crenulate pale yellow crest. Standards spreading or slightly reflexed, 2.5 × 1 cm, obovate with a narrow claw. Style-branches with 1 cm obtuse lobes. Fl. January–February.

Hab.: Stony hills. Moav, Edom.

Area: W. Irano-Turanian (S. Transjordan, N. Syria, S. E. Turkey, N. Iraq, W. Iran).

19. Iris regis-uzziae Feinbr., Notes Roy. Bot. Gard. Edinb. 37:75 (1978). [Plate 171]

Plant 7–15 cm. Bulb oblong-ovoid, 3–4 × 2–3.5 cm; tunics greyish-brown, membranous, prolonged above neck; roots fleshy. Stem short, 1–2(–3)-flowered. Leaves 5–6(–7) in flower-bearing plants, falcate, lanceolate, acute, plicate-canaliculate, many-veined; margin white, conspicuously thickened; lowermost leaf 2–4 cm broad; sheath of uppermost leaf not dilated; in plants not flowering in current season leaves 3, less than 1 cm broad. Spathe-valves pale green, acute. Flowers pale sky-blue, lilac or pale lilac to nearly white, more or less translucent. Falls 4–5 cm; claw winged, 2.5 cm broad, with wings turned upwards and with a median yolk-yellow stripe; limb 1–1.2 cm broad, rounded at tip, with a yolk-yellow crest at centre rising from a yolk-yellow area which is surrounded by a white margin. Standards 2–2.5 × 0.6–0.8 cm, spreading or reflexed, spathulate, often truncate and apiculate at apex. Style-branches 2-lobed; lobes ovate-rounded, obtuse. Capsule 5–7 cm, ellipsoid. Seeds subglobose, rugose, about 4 mm. 2n=22. Fl. January–February.

Hab.: Rocky slopes, Artemisietum herbae-albae and Zygophylletum dumosi, 500–1,000 m. C. Negev; C. Arava Valley; Moav, Edom.

Area: W. Irano-Turanian. Endemic.

3. ROMULEA Maratti

Perennials with corm; tunics coriaceous or membranous, smooth (rarely fibrous), brown. Cataphylls 1–2, sheathing corm and aerial shoot. Stem short. Leaves appearing before flowers, few, all basal, or some cauline; blades of

* Not illustrated.

basal leaves linear, semiterete, 4-grooved, erect or recurved; cauline leaves usually shorter. Inflorescence a single flower or a few flowers in a monochasium; flower sessile or subsessile, subtended by a bract and bracteole (2-valved spathe). Pedicels generally strongly recurved after flowering. Bract herbaceous, glabrous, narrowly membranous-margined; bracteole generally subequal to bract, membranous or with a broad membranous margin; membranous parts often red-dotted. Perianth actinomorphic, generally infundibular (rarely hypocrateriform), with a short hypanthial tube and 6 equal or subequal perianth-segments. Stamens 3, erect, inserted at throat of perianth-tube; filaments free; anthers linear, basifixed. Ovary small; style filiform, erect, with 3 deeply 2-fid branches, stigmas 6. Capsule 3-gonous, dehiscing loculicidally by 3 valves. Seeds numerous, nearly globose or angular, brown.

About 90 species in the Mediterranean and (the majority) in S. Africa.

Literature: A. Béguinot, Revisione monografica del genere *Romulea* Maratti, *Malpighia* 21:49-122, 364-478(1907); 22:377-469(1908); 23:55-117, 185-239, 275-296 (1909). Miriam P. de Vos, The genus *Romulea* in South Africa, *Jour. S. Afr. Bot.*, Suppl. Vol. 9:1-307 (1972).

1. At least lower ¼ of perianth yellow, orange or lemon-coloured on inside   2
- Perianth violet, only base yellow. Seeds minutely rugulose-tuberculate. Plant of maquis and forest from Mt. Carmel and northwards.   **2. R. phoenicia**
2. Tunics of corm membranous. Leaves short, hardly exceeding flowering stem. Plant of Mt. Hermon growing near melting snow.   **3. R. nivalis**
- Tunics coriaceous. Leaves much exceeding flowering stem   3
3. Perianth 2-3 cm, nearly twice as long as bract. Style exceeding anthers.
          1. R. bulbocodium
- Perianth not reaching 2 cm, slightly longer than bract. Style not exceeding anthers. Rare.   **4. R. columnae**

1. Romulea bulbocodium (L.) Seb. & Mauri, Fl. Rom. 17 (1818); Boiss., Fl. 5:115; Post, Fl. 2:587. *Crocus bulbocodium* L., Sp. Pl. 36 (1753). *Ixia bulbocodium* (L.) L., Sp. Pl. ed. 2, 51 (1762); Bot. Mag. 7:t.265 (1794). [Plate 172]

Plant 10-30 cm. Corm 1-1.2(-1.5) cm in diameter; tunics coriaceous. Leaves much longer than inflorescence, grooved on upper face, erect or recurved. Stem 1-6-flowered. Bract 1-2 cm, usually purplish. Perianth 2-3.5 cm, lilac or violet varying in intensity, rarely white; throat and lower ⅓-½ yellow; perianth-segments with 3-5 darker veins. Stamens shorter than style; filaments puberulent. Capsule 1-1.5 cm, oblong-obovoid. Seeds smooth, glossy. Fl. January–March.

Hab.: Batha. Coastal Galilee, Acco Plain, Coast of Carmel, Sharon Plain, ?Philistean Plain; Upper and Lower Galilee, Mt. Carmel, ?Samaria, Judean Mts.; Golan, Gilead.

Area: Mediterranean.

2. Romulea phoenicia Mout., Bull. Soc. Bot. Fr. 100:349 (1953) & Nouv. Fl. Liban Syrie 1:321, t.100 f.6. [Plate 173]

Plant 7–15 cm. Corm 0.8–1.2 cm in diameter; tunics coriaceous. Leaves much longer than inflorescence, usually recurved. Stem 1–3(–4)-flowered. Bract 1.3–1.8 cm; bracteole 1.2–1.7 cm, pale purple. Perianth (1.5–)2–3 cm, violet with 3–5 darker veins, yellow at base of throat. Stamens shorter than style; filaments glabrous. Capsule about 1 cm, oblong-obovoid. Seeds globose, minutely rugulose-tuberculate, not glossy. Fl. January–March(–April).

Hab.: Maquis, pine forest and clearings. Upper Galilee, Mt. Carmel; Golan.

Area: E. Mediterranean (Palestine, Lebanon).

R. phoenicia is close to *R. linaresii* Parl. of Sicily and Greece (incl. islands). According to Mouterde (1953), it differs from *R. linaresii* in the yellow base of its perianth and in its glabrous filaments.

3. Romulea nivalis (Boiss. & Ky.) Klatt, Abh. Naturf. Ges. Halle 12:400 (1882); Boiss., Fl. 5:116; Post, Fl. 2:587. *Trichonema nivale* Boiss. & Ky. in Boiss., Diagn. ser. 2, 4:92 (1859).*

Plant 6–10 cm. Corm 1 cm in diameter; tunics membranous. Leaves generally not exceeding 10 cm, at most slightly overtopping inflorescence during flowering. Stem 1–3-flowered. Bract 1.2 cm. Perianth 2–2.5 cm, about twice as long as bract, orange-yellow in the lower ⅓ or ½, lilac or violet with darker veins above. Stamens shorter than style. Capsule 6–8 mm, ellipsoid. Seeds angular, nearly smooth. Fl. April–May

Hab.: Near snow. Mt. Hermon.

Area: E. Mediterranean (Lebanon, Mt. Hermon, Antilebanon).

4. Romulea columnae Seb. & Mauri, Fl. Rom. 18 (1818); Boiss., Fl. 5:117; Post, Fl. 2:587. [Plate 174]

Plant 5–15(–20) cm. Corm 0.6–1 cm in diameter; tunics coriaceous. Leaves grooved on upper face, long-overtopping inflorescence, mostly recurved. Stem 1–3-flowered. Spathe 1–1.2 cm. Bract 6–13 mm; bracteole nearly entirely membranous. Perianth 1.2–1.5 cm, white or pale blue above throat, lemon-coloured or pale yellow and glabrous at throat, outside greenish suffused with yellow; perianth-segments with 3 darker veins. Stamens not exceeding style. Capsule 0.5–1 cm, ovoid. Seeds subtuberculate. Fl. February–March.

Hab.: Batha. Rare. Mt. Carmel, Samaria, Judean Mts.; Golan.

Area: Mediterranean, Atlantic coast of Europe, Macaronesia.

4. CROCUS L.

Perennials, with subterranean proliferous tunicate corms; tunics membranous, coriaceous or fibrous (of parallel or reticulate fibres); new corm formed every year on top of old one at base of scape. Foliage leaves and short flowering scape enclosed by 3–5 membranous tubular cataphylls. Leaves synanthous or hysteranthous, basal, linear, canaliculate and usually with a white mid-

* Not illustrated.

stripe on upper surface, keeled and 2-canaliculate on lower. Flowers one or several, each on a short subterranean pedicel; pedicel subtended in some species by a membranous sheathing prophyll ("*basal spathe*"). Base of flower surrounded by a tubular bract and usually an inner strap-shaped bracteole, forming a monophyllous or diphyllous "*floral spathe*" or "*proper spathe*". Perianth actinomorphic, infundibular, erect, with a long slender tube and 6 equal or subequal perianth-segments. Stamens 3, free, included; anthers usually longer than filaments, basifixed. Ovary subterranean; style divided above into 3 entire or dissected arms. Capsule chartaceous, oblong or ellipsoid, 3-gonous, dehiscing loculicidally at apex by 3 valves, borne on a lengthened slender pedicel at or above ground-level. Seeds numerous, globose or ovoid, often arillate.

About 90 species in the Euro-Siberian, Mediterranean and Irano-Turanian regions.

The dried style of *C. sativus* L. yields *saffron* used for colouring and flavouring various foods, dishes, etc.

Literature: G. Maw, *A Monograph of the Genus Crocus*, London, 1886. E. A. Bowles, *A Handbook of Crocus and Colchicum for Gardeners*, London, 1924, revised ed. 1952. P. Mouterde, Contribution à l'étude de la flore syrienne et libanaise (suite), *Bull. Soc. Bot. Fr.* 101:420–428 (1954). N. Feinbrun, The genus *Crocus* in Israel and neighbouring countries, *Kew Bull.* 12:269–285 (1957); Chromosome numbers in *Crocus, Genetica* 29:172–192 (1958). C. A. Brighton, B. Mathew & C. J. Marchant, Chromosome counts in the genus *Crocus* (Iridaceae), *Kew Bull.* 28:451–464 (1973). P. Wendelbo & B. Mathew, *Crocus*, in: K. H. Rechinger (ed.), *Flora iranica* 112:2–11, 1975. B. Mathew, *Crocus olivieri* and its allies (Iridaceae), *Kew Bull.* 31:201–208 (1976); *Crocus sativus* and its allies, *Pl. Syst. Evol.* 128:89–103 (1977). N. Feinbrun & A. Shmida, A new review of the genus *Crocus* in Israel and neighbouring countries, *Israel Jour. Bot.* 26:172–189 (1977). C. A. Brighton, Cytology of *Crocus sativus* and its allies, *Pl. Syst. Evol.* 128:137–157 (1977).

1. Perianth bright orange-coloured. Plants growing near northern boundary of country.   **9. C. vitellinus**
- Perianth not orange-coloured   2
2. Anthers white to cream-coloured. Perianth-segments white outside and inside (except for yellow throat).   **1. C. ochroleucus**
- Anthers yellow, orange or dark purple. Perianth-segments variously coloured or, if white inside, marked on outside with a purple or violet midvein or veins or feathering   3
3. Style-arms scarlet (rarely yellow), entire or nearly so (not dissected into branches), cuneate. Leaves (5–)6–17, 1 mm broad or narrower. Corm-tunics finely reticulate-fibrous, silky   4
- Style-arms orange or yellow; each style-arm dissected into several branches. Leaves 3–7, (1–)1.5–3 mm broad. Corm-tunics membranous or of coarse reticulate or parallel fibres   5
4. Style branched in throat of perianth below anther-bases; style-arms long. Perianth-segments 3–7 mm broad, narrowly oblanceolate, with purple-lilac stripes on a whitish ground.   **3. C. moabiticus**

− Style branched above bases of anthers. Perianth-segments (5−)10–12 mm broad, elliptic, often apiculate, lilac or light violet, with darker veins.   **2. C. pallasii**

5(3). Perianth white; outer or all segments with a purple midvein or feathering on outside, especially towards base (rarely outer segments suffused throughout outer surface with purple or violet). Filaments yellow or pale yellow   6

− Perianth lilac, light violet or purplish-pink, with darker veins. Filaments white   7

6. Anthers dark purple, very rarely yellow. Corm-tunics membranous, silky, split at base into narrow strips. Leaves 2–2.5 mm broad, with a broad white mid-stripe. Common, mainly in hill districts.   **8. C. hyemalis**

− Anthers yellow. Corm-tunics of dense parallel fibres dissociating at base. Leaves 1–1.5 mm broad, with a filiform white mid-stripe. Confined mainly to northern districts.   **7. C. aleppicus**

7(5). Corm-tunics reticulate, of fibres interconnected by anastomoses; interspaces rectangular or rhombic, exceeding the fibres in width   8

− Corm-tunics of parallel fibres interconnected by very short anastomoses; interspaces linear or oblong, not exceeding the fibres in width.   **6. C. hermoneus**

8. Ultimate style-branches generally exceeded by anthers; anthers strongly curved; style branching into 3 arms at or below level of anther-bases. Perianth-segments 5–8 mm broad, usually 3-veined near base. Leaves 3–5, 2–3 mm broad. Plants of C. Negev and Edom.   **5. C. damascenus**

− Ultimate style-branches ending at or above level of anther-tips; anthers straight; style branching into 3 arms considerably above anther-bases. Perianth-segments 8–10 mm broad, usually 5–7-veined near base. Leaves (4−)6–7, 1.5–2 mm broad. Plants of Golan and Mt. Hermon.   **4. C. cancellatus**

1. Crocus ochroleucus Boiss. & Gaill. in Boiss., Diagn. ser. 2, 4:93 (1859); Hook. fil., Bot. Mag. 88:t.5297 (1862); Maw, Monogr. 109, t.11; Feinbr., Kew Bull. 12:269; Feinbr. & Shmida, Israel Jour. Bot. 26:173; Boiss., Fl. 5:98; Post, Fl. 2:583. [Plate 175]

Corm small (1–1.7 cm in diameter), depressed-globose; tunics membranous, the outer subcoriaceous, split into strips at base, hardly prolonged above neck. Cataphylls membranous, white. Leaves 5–8, synanthous, 2 mm broad, with a broad white mid-stripe. Basal spathe present. Flowers several, autumnal; perianth white on inner and outer surface; segments (20−)25–30 × 6–10 mm, elliptic, obtuse; throat yellow, bearded. Filaments yellow, somewhat shorter than anthers; anthers white to cream-coloured; pollen white. Style orange or yellow, exceeding anthers, divided into 3 arms dilated towards apex, entire or nearly so. 2n=10. Fl. October–December.

Hab.: *Quercus calliprinos-Q. boissieri* maquis and among rocks, often on basalt soil. Upper Galilee; Golan; Mt. Hermon.

Area: E. Mediterranean (N. Palestine, Lebanon, Syria).

2. Crocus pallasii Goldb., Mém. Soc. Nat. Moscow 5:157 (1817); Mathew, Pl. Syst. Evol. 128:96 (1977); Brighton, *ibid.* 128:137–157; Prodan & Nyárády, in Fl. Romania 11:t.70 ff.4, 4a (1966). *C. sativus* var. *elwesii* Maw, Gard. Chron. ser. 2, 16:430 (1881); Monogr. t.29c (1886). *C. elwesii* (Maw) O. Schwarz, Feddes Repert. 36:74 (1934); Feinbr., Kew Bull. 12:272, f.1 (1957). *C. sativus*

var. *haussknechtii* Boiss. & Reut. ex Maw, Gard. Chron. ser. 2, 16:430 (1881).
C. haussknechtii (Boiss. & Reut. ex Maw) Boiss., Fl. 5:100 (1882). *C. olbanus*
Siehe, Allg. Bot. Zeitschr. 12:1 (1906); Feinbr. & Shmida, *op. cit.*174, f.1C
(1977). *C. thiébautii* Mout. & *C. libanoticus* Mout., Bull. Soc. Bot. Fr. 101:422
(1954). [Plate 176]*

Corm 1.5–2.5 cm in diameter, somewhat depressed; tunics numerous, silky,
finely reticulately fibrous with narrow interspaces, prolonged above neck for
2–3(–6–9) cm. Leaves numerous (6–17), appearing towards end of anthesis,
0.5–1 mm broad, scabridulous or smooth at margin. Basal spathe present.
Flowers several, autumnal; perianth lilac or light violet with darker veins;
throat white or lilac; segments 25–40 × (5–)10–12 mm, elliptic, often apiculate,
5-veined at base; tube white. Filaments white to pale yellow, short; anthers
12–17 mm, yellow. Style scarlet, rarely yellow, divided into 3 arms mostly
above bases of anthers; arms gradually dilated towards apex, ending at level of
anther-tips or somewhat exceeding them. Capsule about 2 cm, oblong-ovoid.
2n=14.** Fl. October–November.

Hab.: Open spaces in maquis areas, batha of *Sarcopoterium spinosum* from
700 m upwards. Upper Galilee, Judean Mts. (rare); Golan (common), Edom.

Area: E. Mediterranean and W. Irano-Turanian (Palestine, Lebanon, Syria,
W., S. and E. Turkey, N. E. Iraq, W. Iran, S. Jugoslavia, S. and E. Bulgaria,
E. Romania, Crimea).

On Mt. Hermon *C. pallasii (C. olbanus)* is very common from 1,800 m up to the
summit (2,814 m) on slopes covered with rubble and dominated by a vegetation of
tragacanthic chamaephytes, such as *Astragalus cruentiflorus* and *Onobrychis cornuta*.
It grows on wind-swept parts where the snow-cover is shallow and unstable; on wind-
sheltered snow-beds it is replaced by *C. hermoneus* subsp. *hermoneus*. From 1,800 m
downwards to 1,200 m, *C. pallasii* becomes less common and occurs in small populations of 1–10 specimens. In N. Golan (1,100 m to 900 m), it is found on basalt soil,
often in clearings of *Quercus calliprinos-Q. boissieri* maquis. It is rather rare in
Upper Galilee, west of Golan. *C. pallasii* includes two forms: one with scarlet styles,
the other with yellow styles. The form with yellow styles is found together with the
scarlet-styled form on Mt. Hermon from an altitude of 1,850 m upwards and predominates at higher elevations. At 2,300 m, 78% of the *Crocus pallasii* population have
been recorded with yellow styles (Feinbr. & Shmida, 1977).

3. Crocus moabiticus Bornm. & Dinsmore, Feddes Repert. 10:383 (1912);
Feinbr., Kew Bull. 12:276; Feinbr. & Shmida, Israel Jour. Bot. 26:176;
Mathew, Pl. Syst. Evol. 128:96; Post, Fl. 2:584. [Plate 177]

Corm about 2 cm in diameter, depressed-globose; tunics numerous, silky,
finely and densely reticulately fibrous with narrow elongated interspaces,
extended along scape for about 5–6 cm. Leaves numerous (10–13), hysteran-

* Drawing reproduced from N. Feinbrun, The genus *Crocus* in Israel and neighbour-
ing countries, *Kew Bull.* 12:269–285 (1957), by permission of Her Majesty's Stationary
Office.
** In some parts of Turkey 2n=12; in Edom and W. Iran 2n=16.

thous, at most 1 mm broad, sparsely ciliolate. Basal spathe present. Flowers several, autumnal; perianth-segments 15–25 × 3–7 mm, narrowly oblanceolate, acute, striped with purple-lilac on a whitish ground. Filaments white or often purple, short; anthers 10–13 mm, yellow. Style scarlet, divided into 3 arms below bases of anthers (usually in throat of perianth); arms dilated towards apex, nearly entire, often equalling perianth-segments in length or even exceeding them. Capsule unknown. 2n=14 (Brighton, unpubl.). Fl. November.

Hab.: Fallow fields. Moav.

Area: Endemic in Transjordan. W. Irano-Turanian.

4. Crocus cancellatus Herbert, Croc. Syn., Bot. Mag. 67:post t. 3864 no. 10 (1841) emend. Feinbr., Kew Bull. 12:278 (1957); Feinbr. & Shmida, Israel Jour. Bot. 26:177, f.1D; Maw, Monogr. 182, t.31b; Boiss., Fl. 5:102 as var. *cilicicus* Maw. [Plate 178]*

Plant 12–15 cm at anthesis. Corm ovoid or depressed; tunics of reticulate fibres (less coarse than in *C. damascenus*) with rectangular or rhombic interspaces, shortly extended (1–2 cm) as bristles along scape. Leaves (4–)6–7, hysteranthous, 1.5–2 mm broad, with a white mid-stripe. Flowers several, autumnal; perianth-segments 20–40 × 8–10 mm, obovate to elliptic, tapering at base, lilac with darker veins, usually 5–7-veined near base and with greyish-purple veins and feathering on outside; tube greyish-purple. Filaments 5–6 mm, white; anthers 12–18 mm, yellow. Style orange, rarely yellow, divided into 3 arms much above anther-bases; each arm further divided into several slender branches; ultimate style-branches ending at or above level of anther-tips. Capsule 2–3 cm, oblong. 2n=8. Fl. November–December.

Hab.: Batha on hills of basalt tuff and scoria (porous basalt rocks), at 1,000–1,100 m. Golan.

Area: E. Mediterranean and W. Irano-Turanian (Lebanon, Syria, S. Turkey, ?Iran).

The rocks of tuff and scoria of volcanic hills, on or around which *C. cancellatus* occurs in N. and C. Golan Heights, are characterized by low water-holding capacity; the soils formed on them are shallow and poor in nutrients. The vegetation consists mainly of hemicryptophytes, e.g., *Teucrium orientale, Carlina hispanica* subsp. *galilaea, Hordeum bulbosum* (Feinbr. & Shmida, *op. cit.* 178).

5. Crocus damascenus Herbert, Bot. Reg. 31: Misc. 1 & t.37 f.1 (1845) emend. Feinbr., Kew Bull. 12:276, 279 (1957); Feinbr. & Shmida, Israel Jour. Bot. 26:178. *C. cancellatus* Herbert var. *damascenus* (Herbert) Maw ex Boiss., Fl. 5:101 (1882); Post, Fl. 2:585. *C. edulis* Boiss. & Bl. in Sched. *C. cancellatus* Herbert forma *damascenus* (Herbert) Mout., Nouv. Fl. Liban Syrie 1:299, t.99 f.1 (1966). [Plate 179]

* The colour photo on t.10 f.3 in Wendelbo & Mathew, Fl. Iran. 112 is not of *C. cancellatus* Herbert.

Plant 15 cm at anthesis. Corm ovoid or depressed, 1.5–2.5 cm in diameter; tunics reticulately and coarsely fibrous with rectangular or rhombic inter-spaces, extended along scape as coarse fibres. Leaves 3–5, hysteranthous, 2–3 mm broad, slightly ciliolate, with a broad white mid-stripe. Flowers several, autumnal; perianth-segments 20–40 × 5–8 mm, narrowly oblanceolate to ellip-tic, lilac with darker veins, usually 3-veined near base, greyish-purple on out-side; tube greyish-purple. Filaments white, about ¼ length of anthers; anthers yellow, strongly incurved; pollen yellow. Style orange, divided into 3 arms at or below anther-bases; each arm further divided into several branches; ulti-mate style-branches distinctly exceeded by anthers or rarely ending at level of anther-tips. Capsule 1.5–2 cm, oblong. 2n−8. Fl. October–December.

Hab.: Hammada, 1,000 m and above, often in Artemisietum herbae-albae. C. Negev; Edom.

Area: W. Irano-Turanian (Palestine, Syria, Iraq, Iran).

After describing *C. damascenus*, the corms of which he received from J. Cartwright, Vice-Consul of Damascus, W. Herbert (Bot. Reg. 31: Misc. 2, 1845) remarked: "... this is probably the plant of which the roots are eaten by the natives". Later, Post (Fl. ed. 1, 770, 1883–1896), who did not distinguish *C. damascenus* from *C. cancellatus*, wrote: "The corm is edible, and sold in the streets of Damascus and other cities".

In the C. Negev this Irano-Turanian species has been collected repeatedly, and it is known at present from several localities in an area of 20–30 square kilometres in the Zin Desert. It is generally connected with Artemisietum herbae-albae and has been found growing together with *Rheum palaestinum* Feinbr. (Feinbr. & Shmida, *op. cit.* 178).

6. Crocus hermoneus Ky. ex Maw, Gard. Chron. ser. 2, 16:559 (1881); Maw, Monogr. 229, t.44; Feinbr. & Shmida, Israel Jour. Bot. 26:178. [Plate 180]

Plant 13–15 cm at anthesis. Corm 1.5–2 cm in diameter, ovoid; tunics dull brown, of dense parallel fibres interconnected by very short anastomoses, dis-sociated at base into free fibres and continued above as cusps along scape for 1.5–3 cm. Leaves hysteranthous (subsp. *hermoneus*) or nearly synanthous (subsp. *palaestinus*), slightly ciliolate. Flowers several, autumnal; perianth-segments (5–)8–9 mm broad, elliptic to oblanceolate, obtuse, lilac to purplish-pink (sometimes ranging to nearly white), with 3 darker veins on outside, usually decurrent onto tube. Filaments 7 mm, white; anthers 12–16 mm, yel-low. Style pale orange to orange, rarely yellow, divided above bases of anthers into 6 or more ultimate branches. Capsule 1–2 cm, oblong-ellipsoid.

1. Plants of high altitudes on Mt. Hermon (1,600–2,500 m), flowering in September-October (rarely November). Corm-tunics extending up to 1.5 cm along scape. Leaves (3–4–)5–7, 2–3 mm broad, appearing in April–June, long after end of flow-ering. Style-branches much exceeding anther-tips (by about 1 cm); ultimate branches usually numerous.   subsp. **hermoneus**
– Plants of hills in S. districts (Judean Mts., Ammon, Moav), flowering in November-December. Corm-tunics extending for 2–3 cm along scape. Leaves 3–4(–5), (1.5–)1.75(–2) mm broad, appearing before end of flowering. Style-branches not

exceeding anther-tips or, more rarely, slightly exceeding them; ultimate branches usually fewer than in subsp. *hermoneus*.   subsp. **palaestinus**

Subsp. **hermoneus**; Feinbr. & Shmida, *op. cit.* 179 (1977). *C. hermoneus* Ky. ex Maw, Gard. Chron. ser. 2, 16:559 (1881); Maw, Monogr. 229, t.44; Bornm., Bull. Herb. Boiss. 7:922–926 (1899); Mout., Bull. Soc. Bot. Fr. 101:423–426 (1954); Boiss., Fl. 5:116; Post, Fl. 2:586. *C. cancellatus* Herbert var. *hermoneus* (Ky.) Mout., Nouv. Fl. Liban Syrie 1:300, t.99 f.3 (1966).* 2n=8. Fl. September–October.

Hab.: Snow-beds on wind-sheltered slopes and dolinas, tragacanthic batha, 1,600–2,800 m. Mt. Hermon.

Area of subspecies: Endemic. E. Mediterranean.

From November or December onwards the snow layer remains more or less undisturbed for several months and melts from April to June.

The snow-beds in which subsp. *hermoneus* is common alternate with wind-swept stretches that are the habitat of *C. pallasii* (Baker) Boiss. Within the above altitudinal range the two taxa are common to about the same degree and flower simultaneously. *C. pallasii*, however, descends to 900 m on Mt. Hermon and extends further S. and W. (Feinbr. & Shmida, *op. cit.* 180).

Subsp. **palaestinus** Feinbr. in Feinbr. & Shmida, Israel Jour. Bot. 26:180 (1977). *C. hermoneus* sensu Feinbr., Kew Bull. 12:280–283, f.3 (1957). *C. hyemalis* Boiss. & Bl. var. *violaceo-splendens* Bornm. & Dinsmore in herb. [Plate 180].** 2n=12. Fl. November–December.

Hab.: Batha among calcareous or basalt rocks. Judean Mts.; S. Golan, Ammon, Moav.

Area of subspecies: Endemic. E. Mediterranean.

The plants vary in intensity of the purplish-pink colour of perianth, in colour of style (from orange to yellow), and to some extent in length of style (somewhat shorter to slightly longer than anthers) (Feinbr. & Shmida, *op. cit.* 180).

7. Crocus aleppicus Baker, Gard. Chron. 1873:609 (1873); Feinbr., Kew Bull. 12:279; Feinbr. & Shmida, Israel Jour. Bot. 26:181. *C. gaillardotii* Boiss. & Bl. ex Maw, Gard. Chron. 1879:234 (1879); Maw, Monogr. 211, t.40; Baker, Handb. Irid. 91 (1892); Boiss., Fl. 5:105; Post, Fl. 2:585. *C. hyemalis* Boiss. & Bl. (? var.) *gaillardotii* Boiss. & Bl. in Boiss., Diagn. ser. 2, 4:93 (1859) *sine statu explicito.* [Plate 181]

Corm ovoid, about 1.5 cm in diameter; tunics fairly thick, dull brown, of dense parallel fibres interconnected by very short anastomoses, dissociated at base into free fibres and extended shortly (up to 1 cm) along scape. Leaves

---

* Not illustrated.

** Drawing reproduced from N. Feinbrun, The genus *Crocus* in Israel and neighbouring countries, *Kew Bull.* 12:269–285 (1957), by permission of Her Majesty's Stationary Office.

(4-)5-7, synanthous, 1-1.5 mm broad, much longer than flowers towards end of flowering, with filiform white mid-stripe. Bract and bracteole well exserted from cataphylls. Flowers several, appearing early in winter; perianth-segments 25-35 × 6-11 mm, elliptic, white, with a purple or violet midvein or feathering on outside towards base (rarely outer segments entirely purplish on outside); tube with dark veins; throat yellow. Filaments short, yellow; anthers yellow or pale yellow. Style orange, often exceeding anthers, much dissected well above bases of anthers into several slender branches, often slightly thickened towards tip. Capsule 1.5-2 cm, ellipsoid. 2n=16. Fl. December-January.

Hab.: Batha, often on basalt soil, up to 1,250 m. Coast of Carmel, kurkar (calcareous sandstone) hills along coast, Upper Galilee; Golan (abundant), Mt. Hermon; Edom.

Area: E. Mediterranean (Palestine, coast of Lebanon, Syria).

In the Golan *Crocus aleppicus* is frequent between 400 m and 1,100 m alt. on marly basalt soil, in clearings of maquis and in a batha of hemicryptophytes. On Mt. Hermon it is found at elevations up to 1,250 m. Druze peasants around Mt. Hermon use the corms for food mixed with meat (Feinbr. & Shmida, *op. cit.* 181).

8. Crocus hyemalis Boiss. & Bl. in Boiss., Diagn. ser. 2, 4:93 (1859) emend. Boiss., Fl. 5:106 (1882); Maw, Monogr. 225, t.43; Feinbr., Kew Bull. 12:283; Feinbr. & Shmida, Israel Jour. Bot. 26:182; Mathew, Kew Bull. 31:207 (1976); Post, Fl. 2:585. *C. hyemalis* Boiss. & Bl. var. *foxii* Maw ex Boiss., Fl. 5:106 (1882); Post, *op. cit.* 586. [Plate 182]

Corm nearly globose, 1.5-2 cm in diameter; tunics membranous, silky, split lengthwise into linear strips at base and shortly extended along scape. Leaves 3-7, synanthous, 2-2.5 mm broad, dark green with a broad white mid-stripe. Bract and bracteole well exserted from cataphylls, unequal; bract tubular; bracteole linear. Flowers several, appearing in early winter; perianth-segments 20-40 × 6-15 mm, elliptic, subacute, white with violet or dark purple veins and longitudinal stripes on outside, especially towards base and on tube; throat yellow to orange. Filaments pale yellow, shorter than anthers; anthers 10-13 mm, dark purple, very rarely yellow; pollen yellow. Style orange, deeply divided well above bases of anthers into 8-15 filiform branches, mostly somewhat exceeding anther-tips. Capsule about 1 cm, shortly ellipsoid. 2n=6 & 6+1-4 B. Fl. November-February.

Hab.: Batha and garigue, mostly on terra rossa. Common. Coastal Galilee, Coast of Carmel, Sharon Plain; Upper and Lower Galilee, Mt. Carmel, Mt. Gilboa, Samaria, Judean Mts.

Area: E. Mediterranean (Palestine, extending along the coast of Lebanon up to Alexandretta).

Very rare single plants with yellow anthers collected in the Judean Mts. and Galilee are apparently mutants of the dark-anthered common form.

Flowers sweet-scented, visited by bees. *C. hyemalis* has not been found on the Golan Heights, where the predominant species is *C. aleppicus*. No record of *C. hyemalis* from the Mediterranean territory of Transjordan has come to our knowledge (Feinbr. & Shmida, *op. cit.* 182).

9. Crocus vitellinus Wahlenb. in J. Berggren, Resor 2, Bihang 59 (1828); Maw, Monogr. 253, t.50; Mathew, Kew Bull. 31:207 (1976); Boiss., Fl. 5:106; Post, Fl. 2:586. *C. syriacus* Boiss. & Gaill. in Boiss., Diagn. ser. 2, 4:94 excl. syn. *C. balansae* J. Gay. [Plate 183]

Corm subglobose, about 1.5 cm in diameter; tunics membranous, maroon, glossy, shortly extended above neck, split at base into flat linear strips. Leaves synanthous, 2–7(–10), 2–3.5 mm broad, shiny-green with a conspicuous silvery-white mid-stripe. Perianth infundibular, orange-coloured, often feathered with dark purple on outside; segments obtuse or rounded, 2–3 × 0.6–0.9 cm; throat glabrous. Filaments and anthers yellow. Style orange to reddish-orange, deeply dissected into numerous branches, somewhat exceeding anthers. Fl. December–January.

Hab.: Rocky slopes. N. E. Galilee near the northern boundary with Lebanon and Syria.

Area: E. Mediterranean (N. E. Palestine, Lebanon, W. Syria, S. Turkey).

Brighton (Kew Bull. 31:215, 1976) counted 2n=8 in *C. vitellinus.*

5. GLADIOLUS L.

Perennials with corm. Corm-tunics reticulate-fibrose. Lower leaves sheath-like, devoid of blade; cauline leaves sheathing at base; blade linear to sword-like. Inflorescence a 1- or 2-sided spike. Flowers each subtended by a bract and a bracteole; bracteole shorter than bract. Perianth zygomorphic, with a more or less curved tube and 6 unequal segments. Stamens 3; filaments attached on tube; anthers linear, basifixed. Ovary 3-locular, many-ovulate; style filiform, 3-lobed with stigmas dilated at apex. Capsule loculicidal, many-seeded. Seeds flattened and winged or globose-angular and wingless.

About 150 species; the majority of species occur in S. Africa, some in the Mediterranean and Irano-Turanian regions and a few in the Euro-Siberian region. Several species are widely cultivated for their showy flowers.

1. Leaves (8–)10–15 mm broad. Flowers pink. Spike lax, 2-sided.   **1. G. italicus**
- Leaves 4–7(–8) mm broad. Flowers dark violet-purple. Spike dense, 1-sided.
                                                                  2. G. atroviolaceus

Plants forming transitions between the two species are found especially in and around Jerusalem. They usually are intermediate in width of leaves and in flower-colour.

1. Gladiolus italicus Mill., Gard. Dict. ed. 8, no. 2 (1768); Becherer, Fortschritte, Ber. Schweiz. Bot. Ges. 74:181–182 (1964); Greuter & Rech. fil., Boissiera 13:165 (1967).*G. segetum* Ker-Gawler, Bot. Mag. 19:t.719 (1804); Czerniakovskaia, in Fl. URSS 4:580, t.36 f.2; Wendelbo & Mathew, Fl. Iran. 112:70, t.8 f.1; Boiss., Fl. 5:139; Post, Fl. 2:603. [Plate 184]

Plant 40–80 cm. Corm globose, 1.5–2 cm in diameter; tunics coarsely fibrous and partly reticulate. Leaves sword-like, mostly 10–15 mm broad, acute, few-veined; veins irregular; median vein thicker than the others, abruptly curved towards leaf-margin in lower part of leaf. Spike lax, 6–16-flowered, 2-sided, somewhat flexuose; bracts lanceolate, the lower nearly equalling flowers. Flowers nearly straight; perianth (3–)4–5(–5.5) cm, pink, campanulate; tube short, slightly curved; segments oblong; upper median segment longer than others. Capsule globose, up to 1 cm. Seeds wingless. Fl. March–April.

Hab.: Cornfields. Acco Plain, Coast of Carmel, Sharon Plain, Philistean Plain; Upper and Lower Galilee, Mt. Carmel, Esdraelon Plain, Samaria, Shefela, Judean Mts., W. and N. Negev; Upper Jordan Valléy; Golan, Gilead, Ammon.

Area: Mediterranean and W. Irano-Turanian.

Literature: J. Galil, Morpho-ecological studies on *Gladiolus segetum* Ker-Gawl., *Israel Jour. Bot.* 18:43–54 (1969).

2. Gladiolus atroviolaceus Boiss., Diagn. ser. 1, 13:14 (1854); Czerniakov-skaia, in Fl. URSS 4:587, t.36 f.5; Wendelbo & Mathew, Fl. Iran. 112:72, t.8 f.3; Boiss., Fl. 5:141. *G. aleppicus* Boiss., *op. cit.* 13 (1854); Post, Fl. 2:604. [Plate 185]

Plant 30–50 cm. Corm ovoid, 1.5–2 cm in diameter; tunics coarsely fibrous. Stem flexuose above. Leaves 3, linear to sword-shaped, 4–7(–8) mm broad, parallel-veined with veins equal in width, sometimes veins irregular, curved towards margin. Spike 4–9-flowered, one-sided, usually rather dense. Perianth (3.5–)4–4.5 cm, dark violet-purple; tube markedly curved; segments obovate; upper median segment not longer than others. Capsule oblong, 1.5–2 cm. Seeds wingless. Fl. March–April.

Hab.: Cornfields; marly calcareous and basalt soils. Samaria, Shefela, Judean Mts., Judean Desert, W. and N. Negev; Lower Jordan Valley; ?Golan, ?Gilead, Ammon, Moav, Edom.

Area: Irano-Turanian.

122. JUNCACEAE

Herbs, mostly perennial, tufted or rhizomatose. Stems mostly erect, devoid of nodes. Leaves spirally or distichously arranged, linear, flat or terete, sheathing at base, sometimes reduced to sheaths only; sheaths open or closed. Inflorescence terminal or pseudolateral, simple or compound, mostly many-flowered; ultimate partial inflorescences monochasial cymes sometimes contracted into heads, or flowers solitary. Flowers small, usually hermaphrodite, actinomorphic, 3-merous, wind-pollinated, mostly protogynous. Perianth of 6 segments in 2 whorls; segments persistent, herbaceous, coriaceous or scarious, rarely coloured. Stamens free, 6 or fewer by reduction of inner whorl; anthers basifixed, introrse. Pistil 1, 3-carpellate; ovary superior, (1–)3-locular, 3- to many-ovulate; style 1; stigmas 3, brush-like. Fruit a loculicidal capsule. Seeds 3 or numerous, sometimes caudate; endosperm starchy; embryo straight, small.

Nine genera and about 320 species in temperate and cold regions all over the globe; in the Tropics restricted to mountainous areas.

Literature: F.G.P. Buchenau, Monographia Juncacearum, *Bot. Jahrb.* 12:1–498 (1890); Juncaceae, in: Engler, *Pflanzenreich* 25 (IV. 36):1–284, 1906. P.F.A. Ascherson & K.O.P.P. Graebner, Juncaceae, *Syn. Mitteleur. Flora* 2(2):414–526, 1904. F. Vierhapper, Juncaceae, in: Engler & Prantl, *Pflanzenfam.* ed. 2, 15a:192–224, 1930. V.I. Kreczetowicz & F.N. Gontscharov, Juncaceae, in: V.L. Komarov (ed.), *Flora URSS* 3:504–576 & 623–631, 1935. R. Maire et al., Flore de l'Afrique du Nord 4:255–305, Paris, 1957. S. Snogerup, Juncaceae, in: K.H. Rechinger (ed.), *Flora iranica* 75:1–35, 1971. T.A. Cope & C.A. Stace, The *Juncus bufonius* L. aggregate in western Europe, *Watsonia* 12:113–128 (1978). P.E. Boissier, *Flora orientalis* 5:346–362, 1882 & 758–759, 1884. G.E. Post, *Flora of Syria, Palestine and Sinai* ed. 2, 2:665–669, 1933.

1. JUNCUS L.

Glabrous perennials, mostly rhizomatous, or annuals. Stems erect, unbranched, with or without cauline leaves. Leaves terete or flat and grass-like, sheathing at base; sheaths mostly open. Inflorescence variously shaped, terminal or pseudolateral with the lower bract simulating prolongation of stem; flowers each subtended by a basal bracteole on pedicel and 2–3 sheathing ones, or bracteoles absent. Perianth-segments 6, 1- or 3-veined. Stamens 6 or fewer; pollen in tetrads. Capsule 3-gonous, 3-locular or 1-locular by incomplete partitions. Seeds numerous, small; testa often forming persistent appendages.

About 225 species, all over the globe but rare in Tropics.

1. Cauline leaves hollow and transversely septate   2
- Cauline leaves absent or, if present, not septate   4
2. Stems with a single long cauline leaf Flowers in dense spherical heads 4–5(–8) mm in diameter. Perianth-segments 2(–2.5) mm.   **9. J. punctorius**
- Stems with 2 or more cauline leaves. Flowers in few-flowered hemispherical heads 6–12(–14) mm in diameter. Perianth-segments 2.5–4.5 mm   3
3. Capsule acute, with a long beak. Perianth-segments acuminate-aristate.   **11. J. fontanesii**
- Capsule with a short mucro. Perianth-segments acute and mucronulate, or the inner obtuse mucronulate.   **10. J. articulatus**
4(1). Perennials   5
- Annuals   9
5. Pith of stem interrupted. Inflorescence subtended by a single leaf-like involucral bract much exceeding inflorescence and seemingly forming a prolongation of stem. All leaves of flowering stems reduced to brown bladeless sheaths.   **4. J. inflexus**

- Pith of stem not interrupted. Leaves and involucral bract not as above   6
6. Stems leafy; cauline leaves 2–4, green, with a ligule between sheath and blade.
                                                                      5. J. subulatus
- Leaves all basal; cauline leaves absent   7
7. Perianth brown; inner perianth-segments broadly hyaline-margined, truncate and
   retuse, with apical auricles.   **3. J. acutus**
- Perianth straw-coloured; inner perianth-segments narrowly hyaline-margined and
   rounded at apex, without apical auricles   8
8. Capsule 3.5–5 mm, tapering towards apex, exceeding perianth by ⅓ or more.
   Plants mainly of desert districts, especially along Rift Valley, up to Hula Plain in
   north.   **2. J. arabicus**
- Capsule 2.5–3.5(–4) mm, obtuse, as long as or slightly longer than perianth.
   Plants of coastal plain.   **1. J. maritimus**
9(4). Flowers in a terminal head. Each head subtended by 2–3 involucral bracts,
   with longest bract overtopping head. Individual flowers not subtended by bracteoles.   **8. J. capitatus**
- Flowers in a lax cymose inflorescence, single and spaced or in dense fan-shaped
   clusters of (2–)3–6. Each flower subtended by 2 bracteoles which are shorter than
   flower   10
10. Perianth-segments 4.5–6(–8) mm. Capsule oblong-ellipsoid.   **6. J. bufonius**
- Outer perianth-segments 2.5–4 mm. Capsule subglobose.   **7. J. sphaerocarpus**

Subgen. 1. JUNCUS. Perennials. Leaves basal, terete, pungent, stem-like, not septate. Inflorescence usually pseudolateral; the 2 lowermost bracts leaf-like, pungent; the lowest forming an apparent prolongation of stem. Involucral bracteoles absent. Seeds with appendages.

1. Juncus maritimus Lam., Encycl. Méth. Bot. 3:264 (1789); Buchenau, Bot. Jahrb. 12:256; Snogerup, Fl. Iran. 75:4, t.1 f.2; Boiss., Fl. 5:354 excl. var. *arabicus*; Post, Fl. 2:666. ?*J. rigidus* Desf., Fl. Atl. 1:312 (1798) *nom. ambig.* [Plate 186]

Perennial, 50–100 cm. Stems numerous, in rows along horizontal creeping rhizomes; pith continuous. Leaf-sheaths brown; blades terete, not septate, pungent. Inflorescence pseudolateral, many-flowered, with more or less erect branches. Lower involucral bract subulate, pungent, shorter to longer than inflorescence. Flowers aggregated in few-flowered heads. Perianth-segments straw-coloured; outer segments somewhat longer, boat-shaped, ovate, acute, short-mucronate; inner segments flat, oblong, rounded and narrowly hyaline-margined at apex. Stamens 6, rarely 3; anthers about twice as long as filaments. Capsule 2.5–3.5(–4) × 1.75–2 mm, broadly ellipsoid to ovoid, mucronate, as long as or slightly longer than perianth, yellow to pale brown. Seeds caudate, with long or short appendages. Fl. June–September.

Hab.: Marshes, mostly saline soil; together with *J. acutus*. Acco Plain, Coast of Carmel, Sharon Plain.

Area: Euro-Siberian (C. and E. Europe) and Mediterranean, slightly extending into the W. Irano-Turanian.

Snogerup in Flora iranica 75:4 (1971) identified *J. rigidus* Desf. (Flora Atlantica 1:312, 1798) with *J. arabicus* (Aschers. & Buchenau) Adamson. This does not seem justified. Desfontaines says in his description of *J. rigidus*: "Nec capsulam nec folia observavi". Thus the main morphological feature to distinguish *J. arabicus* from *J. maritimus* is lacking in his description. On the other hand, the habitat of *J. rigidus* is given as "arenis ad maris litora" which suits *J. maritimus*. In Flore de l'Afrique du Nord by Maire et al. (4:281, 1957), covering the area dealt with in Flora Atlantica, *J. rigidus* Desf. is quoted as synonymous with *J. maritimus* var. *typicus* forma *rigidus* (Desf.) Maire & Weiller, while *J. arabicus* is given as var. *arabicus* Aschers. & Buchenau. *J. rigidus* Desf. is thus regarded here a *nom. ambig.*

J. maritimus which grows in Israel on the coastal plain exhibits variability in the shape of capsule. Some plants have broadly ellipsoid obtuse capsules (typical of *J. maritimus*), but in others capsules are narrowly ovoid and taper in their upper part as in *J. arabicus*; the capsules are, however, only slightly exserted from the perianth (up to ¼ their length). This is suggestive of occasional hybridization between the two species where they meet. According to Snogerup *(op. cit.)*, the chromosome number of both species is 2n=48.

2. **Juncus arabicus** (Aschers. & Buchenau) Adamson, Jour. Linn. Soc. London (Bot.) 50:10–11 (1935). *J. maritimus* Lam. var. *arabicus* Aschers. & Buchenau in Boiss., Fl. 5:354 (1882) excl. syn. (*J. spinosus* Forssk.); Buchenau, Bot. Jahrb. 12:257; Post, Fl. 2:666. Snogerup, Fl. Iran. 75:4, t.1 f.1 as *J. rigidus* Desf. *J. nevskii* V. Krecz. & Gontsch., in Fl. URSS 3:629 (1935). [Plate 187]

Perennial, 75–150 cm, growing in large tufts. Stems numerous; pith continuous. Leaf-sheaths brown; blades terete, not septate, pungent. Inflorescence pseudolateral, many-flowered, up to 35 cm, narrow, with more or less erect branches. Involucral bracts 2; the lower subulate, pungent, shorter to longer than inflorescence. Flowers aggregated in few-flowered clusters. Perianth-segments straw-coloured; the outer somewhat longer, boat-shaped, narrowly ovate, obtuse or acute, often mucronate; the inner flat, oblong, rounded and narrowly hyaline-margined at apex. Stamens 6; anthers 3–5 times as long as filaments. Capsule 3.5–5 × 1.5–1.75(–2) mm, oblong-ovoid, tapering towards apex in its upper ⅓, mucronate, exceeding perianth by ⅓ or more, yellow to pale brown. Seeds caudate, with equal or unequal appendages. Fl. March–December.

Hab.: Saline soil, marshes and along dry wadis. Common mainly in desert districts. E. Samaria; Hula Plain, Upper Jordan Valley, Beit Shean Valley, Lower Jordan Valley, Dead Sea area, Arava Valley; E. Ammon, Moav, Edom. Occasional in Acco Plain, Coast of Carmel, Judean Mts.

Area: Saharo-Arabian and W. Irano-Turanian.

Used for making mats, often the praying mats of Moslems.

3. **Juncus acutus** L., Sp. Pl. 325 (1753); Buchenau, Bot. Jahrb. 12:249; Snogerup, Fl. Iran. 75:5; Boiss., Fl. 5:353 incl. var. *longibracteatus* Buchenau in

Boiss., *op. cit.* 354 (1882); Post, Fl. 2:666. *J. spinosus* Forssk., Fl. Aeg.-Arab. 75 (1775). [Plate 188]

Perennial, 50–150 cm, dark green, in large dense more or less round tufts 0.5–1 m in diameter, forming no creeping rhizomes. Stems numerous, rigid, erect, terete; pith continuous. Leaf-sheaths brown; blades terete, pungent, not septate, usually shorter than stems. Inflorescence pseudolateral, many-flowered, varying in shape and length, compact and short, or lax and elongated, with more or less spreading branches. Lower involucral bract dilated at base, subulate and strongly pungent, shorter to longer than inflorescence. Flowers aggregated in few-flowered clusters; each cluster subtended by several scarious bracts. Perianth-segments brown-maroon, hyaline-margined; the outer nearly boat-shaped, obtuse or acute, short-mucronate; the inner flat, oblong, truncate and retuse, hyaline-margined; the hyaline margin much broadened towards apex, forming auricles. Stamens 6. Capsule hard, brown to maroon, glossy, acute, mucronate, much varying in size and shape. Seeds caudate. Fl. April–September.

Hab.: Marshes, often saline, in the *Inula viscosa–Juncus acutus* association (Eig, Palest. Jour. Bot. Jerusalem ser., 3:231, 1946), on alluvial soils and in depressions between dunes overlaying marshy soil. Common and abundant especially on the coastal plain.

The important differential characters of *J. acutus* used in its delimitation from other species of subgen. *Juncus* are: inner perianth-segments provided with hyaline apical auricles, brown colour of inflorescence and perianth, and general habit of plants which grow in dense and large round tufts.

A great number of infraspecific taxa of *J. acutus* have been described by various authors (cf. Aschers. & Graebn., Syn. Mitteleur. Fl. 2(2):452–455, 1904). Some taxa have been published as distinct species, such as *J. littoralis* C.A. Mey. (1831) from the shores of the island Sara near Baku on the Caspian Sea, and *J. tommasinii* Parl. (1857) from Sicily. *J. littoralis* C.A. Mey. has been accepted at specific rank by Snogerup (Flora iranica 75:6, 1971).

On the basis of the study of abundant material from Israel, we regard *J. acutus* as comprising two subspecies, subsp. *acutus* and subsp. *littoralis*. The two are indistinguishable in general habit and in the variable shape of inflorescence. The main differences between the two taxa are the shape and size of capsule and the number of seeds per capsule, as given below.

Subsp. **acutus**. Subsp. *megalocarpus* Aschers. & Graebn., Syn. Mitteleur. Fl. 2(2):453 (1904). [Plate 188]. Capsule 4–5.5(–6) × 2.5–3 mm, ovoid to conical-ovoid, 3-gonous, turgid, exserted from perianth by more than $\frac{1}{2}$ of its length (mostly by $\frac{2}{3}$). Seeds (40–)70–120 per capsule.

Hab.: Coastal Galilee, Acco Plain, Coast of Carmel, Sharon Plain, Philistean Plain; Upper and Lower Galilee, Mt. Carmel, Samaria, Shefela, ?Judean Mts.; Dan Valley, Hula Plain, Upper Jordan Valley, Beit Shean Valley, Lower Jordan Valley; Golan, E. Ammon, Moav.

Area of subspecies: Mediterranean and W. Irano-Turanian, extending into W. Europe. Also in S. Africa, America and Australia.

Subsp. **littoralis** (C.A. Mey.) Feinbr. comb. nov. Var. *littoralis* (C.A. Mey.) Trautv., Acta Horti Petrop. 5:480 (1878); Aschers. & Graebn., Syn. Mitteleur. Fl. 2(2):454 (1904). *J. littoralis* C.A. Mey., Verz. Pfl. Cauc. 34 (1831). *J. tommasinii* Parl., Fl. Ital. 2:315 (1857). *J. acutus* B. *tommasinii* (Parl.) Aschers. & Graebn., *loc. cit. J. acutus* var. *tommasinii* (Parl.) Arcangeli, Comp. Fl. Ital. 715 (1882); Täckholm & Drar, Fl. Eg. 2:460. *J. acuto-maritimus* E. Meyer in Ledeb., Fl. Ross. 4:234 (1852); Boiss., Fl. 5:362 (1882). [Plate 188]. Capsule 2.5-3.5(-4) × 1.5-2.5 mm, narrower and/or shorter than in subsp. *acutus*, variously shaped and exserted from perianth by at most ½ of its length. Seeds 3-20, rarely up to 40 per capsule.

Hab.: Coastal Galilee, Acco Plain, Coast of Carmel, Sharon Plain, Philistean Plain, W. Negev.

Area of subspecies: Mediterranean and W. Irano-Turanian, extending into S.E. Europe.

In Israel the two subspecies grow in mixed populations along the coastal plain. In the inner plains (Dan Valley, Hula Plain, Upper Jordan Valley, Beit Shean Valley and Lower Jordan Valley) only subsp. *acutus* is found. A similar situation is reported from S. France (Aubouy, 1885, quoted by Buchenau, 1890, and Aschers. & Graebn., 1904) and from Egypt. Täckholm & Drar (Fl. Eg. 2:461, 1950) remarked with regard to *J. acutus* and var. *tommasinii*: "Both the type and the variety (also with transitional forms) are growing in dense tufts, especially on salt lands, moist places, sometimes also on dry sandy localities".

Boissier (Fl. 5:362, 1882), following E. Meyer in Ledebour (Fl. Ross. 4:234, 1852), regarded *J. littoralis* C.A. Mey. as a hybrid between *J. acutus* and *J. maritimus*, named *J. acuto-maritimus* E. Meyer in Ledebour (1852).

It is noteworthy that *J. maritimus* is often found together with *J. acutus* in littoral areas of the Mediterranean. Even on the coast of the Caspian Sea, the type locality of *J. littoralis*, this taxon grows together with *J. acutus* and *J. maritimus*, as noted by C.A. Meyer. Boissier recorded *J. littoralis* C.A. Mey. as a synonym of *J. acuto-maritimus* and remarked: "Inter ambo omnino hybridus videtur". Buchenau (1890) found no morphological basis for the view of E. Meyer and Boissier. It is true that in all forms of var. *littoralis* the shape of the inner perianth-segments is invariably that of *J. acutus*. On the other hand, these plants approach *J. maritimus* in the small size of the capsules, and in their capsules being less exserted from the perianth than in *J. acutus*.

On the coastal plain of Israel plants with fully developed capsules were often collected together with flowering or sterile specimens. Rechinger too (1952:309) lists a specimen "*J. acutus* L. f. *sterilis* (an hybrida?)" from the seashore of Haifa, growing together with typical *J. acutus* and *J. maritimus*. Snogerup identified some of the specimens in our Herbarium, especially from Acco Plain, as hybrids *J. acutus* × *J. littoralis*. The chromosome number 2n=48 is reported by Snogerup for both *J. acutus* and *J. littoralis*.

Subgen. 2. GENUINI Buchenau (1875). Rhizomatous perennials; rhizome horizontal, mostly with short internodes, producing flowering stems and short sterile shoots. Leaves often reduced to basal sheaths. Lowest bract usually large, forming an apparent prolongation of stem; inflorescence therefore pseudolateral. Flowers with involucral bracteoles. Seeds usually without appendages.

4. Juncus inflexus L., Sp. Pl. 326 (1753); Snogerup, Fl. Iran. 75:8. *J. glaucus* Ehrh., Beitr. Naturk. 6:83 (1791); Buchenau, Bot. Jahrb. 12:243; Boiss., Fl. 5:353; Post, Fl. 2:666. [Plate 189]

Perennial, 1–1.5 m, densely tufted. Rhizome branched. Stems rigid, terete, glaucous, smooth when fresh, striate when dry; pith interrupted. Leaves of flowering stems reduced to sheaths; sheaths obtuse, mucronate, dark brown to maroon and glossy at least in lower part. Inflorescence pseudolateral, branched, subtended by a single long subulate pungent bract much exceeding inflorescence. Each flower subtended by 2 bracteoles. Perianth-segments 3–3.5 mm, straw-coloured to brownish, lanceolate, membranous-margined, ending in a subulate point; inner segments somewhat shorter. Stamens 6. Capsule 3-gonous-ovoid, acute, mucronate, as long as to slightly exceeding perianth, brown to maroon, glossy. Seeds not caudate to shortly caudate, reticulate. Fl. April–August.

Hab.: Damp places. Acco Plain; Upper Galilee; Dan Valley, Hula Plain; Golan, Gilead, Edom.

Area: Borealo-Trop. (Europe, Asia, Africa).

Subgen. 3. SUBULATI Buchenau (1885). Rhizomatous perennials. Cauline leaves present, with well-developed blades. Lowest bract of inflorescence much shorter than inflorescence, not forming a prolongation of stem. Seeds with short appendages.

5. Juncus subulatus Forssk., Fl. Aeg.-Arab. 75 (1775); Buchenau, Bot. Jahrb. 12:171; Snogerup, Fl. Iran. 75:8; Boiss., Fl. 5:354; Post, Fl. 2:666. [Plate 190]

Perennial, 50–100 cm. Rhizome horizontal, creeping, with long internodes. Stems numerous, terete; pith continuous. Basal leaves 2–3, reduced to sheaths; cauline leaves 2–4, green, with a well-developed ligule between sheath and blade. Inflorescence terminal, up to 20 cm, lax. Lower involucral bract shorter than inflorescence. Flowers in few-flowered clusters, each flower subtended by 2 bracteoles. Perianth-segments 2.5–3.5 mm, pale green to pale brown; the outer narrowly ovate or lanceolate, acute; the inner broader and somewhat shorter, broadly membranous-margined, obtuse and mucronate. Stamens 6. Capsule 3-gonous-ellipsoid, obtuse, short-mucronate, brown, glossy, as long as perianth to somewhat exceeding it. Seeds not caudate or very shortly caudate. Fl. April–July.

Hab.: Damp places. Acco Plain, Sharon Plain, Philistean Plain; Mt. Carmel, Esdraelon Plain, N. and C. Negev; Dan Valley, Dead Sea area.

Area: Mediterranean, extending into the Irano-Turanian and Saharo-Arabian regions.

Subgen. 4. POIOPHYLLI Buchenau (1875). Annuals, without rhizome, usually caespitose. Leaves all cauline on flowering stems, flat or subterete. Inflorescence terminal. Flowers with 2 bracteoles. Seeds without appendages.

6. Juncus bufonius L., Sp. Pl. 328 (1753); Buchenau, Bot. Jahrb. 12:174; Aschers. & Graebn., Syn. Mitteleur. Fl. 2(2):420; Snogerup, Fl. Iran. 75:16; Boiss., Fl. 5:361; Post, Fl. 2:668. [Plates 191, 192]

Annual, 10–30(–40) cm, many-stemmed, tufted. Stems slender, leafy. Leaves 0.5–2 mm broad, flat to convolute; lower bracts of inflorescence leaf-like. Inflorescence lax, occupying at least the upper ½ of stem; branches dichasial or monochasial, usually not exceeded by involucral bracts. Flowers single (or partly twin), spaced, or 3–6 in dense fan-shaped clusters, each flower subtended by 2 membranous bracteoles that are shorter than flower. Perianthsegments with a pale green mid-stripe and broad hyaline margins, narrowly ovate to lanceolate; outer segments 4.5–6(–8) mm, acute to acuminate, the inner shorter, subacute to acute, sometimes apiculate. Stamens 6. Style usually persisting as a short beak. Capsule oblong-ellipsoid, obtuse, often mucronate, green to maroon, shorter than inner perianth-segments. Seeds ellipsoid, not caudate. Fl. March–June.

A variable species.

Var. **bufonius**. [Plate 191]. Flowers single, spaced, subsessile, mostly in forked spike-like cymes; sometimes part of flowers twin.

Hab.: Habitats with high water-table. Fairly common on the coastal plain. Acco Plain, Coast of Carmel, Sharon Plain, Philistean Plain, W. Negev; Upper and Lower Galilee, Mt. Carmel, Esdraelon Plain, Samaria, Shefela, Judean Mts.; Dan Valley, Hula Plain, Upper Jordan Valley, Beit Shean Valley, Dead Sea area; Golan, Gilead, Ammon, Edom.

Area of var.: as of species.

Var. **congestus** Wahlberg, Fl. Goth. 38 (1820). Var. *fasciculatus* Koch, Syn. Fl. Germ. 732 (1837); Boiss., Fl. 5:361. Var. *hybridus* (Brot.) Parl., Fl. Ital. 2:353 (1857) & (Brot.) Husnot, Bull. Soc. Bot. Fr. 55:50 (1908); Post, Fl. 2:669. *J. hybridus* Brot., Fl. Lusit. 1:513 (1804); Snogerup, Fl. Iran. 75:17; Cope & Stace, Watsonia 12:123, f.4 (1978). [Plate 192]. Flowers (2–)3–6 in dense fan-shaped clusters, occasionally cleistogamous.

Hab.: Less common than var. *bufonius*, in similar habitats, mainly along the coastal plain and the Rift Valley. Acco Plain, Coast of Carmel, Sharon Plain, Philistean Plain; Dan Valley, Hula Plain, Upper Jordan Valley, Dead Sea area; Golan, Gilead.

Area of var.: Mediterranean.

Specimens that are transitional between the 2 varieties have been collected frequently.

Area of species: Euro-Siberian–Boreo-American, Mediterranean and Irano-Turanian; naturalized elsewhere.

Cope & Stace in their recently published article (1978) recognize in W. Europe several species within what they designate the *Juncus bufonius* aggregate. In his treatment of *Juncus* in Flora iranica (1971), Snogerup also distinguished several species in this group. Two of the species treated in these sources concern our Flora: *J. bufonius* L. and *J. hybridus* Brot. These two species are keyed out by Snogerup as follows:

Capsule ovoid. Inner perianth-segments apiculate or acute.   **J. bufonius**
Capsule ellipsoid or prismatic. Inner perianth-segments obtuse or subacute.

                                                                              J. hybridus

Cope & Stace (1978) recognize similar differences between the 2 taxa and add differences in seed characters (obliquely ovoid for *J. bufonius* and barrel-shaped or ovoid for *J. hybridus*).

Studies of extensive herbarium material collected in Israel have not revealed clear differences in the shape of capsules, inner perianth-segments or seeds between the 2 taxa. The main difference lies in the spacing of flowers within the inflorescence, and we distinguish plants with cymes of single or twin flowers from those of flowers in fan-shaped 3–6-flowered clusters. In Israel the two taxa do not differ ecologically or geographically; they are often connected by transition forms which are fairly common in the same districts.

Therefore we see no justification in according these taxa the status of species. and designate them as varieties of *J. bufonius* L., namely var. *bufonius* and var. *congestus* Wahlberg (1820).

Regretfully no data on chromosome numbers from Israel are available for this group of *Juncus*. In the literature, the chromosome numbers given for *J. bufonius* are 2n=108 (also 54, 60, 70, 72, 80, 104, 110), for *J. hybridus* 2n=34.

7. Juncus sphaerocarpus Nees in Funck, Flora (Regensb.) 1:31, 521 (1818); Buchenau, Bot. Jahrb. 12:178; Snogerup, Fl. Iran. 75:19; Boiss., Fl. 5:759. *J. tenageia* auct. non Ehrh. ex L. fil. (1781); Post, Fl. 2:669; Eig, Bull. Inst. Agr. Nat. Hist. Tel-Aviv 4:41 (1926) & 6:53 (1927). [Plate 193]

Annual, 8–25(–40) cm, tufted. Stems slender, flaccid, mostly numerous. Leaves 0.5–1.5 mm broad, flat to convolute, not auriculate. Inflorescence of lax 2-fid cymes, occupying ½ or whole length of plant. Flowers single, spaced, each flower subtended by 2 bracteoles; bracteoles membranous, ovate, mucronate. Perianth-segments 2.5–4 mm, broadly hyaline-margined, the outer narrowly ovate, acuminate and apiculate, the inner somewhat shorter, lanceolate, acute and often apiculate. Stamens 6. Capsule 2–2.5 mm, subglobose, obscurely 3-gonous, obtuse, straw-coloured or pale rust-coloured. Seeds not caudate, ovoid to ellipsoid. Fl. April–June.

Hab.: Damp places. Fairly rare and sporadic. Sharon Plain, Philistean Plain; Esdraelon Plain, Judean Mts.; Golan.

Area: Euro-Siberian, Mediterranean and Irano-Turanian; also in America.

Subgen. 5. JUNCINELLA (Fourr.) V. Krecz. & Gontsch. (1935). Small annuals, caespitose or with a single stem. Leaves and bracts not pungent. Leaves basal on flowering stems; blades flat to subterete. Flowers solitary on stem or in terminal clusters. Bracteoles absent. Seeds without appendages, prominently reticulate.

8. Juncus capitatus Weigel, Obs. Bot. 28, t.2 f.5 (1772); Buchenau, Bot. Jahrb. 12:450; Boiss., Fl. 5:361; Post, Fl. 2:668. [Plate 194]

Annual, 8–15 cm. Stems filiform, leafy at base, each ending in inflorescence. Leaves setaceous, canaliculate, shorter than stem. Inflorescence a terminal dense globose head, about 1 cm in diameter, 3–10-flowered; sometimes 1–2 lateral usually pedunculate heads present in addition. Involucral bracts of head leaf-like, 2–3, unequal, the outer longer than inflorescence. Perianth-segments 4–6 mm, unequal, pale green, broadly hyaline-margined; the outer broadly lanceolate, keeled, acuminate-aristate; the inner ovate, acute, shorter by 1–1.5 mm. Stamens 3. Capsule ellipsoid, obtuse and short-mucronate, long exceeded by outer perianth-segments. Seeds not caudate. Fl. April–May.

Hab.: Damp sandy soils. Rare and sporadic. Sharon Plain; Upper Jordan Valley.

Area: Euro-Siberian, Mediterranean, Trop. Africa.

Subgen. 6. SEPTATI Buchenau (1875). Perennials or rarely annuals; rhizome usually well developed. Leaves and bracts not pungent. All leaves on flowering stems, terete or subterete, hollow, with distinct transverse septa. Flowers in heads. Bracteoles absent. Stamens shorter than perianth-segments; filaments more or less rigid.

9. Juncus punctorius L. fil., Suppl. 208 (1781); Buchenau, Bot. Jahrb. 12:277; Snogerup, Fl. Iran. 75:21; Boiss., Fl. 5:357; Post, Fl. 2:667; incl. var. *exaltatus* (Decne.) Buchenau, Abh. Nat. Ver. Bremen 4:429 (1875). *J. exaltatus* Decne., Ann. Sci. Nat. Bot. ser. 2, 2:16 (1834). [Plate 195]

Perennial, 50–200 cm. Rhizome thick, creeping, with short internodes. Stem terete to subcompressed, with 3 obtuse sheaths at base and 1 cauline leaf with terete, transversely septate, usually erect leaf-blade. Inflorescence compound, with numerous dense subglobose, many-flowered heads 4–5(–8) mm in diameter. Perianth-segments 2(–2.5) mm, equal in length, linear-lanceolate, acute to apiculate, narrowly hyaline-margined; the outer boat-shaped, the inner flat. Stamens 6. Style and stigmas long-exserted. Capsule 2.5 mm, 3-gonous-ovoid, more or less abruptly ending in a 0.5 mm beak, dark brown, glossy, about as long as perianth. Seeds ovoid, 2-apiculate. Fl. April–July.

Hab.: Marshes, ditches, in water. Rare. ?Golan, Gilead, Moav, Edom.

Area: Saharo-Arabian (S. W. Asia, Sinai, N. Africa); also S. and E. Africa.

10. Juncus articulatus L., Sp. Pl. 327 (1753); Snogerup, Fl. Iran. 75:22. *J. lampocarpus* Ehrh. ex Hoffm., Deutschl. Fl. ed. 2, 1:166 (1800); Reichenb., Icon. Fl. Germ. 9:t.405 (1847; "*lamprocarpus*"); Buchenau, Bot. Jahrb. 12:376; Boiss., Fl. 5:358; Post, Fl. 2:667. [Plate 196]

Perennial, 10–60 cm, tufted or with a horizontal or oblique, mostly short rhizome. Stems with continuous pith. Basal leaves reduced to sheaths; cauline leaves 2–3 or more, with terete or compressed, transversely septate blades. Inflorescences more or less compound, with dense hemispherical 4–20(–30)-flowered heads, 6–10 mm in diameter. Lower bracts of the inflorescence shorter than inflorescence. Perianth-segments 2.5–3 mm, more or less equal in length, ovate to broadly lanceolate, pale to dark brown, hyaline-margined; the outer keeled, the inner more or less flat, all acute and mucronulate, or the inner obtuse. Stamens 6. Capsule 3–3.5(–4) mm, somewhat exceeding perianth, 3-gonous-ovoid to -ellipsoid, obtuse, abruptly contracted at apex and mucronate. Seeds ovoid. Fl. June–August.

Hab.: Marshes. Fairly rare. Acco Plain, Sharon Plain; Judean Mts.; Golan.

Area: Euro-Siberian, Mediterranean and Irano-Turanian; also S. Africa, S. Australia, New Zealand, N. America.

11. Juncus fontanesii J. Gay in Laharpe, Monogr. Jonc. (1825) preprinted from Mém. Soc. Hist. Nat. Paris 3:130 (1827) subsp. **pyramidatus** (Laharpe) Snogerup, Fl. Iran. 75:25 (1971); Post, Fl. 2:668 as var. *pyramidatus* (Laharpe) Buchenau. *J. pyramidatus* Laharpe, *op. cit.* 128 (1825); Boiss., Fl. 5:359. [Plate 197]

Perennial, 25–60(–80) cm, with short rhizomes, tufted, or sometimes with long creeping stems. Stems 1–2 mm in diameter, with continuous pith. Basal and lower cauline leaves reduced to sheaths; 2–3 cauline leaves, with compressed, hollow, transversely septate blades. Inflorescence more or less compound, with hemispherical to subglobose 6–20-flowered heads, (8–)10–12(–14) mm in diameter, at maturity appearing echinate owing to diverging long-beaked capsules and acuminate perianth-segments. Perianth-segments 3.5–4.5 mm, often reddish-brown, lanceolate, acuminate-aristulate, more or less striate, narrowly hyaline-margined; the outer nearly keeled, the inner flat, often slightly longer than the outer. Stamens 6. Capsule 4.5–5(–6) mm, narrowly pyramidal, 3-quetrous, reddish-brown, glossy, gradually tapering to a long beak, exceeding perianth often by ⅓ to ½. Seeds obovoid, shortly apiculate. Fl. February–September.

Hab.: Marshes and banks of streams. Fairly common. Acco Plain, Sharon Plain, Philistean Plain; Upper and Lower Galilee, Esdraelon Plain, Samaria, Judean Mts.; Dan Valley, Hula Plain; Golan, Ammon, Moav, Edom.

Area of subspecies: E. Mediterranean, extending East to mountains of N. Iraq.

The known monographer of Juncaceae, F. Buchenau (Bot. Jahrb. 1:140–141, 1881) at first accepted Ascherson's view that *J. pyramidatus* Laharpe is synonymous with the more widely distributed *J. fontanesii* J. Gay. However, later (Bot. Jahrb. 7:168, 1885 & 12:329–330, 1890) Buchenau distinguished var. *pyramidatus* (Laharpe) Buchenau and var. *kotschyi* (Boiss.) Buchenau as separate taxa within *J. fontanesii*. Boissier (Fl. 5:359) regarded *J. pyramidatus* as a distinct Saharo-Arabian and Irano-Turanian species, differing from the W. Mediterranean *J. fontanesii* J. Gay mainly in its larger fruiting heads. According to Buchenau, heads of *J. fontanesii* are 8-10 mm in diameter and those of var. *pyramidatus* 10-12(-14) mm in diameter. Snogerup (1971) assigned the above two varieties the rank of subspecies: subsp. *pyramidatus* (Laharpe) Snogerup and subsp. *kotschyi* (Boiss.) Snogerup.

123. GRAMINEAE
(Poaceae)

Annuals or perennials, herbaceous, rarely woody, often rhizomatous or stoloniferous. Flowering stems *(culms)* mostly cylindrical, mostly with hollow internodes and solid nodes. Leaves alternate, distichous, consisting of *sheath, ligule* and *blade*; sheath surrounding the stem with free or more rarely connate margins, sometimes with tooth-like *auricles* at mouth; ligule, along inner side of junction between sheath and blade, membranous or reduced to a fringe of hairs, rarely absent. Inflorescence a spike, panicle, raceme or false raceme (in which a sessile spikelet is accompanied by a pedicellate one). The elementary unit of the inflorescence is the *spikelet*, consisting of one to many flowers *(florets)* alternately borne on a slender axis *(rachilla)* and subtended by (0-1-)2(-3) bracts *(glumes)*; each floret subtended by a lower flowering bract *(lemma)* and an upper flowering bract *(palea)*; lemma often with a terminal or dorsal awn; palea usually membranous, 2-veined and 2-keeled; lemma or spikelet with a thickened, sometimes elongate and pungent base *(callus)*. Florets usually hermaphrodite, composed of 1-3(-6) stamens and a unilocular superior ovary with 2 hairy to plumose stigmas, which are sessile or borne on more or less elongated styles; outside the whorl of stamens are (0-)2(-3) small fleshy scales *(lodicules)* which play a role in opening of bracts during anthesis. Fruit a *caryopsis,* consisting of a single seed and mostly adherent pericarp; caryopsis sometimes adherent to lemma and palea; seed with straight embryo lying along one side of copious mostly starchy endosperm.

One of the largest families of flowering plants comprising 500-600 genera and 8,000-10,000 species widely distributed in nearly all parts of the globe and including the most important agricultural plants, such as wheat, rice, maize, barley, oats.

Literature: O. Stapf, in: D. Prain (ed.), *Flora of Tropical Africa,* 9:1-768, 1917-1920; O. Stapf & C.E. Hubbard, *op. cit.* 769-1132, 1930-1934. R.J. Roshevits, Gramineae, in: V.L. Komarov (ed.), *Flora URSS* 2:1-778, Leningrad, 1934. R. Pilger, Gramineae III, Unterfam. Panicoideae, in: Engler & Prantl, *Pflanzenfam.* ed. 2, 14 e, Leipzig & Berlin, 1940. V. & G. Täckholm & M. Drar, *Flora of Egypt* 1:125-557, Cairo, 1941. A.S. Hitchcock, *Manual of the Grasses of the United States,* USDA Misc. Publ. no. 200, ed. 2, Washington, 1950. R. Maire, *Flore de l'Afrique du Nord* 1, 1952; 2, 1953; 3, 1955, Paris. C.E. Hubbard, *Grasses,* Penguin Books, 1954. N.L. Bor, *The Grasses of Burma, Ceylon, India and Pakistan,* Oxford, 1960. H. Jacques-Félix, *Les Graminées (Poaceae) d'Afrique Tropicale* 1:1-345, Inst. Res. Agron. Trop., Bull. Sci.

8, Paris, 1962. P. Mouterde, *Nouvelle Flore du Liban et de la Syrie* 1, Text et Atlas, Beyrouth, 1966. N.N. Tzvelev, The system of the grasses (Poaceae) indigenous to the USSR, *Bot. Zhur.* 53:301–312 (1968). N.L. Bor, Gramineae, in: C.C. Townsend, E. Guest & Ali al Rawi (eds.), *Flora of Iraq* 9, Baghdad, 1968. N.L. Bor, Gramineae, in: K.H. Rechinger (ed.), *Flora iranica* 70, Graz, 1970. C.W. Clayton, Gramineae, in: F.N. Hepper (ed.), *Flora of West Tropical Africa* ed. 2, 3:349–512, 1972. V. Täckholm, *Students' Flora of Egypt* ed. 2, 609–762, Beirut, 1974. N.N. Tzvelev, *Poaceae URSS*, Leningrad, 1976 (in Russian). T.G. Tutin et al. (eds.), Gramineae, in: *Flora europaea* 5:118–267, Cambridge, 1980. P.E. Boissier, *Flora orientalis* 5:432–692, 1884. G.E. Post, *Flora of Syria, Palestine and Sinai* ed. 2, 2:687–795, 1933.

Synopsis of Tribes and Genera
Tribes in accordance with N.N. Tzvelev (Poaceae URSS, 1976)

Subfam. Pooideae A. Br.

Tribe 1. ORYZEAE Dumort.
1. Leersia

Tribe 2. BRACHYPODIEAE C.O. Harz
2. Brachypodium

Tribe 3. TRITICEAE Dumort.
3. Elymus
4. Crithopsis
5. Eremopyrum
6. Heteranthelium
7. Aegilops
8. Triticum
9. Secale
10. Hordeum
11. Taeniatherum

Tribe 4. BROMEAE Dumort.
12. Bromus
13. Boissiera

Tribe 5. AVENEAE Dumort.
14. Avena
15. Arrhenatherum
16. Gaudinia
17. Pilgerochloa
18. Lophochloa
19. Avellinia
20. Trisetaria
21. Aira
22. Antinoria
23. Corynephorus
24. Holcus
25. Milium
26. Ammophila
27. Lagurus

28. Polypogon
29. Gastridium
30. Triplachne

Tribe 6. PHALARIDEAE Kunth
31. Phalaris

Tribe 7. PHLEEAE Dumort.
32. Beckmannia
33. Phleum
34. Rhizocephalus
35. Alopecurus
36. Cornucopiae

Tribe 8. POEAE R. Br.
37. Festuca
38. Lolium
39. Vulpia
40. Loliolum
41. Catapodium
42. Ctenopsis
43. Cutandia
44. Sphenopus
45. Psilurus
46. Poa
47. Catabrosa
48. Puccinellia
49. Sclerochloa
50. Dactylis
51. Cynosurus
52. Lamarckia
53. Briza

Tribe 9. MONERMEAE C.E. Hubbard
54. Parapholis
55. Monerma

Tribe 10. SESLERIEAE Koch
  56. Echinaria
  57. Ammochloa
Tribe 11. MELICEAE Reichenb.
  58. Melica
Tribe 12. STIPEAE Dumort.
  59. Stipa
  60. Piptatherum
Tribe 13. ARUNDINEAE Dumort.
  61. Arundo
  62. Phragmites
Tribe 14. DANTHONIEAE (G. Beck)
  C.E. Hubbard
  63. Asthenatherum
  64. Schismus
Tribe 15. ARISTIDEAE C.E. Hubbard
  ex Bor
  65. Aristida
  66. Stipagrostis
Tribe 16. AELUROPODEAE Nevski ex
  Bor
  67. Aeluropus
Tribe 17. PAPPOPHOREAE Kunth
  68. Enneapogon
Tribe 18. CYNODONTEAE Dumort.
  (comprising Eragrostideae and
  Chlorideae auct.)
  69. Eragrostis
  70. Eleusine
  71. Dactyloctenium
  72. Desmostachya

  73. Dinebra
  74. Diplachne
  75. Tetrapogon
  76. Chloris
  77. Cynodon
  78. Sporobolus
  79. Crypsis
Tribe 19. PANICEAE R. Br.
  80. Panicum
  81. Echinochloa
  82. Brachiaria
  83. Paspalum
  84. Paspalidium
  85. Digitaria
  86. Setaria
  87. Pennisetum
  88. Cenchrus
  89. Tricholaena
  90. Anthephora
Tribe 20. ANDROPOGONEAE
  Dumort.
  91. Hemarthria
  92. Lasiurus
  93. Imperata
  94. Saccharum
  95. Sorghum
  96. Dichanthium
  97. Eremopogon
  98. Andropogon
  99. Hyparrhenia
  100. Cymbopogon

1. All or nearly all spikelets proliferating (viviparous), i.e. with florets transformed into leafy buds and variously developed lemmas; lemmas keeled. Inflorescence a panicle.   **46. Poa**
- Spikelets not proliferating   2
2. Spikelets enclosed in a burr-like involucre of more or less fused bristles or flattened spines; burrs arranged in a spike-like inflorescence, deciduous at maturity, ovoid or globose, 10–15 mm broad, each enclosing only few spikelets.
                                                              88. Cenchrus
- Spikelets not enclosed in a burr-like involucre   3
3. Spikelets subtended at base by an involucre of long scabrous, feathery or ciliate bristles (transformed sterile branchlets). Inflorescence a cylindrical spike-like panicle   4
- Spikelets not subtended by bristles as in above   5

4. Annuals. Spikelets falling at maturity; bristles persistent, retrorsely or antrorsely scabrous.   **86. Setaria**
- Tufted perennials. Spikelets falling at maturity together with bristles; bristles, at least inner ones, plumose or ciliate.   **87. Pennisetum**

5(3). Inflorescence appearing as a single 4-ranked terminal hairy spike; its upper part sometimes split into two 2-ranked spikes; in fact, inflorescence formed by 2 spikes united at their inner faces by means of long hairs. Glumes hyaline, persistent; lemmas long-hairy and long-awned. Plants of desert rocks.

75. Tetrapogon
- Plants not as above   6

6 Inflorescence *digitate*, i.e. consisting of 2–10 spikes or spike-like false racemes, borne at tip of culm or branch, or *subdigitate* with spikes or false racemes borne along rachis to 1–1.5 cm below tip   7
- Inflorescence neither digitate nor subdigitate   17

7. Lemmas and/or glumes awned or long-mucronate   8
- Lemmas and glumes awnless and not long-mucronate   13

8. Spikelets 2(-3) on each node of rachis; one spikelet of each group sessile (or short-pedicellate), the other(s) pedicellate   9
- Spikelets solitary on each node of rachis   12

9. Inflorescence consisting of a single terminal raceme-pair, or of a single group of 3–9 subdigitately arranged racemes   10
- Inflorescence a panicle of several lateral raceme-pairs, each pair on a peduncle subtended by a sheath or small spathe (spatheole)   11

10. Inflorescence a single terminal raceme-pair. Nodes of culms not long-bearded. Upper glume of sessile spikelet with a slender awn.   **98. Andropogon**
- Inflorescence subdigitate, of several racemes. Nodes of culms long-bearded. Glumes awnless.   **96. Dichanthium**

11(9). Exserted portion of awn about as long as spikelet; awned lemma of upper floret (upper lemma) ciliate. Rare plants of S. Negev and Arava Valley.

100. Cymbopogon
- Exserted portion of awn several times as long as spikelet; awned upper lemma glabrous. Very common plants.   **99. Hyparrhenia**

12(8). Spikelets with 1 fertile floret; lemma long-ciliate along marginal veins in its upper part and with awn nearly as long as or much longer than lemma; upper glume ending in a straight mucro or short awn.   **76. Chloris**
- Spikelet with 3 or more fertile florets; lemma not as above; upper glume ending in a curved mucro or awn.   **71. Dactyloctenium**

13(7). Lower glume as long as spikelet, coriaceous; upper glume nearly as long as lower glume, more delicate in texture, adhering to concavity of rachis; racemes glabrous.   **91. Hemarthria**
- Glumes not as above   14

14. Lower glume minute, scale-like, or absent; lower floret sterile, reduced to a lemma (lower lemma) which is as long as spikelet   15
- Lower glume nearly as long as the upper or at least ½ its length; lower floret fertile   16

15. Stoloniferous perennials. Racemes 2, rarely 3 or 1, usually 3–6 cm; lemma of upper floret with involute margins.   **83. Paspalum**
- Annual weeds. Racemes usually more than 2, 5–10 cm; lemma of upper floret with thin flat margins.   **85. Digitaria**

16(14). Spikelets 2- to several-flowered. Annuals, introduced as casual weeds from tropical countries.   **70. Eleusine**

- Spikelets with 1 fertile floret; rachilla prolonged beyond fertile floret. Perennials, with scaly rhizomes and creeping stolons, very common.   **77. Cynodon**

17(6). Inflorescence a short and dense panicle seated in a cup-shaped involucre, or closely subtended by an involucre formed by broadened sheaths of 2–3 uppermost leaves; spikelets 1-flowered   18

- Inflorescence not as above and not subtended by an involucre as above   19

18. Involucre cup-shaped.   **36. Cornucopiae**

- Involucre formed by 2–3 broadened sheaths of uppermost leaves.   **79. Crypsis**

19(17). Inflorescence a very dense, ovoid or globose head; spikelets sessile or subsessile   20

- Inflorescence differently shaped or if head-like, spikelets distinctly pedicellate 24

20. Lemma with a single dorsal awn and 2 shorter bristles or aristulae; glumes feather-like; head long-bristly.   **27. Lagurus**

- Lemma awnless or with 5–7 awns; glumes not as above   21

21. Mature inflorescence prickly; lemma with 5–7 unequal recurved lobes or awns, prickly at maturity.   **56. Echinaria**

- Mature inflorescence not prickly; lemma not as in above   22

22. Small annuals. Heads overtopped by leaves. Lemmas 5-veined   23

- Low stoloniferous perennials. Heads not overtopped by leaves. Lemmas 7-11-veined.   **67. Aeluropus**

23. Spikelets many-flowered, glabrous. Plants of dunes and sandy soils.   **57. Ammochloa**

- Spikelets 1-flowered, with clavate hairs on lower part of glumes and lemmas. Plants of steppes and deserts.   **34. Rhizocephalus**

24(19). Tufted perennials of desert rocks. Inflorescence spike-like, cylindrical, 4–7 × 0.2–0.3 cm, composed of subsessile deciduous clusters, becoming black at maturity; rachis angular, slender, wavy, with branches reduced to minute stumps; clusters consisting of 4–5 spikelets; spikelets dorsally compressed; their lower glumes smooth, forming a false involucre to the cluster with gaps between their bases.   **90. Anthephora**

- Plants with a different set of characters   25

25. Spikelets arranged in a true spike or in a spike-like raceme with pedicels not exceeding 2 mm; spike or raceme solitary, or in some cases several to many spikes or racemes are borne on a central rachis and form a compound inflorescence   26

- Spikelets arranged in a panicle with long or short branches and pedicels, or in a raceme with pedicels longer than 2 mm; in some plants branches of panicle much reduced and present only on lower nodes of rachis   59

26. Spike or raceme solitary   27

- Spikes or racemes in compound inflorescences, sessile or short-pedunculate, spread along a central rachis or gathered into dense, globose or oblong heads 51

27. Glumes connate by their margins for ¼ to ½ of their length; lemma with a dorsal awn; spikelets 1-flowered; pedicels cupuliform at tip. Inflorescence with branches bearing single spikelets on very short pedicels.   **35. Alopecurus**

- Glumes not connate or connate just at base. Plants with a different set of characters   28

28. Spikelets solitary at each node of rachis, rarely 2 or 3 at lower nodes   29
 - Spikelets in groups of 2 or 3 at each node of rachis   46
29. Lateral spikelets with 1 glume (upper glume); terminal spikelets with 2 glumes   30
 - Lateral spikelets with 2 glumes; lower glume sometimes minute   32
30. Glume not longer than 2 mm, about ¼ length of spikelet (excl. awn); lemma awned.   **45. Psilurus**
 - Glume longer   31
31. Spikelets 2- to many-flowered.   **38. Lolium**
 - Spikelets 1-flowered.   **55. Monerma**
32(29). Lemma with 1 dorsal or subterminal awn (or 3–5 awns); apex of lemma often 2-lobed or 2-dentate   33
 - Lemma with a terminal awn or awns, or lemma awnless   34
33. Spikelets sessile; awn of lemma distinctly geniculate at maturity, twisted below bend.   **16. Gaudinia**
 - Spikelets pedicellate; awn or awns not geniculate at maturity, straight or recurved, only slightly twisted at base.   **12. Bromus**
34(32). Lower glume at most ⅓ length of upper, sometimes minute; upper glume somewhat shorter than spikelet   35
 - Glumes equal in length or, if unequal, lower glume at least ½ length of upper   36
35. Lemma with a terminal awn at least ½ length of lemma proper.   **39. Vulpia**
 - Lemma mucronate but not awned.   **42. Ctenopsis**
36(34). Spikelets 1-flowered, awnless   37
 - Spikelets 2- to many-flowered, awned or awnless   38
37. Inflorescence a true spike with spikelets sessile in concavities of the spike-rachis; glumes seated side by side in front of each lateral spikelet.   **54. Parapholis**
 - Inflorescene with laterally compressed spikelets borne on very short pedicels; glumes boat-shaped, keeled, not seated as above.   **79. Crypsis**
38(36). Glumes subulate; rachis of spike breaking into a number of sections, each of which bears 1 or 2 fertile spikelets and 1 to several sterile spikelets reduced to empty glumes; spike very bristly; lemma papillose and long-awned.   **6. Heteranthelium**
 - Spike with a different set of characters   39
39. Spike 6–14 cm (excl. awns), long-awned; spikelets with 2 hermaphrodite florets; lemma 15–24 mm, pectinately spinulose-ciliate on the keel, ending in an awn much longer than lemma proper. Perennials of rocky mountain slopes in northern districts.   **9. Secale**
 - Lemma not spinulose-ciliate. Plants with a different set of characters   40
40. Glumes very unequal; longer glume less than ½ length of spikelet; lemma with a straight terminal awn. Spikelets 1.5–3.5 cm (excl. awns).   **2. Brachypodium**
 - Plants with a different set of characters   41
41. Spikelets in 2 rows on one side of a flattened rachis, oriented with their narrow side towards rachis. Annuals   42
 - Spikelets not oriented as above   43
42. Glumes considerably shorter than spikelet; lemma muticous.   **41. Catapodium**
 - Glumes about as long as spikelet; lemma short-awned.   **40. Loliolum**
43(41). Glumes and lemmas acuminate or tapering to a short awn. Spike 1.5–4 × 1.25 cm; rachis fragile at maturity.   **5. Eremopyrum**

- Plants with a different set of characters   44
44. Glumes symmetrical. Tall perennials of marshes, maquis or seashore. **3. Elymus**
- Glumes asymmetrical. Annuals   45
45. Glumes boat-shaped with 1 or 2 keels; keels prominent from base to apex. Culti-
    vated plants with a tough spike-rachis or wild plants with a fragile rachis; in
    wild plants spikelet densely bearded at base.   **8. Triticum**
- Glumes rounded on back, not keeled, rarely with one slightly prominent keel
    and then base of spikelet not bearded.   **7. Aegilops**
46(28). Spikelets 3 at each node of rachis; lateral spikelets of triplet generally stami-
    nate or neuter, short-pedicellate; median spikelet hermaphrodite, sessile or sub-
    sessile; lemma in at least one of the spikelets long-awned.   **10. Hordeum**
- Spikelets 2 at each node of rachis, or if 3, lemmas not awned   47
47. Spikelets awnless; back of upper glume turned towards rachis.   **91. Hemarthria**
- Spikelets awned or lower glume long-caudate   48
48. One of the pair of spikelets sessile, the other pedicellate   49
- Both spikelets sessile; glumes subulate or strap-shaped   50
49. Lower glume of sessile spikelet with a roundish pit above middle; lemma bear-
    ing a long geniculate awn.   **97. Eremopogon**
- Lower glume of sessile spikelet devoid of a pit; lemma not awned. **92. Lasiurus**
50(48). Awns 6–10 cm; rachis tough.   **11. Taeniatherum**
- Awns at most 3 cm; rachis fragile.   **4. Crithopsis**
51(26). Spikelets 2-flowered, dorsally compressed; lower floret staminate or reduced
    to a lemma (lower lemma); upper floret hermaphrodite; upper glume about as
    long as lower lemma; spikelets falling entire at maturity   52
- Spikelets 1- to many-flowered, laterally compressed and, if 2-flowered, both
    florets or lower one hermaphrodite   55
52. Ligule membranous. Lower glume absent; upper glume and lower lemma den-
    sely and softly long-ciliate at margin, broadly ovate and apiculate; racemes 5–9
    cm; spikelets 2.5–3.5 mm.   **83. Paspalum**
- Ligule present as row of cilia or absent. Glumes and lemmas not as in above 53
53. Ligule absent altogether or present in upper leaves only. Lower glume apiculate
    or cuspidate, veined; lower lemma apiculate, cuspidate or awned.
                                                                81. Echinochloa
- Ligule present as row of dense cilia in all leaves. Lower glume veinless or ob-
    scurely veined, sometimes minute   54
54. Lower glume abaxial (turned away from rachis); spikelets glabrous, ovate in
    outline, less than 2 times as long as broad, in 2 regular rows on a stiff rachis.
                                                                84. Paspalidium
- Lower glume adaxial (turned towards rachis); spikelets glabrous or pubescent,
    oblong-lanceolate or oblong-ovate in outline, more than 2 times as long as
    broad, mostly in irregular rows.   **82. Brachiaria**
55(51). Spikelets subcircular or obovate in outline, 1-2-flowered; glumes equal,
    slightly shorter than florets; spikes appressed to main rachis. **32. Beckmannia**
- Plants with a different set of characters   56
56. Glumes narrow, acuminate-aristate, overtopping florets; spikes spreading from
    main rachis or reflexed. Annuals.   **73. Dinebra**
- Glumes acute or obtuse, shorter than spikelet. Perennials   57
57. Inflorescence consisting of 4–8(–15) cm spike-like racemes; spikelets spaced

along rachis of raceme; lemma emarginate at apex, with midvein excurrent into a mucro from the sinus.   **74. Diplachne**
- Inflorescence consisting of 1–2.5 cm spikes or racemes; lemma entire at apex   58
58. Inflorescence 20–40 cm, dense; spikes very numerous, close on main axis; lemma faintly 3-veined. Leaf-blades very long (30–60 cm).   **72. Desmostachya**
- Inflorescence much shorter; spikes in terminal heads or spaced along main rachis; lemma distinctly 7–11-veined. Leaf-blades short.   **67. Aeluropus**
59(25). Inflorescence a one-sided panicle with spikelets conspicuously of 2 kinds, fertile and sterile; each branch bearing one or a few fertile and several sterile spikelets (empty glumes and lemmas); panicle awned, at most 4 times as long as broad (excl. awns); width not exceeding 2.5 cm   60
- Inflorescence differing from above; spikelets not of 2 kinds or, if fertile and sterile spikelets on same branch, panicle not awned   61
60. Glumes and lemmas of sterile spikelets awnless and obtuse.   **52. Lamarckia**
- Sterile glumes and lemmas extended into long awns.   **51. Cynosurus**
61(59). Spikelets in pairs (rarely in threes) on each node of ultimate branches; one spikelet of each pair sessile, the other(s) pedicellate; glumes awnless, firmer than mostly hyaline lemma   62
- Spikelets arranged differently   64
62. Spikelets of each pair similar, hermaphrodite; inflorescence long-hairy   ·63
- Spikelets of each pair dissimilar, the sessile hermaphrodite, the pedicellate staminate or neuter; inflorescence not long-hairy.   **95. Sorghum**
63. Inflorescence a terminal spike-like dense panicle about 2 cm broad.   **93. Imperata**
- Inflorescence broader, very long (20–60 cm), of numerous very fragile racemes.   **94. Saccharum**
64(61). Tall reeds (1.5–6 m) of river-banks and other moist habitats, with large feathery decompound panicles. Lemma or rachilla bearing long soft hairs that envelop florets   65
- Plants with a different set of characters   66
65. Glumes more or less equal; lemma covered with long hairs.   **61. Arundo**
- Lower glumes ½ length of upper; rachilla bearing long hairs.   **62. Phragmites**
66(64). At least one lemma in each spikelet with 7–9 awns; spikelets at most 6-flowered. Plants of steppes and deserts   67
- Lemma awnless, or with 1–3 awns, or, if with up to 5 awns, spikelets with more than 6 florets   68
67. Lemma (excl. awns) about 2 mm; glumes much longer than lemma (excl. awns).   **68. Enneapogon**
- Lemma much longer than above; glumes not longer than lemma (excl. awns).   **13. Boissiera**
68(66). Glumes awned; awn 3–4 times as long as spikelet; spikelets 1-flowered. Annuals.   **28. Polypogon**
- Glumes not awned or, if awned, awns shorter or spikelet not 1-flowered; rarely glumes absent   69
69. Lemma (or some of lemmas) bearing an awn which is as long as to longer than lemma proper, rarely ⅔ its length (sometimes awn concealed between glumes!)   70
- Lemma awnless or provided with an awn considerably shorter than lemma proper (at most ½ its length)   91

70. Awn of lemma divided into 3 branches   71
  - Awn not branched   72
71. Branches of awn glabrous or scabrous.   **65. Aristida**
  - Branches of awn (all or median branch only) feathery.   **66. Stipagrostis**
72(70). Glumes connate by their margins at least in lower part; spikelets 1-flowered.
                                                                  35. Alopecurus
  - Glumes not connate   73
73. Spikelets arranged along one side of flattened rachis, 2-flowered; only upper
    floret fertile; glumes very unequal, lower glume about as long as broad.
                                                                 81. Echinochloa
  - Spikelets not arranged as above, 1- to many-flowered; lower glume not as above
                                                                                  74
74. Spikelets 1-flowered   75
  - Spikelets 2- to several-flowered   80
75. Awn of lemma 5–25 cm.   **59. Stipa**
  - Awn shorter   76
76. Annuals. Inflorescence spike-like   77
  - Tall perennials (50–100 cm). Inflorescence an effuse or contracted panicle, not
    spike-like; lemma indurated at maturity   79
77. Lemma with 1 awn forming prolongation of its midvein; glumes firm, glossy
    and ventricose at base.   **29. Gastridium**
  - Lemma with a geniculate dorsal awn and with 2 bristle-like cusps forming
    prolongation of 2 lateral veins; glumes membranous, not inflated at base   78
78. Dorsal awn arising from lemma above its middle.   **20. Trisetaria**
  - Dorsal awn arising from near base of lemma.   **30. Triplachne**
79(76). Lemma or caryopsis glabrous or, if hairy, ligule 4–10 mm; panicle effuse.
                                                                  60. Piptatherum
  - Lemma hairy; ligule shorter or absent; panicle contracted.   **59. Stipa**
80(74). Glumes awned; awn as long as to ¾ length of glume proper   81
  - Glumes awnless or with awn much shorter than glume proper   82
81. Spikelets 2-flowered; lower floret hermaphrodite, awnless, the upper staminate,
    ending in a short flexuose awn.   **24. Holcus**
  - Spikelets 4–9-flowered, each floret with an awned lemma.   **39. Vulpia**
82(80). Awn of lemma 2.5–3.5 mm, hardly exserted from glumes, articulated near
    middle and with a hairy ring at joint, brownish and twisted below middle, pale
    and club-shaped above, arising at or near base of lemma; branches of panicle
    capillary.   **23. Corynephorus**
  - Awn of lemma not as above   83
83. Perennial of sand dunes. Lemma 2-lobed, with a straight awn in the sinus and
    veins ending in tufts of hairs as long as lobes; length of lemma (excl. awn) not
    exceeding 5 mm; width of panicle at most 1.5 cm; spikelets 2–3-flowered; florets
    hermaphrodite. Ligule a fringe of cilia.   **63. Asthenatherum**
  - Plants not as above. Ligule membranous or absent   84
84. Perennial of hill districts; culms with 1 or 2 globose tubers at base. Spikelets
    about 1 cm, 2-flowered; lower floret staminate with a long geniculate awn aris-
    ing from base of lemma; upper floret hermaphrodite with a shorter awn, arising
    from upper part of lemma.   **15. Arrhenatherum**
  - Plants not as above   85

85. Spikelets about 2 mm (excl. awns); glumes subequal in length and width; panicle very lax, ovate in outline.   **21. Aira**
 - Spikelets larger   86
86. Awn of lemma apical; glumes very unequal; spikelets (excl. awns) 6–10 mm.
                                                                         39. Vulpia
 - Awn of lemma dorsal or arising in or somewhat below sinus between 2 apical
teeth   87
87. Spikelets (excl. awns) 1 cm or longer   88
 - Spikelets (excl. awns) not exceeding 8 mm   90
88. Awn of lemma dorsal, geniculate and twisted below knee, arising at about middle of lemma   89
 - Awn(s) of lemma not geniculate, arising in or below sinus between apical lobes
or teeth of lemma.   **12. Bromus**
89. Glumes nearly equal or, if unequal, lower glume at least ½ length of upper;
pedicels flexuose, often curved under spikelet; spikelets usually nodding.
                                                                          14. Avena
 - Glumes very unequal; lower glume less than ½ length of upper; pedicels
straight, rigid, club-shaped; spikelets not nodding.   **17. Pilgerochloa**
90(87). Lemma with a dorsal geniculate awn arising below sinus between 2 lobes;
each lobe ending in a bristle; spikelets 1–2-flowered; glumes longer than florets
(excl. awns) or as long as florets. (In some plants of *Trisetaria koelerioides* lemmas awnless and devoid of apical bristles).   **20. Trisetaria**
 - Lemma with a straight awn arising from sinus between 2 lobes or teeth which
are acute or obtuse and not ending in a bristle; spikelets 3- to several-flowered;
glumes somewhat shorter than spikelet.   **18. Lophochloa**
91(69). Glumes absent; spikelet 1-flowered, laterally compressed, 3 mm; lemma
keeled, ciliate; stamens 6. Rare plant of humid places.   **1. Leersia**
 - Glumes 2 or 1; stamens 1–3   92
92. Spikelets 2-flowered; lower floret staminate or reduced to a lower lemma; upper
floret hermaphrodite; upper glume about as long as lower lemma; spikelets
deciduous as a whole at maturity   93
 - Spikelets not as above, 1- to many-flowered and, if 2-flowered, both florets or
lower one hermaphrodite   95
93. Inflorescence compound, of several racemes or spikes borne along central rachis;
lower glume adaxial (turned towards rachis). Nodes of culm densely bearded.
                                                                     82. Brachiaria
 - Inflorescence an effuse or compact panicle; lower glume abaxial (turned away
from rachis)   94
94. Glumes and lower lemma covered with conspicuous long silky hairs.
                                                                    89. Tricholaena
 - Glumes and lower lemma glabrous or short-hairy.   **80. Panicum**
95(92). Spikelet with 1 or 2 sterile lemmas forming an apical clavate or cylindrical
structure above the 1 or 2 hermaphrodite florets. Perennials; leaf-sheaths tubular
to mouth.   **58. Melica**
 - Spikelets not as above. Plants with a different set of characters   96
96. Spikelets containing 1 hermaphrodite (fertile) floret   97
 - Spikelets containing 2 or more hermaphrodite florets   104
97. Hermaphrodite floret subtended at base by 1 or 2 minute scales (reduced lemmas

of sterile florets), at maturity falling off from persistent glumes together with
basal scales; glumes keeled, often winged; spikelets strongly compressed.

31. Phalaris

- Hermaphrodite floret devoid of basal scales; sometimes prolongation of rachilla
  bearing a reduced upper floret   98
98. Tall rhizomatous perennial of sand dunes. Panicle spike-like, cylindrical, 7–25
    cm long, up to 2.5 cm thick; glumes 8 mm or longer; lemma about as long as
    glumes; callus long-bearded.   **26. Ammophila**
- Plants with a different set of characters   99
99. Panicle very lax, ovate in outline, 5–15 cm broad; branches very long, widely
    spreading; pedicels much longer than spikelets.   **25. Milium**
- Panicle narrower, often spike-like; branches not as above; pedicels not longer
  than spikelets   100
100. Ligule a fringe of hairs   101
- Ligule membranous or hyaline   102
101. Perennials. Glumes distinctly unequal; ripe caryopsis falling from lemma and
     palea, red in colour.   **78. Sporobolus**
- Annuals. Glumes subequal; caryopsis not as above.   **79. Crypsis**
102(100). Glumes unequal, ventricose at base.   **29. Gastridium**
- Glumes equal or subequal, not as above   103
103. Glumes abruptly ending in a short awn or a short incurved mucro, firmly com-
     pressed and keeled; panicle very dense, cylindrical or ovoid.   **33. Phleum**
- Glumes muticous, not as above; panicle pyramidal-oblong, usually interrupted,
  lobed.   **28. Polypogon**
104(96). Perennials with stiff culms, 50–80 cm; panicle 3–12 cm, contracted except
         during anthesis; spikelets 5–9 mm, in dense, strongly compressed, sessile or sub-
         sessile clusters, 3 or more to a cluster.   **50. Dactylis**
- Spikelets not in clusters as above. Plants with a different set of characters   105
105. Spikelets not longer or slightly longer than glumes   106
- Spikelets considerably longer than glumes   110
106. Inflorescence an effuse panicle; spikelets 1–1.5 mm.   **22. Antinoria**
- Inflorescence more or less spike-like; spikelets longer   107
107. Lower glume subulate above and about ¼ length of lanceolate upper glume,
     which is nearly as long as spikelet; lemmas linear-lanceolate, very narrow, 2-fid
     into subulate teeth and with an awn in the sinus; palea much shorter than
     lemma.   **19. Avellinia**
- Glumes and lemmas not as above   108
108. Ligule a fringe of hairs. Lemma 7–9-veined; glumes strongly veined; lower
     glume 5–7-veined.   **64. Schismus**
- Ligule membranous. Lemma 5-veined; lower glume 1-veined   109
109. Upper glume ovate, about twice as long as broad; lemma 2-lobed or emarginate,
     awnless or sometimes with a short straight awn in sinus between lobes.

18. Lophochloa

- Upper glume lanceolate, at least 4 times as long as broad; lemma usually with 2
  apical bristles and with a dorsal geniculate awn.   **20. Trisetaria**
110(105). Lemma suborbicular to broadly ovate, cordate at base, 5–9-veined; glumes
          deeply concave, suborbicular when flattened; spikelets broadly ovate.   **53. Briza**
- Lemma and glumes not as above   111

111. Glumes membranous; lower glume minute, veinless, obovate and nearly as long as broad or rarely ovate and eroded at tip; spikelets 2–3(–4) mm; panicle effuse or rarely contracted   112
  – Plants with a different set of characters   113
112. Panicle 6 × 4 cm; branches in pairs at each node of rachis, widely spreading, capillary, di- and trichotomously divided and redivided; pedicels of spikelets club-shaped. Annuals of arid habitats.   **44. Sphenopus**
  – Panicle with several branches at nodes of rachis; pedicels not club-shaped. Rhizomatous perennials of humid habitats.   **47. Catabrosa**
113(111). Lemma 3-veined; upper glume 1-veined   114
  – Lemma 5-veined; upper glume 3–9-veined, rarely 1-veined   115
114. Ligule membranous or hyaline, about as long as broad. Branches and pedicels of panicle stout, rigid.   **43. Cutandia**
  – Ligule reduced to a rim of hairs or, if membranous, very short and distally ciliate. Branches and pedicels of panicle slender, not rigid.   **69. Eragrostis**
115(113). Annuals   116
  – Perennials   119
116. Lemma 2-lobed at apex, often with a mucro or short awn in the sinus; panicle dense, spike-like or head-like.   **18. Lophochloa**
  – Lemma not as above   117
117. Branches and pedicels of panicle filiform; lemma keeled, keel and veins hairy.
                                                                         46. Poa
  – Panicle and lemma not as above   118
118. Glumes distinctly unequal in length, very obtuse, the upper 5–9-veined, broadly hyaline-margined; inflorescence a dense panicle, 1.5–3.5 cm.   **49. Sclerochloa**
  – Glumes nearly equal in length, the upper 1–3-veined; inflorescence usually longer, a panicle or a spike-like raceme.   **41. Catapodium**
119(115). Culms bulbous at base or, if not, lemma with a tuft of long wavy wool at base; pedicels and branches filiform.   **46. Poa**
  – Culms not bulbous at base; lemma without wool at base   120
120. Lemma very obtuse to truncate; hilum punctiform.   **48. Puccinellia**
  – Lemma acute to acuminate, muticous, mucronate or awned; hilum linear, long
                                                                              121
121. Leaves with falcate auricles. Caryopsis hairy at apex, but without a hairy appendage. Spikelets about 1 cm; lemma 6–7 mm.   **37. Festuca**
  – Leaves without auricles. Caryopsis with a hairy apical appendage. Spikelets 2–3(–4) cm; lemma 9–15 mm.   **12. Bromus**

Tribe 1. ORYZEAE Dumort. (1823). Annuals or perennials. Leaf-sheaths split nearly to base, without auricles; ligule membranous, lacerate. Inflorescence a contracted or, more rarely, lax panicle. Spikelets all alike, generally hermaphrodite, 1–3-flowered; the terminal floret hermaphrodite, the 2 lower florets, when present, empty and scale-like. Glumes 2, very short, often rudimentary; lemma of terminal floret coriaceous or scarious, 5-veined and often with a terminal awn; palea 3-veined. Lodicules 2, entire or 2-lobed. Stamens 6, rarely 1–3. Caryopsis ellipsoid; embryo shorter than caryopsis; hilum linear. Chromosomes small; x=12.

Nine genera, mainly in trop. and subtrop. regions but also in temperate parts of both hemispheres.

1. LEERSIA Swartz

Perennial aquatic herbs. Leaf-blades flat, more or less scabrous; ligule membranous. Inflorescence a panicle with slender branches and subsessile or shortly pedicellate spikelets. Spikelets hermaphrodite, strongly laterally compressed, with 2 lower florets rudimentary, the upper hermaphrodite. Glumes absent; lemma of hermaphrodite floret scarious, rigid, strongly compressed, boat-shaped, sharply keeled, 5-veined, awnless, tightly enclosing palea; palea as long as lemma, but narrower, 3-veined. Lodicules 2. Stamens 6 or 3, rarely 1. Stigmas plumose. Caryopsis obovate, strongly compressed; hilum linear.

Fifteen species of trop. and subtrop. regions.

1. Leersia hexandra Swartz, Nov. Gen. Sp. Pl. 21 (1788); Boiss., Fl. 5:469; Post, Fl. 2:712. [Plate 198]

Plant up to 1 m. Culms stoloniferous at base, rooting at nodes, then ascending, slender, smooth and glabrous, bearded at nodes. Leaf-blades narrowly linear-acuminate, glabrous, more or less scabrous on veins and margins; ligule truncate, lacerate. Panicle erect; branches few, scarcely spreading. Spikelets 3.5–4 mm, obliquely ovate-oblong, unilaterally imbricate, on very short pedicels, caducous; lemma pectinate-ciliate or -aculeolate along keel and margins. Stamens 6. Fl. June–September.

Hab.: Marshes and wet sites. Sharon Plain, Philistean Plain; Dan Valley, Hula Plain; Golan.

Area: Mediterranean, Irano-Turanian, Saharo-Arabian and Trop.

Tribe 2. BRACHYPODIEAE C.O. Harz (1880). Annuals or perennials. Leaf-blades flat; sheaths usually split nearly to base, without auricles; ligules membranous, short, densely short-hairy at back. Inflorescence a terminal spike-like 1–10-spiculate raceme. Spikelets all alike, sessile or shortly pedicellate, several- to many-flowered, their narrow side turned towards rachis; rachilla disarticulating above glumes and between florets. Glumes 2, unequal, coriaceous, shorter than lower florets, persistent; lemma (5–)7–9-veined, coriaceous, not keeled, usually with a straight terminal awn; palea 2-veined, 2-keeled, pectinate-ciliate. Lodicules 2, broadened in their upper part, ciliate towards apex. Stamens 3. Caryopsis narrowly ellipsoid, grooved, with a hairy apical appendage, adherent to lemma and palea; embryo small; hilum linear, about as long as caryopsis. Chromosomes small; x=5, 7, 9.

One genus in subtrop. and temperate regions of Eurasia and Africa and in mountain zones of Tropics of both hemispheres.

2. Brachypodium Beauv.

Annuals or perennials, often with branched rhizomes. Leaf-blades flat; ligule membranous, short. Inflorescence a terminal raceme, 1–10-spiculate, with distichous subsessile or shortly pedicellate spikelets whose narrow side is turned towards rachis. Spikelets several- to many-flowered; florets hermaphrodite or the uppermost staminate; rachilla fragile. Glumes unequal, persistent, shorter than spikelet and than lowest floret, many-veined; veins elevated; lemma coriaceous-membranous, rounded on back, lanceolate, with a straight terminal awn or mucro; palea truncate, shorter than lemma, pectinate-ciliate along keels in upper part. Lodicules 2, ciliate towards tip. Stamens 3. Caryopsis oblong or narrowly elliptic, with a hairy apical appendage, enveloped by lemma and palea; embryo very small; hilum linear, as long as caryopsis.

About 20 species in temperate regions and in mountains of Tropics.

1. Annual. Racemes 1–4(–6)-spiculate; anthers not exceeding 1 mm. Plants 15–25(–40) cm.   **2. B. distachyon**
- Rhizomatous perennial. Racemes 6–8(–15)-spiculate; anthers 3.5–5.5 mm. Plants 50–80 cm.   **1. B. pinnatum**

Sect. 1. BRACHYPODIUM. Perennial. Raceme usually elongate, 6–15-spiculate. Anthers at least 2.5 mm.

1. Brachypodium pinnatum (L.) Beauv., Agrost. 101, 155 (1812); Bor, Fl. Iraq 9:167, t.57; Boiss., Fl. 5:658; Post, Fl. 2:778. *Bromus pinnatus* L., Sp. Pl. 78 (1753). [Plate 199]

Perennial, 50–80 cm, with creeping rhizome. Culms erect, rigid, not branched. Raceme 7–20 cm, erect or nodding, 6–8(–15)-spiculate. Spikelets 2–3.5 cm (excl. awns), many-flowered, spaced on raceme; rachilla fragile. Glumes oblong-lanceolate, acute, mucronate, with elevated veins; lower glume 4–7 mm, the upper 6–9 mm; lemma 6–10 mm (excl. awn), 7-veined, glabrous, ending in an awn 1–5 mm. Anthers 3.5–5.5 mm. Fl. May–June.

Hab.: Maquis and garigue. Upper Galilee, Mt. Carmel.

Area: Euro-Siberian, Mediterranean and Irano-Turanian.

Sect. 2. TRACHYNIA (Link) Nyman, Consp. 843 (1882). Annual. Racemes short, stiff, few-spiculate. Anthers at most 1 mm.

2. Brachypodium distachyon (L.) Beauv., Agrost. 101, 155 (1812); Boiss., Fl. 5:657; Post, Fl. 2:777. *Bromus distachyos* L., Cent. Pl. 2:8 (1756). *Trachynia distachya* (L.) Link, Hort. Berol. 1:43 (1827). [Plate 200]

Annual, 15–25(–40) cm, glaucous. Culms fasciculate, geniculately ascending, rarely single, erect. Leaves glabrous or more or less pilose. Raceme 2–4 cm. Spikelets appressed to axis, 1.5–3.5 cm (excl. awns), 10–14-flowered, imbricate, glabrous or hairy; rachilla fragile except in lower part of spikelet. Glumes and lemmas rounded on back; glumes lanceolate, with elevated veins;

lower glume 6–7 mm, the upper 10–11 mm; lemma 8–10 mm, with elevated veins; awn up to 1½ times as long as lemma proper in upper florets. Fl. February–May.

Var. **distachyon.** Spikelets glabrous.

Hab.: Batha, fields and roadsides. Acco Plain, Coast of Carmel, Sharon Plain, Philistean Plain; Upper and Lower Galilee, Mt. Carmel, Esdraelon Plain, Mt. Gilboa, Samaria, Shefela, Judean Mts., Judean Desert, N. and C. Negev; Upper Jordan Valley, Dead Sea area; ?Golan.

Var. **hispidum** Pamp., Bull. Soc. Bot. Ital. 1914:11 (1914). *Trachynia distachya* var. *hispida* (Pamp.) Bor, Fl. Iraq 9:170 (1968). Spikelets more or less hairy.

Hab.: As above. Sharon Plain, Philistean Plain; Upper and Lower Galilee, Mt. Carmel, Esdraelon Plain, Shefela, Judean Mts., Judean Desert, N. Negev; Upper Jordan Valley, Dead Sea area; Golan, Gilead, Ammon, Moav, Edom.

Area of species: Mediterranean and Irano-Turanian, extending into the Saharo-Arabian.

Variation in indumentum not confined to spikelets but also found in leaves.

Tribe 3. TRITICEAE Dumort. (1823). Annuals or perennials. Leaf-blades flat; sheaths split nearly to base (on vegetative shoots often tubular); auricles often present; ligule membranous. Inflorescence a terminal spike or spike-like raceme. Spikelets sessile or subsessile, alternating on opposite sides of rachis, solitary or in groups of 2–3, hermaphrodite or some staminate or neuter, 1- to many-flowered; rachis of spike fragile or tough. Glumes 2, coriaceous, strongly veined; lemma 5–13-veined, awnless or awned; awn terminal, neither twisted nor geniculate; palea 2-veined, 2-keeled. Lodicules 2, entire or dentate, mostly ciliate or hairy at margin. Stamens 3. Ovary hairy, with nearly sessile plumose stigmas and hairy apical appendage. Caryopsis free or adherent to lemma and palea; embryo varying in size; hilum linear, long. Chromosomes large; x=7.

3. ELYMUS L.

Perennials, tufted or rhizomatous. Leaves flat or rolled, auriculate; ligule a narrow membranous rim. Inflorescence a terminal spike; rachis mostly tough. Spikelets solitary at each node of rachis, rarely in pairs (not in our region), oriented with their broad side towards rachis, several- to many-flowered; rachilla disarticulating above glumes and between florets, sometimes spikelets falling entire. Glumes more or less equal, with 1 to several distinct veins, shortly awned or awnless, acute or obtuse; lemma lanceolate, 5-veined; palea 2-keeled. Lodicules 2, ciliate in upper half. Stamens 3. Caryopsis fusiform with a hairy apical appendage; hilum linear, as long as caryopsis. x=7.
About 100 species in extratropical regions of both hemispheres.

Literature: H. Runemark & W.K. Heneen, *Elymus* and *Agropyron,* a problem of generic delimitation, *Bot. Not.* 121:51-79 (1968). A. Melderis, *Elymus, Jour. Linn. Soc. London (Bot.)* 76:369-384 (1978).

1. Lemma awned; awn 1.5-3.5 cm. Plant of maquis.   **1. E. panormitanus**
- Lemma muticous or shortly awned   2
2. Glumes 10-18 mm, at least ⅔ length of spikelet in middle of spike. Plant of sandy soils near seashore.   **3. E. farctus**
- Glumes 6-8(-10) mm, about ½ length of spikelet in middle of spike or shorter. Plant of salt marshes.   **2. E. elongatus**

1. Elymus panormitanus (Parl.) Tzvelev, Sched. Herb. Fl. URSS 18:27 (1970). *Agropyron panormitanum* Parl., Pl. Rar. Sic. fasc. 2.20 (1840); Boiss., Fl. 5:663; Post, Fl. 2:778. [Plate 201]

Perennial, 80-100 cm, with short rhizome. Culms erect, with brown nodes. Leaves flat. Spike 10-20 cm (excl. awns), with loosely imbricated spikelets; rachis tough, scabrous at angles. Spikelets 3-5-flowered, appressed to rachis. Glumes 15-20 mm, lanceolate, prominently 7-9-veined and sulcate, gradually tapering to a mucro or short awn; lemmas shorter than glumes, veined towards tip, scabridulous in upper half, ending in a straight 1.5-3.5 cm awn; callus short, appressed-hairy. Anthers about 5 mm. Fl. May-June.

Hab.: Maquis. Upper Galilee; Golan.

Area: S. Mediterranean, extending into the W. Irano-Turanian (S. Spain, S. Italy, Jugoslavia, Romania, S. E. Russia, Turkey, Lebanon, Palestine, Syria, Iraq, Iran).

2. Elymus elongatus (Host) Runemark, Hereditas 70:156 (1972). *Triticum elongatum* Host, Gram. Austr. 2:18, t.23 (1802). *Agropyron elongatum* (Host) Beauv., Agrost. 102 (1812); Boiss., Fl. 5:665; Post, Fl. 2:779; incl. var. *multiflorum* Eig, Bull. Inst. Agr. Nat. Hist. Tel-Aviv 6:69 (1927) et var. *haifensis* Melderis in Rech. fil. (under *Elytrigia),* Ark. Bot. ser. 2, 2(5):304 (1952). [Plate 202]

Perennial, 30-80(-100) cm, tufted. Culms glabrous. Leaves mostly rolled, erect, rigid, ending in a sharp point. Spike 10-25 cm; rachis tough, scabridulous at angles. Spikelets 10-15(-25) mm, 6-13(-25)-flowered, appressed to rachis, laterally compressed, awnless. Glumes 6-8(-10) mm, 5-9-veined, obtuse or truncate; lemma 7-10 mm, obtuse, emarginate at apex, glabrous. Anthers 4-4.5 mm. 2n=14. Fl. April-July.

Hab.: Salt marshes. Acco Plain, Coast of Carmel, Sharon Plain, Philistean Plain; Esdraelon Plain; Golan.

Area: Mediterranean.

"... mostly found amid the tufts of *Juncus acutus* L. and *Juncus maritimus* Lam. which protect it from the herds of the Bedouin camping in the marshes" (A. Eig, *loc. cit.*).

3. Elymus farctus (Viv.) Runemark ex Melderis, Jour. Linn. Soc. London (Bot.) 76:382 (1978) subsp. **farctus.** *Triticum farctum* Viv., Ann. Bot. (Konig &

Sims) 1(2):159 (1804). *Agropyron farctum* Viv., Fl. Ital. Fragm. 1:28, t.26 (1808). *Agropyron junceum* (L.) Beauv., Agrost. 102, 146 (1812) subsp. *mediterraneum* Simonet & Guinochet, Bull. Soc. Bot. Fr. 85:176 (1938). Boiss., Fl. 5:665; Post, Fl. 2:779. [Plate 203]

Perennial, 50–80 cm, with long creeping rhizome. Culms rigid, fairly thick. Leaves usually rolled, rigid, glaucous, with elevated veins. Spike 15–35 cm, erect, 8–12(–15)-spiculate, usually longer than leaves; rachis more or less fragile, breaking at maturity above each spikelet, glabrous at angles. Spikelets 10–25 mm, 5–9-flowered, glabrous, appressed to rachis, laterally compressed, awnless. Glumes 10–18 mm, lanceolate to oblong, obtuse, 6–12-veined, asymmetrically keeled, as long as or somewhat shorter than spikelet; lemma 10–20 mm, obtuse, keeled towards apex; palea spinulose-ciliate along keels. Anthers (7–)10–12 mm. 2n=42, 56. Fl. April–June.

Hab.: Seashores, sandy soil or sandy loam. Coastal Galilee, Acco Plain, Coast of Carmel, Sharon Plain.

Area of subspecies: Mediterranean.

This species was first recorded from Palestine by Eig (*op. cit.* 67, 1927).

The following additional subspecies are not found in Palestine: subsp. *borealiatlanticus* (Simonet & Guinochet) Melderis (2n=28) of N. and W. Europe, subsp. *bessarabicus* (Săvul. & Rayss) Melderis (2n=14) of the Black Sea coast from Bulgaria to Crimea and subsp. *rechingeri* (Runemark) Melderis (2n=14) of the Aegean.

4. CRITHOPSIS Jaub. & Spach

Annuals. Leaves flat, long-auriculate; ligule very short. Inflorescence a terminal spike, dense, bristly, with spikelets in pairs at each node of rachis; rachis fragile, densely hairy. Spikelets 2-flowered; lower floret hermaphrodite, the upper neuter, long-pedicellate; rachilla flattened, disarticulating above glumes. Glumes equal, erect, linear-subulate, tapering to an awn, coriaceous, 3-veined, scabrous; lemma coriaceous at maturity, flat on back, involute, 5-veined, tapering to a long flat awn equal in length to awn of glume; palea 2-keeled, ciliate on keels. Lodicules 2, ciliate above. Stamens 3. Caryopsis fusiform, with a hairy apical appendage, firmly adherent to lemma and palea; hilum linear, as long as caryopsis.

A monotypic genus.

1. Crithopsis delileana (Schult. & Schult. fil.) Roshev., Zlaki 319 in obs. (1937); Bor, Fl. Iraq 9:226, t.77. *Elymus delileanus* Schult. & Schult. fil., Mantissa 2:424 (1824); Boiss., Fl. 5:692. *C. rhachitricha* Jaub. & Spach, Ill. Pl. Or. 4:30, t.321 (1851). *E. geniculatus* Del., Fl. Eg. 174 (1813–1814), t.13 f.1 (1826) non Curt. (1790); Post, Fl. 2:795. [Plate 204]

Annual, 10–30 cm. Culms geniculate-ascending. Spike 2–5 cm (excl. awns), oblong; rachis fragile, flat, ciliate, with dense 2 mm long bristles. Awns of spikelet nearly equal. 2n=14 (Bowden, Canad. Journ. Gen. Cyt. 8:131, 1966). Fl. March–May.

Hab.: Steppes and batha. Sharon Plain; Upper Galilee, Mt. Carmel, Mt. Gilboa, Samaria, Judean Mts., Judean Desert, N. and C. Negev; Upper Jordan Valley, Beit Shean Valley, Lower Jordan Valley, Dead Sea area; Golan, Gilead, Ammon, Moav, Edom.

Area: Mediterranean and Irano-Turanian.

5. EREMOPYRUM (Ledeb.) Jaub. & Spach

Annuals. Leaves flat, auriculate; ligule short. Inflorescence a short, compact, laterally compressed terminal spike; rachis generally fragile, with very short internodes. Spikelets solitary, seated distichously at a wide angle to rachis, laterally compressed, 2- to several-flowered; lower 1-2 florets hermaphrodite, the upper rudimentary. Glumes subequal, coriaceous, keeled, narrow, taper-ing and ending in a short point or awn; lemma coriaceous, keeled, muticous or drawn out into a slender awn; palea shorter than lemma, membranous, 2-keeled, scabrous on keels. Lodicules 2, obliquely ovate, fimbriate or ciliate above. Stamens 3. Caryopsis oblong, deeply grooved on adaxial side, with a hairy apical appendage; hilum linear, long.

Eight species in the Mediterranean and Irano-Turanian regions, extending to N. W. India.

1. Keels of palea produced above into 2 teeth or short awns with a deep cleft between them; glumes and lemmas long-pilose.   **2. E. distans**
- Keels of palea ending in 2 minute teeth; glumes and lemmas glabrous or appressed-puberulent, rarely long-pilose.   **1. E. bonaepartis**

1. Eremopyrum bonaepartis (Spreng.) Nevski, Acta Inst. Bot. Acad. Sci. URSS ser. 1, 1:18 (1933); Nevski in Fl. URSS 2:663 (1934); Bor, Fl. Iran. 70:189; Bor, Fl. Iraq 9:229, t.78. *Triticum bonaepartis* Spreng., Erst. Nachtr. Bot. Gart. Halle 40 (1801). *T. squarrosum* Roth, Neue Beitr. Bot. 128 (1802). *Agropyrum squarrosum* (Roth) Boiss., Fl. 5:668 (1884); Post, Fl. 2:789. [Plate 205]

Annual, 10-30 cm. Culms geniculate below, puberulent below spike. Spike 2-3.5(-4) cm, oblong, long-exserted; rachis fragile. Spikelets crowded, 3-5-flowered, glabrous, scabrous or sparsely pilose. Glumes shorter than spikelet, lanceolate, acute and ending in a sharp point, keeled, scabrous along upper part of keel; lemma lanceolate, acute, in upper spikelets often shortly aristate; keels of palea ending in 2 minute teeth. Fl. March–May.

Var. **bonaepartis.** [Plate 205]. Lemmas glabrous or with short appressed hairs.
   Hab.: Steppes and fields. C. Negev; E. Gilead, E. Ammon, Moav, Edom.

Var. **sublanuginosum** (Drob.) Melderis in Rech. fil., Ark. Bot. ser. 2, 2(5):305 (1952). *Agropyron orientale* (L.) Roem. & Schult. var. *sublanuginosum* Drob., Trav. Mus. Bot. Acad. Sci. Pétersb. 16:135 (1916). Lemmas long-pilose.

Hab.: As above. E. Ammon.

Area of species: Irano-Turanian.

2. Eremopyrum distans (C. Koch) Nevski, Acta Inst. Bot. Acad. Sci. URSS ser. 1, 1:18 (1933) in obs.; Nevski, Fl. URSS 2:665 (1934); Bor, Fl. Iraq 9:234, t.80 (1968). *Agropyron distans* C. Koch, Linnaea 21:426 (1848). *E. orientale* (L.) Jaub. & Spach var. *lasianthum* (Boiss.) Maire, Bull. Soc. Hist. Nat. Afr. Nord 20:207 (1929). *A. lasianthum* Boiss., Diagn. ser. 1, 13:68 (1854). *A. orientale* (L.) Roem. & Schult. var. *lasianthum* (Boiss.) Boiss., Fl. 5:668 (1884). [Plate 206]

Annual, 5–15 cm. Culms geniculate, smooth and glabrous, puberulent below spike. Spike 2–4 cm, oblong, enclosed at base by leaf-sheath; rachis fragile. Spikelets crowded, 3–5(–6)-flowered, with densely pilose glumes and lemmas. Glumes (incl. awns) as long as spikelet, narrow, subulate-aristate; awn of glumes 4–6 mm; lemma of lower florets broader than glume, rounded on back, keeled above, aristate; awn of lemma 6 mm; keels of palea produced above into 2 teeth or short awns with a deep cleft between them. Fl. April–May.

Hab.: Hammada. C. Negev; E. Gilead, E. Ammon, Edom.

Area: Irano-Turanian.

6. HETERANTHELIUM Hochst. ex Jaub. & Spach

Annuals. Leaves flat, auriculate; ligule short. Inflorescence an awned and bristly terminal spike; rachis only partially fragile, breaking up into a number of sections; each section consists of at least 3 fused rachis-segments diminishing in length from below upwards and with a spikelet at each joint; the lowest spikelet fertile, the next above smaller, fertile or sterile with a reduced lemma, and succeeding spikelet(s) reduced to a bunch of awned glumes. Fertile spikelet: glumes 2, subequal, subulate, curved at base; florets 3 to several; lemma lanceolate-oblong, papillose and pilose, 5–7-veined, each ending in a long flattened scabrous awn long-pilose on one side; palea 2-keeled, with veins produced as 2 mucros; rachilla prolonged and crowned by a tuft of reduced lemmas. Lodicules 2, ciliate. Stamens 3. Caryopsis oblong, laterally compressed, with a hairy apical appendage; embryo $\frac{1}{4}$–$\frac{1}{3}$ length of caryopsis; hilum linear, as long as caryopsis.

A monotypic genus.

1. Heteranthelium piliferum (?Banks & Sol.) Hochst. ex Jaub. & Spach, Ill. Pl. Or. 4:24, t.318 (1851); Hochst. in Kotschy, Pl. Alepp. exsicc. no. 130 (1843) *nom. nud.*; Bor, Fl. Iraq 9:242, t.84; Boiss., Fl. 5:672; Post, Fl. 2:781. *?Elymus pilifer* Banks & Sol. in Russ., Nat. Hist. Aleppo ed. 2, 2:244 (1794). [Plate 207]

Annual, 5–20(–30) cm. Leaf-blades minutely hairy; sheaths slightly inflated, retrorsely villose. Spike 3–5 cm (incl. awns), bristly, often purplish. Fl. April–May.

Hab.: Steppes. Judean Desert, N. Negev; Gilead, Edom.
Area: Irano-Turanian.

7. Aegilops L.

Annuals, mostly autogamous. Leaves flat, rarely rolled, auriculate; ligule short. Inflorescence a terminal spike, linear or ovate to lanceolate in outline, wholly or partly awned or awnless, mostly scabrous, rarely hairy, 2–20-spiculate, generally with 1–4 vestigial spikelets at base; spike breaking off at maturity above vestigial spikelets and falling as a unit, or disarticulating into spikelets. Spikelets solitary at nodes of flattened rachis, hermaphrodite or the upper neuter. Florets 2–8, the upper often staminate or neuter. Glumes equal, coriaceous, awned, dentate or muticous, veined, rounded on back, rarely keeled; awns veined, scabrous; lemma papery or membranous, firmer and distinctly veined towards tip, dentate or awned; palea 2-keeled. Lodicules 2, cuneate, fimbriate on upper margin. Stamens 3. Caryopsis compressed, mostly oblong, with a hairy apical appendage; embryo about $\frac{1}{5}$ length of caryopsis; hilum linear, as long as caryopsis. x=7.

About 25 Mediterranean and Irano-Turanian species.

Literature: A. Eig, A second contribution to the knowledge of the Flora of Palestine, *Bull. Inst. Agr. Nat. Hist. Tel-Aviv* 6:70–77 (1927). P.M. Zhukovsky, Specierum generis Aegilopsis L. revisio critica, *Bull. Appl. Bot. Pl.-Breed. (Leningrad)* 18:417–609 (1928). A. Eig, Notes sur le genre *Aegilops, Bull. Soc. Bot. Genève* Ser. 2, 19 (2):322–333 (1927–1928); Monographisch-kritische Uebersicht der Gattung *Aegilops, Feddes Repert., Beih.* 55:1–228 (1929); Contribution à la connaissance de l'*Aegilops kotschyi* Boiss., *Bull. Appl. Bot. Pl.-Breed.* 24(2):395–397 (1931); *Aegilops*, in E. Hannig & H. Winkler, *Die Pflanzenareale*, 4 Reihe, 4, Karte 38–41, Jena, 1936. W.M. Bowden, The taxonomy and nomenclature of the wheats, barleys and ryes and their wild relatives, *Canad. Jour. Bot.* 37:637–684 (1959). M.S. Chennaveeraiah, Karyomorphological and cytotaxonomic studies in *Aegilops, Acta Horti Gothob.* 23:85–178(1960). K. Hammer, Zur Taxonomie und Nomenklatur der Gattung *Aegilops* L., *Feddes Repert.* 91:225–258 (1980); Vorarbeiten zur monographischen Darstellung von Wildpflanzensorten: *Aegilops* L., *Kulturpflanze* 28:33–180 (1980).

1. Glumes awned at least in terminal spikelet of spike or ending in 0.5 cm or longer teeth   2
   - Glumes awnless in all spikelets, often ending in 1–2(–3) short teeth   6
2. Caryopsis adherent to lemma and palea; veins of glumes slender, subequal in width, elevated and parallel, separated by usually equal interspaces; vestigial spikelets at base of spike mostly 3   3
   - Caryopsis free; veins of glumes very unequal in width, many of them very broad and flattened, separated by unequal interspaces; vestigial spikelets at base of spike mostly 1 or 2, rarely 3   4
3. Plant of deserts and steppes. Awns of glumes generally 3, more or less equal in length and width; lemmas of lower spikelets awned; glumes 2.5–4 mm broad.

                                                                                8. A. kotschyi
   - Plant common in most parts of country, rare in Negev and other desert districts.

Awns of glumes 2 or 3, mostly unequal in width and length and often short or tooth-like in part of spikelets; lemmas of lower spikelets mostly dentate, rarely awned and then awns unequal; glumes 4-6 mm broad. **7. A. peregrina**

4(2). Vestigial spikelets (remaining at tip of culm after falling of spike) 3, rarely 2; spike 3.5-6 cm (excl. awns). **9. A. triuncialis**

- Vestigial spikelets 1, rarely 2; spike 1.5-2.5(-3) cm (excl. awns) 5

5. Awns of terminal spikelet distinctly longer than awns of lateral spikelets; glumes of terminal spikelet 3-awned. Not common. **10. A. biuncialis**

- Awns of terminal spikelet shorter than or at most as long as awns of lateral spikelets; glumes of terminal spikelet 4-5-awned. Common. **11. A. geniculata**

6(1). Glumes of lateral spikelets truncate, muticous or with a small lateral tooth, scabrous, not hairy; apex of glumes bordered by a thick vein. Plants of alluvial soils. **5. A. speltoides**

- Glumes of lateral spikelets 2(-4)-dentate or emarginate, or if truncate, glumes densely hairy 7

7. Spikes densely appressed-hairy; lemmas of uppermost spikelet (rarely of 2 upper spikelets) awned, those of lateral spikelets awnless; spike disarticulating into single spikelets, each spikelet with a laterally adjoining rachis-internode. **6. A. crassa**

- Spikes scabrous, not hairy 8

8. Terminal spikelet with 2 long and strong awns of lemmas; at least one of the awns (6-)7 cm or longer; lateral spikelets awnless; spike falling as a unit at maturity 9

- Spike awnless or lemmas of lateral and terminal spikelets awned (and then often lowest spikelets awnless); awns usually shorter than above 10

9. Lemmas of terminal spikelet with 2 somewhat unequal awns; the shorter awn not less than ¾ length of the longer. Anthers (4-)5-6 mm. Plant of sandy soils, rarely of grey steppe soil. **3. A. longissima***

- Lemmas of terminal spikelet with 2 distinctly unequal awns; the shorter usually less than ½ length of the longer. Anthers 2-3 mm. Plant of terra rossa or terra rossa sometimes mixed with loess. **4. A. searsii**

10(8). Spike falling at maturity as a unit; glumes 4-6 mm broad. **7. A. peregrina**

- Spike disarticulating into single spikelets; glumes 1.5-2.5 mm broad 11

11. Plant of coastal plain from Philistean Plain northwards, (40-)50-70(-100) cm. Spike 7-13 cm (excl. awns); spikelets 8-13 mm (excl. awns); tip of lemma awned and 1-2-dentate at base of awn (in var. *mutica* awns in uppermost spikelets only).
2. A. sharonensis*

- Plant of Negev and Edom, (15-)20-35(-45) cm. Spike 4-5(-8) cm (excl. awns); spikelets 5.5-8.5 mm (excl. awns); tip of lemma awned but devoid of teeth (in var. *anathera* awns found in 1-2 uppermost spikelets only). **1. A. bicornis**

---

* Hanna Ankori & D. Zohary (Cytologia 27:314-324, 1962) described hybrid swarms between *A. sharonensis* and *A. longissima* in disturbed sites in the vicinity of pure populations of both species in the Sharon Plain. Members of the hybrid populations were intermediate between the 2 species with respect to presence and length of awns in lateral spikelets, several morphological characters of the lemma, cross-section of the awn, etc. The chromosomes of the 2 species are known to differ in a reciprocal translocation.

Sect. 1. SITOPSIS (Jaub. & Spach) Zhuk., *op. cit.* 466 (1928). Sect. *Platystachys*
Eig, Monogr. 68 (1929). Spike long, linear, slender, 5–20-spiculate, mostly
awned, at maturity falling as a unit or disarticulating into spikelets, each
spikelet with a rachis-internode immediately below it. Spikelets linear, more
or less equal in size, 5–6 times as long as broad. Glumes awnless; all lemmas
or only those in terminal spikelet 1-awned. Caryopsis adherent to lemma and
palea.

1. Aegilops bicornis (Forssk.) Jaub. & Spach, Ill. Pl. Or. 4:10, t.309 (1850);
Fig, Monogr. 72, t.3 a, f, g, h; Hammer, Feddes Repert. 91:231 (1980); Boiss.,
Fl. 5:677 p.p.; Post, Fl. 2:780. *Triticum bicorne* Forssk., Fl. Aeg.-Arab. 26
(1775). [Plate 208]

Annual, 15–45 cm. Spike 4–5(–8) cm (excl. awns), many-spiculate, distichous, mostly awned, at maturity disarticulating into spikelets, each spikelet
falling with rachis-internode immediately below it; a few lower spikelets often
remain attached at tip of culm. Spikelets 5.5–8.5 mm (excl. awns), mostly
3-flowered (2 florets fertile); the attached rachis-internode ½ length of spikelet. Glumes 4.5–5.5 mm, 5-veined, (1–)2-dentate at tip, emarginate or with an
angle between teeth, asymmetrical, keeled; veins elevated, slender, parallel,
scabrous; lemma 4–6(–7.5) mm, boat-shaped, veined and faintly keeled in
upper part, mostly ending in a slender awn (in lower spikelets awn short or
absent); awn usually not flanked by lateral teeth. Caryopsis adherent to
lemma and palea. 2n=14. Fl. March–May.

Var. **bicornis.** Var. *typica* Eig, Monogr. 73 (1929). [Plate 208]. All spikelets
awned, sometimes lowest awnless.
Hab.: Sands. W., N. and C. Negev; Edom.

Var. **anathera** Eig, Bull. Soc. Bot. Genève ser. 2, 19(2):325 (1927–1928). Var.
mutica (Aschers.) Eig, Monogr. 73 (1929). *Triticum bicorne* B. *muticum*
Aschers., Magyar Bot. Lapok 1, 6:10 (1902). All spikelets except 1–3 upper-
most awnless.
Hab.: Rare and sporadic; together with var. *bicornis.*
Area of species: Saharo-Arabian.

2. Aegilops sharonensis Eig, Notizbl. Bot. Gart. Berlin 10:489 (1928);
Monogr. 73, t.3 b–e (1929); Post, Fl. 2:788. *A. longissima* Schweinf. &
Muschler emend. Eig, *op. cit.* (1927) subsp. *sharonensis* (Eig) Hammer, Feddes
Repert. 91:231 (1980). [Plate 209]

Annual, (40–)50–70(–100) cm. Spike 7–13 cm (excl. awns), 7–15-spiculate,
usually distinctly distichous, more or less flattened, awned (except in var.
mutica), disarticulating at maturity into spikelets, each spikelet with rachis-
internode immediately below it; the lowermost spikelet (or few lowest spikelets) remains attached at tip of culm. Spikelets 8–13 mm (excl. awns), 3–5-
flowered; attached rachis-internode shorter than spikelets and curved. Glumes
6–7 mm, 6–7-veined, 2-dentate at tip, emarginate or with an angle between

teeth; veins elevated, slender, parallel, scabrous; lemma 8–11 mm, boat-shaped, 5-veined in upper part, ending in a slender awn flanked by 1–2 small teeth; awn 3–8 cm long, scabrous, longest towards apex of spike; 1–3 lower-most spikelets usually almost awnless. Caryopsis adherent to lemma and palea. 2n=14. Fl. March–June.

Var. **sharonensis**. Var. *typica* Eig, Monogr. 75, t.3 b, d, e. Lemmas awned in all spikelets except in the lowest.
    Hab.: Sandy soils. Acco Plain, Sharon Plain, Philistean Plain.

Var. **mutica** (Post) Eig, Monogr. 75, t.3 c (1929). *A. bicornis* (Forssk.) Jaub. & Spach var. *mutica* Post, Fl. ed. 1, 901 (1883–1896). Lemmas awnless in all spikelets, or awned in uppermost spikelets only.
    Hab.: Rare. Acco Plain, Sharon Plain.
    Area of species: E. Mediterranean. Endemic to the coastal plain of Palestine and Lebanon south of the Litani River.

    Literature: Hanna Ankori & D. Zohary, Natural hybridization between *Aegilops sharonensis* and *A. longissima, Cytologia* 27:314–324 (1962).

3. Aegilops longissima Schweinf. & Muschler in Muschler, Man. Fl. Eg. 1:156 (1912) emend. Eig, Bull. Inst. Agr. Nat. Hist. Tel-Aviv 6:73 (1927); Eig, Monogr. 79, t.2 c (excl. some specimens from Transjordan); Post, Fl. 2:788. *A. longissima* subsp. *longissima*; Hammer, Feddes Repert. 91:231 (1980). *Triti-cum longissimum* (Schweinf. & Muschler) Bowden, Canad. Jour. Bot. 37:666 (1959). [Plate 210]
    Annual, (40–)50–100 cm. Spike 10–20 × 0.2–0.3 cm (excl. awns), 10–20-spiculate, narrowly linear, the greater part of spike falling at maturity; lower spikelets remaining on culm or falling later singly or in 2–3 pieces. Spikelets mostly 12–14 mm, 3–5-flowered (fertile florets usually 2); rachis-internodes nearly as long as spikelets in middle of spike. Glumes 6–8 mm, 6–7-veined, 2-dentate at tip (in terminal spikelet usually 3-dentate), with an angle or sinus between teeth; veins elevated, slender, parallel, scabrous; lemma in lateral spikelets awnless, keeled, folded in upper $\frac{1}{3}$, scabrous; in terminal spikelet lemma in each of the 2 fertile florets tapering into a 7–12 cm scabrous awn flanked at base by 1–2 unequal short aristulate teeth; the 2 awns somewhat unequal; the shorter mostly not less than $\frac{3}{4}$ length of the longer. Anthers (4–)5–6 mm. Caryopsis adherent to lemma and palea. 2n=14. Fl. March–June.
    Hab.: Sandy soils derived from the sandstone of the coastal plain or from Nubian sandstone, rarely loess or grey calcareous soil. Acco Plain, Sharon Plain, Philistean Plain; W. and N. Negev; Gilead, Ammon.
    Area: E. Mediterranean, extending into the Saharo-Arabian (Palestine, Sinai, Egypt).

4. Aegilops searsii Feldman & Kislev ex Hammer, Feddes Repert. 91:231 (1980); Feldman & Kislev, Israel Jour. Bot. 26:193, ff.1–3 (1977) *nom. invalid.*

Triticum searsii Feldman & Kislev, *op. cit.* 91 (1977) *nom. invalid.* [Plate 211]

Annual, 10–40(–50) cm, close to *A. longissima,* with a spike awned only at tip and falling as a unit at maturity. Differs from *A. longissima* mainly in habitat and in the following characters: spike shorter, 5–11 cm (excl. awns), mostly with its lower part enclosed in uppermost leaf-sheath; spikelets fewer, 8–12; rachis-internodes more or less as long as spikelets; fertile florets generally 2 per spikelet in middle of spike, 1 in terminal spikelet; awns of lemma in terminal spikelet very unequal, the longer 8–10 cm, the shorter usually less than ½ their length; awns flanked at base by 1–2 unequal short-aristulate teeth; middle tooth of glumes in terminal spikelet sometimes lengthened into an up to 1 cm long awn. Anthers 2–3 mm. Caryopsis more or less free. 2n=14. Fl. April–June.

Hab.: More or less degraded batha (Sarcopoterietum spinosi semistepposum), fallow fields, terra rossa sometimes mixed with loess. Samaria, Shefela, Judean Mts.; Golan, Gilead, Ammon, Moav.

Area: W. Irano-Turanian (Palestine, E. Lebanon, Syria).

5. **Aegilops speltoides** Tausch, Flora (Regensb.) 20:108 (1837); Eig, Bull. Inst. Agr. Nat. Hist. Tel-Aviv 6:72–73 (1927), incl. var. *macrostachys* Eig, *op. cit.* 73 (1927). *Triticum speltoides* (Tausch) Gren. ex Richter, Pl. Eur. 1:129 (1890); Hammer, Feddes Repert. 91:230 (1980). [Plate 212]

Annual, 40–70 cm. Spike (6–)7–11(–15) cm (excl. awns), mostly 7–10-spiculate, scabrous, disarticulating at maturity or falling as a unit. Spikelets mostly 11–15 mm, 4–6-flowered, sessile in hollows of rachis; vestigial spikelet at base of spike 1 or none. Glumes 5–7 mm, 7–10-veined, obliquely truncate at apex, asymmetrical, somewhat keeled, with parallel, scabrous, elevated veins, unequal in thickness; glumes of lateral spikelets truncate with a thickened elevated vein across tip, with a small tooth at end of keel and a narrow hyaline membranous margin beyond vein; glumes of terminal spikelet somewhat longer, rounded at apex and devoid of tooth; lemma 7–10 mm, boat-shaped, awned or muticous, keeled and 5-veined in its upper part. Caryopsis adherent to lemma and palea. 2n=14. Fl. April–June.

Hab.: Humid alluvial soil. Sharon Plain, Philistean Plain; Mt. Carmel, Esdraelon Plain.

Var. **speltoides.** *A. speltoides* Tausch, *loc. cit.*; Eig, Monogr. 83, t.2 e; Post, Fl. 2:789. *A. speltoides* subsp. *speltoides*; Hammer, Feddes Repert. 91:230 (1980). *A. speltoides* var. *aucheri* (Boiss.) Fiori in Fiori & Paoletti, Fl. Anal. Ital. 4, App.:32 (1907); Eig, *op. cit.* 73 (1927). *A. aucheri* Boiss., Diagn. ser. 1, 5:74 (1844); Boiss., Fl. 5:678 excl. var. *polyathera* Boiss., *loc. cit.* (1884). [Plate 212]. Spike falling at maturity as a unit. Awns present on 2 lowest lemmas of terminal spikelet only, stout, about 7 cm; rarely awns also on some lateral spikelets and then shorter and thinner.

Var. **ligustica** (Savign.) Fiori in Fiori & Paoletti, Fl. Anal. Ital. 4, App.:33 (1907). *A. ligustica* (Savign.) Coss., Bull. Soc. Bot. Fr. 11:164 (1864); Eig,

Monogr. 81, t.2 d; Post, Fl. 2:788. *A. speltoides* subsp. *ligustica* (Savign.) Zhuk., *op. cit.* 530 (1928); Hammer, *loc. cit. Agropyrum ligusticum* Savign., Diar. Atti Congr. Sc. Genov. 1846:138 (1847). [Plate 212]. Spike disarticulating into spikelets, each spikelet with the rachis-internode immediately below it. All spikelets awned; awns of terminal spikelet somewhat stouter than those of lateral ones.

Area of species: E. Mediterranean extending into the Irano-Turanian.

Literature: D. Zohary & D. Imber, Genetic dimorphism in fruit types in *Aegilops speltoides*, Heredity 18:223–231 (1963).

Sect. 2. VERTEBRATA Zhuk., *op. cit.* 450 (1928) emend. Kihara, Züchter 12:61 (1940). Sect. *Pachystachys* Eig, Monogr. 84 (1929). Spike more or less long, cylindrical, mostly thick, mostly awned, at maturity disarticulating into spikelets, each spikelet with the laterally adjacent rachis-internode. Spikelets more or less ventricose or linear. Glumes mostly awnless; lemmas of several upper spikelets often 1-awned; awns broad and mostly flat. Caryopsis adherent to lemma and palea.

6. Aegilops crassa Boiss., Diagn. ser. 1, 7:129 (1846) subsp. **vavilovii** Zhuk., Bull. Appl. Bot. Pl.-Breed. (Leningrad) 18:554 (1928); Hammer, Feddes Repert. 91:234 (1980). *A. crassa* var. *palaestina* Eig, Bull. Soc. Bot. Genève ser. 2, 19:326, ff. a, b (1927–1928); Monogr. 91; Post, Fl. 2:787. [Plate 213]

Annual, 20–40 cm. Spike 10–15 × 5–6 mm (excl. awns), appressed-hairy, cylindrical or slightly zigzag-shaped, tapering towards tip, 7–11-spiculate, disarticulating into spikelets at maturity, each spikelet with the laterally adjoining rachis-internode. Spikelets 12–14 mm, 4–5-flowered, linear; vestigial spikelets 1–2 or none. Glumes nearly truncate, densely hairy, membranous at tip, with (1–)2(–4) teeth (or a weak awn) and a shallow sinus between them and with veins equal in width; lemma thickened in upper part and usually keeled, ending in a tooth; lemma of terminal spikelet ending in a 5–8 cm awn; awn flattened at base and arcuate (rarely some lateral spikelets awned). 2n=28, 42. Fl. April–May.

Hab.: Very rare. Jerusalem, W. and C. Negev.

Area of subspecies: W. Irano-Turanian (Palestine, Syria).

Area of species: Irano-Turanian (Palestine, Syria, Iraq, Iran, ?Turkey, Transcaucasia, Afghanistan, C. Asia).

Sect. 3. AEGILOPS. Sect. *Pleionathera* Eig, Monogr. 117 (1929). Spikes short, few-spiculate, more or less ovate, elliptic or lanceolate in outline, rarely elongate-linear, mostly awned, at maturity falling as a unit. Spikelets 2–5, ventricose or elliptic, rarely linear. Awns always more than 1, on glumes stronger than on lemmas, rarely spike muticous. Vestigial spikelets at base of spike present. Caryopsis free or adherent.

7. Aegilops peregrina (Hackel) Maire & Weiller in Maire, Fl. Afr. Nord 3:358

(1955); Bor, Fl. Iraq 9:186; Hammer, Feddes Repert. 9:236 (1980). *A. pere-grina* (Hackel) Eig, Monogr. 121 (1929) *nom. invalid. A. peregrina* (Hackel) Melderis, Ark. Bot. ser. 2, 5(1):71 (1959). *Triticum peregrinum* Hackel in J. Fraser, Ann. Scott. Nat. Hist. 62:101–103 (1907). *A. variabilis* Eig, Monogr., Feddes Repert. Beih. 55:121, tt. 9, 10, 11 (1929); Post, Fl. 2:785. [Plate 214]

Annual, 15–40 cm. Spike 1.5–7.5 cm (excl. awns), falling as a unit at maturity, widely varying in size and shape (ovate to linear in outline), in length of awns (awned or dentate) and in ratio of spikelet to rachis-internode; vestigial spikelets 3, rarely (1–)2–4. Spikelets (2–)3–5(–7), (3–)4–5(–6)-flowered. Glumes 6–8 × 4–6 mm, scabrous, with slender parallel elevated veins more or less equal in width; awns of glumes 2 or 3, generally unequal in width and length, spreading at maturity; lemma of lower spikelets dentate, rarely awned and, if so, awns unequal. Caryopsis adherent to lemma and palea. 2n=28. Fl. March–May.

"This is a species of the most extraordinary variability and for this very reason was called *A. variabilis* by Eig, although he was perfectly aware that Hackel had already given the name *Triticum peregrinum* to a specimen of this species which had been casually introduced into Scotland. Eig says that he prefers to give a new name to the species because Hackel's specimen was atypical and therefore the specific epithet must be abandoned. Maire & Weiller seem to be the first authors to have validly published the combination" (Bor, Fl. Iran. 70:197,1970).

1. Spike 1.5–4 cm (excl. awns); lower spikelets distinctly longer than adjacent rachis-internode.   subsp. **peregrina**
–   Spike 3.5–7.5 cm (excl. awns); spikelets about as long as adjacent rachis-internode.
                                                                                              subsp. **cylindrostachys**

Subsp. **peregrina.** *A. variabilis* subsp. *eu-variabilis* Eig & Feinbr. in Eig, Monogr. 123, tt.9 a, 10 (1929). [Plate 214]. Spike 1.5–4 cm (excl. awns), broadly ovate to lanceolate in outline, 3–4(–6)-spiculate; lower spikelets longer than adjacent rachis-internode.

Most variable in length of awns. Awned variants predominate; awns present on glumes and mostly absent on lemmas. Variants with dentate glumes in all or in a part of spikelets occur more rarely. Eig & Feinbrun in Eig (*op. cit.*) described several varieties within subsp. *eu-variabilis*.

Hab.: Batha and fallow fields. Common. Coast of Carmel, Sharon Plain, Philistean Plain; Upper and Lower Galilee, Mt. Carmel, Samaria (incl. E. Samaria), Shefela, Judean Mts.; Upper Jordan Valley, Dead Sea area; Golan, Gilead, Ammon.

Subsp. **cylindrostachys** (Eig & Feinbr.) Hammer, Feddes Repert. 91:237 (1980). *A. variabilis* Eig subsp. *cylindrostachys* Eig & Feinbr. in Eig, Monogr. 125, tt.9 b, 11 (1929). [Plate 214]. Spike 3.5–7.5 cm (excl. awns), narrowly lanceolate to linear, (3–)5(–7)-spiculate; rachis-internodes generally about as long as spikelets.

Awned variants rare; when present, awns on glumes short; awns on lemmas present only in terminal spikelet, or very short, or lemma dentate. Eig &

Feinbrun in Eig (*op. cit.*) described 3 varieties within subsp. *cylindrostachys*.

Hab.: Batha and fallow fields. Less common than subsp. *peregrina*, especially in hill districts. Sharon Plain, Philistean Plain; Mt. Carmel, Shefela, Judean Mts., N. and W. Negev; Lower Jordan Valley; Gilead.

Area of species: S. Mediterranean (Morocco, Algeria, Libya, Egypt, Palestine, Lebanon, Syria, Cyprus, Crete, S. Greece, S. Italy).

Tutin & Humphries in Flora europaea (5:201) mistakenly cite *A. peregrina* (Hackel) Maire & Weiller as a synonym of *A. uniaristata* Vis. In fact, *A. uniaristata* is not even close to *A. peregrina* and belongs to Sect. *Comopyrum* (Jaub. & Spach) Zhuk., 1928 (Sect. *Macrathera* Eig, 1929). The difference between the two species is clearly seen in the type of glume venation. In *A. uniaristata* veins on glumes are broad and unequal in width, and resemble those of *A. geniculata*. In contrast, veins on glumes in *A. peregrina* are thin and equal in width and peculiar to *A. peregrina* and *A. kotschyi*. Geographically, *A. uniaristata* is confined to parts of the Balkan peninsula and to W. Anatolia. The areas of the two species overlap only in some parts of S. Greece.

8. Aegilops kotschyi Boiss., Diagn. ser. 1, 7:129 (1846); Eig, Monogr. 127, t.12; Bor, Fl. Iraq 9:184; Hammer, Feddes Repert. 91:237 (1980); Post, Fl. 2:785. *A. triuncialis* L. var. *kotschyi* (Boiss.) Boiss., Fl. 5:674 (1884). *Triticum kotschyi* (Boiss.) Bowden, Canad. Jour. Bot. 37:675 (1959). [Plate 215]

Annual, 15–30(–40) cm. Spike 2–3(–4) cm (excl. awns), awned, lanceolate to ovate-lanceolate, falling as a unit at maturity; 2 lower rachis-internodes mostly shorter than adjacent spikelet; vestigial spikelets at base of spike 3, rarely 2 or 4. Spikelets (3–)4–5(–6), mostly narrow, the 2 lower spikelets mostly 3–4-flowered, with 2–3 fertile florets; upper spikelets only partly fertile; glumes and lemmas awned; awns mostly 8–14 to a spikelet, often purplish. Glumes 5–7 × 2.5–4 mm, often distinctly shorter than flowers, with slender elevated scabrous parallel veins, equal in width and more or less equally spaced; awns of glumes 2.5–3.5 cm, more or less equal, mostly 3, rarely 2 (the middle one then lacking or reduced to a tooth); lemma 1–3-awned; awns unequal and shorter than awns of glumes. Caryopsis adherent to lemma and palea. 2n=28. Fl. March–April.

Var. **kotschyi**. Incl. var. *typica* Eig, Monogr. 128, t.12b (1929) & var. *palaestina* Eig, Monogr. 128, t.12 a,e (1929). [Plate 215]. Spike ovate-lanceolate to lanceolate in outline, awned.

Hab.: Steppes. Common. E. Samaria, Judean Mts. (rare), Judean Desert, W., N. and C. Negev; Lower Jordan Va'ley, Dead Sea area, Arava Valley; ?Golan, E. Gilead, E. Ammon, Moav, Edom.

Var. **brachyathera** Eig, Bull. Appl. Bot. Pl.-Breed. (Leningrad) 24, 2:395–397 (1931); Post, Fl. 2:785 (err. as var. *anathera* Eig). Spike narrowly lanceolate; spikelets all awnless or the upper shortly and irregularly awned.

Hab.: Nubian sandstone hills. Together with var. *kotschyi*. Rare. Edom.

Area of species: Irano-Turanian and Saharo-Arabian (Palestine, Syria, Cyprus, Turkey, Transcaucasia, Iraq, Iran, Kuwait, Egypt, Libya).

A. kotschyi is fairly uniform in Palestine and seems to be so over the greater part of its area. Following Bor (1968), we designate the predominant local taxon as var. *kotschyi*. It is identical with var. *palaestina* Eig. The only other variant of *A. kotschyi* in Palestine is var. *brachyathera* Eig (1931), a few specimens of which were collected in Edom together with the typical form.

Eig (Monogr. 128) refers to var. *palaestina* as a taxon common all over the southern part of the general area of *A. kotschyi*. Another taxon, var. *leptostachya* (Bornm.) Eig, *loc. cit.*, characterized by linear spikes, occurs in Iraq. Var. *hirta* Eig appears together with var. *kotschyi* in Iraq and the Caucasus.

9. Aegilops triuncialis L., Sp. Pl. 1051 (1753); Eig, Monogr. 130, t.13 a–e; Hammer, Feddes Repert. 91:238 (1980); Boiss., Fl. 5.671; Post, Fl 2:784. [Plate 216]

Annual, 20–45 cm. Spike 3.5–6 cm (excl. awns), mostly awned, linear, falling as a unit at maturity, rarely disarticulating into spikelets; rachis-internodes mostly about as long as adjacent spikelet; vestigial spikelets at base of spike 3, rarely 2. Spikelets mostly 4–5, narrowly elliptic. Glumes of lower spikelets 7–10 mm, with veins unequal in width and unequally spaced, scabrous or hairy, some of them flattened; awns of glumes unequal in length; those of terminal spikelet 3 in number, mostly 5–7 cm, distinctly longer than awns of lateral spikelets; lemma in lateral spikelets mostly awnless, in terminal spikelet 1-awned, sometimes with 1–2 shorter awns in addition. Caryopsis free. 2n=28. Fl. May–June.

Hab.: Batha. Rare. Upper Galilee; Judean Mts.; Golan.

Area: Mediterranean and W. Irano-Turanian.

10. Aegilops biuncialis Vis., Fl. Dalm. 1:t.1 f.2 *nom. et figura* (1842); Fl. Dalm. 3:344 (1852) *descr.*; Tzvelev, Poaceae URSS 159 (1976); Eig, Monogr. 135, t.14 d–g; Post, Fl. 2:784. *A. lorentii* Hochst., Flora (Regensb.) 28:25 (1845); Bor, Fl. Iraq 9:176; Hammer, Feddes Repert. 91:239 (1980). *A. ovata* var. *lorentii* (Hochst.) Boiss., Fl. 5:674 (1884). [Plate 217] See p. 414.

Annual, 15–40(–50) cm. Spike 2–3 cm (excl. awns), awned, falling as a unit at maturity; lower rachis-internode shorter than adjacent spikelet; vestigial spikelets at base of spike 1, rarely 2. Spikelets mostly 2, rarely 3–4, 4–5-flowered; the 2 lower florets fertile. Glumes 8–10 mm, awned, scabridulous, with veins unequal in width, unequally spaced, some of them flattened; awns of glumes equal in width, in terminal spikelet 4–7 cm, 3 in number, in lateral spikelet(s) shorter, 2 or 3 in number; lemma dentate and often shortly awned. Caryopsis free. 2n=28. Fl. April–June.

Hab.: Batha, stony hillslopes, degraded maquis. Upper Galilee, Judean Mts.; Golan, Gilead.

Area: Mediterranean, extending into the W. Irano-Turanian.

11. Aegilops geniculata Roth, Bot. Abh. 45 (1787); Tzvelev, Nov. Syst. Pl. Vasc. (Leningrad) 8:64 (1971); Hammer, Feddes Repert. 91:240 (1980). *A.*

ovata auct. non L.; Eig, Monogr. 141, t.15 a-f; Boiss., Fl. 5:673 excl. vars.; Post, Fl. 2:783. [Plate 218]

Annual, 15-40 cm. Spike 1.5-2.5 cm × 4-9 mm (excl. awns), awned, falling as a unit at maturity; lower rachis-internode shorter than adjacent spikelet; vestigial spikelets at base of spike 1, rarely 2 or none. Spikelets 3(-4), urceolate, the lower spikelets hermaphrodite, the upper sterile; glumes and lemmas awned; awns widely spreading at maturity. Glumes 7-8 × 5-6 mm, with veins unequal in width and unequally spaced, scabrous or hairy, some of them flattened; awns of glumes mostly 3-4(-5) in number, 2-3.5 cm, equal in width, in terminal spikelets 4-5 in number, as long as or shorter than in lateral spikelets; awns of lemma (1-)2(-3) in number, unequal on same lemma and mostly shorter than awns of glumes. Caryopsis free. 2n=28. Fl. March-May.

Hab.: Batha and fallow fields. Fairly common. Coast of Carmel; Upper and Lower Galilee, Mt. Carmel, Samaria, Judean Mts., W. Judean Desert; Dan Valley; Golan, Gilead, Ammon.

Area: Mediterranean.

8. Triticum L.

Annuals. Leaves flat, shortly auriculate; ligule membranous. Inflorescence a terminal spike. Spikelets solitary at nodes of fragile or tough shortly hairy rachis, 2-6(-9)-flowered, laterally compressed; florets hermaphrodite, self-pollinated, 1-2 uppermost usually sterile. Glumes subequal, ovate or oblong, ventricose, mostly shorter than spikelet, coriaceous, asymmetrical, veined, more or less keeled, 1-2-dentate, apiculate or awned; lemma subventricose, boat-shaped, coriaceous, keeled towards apex; apex entire, 1-2-dentate or awned; callus very short; palea membranous, 2-veined, 2-keeled, ciliate along keels. Lodicules 2, ciliate on upper margin. Stamens 3. Caryopsis free or adherent to lemma and palea, oblong-elliptic, hairy at apex, deeply grooved along adaxial side; embryo about ⅕ length of caryopsis; hilum linear, as long as caryopsis. x=7.

About 15 species in the Mediterranean and Irano-Turanian regions.

Literature: F. Körnicke, *Die Arten und Varietäten des Getreides. Handbuch des Getreidebaues* 1, Bonn, 1885. J. Percival, *The Wheat Plant*, London, 1921. C.A. Flaksberger, *Kulturnaya Flora SSSR. (1) Khlebnye Zlaki. Psheniča*. Moskwa & Leningrad, 1935. N.L. Bor, in: C.C. Townsend et al. (eds.), *Flora of Iraq* 9:194-208, Baghdad, 1968. M. Feldman, Wheats, in: N.W. Simmonds (ed.), *Evolution of Crop Plants* 120-128, London and New York, 1976. D. Zohary, Wild genetic resources of crops in Israel, *Israel Jour. Bot.* 32:97-100 (1983). See p. 414.

Cultivated wheat species are diploid, tetraploid and hexaploid. They are subdivided into 3 groups according to their genomic constitution. The main species are:

(a) Diploid einkorn wheat, *T. monococcum* L. (genomic constitution AA), which is cultivated today on a small scale as a cereal in the Balkans and in Anatolia. Its wild progenitor is the E. Mediterranean *T. boeoticum* Boiss. emend. Schiem. which is also found on Mt. Hermon.

(b) Tetraploid emmer wheat, *T. dicoccon* (Schrank) Schübler, and hard wheat, *T. durum* Desf. (genomic constitution AABB). Today *T. dicoccon* is a relict cereal while *T. durum* is cultivated to a considerable extent in the Mediterranean region, mainly for macaroni and similar products. The wild progenitor of this group is *T. dicoccoides*.

(c) Hexaploid free-threshing soft or common bread wheat, *T. aestivum* L. (genomic constitution AABBDD). Its numerous varieties constitute the bulk of the present world production. No wild hexaploid wheat occurs in nature; *T. dicoccoides* is regarded as the donor of the AABB genome to *T. aestivum* via cultivated emmer.

1. Triticum dicoccoides (Koern. ex Aschers. & Graebn.) Aaronsohn, Verh. Zool.-Bot. Ges. Wien 59:491 (1910); Bor, Fl. Iraq 9:204. *T. vulgare* Vill. var. *dicoccoides* Koern., Verh. Nat. Ver. Pr. Rheinl. 46:21 (1889) *nom. nud. T. dicoccon* (Schrank) Schübler var. *dicoccoides* Koern. ex Aschers. & Graebn., Syn. Mitteleur. Fl. 2(1):679 (1901). *T. turgidum* L. var. *dicoccoides* (Koern. in Schweinf.) Bowden, Canad. Jour. Bot. 37:671 (1959). [Plate 219]

Annual, 100–150 cm. Culms glabrous; nodes sometimes hairy. Spike 3–10 × 1.5 cm (excl. awns), laterally compressed, dense; rachis fragile, ciliate on margins, with a tuft of hairs up to 5 mm at each node. Spikelets 14–15 mm, imbricate, appressed to rachis, glabrous or hairy, straw-coloured or russet to black; florets 2, rarely 1. Glumes sharply keeled, with a midvein produced as a sharp tooth, and with a lateral vein ending in a weak obtuse tooth; lemma awned; awn 15–20 cm, flattened, straight. Caryopsis 9–11 × 1.7–2.5 mm, adherent to lemma and palea. 2n=28. Fl. April–May.

Hab.: Rocky places and soils developed on basalts and hard limestones. Herbaceous cover in the Quercetum ithaburensis belt. Upper Galilee, Mt. Carmel, Esdraelon Plain, Samaria, Judean Mts.; Upper Jordan Valley; Golan, Gilead.

Area: E. Mediterranean (Palestine, Lebanon, Syria, Transcaucasia, Iran).

T. dicoccoides was discovered in Antilebanon by T. Kotschy in 1855, and rediscovered in 1906 in several locations in Palestine (Mt. Hermon, Upper Galilee near Rosh Pinna, Moav) by A. Aaronsohn.

9. Secale L.

Annuals or perennials. Leaves flat or rolled, auriculate; ligule membranous, short. Inflorescence a dense terminal spike, laterally compressed, with solitary spikelets at each node of fragile or tough rachis. Spikelets 2–3-flowered, laterally compressed, appressed to rachis; florets hermaphrodite, cross-pollinated; rachilla prolonged and ending in rudiment of third floret. Glumes subequal, coriaceous, subulate-lanceolate, 1-veined, keeled, awnless or tapering to a straight awn; lemma lanceolate, coriaceous, 5-veined, prominently keeled, pectinately ciliate along keel and margins and ending in a long straight scabrous awn; callus short, bearded; palea nearly as long as lemma, membranous, 2-keeled and aculeolate-ciliate at keels in upper part. Lodicules 2, cuneate, fimbriate on rounded apex. Stamens 3. Caryopsis oblong, hairy at apex, free; hilum linear, as long as caryopsis. x=7.

Five species in temperate regions.

Secale cereale L., rye, is a temperate cereal grown as a grain crop and in many areas as forage. Rye is an important food plant especially in N. and E. Europe.

Literature: R. J. Roshevitz, A monograph of the wild, weedy and cultivated species of rye, *Acta Inst. Bot. Acad. Sci. URSS* (Ser. 1) 6:105–163 (1947). D. Zohary, Origin of South-West Asiatic cereals: wheats, barley, oats and rye, in: P.H. Davis et al. (eds.), *Plant Life of South-West Asia*, 253–258, Bot. Soc. Edinburgh, 1971. G.M. Evans, Rye, in: N.W. Simmonds (ed.), *Evolution of Crop Plants*, 108–111, London and New York, 1976.

1. Secale montanum Guss., Ind. Sem. Horto Boccad. 11 (1825); Fl. Sic. Prodr. 1:145 (1827); Bor, Fl. Iraq 9:262, t.90; Boiss., Fl. 5:670; Post, Fl. 2:780. [Plate 220]

Perennial, 40–100 cm, glaucous, tufted. Culms glabrous below spike. Auricles of leaves short, falcate. Spike 6–14 cm (excl. awns), somewhat nodding; rachis densely hairy along margins, fragile, disarticulating below spikelets into joints; joints 2.5–3.5 mm long, bearded at nodes and angles. Spikelets 10–11 mm (excl. awns), 2-flowered. Glumes 15–23 mm (incl. awns), scabrous on keel; lemma 15–24 mm (excl. awn), ending in a scabrous awn usually much longer than lemma proper. 2n=14. Fl. June–July.

Hab.: Rocky mountain slopes. Golan.

Area: Irano-Turanian and S. Mediterranean (Iran, Iraq, C. and E. Turkey, Lebanon, Mt. Hermon, Greece, Jugoslavia, Sicily, S. E. Spain, Morocco).

S. montanum is regarded as the wild ancestor of cultivated rye.

10. HORDEUM L.

Annuals or perennials. Leaves flat, generally auriculate; ligule short. Inflorescence a terminal bristly spike. Spikelets 1(–2)-flowered, in triplets; triplets seated alternately in cavities of rachis in 2 longitudinal rows (in cultivated species in 6 or 4 rows); rachis fragile in wild species, the three spikelets falling together; median spikelet hermaphrodite, sessile or pedicellate; the two lateral spikelets staminate or neuter, pedicellate (in some cultivated species lateral spikelets hermaphrodite); in the median spikelet rachilla prolonged as a bristle beyond floret. Florets self-pollinated. Glumes narrow, often linear-subulate, ending in an awn, standing in front of spikelets; lemma ovate, 5-veined, muticous or ending in a long awn; palea membranous, 2-keeled. Lodicules 2, cuneate, ciliate above. Stamens 3. Caryopsis adhering to palea and lemma, furrowed on adaxial side, with a hairy apical appendage; hilum linear, as long as caryopsis. x=7.

About 25 species in subtropical and temperate regions of both hemispheres and partly in mountains of the Tropics.

The cultivated species *H. vulgare* L. is of major economic importance.

Literature: S. A. Nevski, A monograph of the genus *Hordeum* L., *Acta Inst. Bot. Acad. Sci. URSS* (Ser. 1) 5:64–255 (1941). G. Covas, Taxonomic observations of the N. American species of *Hordeum, Madroño* 10:1–21 (1949). D. Zohary, Is *Hordeum agriocriton* Åberg the ancestor of six-rowed cultivated barley?, *Evolution* 13:279–290 (1959), Studies on the origin of cultivated barley, *Bull. Res. Counc. Israel* 9D:21–42 (1960). Tana Tovia & D. Zohary, Natural hybridization between *Hordeum vulgare* and *H. spontaneum*. Corroborative evidence from progeny tests and artificial hybrids, *Bull. Res. Counc. Israel* 11D:43–46 (1962). J. R. Harlan & D. Zohary, Distribution of wild wheats and barley, *Science* 153:1074–1080 (1966). D. Zohary, Origin of South-West Asiatic cereals: wheats, barley, oats and rye, in: P. H. Davis et al. (eds.), *Plant Life in South-West Asia*, 245–247, Bot. Soc. Edinburgh, 1971. D. Zohary, The origin of cultivated cereals and pulses in the Near East, in: J. Wahrman & K. R. Lewis (eds.), *Chromosomes Today* 4:313–315, 1973. J. R. Harlan, Barley, in: N. W. Simmonds (ed.), *Evolution of Crop Plants*, 93–98, London and New York, 1976. T. A. Booth & A. J. Richards, Studies in the *Hordeum murinum* aggregate. I. Morphology, *Journ. Linn. Soc. London (Bot.)* 72:149–159 (1976).

1. Rachis of spike fragile. Wild species   2
- Rachis of spike tough. Cultivated species.   **H. vulgare**
2. Awn of lemma in median spikelet of triplet stout, 8–14(–20) cm (rarely lemmas of lateral spikelets aristate as in median spikelet).   **3. H. spontaneum**
- Awn of lemma of median spikelet much shorter   3
3. Perennial, 60–100 cm. Lemma of lateral spikelets acute or acuminate, not aristate 4
- Annual, 10–50(–60) cm. Lemma of lateral spikelets aristate   5
4. Base of culm thickened, tuber-like. Inner glumes of lateral spikelets linear-lanceolate, outer glumes setaceous.   **1. H. bulbosum**
- Base of culm not tuber-like. Glumes of all spikelets setaceous.   **2. H. secalinum**
5(3). Floret of median spikelet sessile. Glumes of median spikelet not ciliate   6
- Floret of median spikelet pedicellate. Glumes of median spikelet long-ciliate.
                                                                        4. H. glaucum
6. Inner (upper) glume of lateral spikelets expanded into a wing on one side below; width of its lower part 0.6–1.4 mm; outer glume bristle-like.   **5. H. marinum**
- Both glumes of lateral spikelets subulate, or the inner slightly widened below; width of its lower part 0.2–0.4 mm.   **6. H. hystrix**

1. Hordeum bulbosum L., Cent. Pl. 2:8 (1756); Bor, Fl. Iraq 9:245; Boiss., Fl. 5:688; Post, Fl. 2:794. [Plate 221]

Perennial, 60–100 cm, tufted. Lowermost internode of culms thickened, tuber-like, covered with old leaf-sheaths. Leaves long, with well-developed auricles. Spike 5–17 × 0.6–0.8 cm (excl. awns), elongate, dense, usually nodding, yellowish or purplish; rachis disarticulating rather late at maturity. Spikelets 1-flowered; prolongation of rachilla more than ½ length of lemma.

Median spikelet subsessile, hermaphrodite, aristate; glumes lanceolate and ciliate at base, ending in a fine setaceous awn exceeding lemma; lemma oblong,
attenuate into a slender straight awn 2–3 times as long as glumes (incl. awns).
Lateral spikelets pedicellate, staminate, muticous; outer glumes setaceous, the
inner linear-lanceolate, each ending in a fine awn, exceeding glumes of
median spikelet; lemma lanceolate, awnless, somewhat longer than in median
spikelet. 2n=14, 28 (D. Zohary, unpubl.). Fl. March–May.

Hab.: Batha, fallow fields. Common in the hills. Acco Plain, Sharon Plain,
Philistean Plain; Upper and Lower Galilee, Mt. Carmel, Esdraelon Plain, Mt.
Gilboa, Samaria, Shefela, Judean Mts., Judean Desert, W., N. and C. Negev;
Upper Jordan Valley, Beit Shean Valley; Golan, Gilead, Ammon, Moav,
Edom.

Area: Mediterranean and Irano-Turanian.

2. Hordeum secalinum Schreb., Spicil. Fl. Lips. 148 (1771); Boiss., Fl. 5:687;
Post, Fl. 2:794 as *H. nodosum* auct. non L. [Plate 222]

Perennial, 60–100 cm. Differs from *H. bulbosum* L. in culms not thickened
at base, leaves not auriculate or with very small auricles in lower leaves, spikes
generally smaller; glumes of all spikelets setaceous. Fl. summer.

Hab.: N. Golan. Rare.

Area: Mediterranean and Euro-Siberian.

3. Hordeum spontaneum C. Koch, Linnaea 21:430 (1848) emend. Bachteev,
Bot. Zhur. 47:846 (1962); Bor, Fl. Iraq 9:252, t.87. *H. ithaburense* Boiss.,
Diagn. ser. 1, 13:70 (1854); Boiss., Fl. 5:686. *H. spontaneum* var. *ithaburense*
(Boiss.) Nábělek, Publ. Fac. Sci. Univ. Masaryk 111:32 (1929); Post, Fl. 2:793.
H. vulgare L. subsp. *spontaneum* (C. Koch) Thell., Fl. Advent. Montpellier
160 (1912). [Plate 223]

Annual, 50–80(–100) cm. Leaves long, glabrous or sparsely hairy, with well-
developed auricles. Spike 4–10 × 0.7–1 cm (excl. awns), elongate, dense, distichous; rachis densely silky-villose along margins, fragile, readily disarticulating. Spikelets 1-flowered; prolongation of rachilla more than ½ length of
lemma. Glumes linear-lanceolate, silky-villose, each ending in a fine awn
somewhat longer than glume proper. Median spikelet hermaphrodite, sessile;
lemma lanceolate, ending in a long tough awn; awn stout, flattened, scabrous, straight, 8–14(–20) cm long, about 1.5 mm broad at base. Lateral
spikelets staminate, neuter or vestigial, shortly pedicellate; lemma awnless,
rounded-obtuse at apex, rarely mucronate. 2n=14. Fl. April–May.

Hab.: Batha and fallow fields. Common and locally abundant. Sharon
Plain, Philistean Plain; Upper and Lower Galilee, Mt. Carmel, Esdraelon
Plain, Mt. Gilboa, Samaria, Shefela, Judean Mts., Judean Desert, N., C. and
S. Negev; Dan Valley, Upper Jordan Valley, Dead Sea area; Golan, Gilead,
Ammon, Moav, Edom.

Area: E. Mediterranean and Irano-Turanian.

H. spontaneum C. Koch is the wild progenitor of cultivated *H. vulgare* L. (2n=14). Apart from its ecological features, cultivated barley differs from its wild progenitor primarily in the gene controlling brittleness of the rachis. Six-rowed cultivated varieties additionally differ in a gene changing the two-rowed spike into a six-rowed spike, and naked grain varieties in a gene changing hulled grain to naked grain.

Wild brittle six-rowed barley plants have been described by Åberg (1938, 1940) from Tibet as *H. agriocrithon* which was regarded for some time as the wild progenitor of cultivated six-rowed barley. D. Zohary (1960, 1962) demonstrated that such forms are secondary hybrid derivatives resulting from occasional spontaneous hybridization between six-rowed *H. vulgare* and the wild *H. spontaneum*. Agriocrithon-like plants have been found in Israel on sites where stands of *H. spontaneum* grow in close proximity to cultivated barley.

4. Hordeum glaucum Steud., Syn. Pl. Glum. 1:352 (1854); Bor, Fl. Iraq 9:248. *H. stebbinsii* Covas, Madroño 10:18 (1949). *H. murinum* auct. non L.; Boiss., Fl. 5:686; Post, Fl. 2:793, incl. var. *violaceospicum* Eig & var. *fuscospicum* Eig, Bull. Inst. Agr. Nat. Hist. Tel-Aviv 6:69, 70 (1927). [Plate 224]

Annual, 10–50 cm. Leaves glaucous; sheaths glabrous; blades sparsely pilose; auricles 1–3 mm. Spike 3–6(–9) cm (excl. awns), very dense; internodes of rachis ciliate at margins. Spikelets 1-flowered. Median spikelet sessile; its glume linear-lanceolate, long-ciliate on both margins; floret pedicellate; lemma glabrous, its awn 1–2.5 cm; anthers 0.2–0.5(–1) mm, included at anthesis; prolongation of rachilla slender, scabrous, 3–6 mm. Lateral spikelets shortly pedicellate, neuter, rarely staminate; inner (upper) glumes similar to those of median spikelet, linear-lanceolate, ciliate on both margins; outer (lower) glumes linear-subulate, narrower than inner (upper) ones; lemma longer than in median spikelet, tapering to a longer awn; prolongation of rachilla of lateral spikelets 2–3(–4) mm, stout, hairy, orange-coloured at maturity. 2n=14, 28. Fl. February–April.

Hab.: Ruderal places, roadsides, near habitations. Common. Acco Plain, Coast of Carmel, Sharon Plain, Philistean Plain; Upper and Lower Galilee, Mt. Carmel, Esdraelon Plain, Samaria, Shefela, Judean Mts., Judean Desert, W., N. and C. Negev; Hula Plain, Upper and Lower Jordan Valley, Dead Sea area, Arava Valley; Golan, Gilead, Ammon, Moav, Edom.

Area: Mainly Mediterranean and W. Irano-Turanian.

The so-called *Hordeum murinum* aggregate treated by various authors in recent publications comprises three taxa: 1. *H. murinum* L., 2. *H. leporinum* Link and 3. *H. glaucum* Steud. [respectively, 1. *H. murinum* subsp. *murinum*, 2. subsp. *leporinum* (Link) Arcang. and 3. subsp. *glaucum* (Steud.) Tzvelev]. The main characters by which the three taxa can be separated are pointed out in the following key.

1. Lemma of lateral spikelets as long as lemma of median spikelet. Plant of Central and Atlantic Europe and part of the Mediterranean. Described from Europe.
H. murinum
– Lemma of lateral spikelets distinctly longer than lemma of median spikelet   2
2. Anthers in median spikelet 0.7–1.4 mm, usually exserted from floret; prolongation

of rachilla of lateral spikelets slender, scabrous, not orange-coloured nor yellow. Mainly Mediterranean. Described from Greece.   **H. leporinum**
- Anthers in median spikelet 0.2–0.5 mm, rarely reaching 1 mm, not exserted from floret (found under palea after flowering, on top of developing caryopsis); prolongation of rachilla of lateral spikelets stout, hairy, orange-coloured or yellow. Mediterranean and Irano-Turanian. Described from the Sinai Peninsula.   **H. glaucum**

Our plants of the *murinum* complex all belong to *H. glaucum* Steud. They are common ruderal plants growing throughout the country. D. Zohary (unpubl., personal communication) and Y. Waisel (1960, MSc. Thesis, Hebrew Univ.) found in Israel both diploids (2n=14) and tetraploids (2n=28). Diploids are apparently common in more arid parts of the country; they have small included anthers. Tetraploid plants apparently have somewhat longer anthers and somewhat longer spikes.

5. Hordeum marinum Huds., Fl. Angl. ed. 2, 1:57 (1778); Bor, Fl. Iraq 9:250, t.86; Post, Fl. 2:794. *H. maritimum* Stokes in With., Arr. Brit. Pl. ed. 2, 1:127 (1787); Boiss., Fl. 5:687. [Plate 225]

Annual, 10–40(–60) cm. Leaves sparsely pilose, sometimes glabrous; auricles short. Spike 2–5 cm (excl. awns), very bristly. Spikelets 1-flowered, glabrous or hairy. Median spikelet of triplet sessile, the lateral ones pedicellate, shorter than median spikelet; floret of median spikelet sessile. Glumes scabrous (not ciliate), 2–3 times as long as floret, bristle-like, except the inner (upper) glume of each lateral spikelet which is expanded into a wing in its lower part, 0.6–1.4 mm broad. Lemma of lateral spikelets awned. Fl. April–May.

Hab.: Saline marshes, banks of rivers, alluvial soils. Acco Plain, Coast of Carmel, Sharon Plain, Philistean Plain, W. Negev; Upper and Lower Galilee, Mt. Carmel, Esdraelon Plain, Samaria; Hula Plain, Upper Jordan Valley.

Area: Mediterranean and Irano-Turanian, extending into the Euro-Siberian region and S. Africa.

6. Hordeum hystrix Roth, Catalecta Bot. 1:23 (1797); Humphries, Hordeum, in Fl. Eur. 5:205. *H. geniculatum* All., Fl. Pedem. 2:259, t.91 f.3 (1785) *nom. ambig.*; Bor, Fl. Iraq 9:246, t.85. *H. maritimum* Stokes subsp. *gussonianum* (Parl.) Aschers. & Graebn., Syn. Mitteleur. Fl. 2(1):737 (1902). [Plate 226]

Annual, up to 30 cm, glabrous. Auricles very short. Spike 1.5–4(–5) cm (excl. awns). Differs from *H. marinum* mainly in lateral spikelets having equal bristle-like glumes, or one inner glume slightly broader in the lower part (not reaching 0.5 mm in width). Fl. March–May.

Hab.: Saline marshes, swampy ground, sandy sites. Acco Plain, Coast of Carmel, Sharon Plain, Philistean Plain, W. Negev; Upper Galilee, Mt. Carmel, Esdraelon Plain, Judean Mts.; Hula Plain; Golan.

Area: Mediterranean and Irano-Turanian, extending into the Euro-Siberian.

11. Taeniatherum Nevski

Annuals. Leaves flat or rolled, shortly auriculate; ligule very short. Inflorescence a dense spike; rachis tough, not disarticulating at maturity. Spikelets in

pairs at nodes of rachis, in 2 rows, all alike, 2-flowered; lower floret hermaphrodite, the upper reduced to a scale-like rudiment. Glumes equal, very shortly
connate at base with each other and with glumes of neighbouring spikelet,
rigid, subulate, each passing into a slender scabrous awn; lemma lanceolate,
coriaceous, 5-veined, ending in a long scabrous awn; callus short, flat; palea
nearly as long as lemma, obtuse or retuse. Lodicules 2, ciliate in upper half,
with lateral tooth. Stamens 3. Caryopsis oblong, adhering to lemma and
palea, with a hairy apical appendage.

Two or three Irano-Turanian and Mediterranean species.

1. **Taeniatherum crinitum** (Schreb.) Nevski, Acta Univ. As. Med. ser. 3b, Bot.
fasc. 17:38 (1934); Bor, Fl. Iraq 9:264, t.91. *T. caput-medusae* (L.) Nevski var.
crinitum (Schreb.) Humphries, Jour. Linn. Soc. London (Bot.) 76:343 (1978).
Elymus crinitus Schreb., Beschr. Gräser 2:15 (1777); Post, Fl. 2:795. *E. caput-
medusae* Boiss., Fl. 5:691 non L. *E. caput-medusae* L. subsp. *crinitus*
(Schreb.) Maire & Weiller, Fl. Afr. Nord 3:393 (1955). [Plate 227]

Annual, 10–30 cm. Culms glabrous. Leaf-blades sparsely pilose. Spike 3–7
cm (excl. awns), erect, often dark purple at maturity. Glumes 2–3 cm (incl.
awns), somewhat curved at base, erecto-patent; lemma 8–12 mm, minutely
papillose and scabrous; awn stout, in upper and middle part of spike 6–10 cm,
0.8–1.0 mm broad at base, curved outwards. Fl. May–July.

Hab.: Steppes and batha; Nubian sandstone, basalt, calcareous soil. Upper
Galilee, Judean Mts., Judean Desert; Golan, Gilead, Ammon.

Area: Irano-Turanian.

Tribe 4. BROMEAE Dumort. (1823). Annuals or perennials. Leaf-blades flat;
sheaths tubular nearly to mouth, soon splitting, with or without auricles;
ligule membranous. Inflorescence a panicle, lax and often nodding, or dense
and erect. Spikelets all alike, several- to many-flowered, hermaphrodite;
rachilla disarticulating above glumes and between florets. Glumes 2, shorter
than spikelet, unequal; lemma keeled or rounded on back, 2-fid at hyaline tip,
with a subterminal awn, rarely 3–9 awns, or awnless; awns straight or
recurved; palea hyaline, 2-veined, 2-keeled. Lodicules 2, mostly tapering
above, not ciliate. Stamens 3 or 2. Caryopsis ellipsoid, flattened at back,
grooved on ventral side, with a hairy apical appendage; embryo small; hilum
linear, as long as caryopsis. Chromosomes small to large; x=7.

12. BROMUS L.

Annuals or perennials. Leaf-blades flat; sheaths tubular, soon splitting; ligule
membranous. Inflorescence a panicle, contracted or effuse, erect or nodding,
in some species raceme-like. Spikelets pedicellate, several- to many-flowered,
hermaphrodite, mostly compressed; rachilla disarticulating above glumes and

between florets. Glumes unequal, acute, membranous-margined, persistent; the lower 1-7-veined, the upper 3-9-veined; lemma keeled or rounded on back, 5-13-veined, 2-fid at apex (very rarely entire), with 1-3(-5) awns arising in sinus between apical teeth or lobes, rarely awnless or mucronate; awns straight or recurved; palea 2-keeled, rounded at apex, with spaced cilia on keels. Lodicules 2, glabrous. Stamens 3 or 2. Caryopsis flattened at back, oblong, with a hairy apical appendage, adhering to lemma and palea; embryo small; hilum linear, as long as caryopsis. x=7.

About 50 species, mainly in temperate regions, with some species in trop. mountains.

Literature: A. de Cugnac & A. Camus, Sur quelques espèces de *Bromus* et leurs hybrides. I., *Bull. Soc. Bot. Fr.* 78:327-341 (1931). Z. Ovadiahu-Yavin, Cytotaxonomy of the genus *Bromus* of Palestine, *Israel Jour. Bot.* 18:195-216 (1969). H. Scholz, Zur Systematik der Gattung *Bromus* (Gramineae), *Willdenowia* 6:139-159 (1970). P.M. Smith, Taxonomy and nomenclature of the Brome-grasses (*Bromus* L. s.l.), *Notes Roy. Bot. Gard. Edinb.* 30:361-375 (1970); Serology and species relationships in annual *Bromus* (*Bromus* L. sect. *Bromus*), *Ann. Bot.* (*Konig & Sims*) 36:1-30 (1972). H. Scholz, Synaptospermie und Heterodiasporie in der Gattung *Bromus* (Gramineae), *Willdenowia* 8:341-350 (1978).

Infraspecific variability in *Bromus* displays noteworthy features. Parallel variation in indumentum of glumes and/or lemmas is found in the majority of local species belonging to sections *Bromus* and *Pnigma*. Species of section *Bromus* (*japonicus*, *danthoniae*, *lanceolatus*, *alopecuros*, *fasciculatus*, *rubens* and *madritensis*) are each represented by two varieties which differ in the indumentum, especially of glumes and lemmas. With the exception of one or two cases, the two varieties of each species grow together, i.e. are similar in their ecological requirements.

1. Lemma keeled on back, muticous or mucronate or with a weak 1-3 mm awn. Spikelets strongly compressed; panicle very lax, 15-30(-40) cm. Tall plants.
                                                                    16. B. catharticus
- Lemma rounded on back, awned; awns exceeding 3 mm   2
2. Tufted perennials with fibrous or reticulate remnants of preceding years' leaf-sheaths at base. Awn shorter than lemma proper   3
- Annuals devoid of remnants of leaf-sheaths as above. Awn rarely shorter than lemma   4
3. Leaf-sheaths of preceding years disintegrated into parallel fibres or strips. Plants of Mediterranean hill districts.   **1. B. syriacus**
- Leaf-sheaths of preceding years turned into reticulate fibres. Plants of altitudes above 900 m.   **2. B. tomentellus**
4(2). Spikelets (excl. awns) lanceolate or oblong, narrower at apex than at middle; lower glume 3-7-veined, the upper 5-9-veined   5
- Spikelets cuneate, broader at apex than at middle; lower glume 1-veined, the upper 3-veined   10
5. Upper florets in spikelet with 3(-5)-awned lemmas; lemmas of lower florets 1-awned; lateral awns in upper florets straight, shorter and weaker than dorsal (median) awn; dorsal awn curved, spreading. Plants of steppes and hammadas.
                                                                    5. B. danthoniae

- All lemmas 1-awned   6
6. Awns erect and straight also at maturity, about as long as or shorter than lemma. Hydrophytes.   **3. B. brachystachys**
- Awns spreading at maturity, mostly as long as or longer than lemma. Plants of other habitats   7
7. Panicle obovate to elliptic in outline, very dense, 2–5(–6) × 1–2.5 cm (excl. awns); spikelets numerous, shortly pedicellate, more or less erect, 10–20 mm (excl. awns), narrow.   **8. B. scoparius**
- Panicle not as above   8
8. Panicle lax; branches and pedicels capillary, spreading or nodding, as long as or longer than spikelets; spikelets nodding at maturity.   **4. B. japonicus**
- Panicle usually raceme-like or spike-like; branches and pedicels not capillary, mostly erect; spikelets erect or slightly spreading, not nodding   9
9. Apical teeth of lemma acuminate-aristulate, narrow. Longest pedicels (and eventually branches) not exceeding 4 mm, erect; spikelets borne singly on rachis of inflorescence; rarely pedicels up to 10 mm and then spikelets in groups of 2 or 3.
                                                   7. B. alopecuros subsp. **caroli-henrici**
- Apical teeth of lemma acute, not acuminate-aristulate. Shortest pedicels and branches 5 mm.   **6. B. lanceolatus**
10(4). Panicle one-sided; branches capillary, fairly long, all curved to one side; spikelets (incl. awns) up to 40 × 4 mm. Lemma 3–3.5 mm broad.   **9. B. tectorum**
- Plants with a different set of characters   11
11. Callus at base of floret cuneate, longer than broad, pointed at tip; callus-scar oblong. Culm tomentose below panicle; branches and pedicels densely hispidulous. Plant of sandy soils.   **15. B. rigidus**
- Callus at base of floret rounded, not longer than broad. Culm mostly glabrous; branches and pedicels scabridulous or glabrous, more rarely hispidulous   12
12. Spikelets (incl. awns) 5–9 cm; at least part of panicle-branches or pedicels 2–4 cm; panicle effuse; internodes of panicle-rachis 2 cm or longer   13
- Spikelets not exceeding 4 cm (incl. awns) or, if reaching 5 cm, panicle-branches or pedicels not exceeding 1 cm; panicle contracted; internodes of panicle-rachis at most 1 cm   15
13. Lemma 3–3.5 mm broad; awns 35–40 mm; spikelets 70–90 mm (incl. awns).
                                                                   14. B. diandrus
- Lemma 1.5–2.5 mm broad; awns 15–30 mm; spikelets 35–65 mm (incl. awns), rarely 70 mm   14
14. Many panicle-branches or pedicels as long as or longer than spikelet; spikelets 50–65(–70) mm. Lemma 2–2.5 mm broad.   **13. B. sterilis**
- The majority of panicle-branches or pedicels shorter than spikelet; spikelets 35–50 mm. Lemma 1.5–2 mm broad.   **12. B. madritensis** subsp. **madritensis**
15(12). Panicle densely contracted, obovoid or cuneate; internodes of panicle-rachis not exceeding 2 mm, only lowermost internode sometimes up to 5–7 mm   16
- Panicle not as above; at least two lower internodes of panicle-rachis 5 mm or longer.   **12. B. madritensis** subsp. **delilei**
16. Panicle very compact, obovoid, of numerous spikelets. Awns straight even at maturity.   **11. B. rubens**
- Panicle cuneiform, of several spikelets (1–8, rarely 10). Awns widely spreading after maturity.   **10. B. fasciculatus**

Sect. 1. PNIGMA Dumort., Obs. Gram. Belg. 117 (1824). *Bromopsis* Fourr. *pro gen*. Perennials, tufted. Spikelets more or less lanceolate, narrowed towards apex during anthesis. Lower glume 1-veined, the upper 3-veined. Awn straight, usually shorter than lemma, rarely absent.

1. Bromus syriacus Boiss. & Bl. in Boiss., Diagn. ser. 2, 4:139 (1859); Yavin, *op. cit*. 198. *B. erectus* Huds. var. *syriacus* (Boiss. & Bl.) Boiss., Fl. 5:644; Post, Fl. 2:771. *Bromopsis syriaca* (Boiss. & Bl.) Holub, Folia Geobot. & Phytotax. Praha 8:169 (1973). [Plate 228]

Tufted perennial, 60–100 cm. Shoots fertile or sterile; leaf-sheaths of preceding years disintegrating into parallel fibres or strips. Culms erect, leafy; leaf-blades long, flat, flaccid. Panicle 10–20 cm, lax, ovoid-oblong, erect or drooping; branches capillary, scabridulous, up to 6 cm, 1–2-spiculate, more or less erect, the lower spreading. Spikelets (15-)20–30 mm (excl. awns), 4–8-flowered, oblong-lanceolate, compressed. Glumes scarious, hyaline-margined, keeled; the lower 5–6 mm, the upper 9 mm; lemma 9–11 mm (excl. awns), hyaline-margined, 5–7-veined (with 3 veins prominent); apical teeth very short, hyaline; awn straight, $\frac{1}{3}$–$\frac{1}{2}$ length of lemma proper. Stamens 3. Fl. April–June.

Hab.: Maquis, garigue and batha, terra rossa on dolomite. Upper Galilee; Dan Valley; less common on Mt. Carmel, Samaria, Judean Mts.

Area: E. Mediterranean (Palestine, Lebanon, Syria).

2. Bromus tomentellus Boiss., Diagn. ser. 1, 7:126 (1846); Bor, Fl. Iraq 9:162, t.55; Yavin, *op. cit*. 198; Boiss., Fl. 5:646; Post, Fl. 2:771. Incl. forma *brachyathera* Opphr. in Opphr. & Evenari, Florula Cisiord., Bull. Soc. Bot. Genève ser. 2, 31:163 (1941). *Bromopsis tomentella* (Boiss.) Holub, Folia Geobot. & Phytotax. Praha 8:169 (1973). [Plate 229]

Tufted perennial, (20-)50–60 cm. Rhizome thick, oblique, with several shoots. Bases of fertile and sterile shoots tightly clothed with a dense net of reticulate fibres. Culms erect, leafy. Leaves densely velutinous and ciliate. Panicle 5–12 cm (incl. awns), lax, few-spiculate, erect; branches up to 3 cm, 1–2 at nodes of rachis, 1-spiculate, as long as spikelet or longer. Spikelets 20–25 mm (excl. awns), 6–9-flowered, oblong-lanceolate, compressed; florets spaced on rachilla. Glumes and lemmas membranous, keeled; lemma 10–15 × 3.5–4 mm, scabridulous; apical teeth short, hyaline; awn straight, erect or spreading, shorter than lemma. Stamens 3; anthers 5–6 mm. Fl. May–July.

Hab.: Mountain steppes. ?Golan, Edom.

Area: W. Irano-Turanian and E. Mediterranean.

Sect. 2. BROMUS. Annuals. Spikelets lanceolate, narrowed towards apex even after flowering. Lower glume 3–7-veined, the upper 5–9-veined. Awn straight, spreading or recurved, about as long as lemma.

3. Bromus brachystachys Hornung, Flora (Regensb.) 16:417, t.1 (1833); Bor,

Fl. Iraq 9:134, t.45; Bor, Fl. Iran. 70:111–112; Boiss., Fl. 5:654; Post, Fl. 2:775–776; incl. var. *longipes* Post, Fl. ed. 1, 894 (1883–1896). *B. aegyptiacus* Tausch, Flora (Regensb.) 20:124 (1837) subsp. *palaestinus* Melderis, Ark. Bot. ser. 2, 2(5):300, ff.2, 3 (1952). *B. palaestinus* (Melderis) Melderis in Mout., Nouv. Fl. Liban Syrie 1:128 (1966). [Plate 230]

Annual, 25–60 cm. Panicle lax or contracted, 12–20 cm, ovate-oblong to oblong; branches in fascicles at nodes of rachis, erecto-patent to erect, the short ones about as long as spikelet, the longest 3–6 cm, sometimes up to 10 cm (var. *longipes* Post); generally branches 1–3-spiculate. Spikelets 10–15(–17) × 4–5 mm (excl. awns), ovate to ovate-lanceolate, usually (5–)8–10-flowered. Lemma 5–6.5 mm, 7-veined, often minutely appressed-hispidulous; apical teeth short, triangular; awn 6–7 mm (rarely shorter), erect, straight, arising just below apex of lemma; palea shorter than lemma. 2n=14. Fl. March–May.

Hab.: Marshes, river banks, etc., heavy, usually slightly saline soil. Acco Plain, Coast of Carmel, Sharon Plain, Philistean Plain; Upper Galilee, Esdraelon Plain, Samaria, Shefela, Judean Mts.; Dan Valley, Hula Plain, Upper Jordan Valley, Beit Shean Valley, Lower Jordan Valley, Dead Sea area; Golan, Gilead, Ammon, Moav.

Area: E. Mediterranean and W. Irano-Turanian (Egypt, Palestine, Lebanon, Syria, Turkey, Iraq, Iran, Afghanistan); described from a casual plant in Germany.

4. Bromus japonicus Thunb., Fl. Jap. 52, t.11 (1784); Bor, Fl. Iran. 70:112; Yavin, *op. cit.* 207; Post, Fl. 2:776. [Plate 231]

Annual, 20–80 cm. Panicle lax, up to 15 cm; branches and pedicels capillary, spreading and finally nodding, the longer branches 3–7 cm. Spikelets 15–30 × 4–7 mm (excl. awns), oblong-lanceolate, 8–12-flowered, glabrous (var. *japonicus*) or softly pilose [var. *velutinus* (Koch) Bornm., Bot. Jahrb. 61 Beiblatt 140:176, 1927]. Lemma about 9 mm, membranous-margined, broadly elliptic, broadest slightly below awn insertion; apical teeth short, triangular; awn arising at about ⅓ down from apex of lemma, slightly bent, spreading at maturity; awn in lower florets up to 4 mm (sometimes obsolete), in upper florets up to 12 mm. Stamens 3. 2n=14. Fl. April–June.

Hab.: Batha and fallow fields, on various soils. Sharon Plain, Philistean Plain; Upper and Lower Galilee, Mt. Carmel, Shefela, Samaria, Judean Mts., C. Negev; Upper Jordan Valley, Arava Valley; Golan, Gilead.

Area: Euro-Siberian, Mediterranean and Irano-Turanian.

5. Bromus danthoniae Trin. in C.A. Mey., Verz. Pfl. Cauc. 24 (1831); Bor, Fl. Iraq 9:136, t.47; Yavin, *op. cit.* 209; Boiss., Fl. 5:652 as *B. macrostachys* Desf. var. *triaristatus* Hackel (1879). *B. lanceolatus* Roth var. *danthoniae* (Trin.) Dinsmore in Post, Fl. 2:775 (1933). [Plate 232]

Annual, 10–20(–35) cm. Panicle raceme-like, 3–5(–10) cm, 2–5-spiculate; branches and pedicels appressed to rachis, much shorter than spikelet, 1-spiculate. Spikelets 20–25(–30) × 4–5 (–7) mm (excl. awns), oblong-lanceolate,

10–20-flowered, pale green turning straw-coloured or purplish, glabrous to scabrous (var. *danthoniae*) or pubescent (var. *lanuginosus* Roshev. in Fed-tschenko, Fl. Turkmen. 1:166, 1932). Lemma 10–12 mm, hyaline at margins and apex, 3(–5)-awned in upper florets, 1-awned in lower florets; awns mostly purple; middle awn longest, arising ⅓ down from apex of lemma, recurved and spreading; lateral awns straight, much shorter and weaker; apical teeth minute. Stamens 3. 2n=14. Fl. April–May.

Hab.: Steppes and hammadas. N. and C. Negev; E. Gilead, Ammon, Moav, Edom.

Area: Irano-Turanian.

6. **Bromus lanceolatus** Roth, Catalecta Bot. 1:18 (1797); Yavin, *op. cit.* 209; Post, Fl. 2:774. *B. macrostachys* Desf., Fl. Atl. 1:96, t.19 f.2 (1798); Boiss., Fl. 5:652. [Plate 233]

Annual, 15–60 cm. Panicle raceme-like, erect, mostly dense, of several spikelets, or sometimes more lax, with more numerous spikelets and with erect or sometimes slightly spreading branches; branches 0.5–2.5 cm, 1–2-spiculate. Spikelets 20–40(–50) × 4–6 mm (excl. awns), oblong-lanceolate, 8–20-flowered, greenish, mostly softly woolly (var. *lanatus* Kerguélen), rarely glabrous (var. *lanceolatus*). Lemma 12–15(–20) mm, lanceolate, more or less prominently veined; apical teeth triangular, acute; awn 15–25 mm, somewhat bent and spreading at maturity, arising about ¼–⅓ down from apex of lemma. Stamens 3. Fl. March–May.

Var. **lanceolatus.** Spikelets glabrous.
Hab.: Rare. Esdraelon Plain, Shefela, Judean Mts.

Var. **lanatus** Kerguélen, Bull. Soc. Bot. Fr. 124:340 (1977). *B. lanuginosus* Poir. in Lam., Encycl. Méth. Bot. Suppl. 1:703 (1811) excl. var. β (=*B. divaricatus* Rohde in Loisel., 1910).* Var. *lanuginosus* (Poir.) Dinsmore in Post, Fl. 2:774 (1933) *nom. illeg. B. macrostachys* Desf. var. *lanuginosus* (Poir.) Coss. & Durieu, Expl. Sci. Algérie, Bot. II, 162 (1855). *B. oxyphloeus* Paine, Palest. Explor. Soc. Third Statement 128 (1875). [Plate 233]. Spikelets softly woolly. 2n=28.

Hab.: Batha, fields and roadsides. Fairly common. Sharon Plain, Philistean Plain; Upper and Lower Galilee, Mt. Carmel, Mt. Gilboa, Samaria, Shefela, Judean Mts.; Upper and Lower Jordan Valley; Golan, Gilead, Ammon, Moav, Edom.

Area of species: Mediterranean and Irano-Turanian.

Hybridization appears to occur between var. *lanatus* and var. *lanceolatus*. A specimen in the Herbarium of the Hebrew University, collected in the Esdraelon Plain, displays glabrous glumes and woolly lemmas.

---

* "L'épithète '*lanuginosus*' est illégitime et superflue, donc à rejeter, typifiée par *Bromus divaricatus* qui est à épillets glabres!" (Kerguélen, *loc. cit.*).

7. Bromus alopecuros Poir., Voy. Barb. 2:100 (1789) subsp. **caroli-henrici** (Greuter) P.M. Smith, Jour. Linn. Soc. London (Bot.) 76:360 (1978). *B. caroli-henrici* Greuter, Ann. Naturhist. Mus. (Wien) 75:83–89 (1971). Yavin, *op. cit.* 209 & Boiss., Fl. 5:650 & Post, Fl. 2:774 as *B. alopecuros*. [Plate 234]

Annual, 25–40 cm. Panicle raceme-like or spike-like, stiffly erect, with spikelets borne singly at nodes of rachis on stiff, very short pedicels (0.5–2 mm); rarely spikelets in groups of 2 or 3 at lower nodes of axis, and then pedicels 3–10 mm. Spikelets 20–40(–50) × 4–6 mm (excl. awns), lanceolate, (5–)8–12-flowered, erect to slightly spreading, varying in indumentum from glabrous or puberulent to villose. Lemma 11–15(–17) mm (excl. awns), 9-veined, gradually acuminate; veins slender; apical teeth acuminate aristulate; awn arising about ⅓ down from apex of lemma, 1½ times as long as lemma, curved outwards and spreading. 2n=14. Fl. February–May.

Hab.: Various soils in forest, maquis, batha, also in wet places. Coastal Galilee, Acco Plain, Sharon Plain, Philistean Plain; Upper and Lower Galilee, Mt. Carmel, Esdraelon Plain, Mt. Gilboa, Samaria, Shefela, Judean Mts., N. and W. Negev; Hula Plain; Upper and Lower Jordan Valley; Golan, Ammon.

Area of subspecies: E. Mediterranean (Palestine, Lebanon, Syria, Cyprus, Turkey, Crete, Cyclades, Sporades).

Plants with glabrous and hairy spikelets usually grow together. They have been designated respectively as var. *calvus* Halácsy (1904) and var. *poiretianus* Maire & Weiller (1955) of *B. alopecuros* subsp. *alopecuros*.

According to Greuter (1971), *B. alopecuros* differs from *B. caroli-henrici* in shorter spikelets and florets and especially in the apical teeth of lemma being short, broadly 3-angular and acute, not long and acuminate-aristulate. In addition, the spikelets in *B. alopecuros* are as a rule (rarely in *B. caroli-henrici*) borne in groups of 2 or 3 on nodes of axis. The area of *B. alopecuros* extends from the mainland of Greece westwards.

8. Bromus scoparius L., Cent. Pl. 1:6 (1755); Amoen. Acad. 4:266 (1759); Yavin, *op. cit.* 207; Boiss., Fl. 5:651; Post, Fl. 2:773. Incl. var. *multiflorus* Opphr. in Opphr. & Evenari, Florula Cisiord., Bull. Soc. Bot. Genève ser. 2, 31:165 (1941). [Plate 235]

Annual, 10–40 cm. Panicle 2–5(–6) × 1.2 cm (excl. awns), erect, very dense, obovate to elliptic in outline, obtuse; branches and pedicels very short. Spikelets numerous, 10–20 × 2–3 mm (excl. awns), 6–8-flowered, more or less erect, subcompressed, glabrous (var. *scoparius*=var. *psilostachys* Halácsy, Consp. Fl. Graec. 3:399, 1904) or pubescent (var. *villiglumis* Maire & Weiller in Maire, Fl. Afr. Nord 3:259, 1955). Lemma 8–9 mm, oblong-lanceolate, hyaline-margined, more or less prominently veined; apical teeth short, acute; awns as long as or longer than lemma, spreading at maturity, arising about ⅓ down from apex of lemma; palea slightly shorter than lemma. Stamens 3. 2n=14. Fl. February–May.

Hab.: Fields and roadsides. Very common. Coastal Galilee, Acco Plain,

Coast of Carmel, Sharon Plain, Philistean Plain; Upper and Lower Galilee, Mt. Carmel, Esdraelon Plain, Samaria, Shefela, Judean Mts., Judean Desert, W., N. and C. Negev; Dan Valley, Hula Plain, Upper Jordan Valley, Beit Shean Valley, Lower Jordan Valley, Dead Sea area; Golan, Gilead, Ammon, Edom.

Area: Mediterranean and Irano-Turanian, extending into the Euro-Siberian.

Var. *scoparius* and var. *villiglumis* grow together in the same populations.

Sect. 3. GENEA Dumort., Obs. Gram. Belg. 116 (1824). Annuals. Spikelets lanceolate only when young, later cuneate, broadened towards apex. Lower glume 1-veined, the upper 3-veined. Awn longer than lemma, usually flattened, rough.

9. Bromus tectorum L., Sp. Pl. 77 (1753); Bor, Fl. Iraq 9:160; Yavin, *op. cit.* 198; Boiss., Fl. 5:647; Post, Fl. 2:771. Incl. var. *longipilus* Borbás in Nábělek, Publ. Fac. Sci. Univ. Masaryk 111:22 (1929) & var. *širjaevii* Podp. in Nábělek, *loc. cit.* [Plate 236]

Annual, 10-40 cm. Leaves flat, soft; ligule up to 5 mm, lacerate at tip. Panicle 5-10(-12) cm (incl. awns), turned to one side, with clusters of crowded nodding spikelets; branches and pedicels capillary, thickened under spikelet, flexuose and bent sidewards, antrorsely hispidulous. Spikelets 30-40 mm (incl. awns), narrowly cuneate, 5-9-flowered. Glumes and lemmas silvery-hyaline at margins and apex; lemma 7-veined, glabrous (var. *tectorum*) or hairy (var. *hirsutus* Regel, Acta Horti Petrop. 7:600, 1880); apical teeth about 1 mm, hyaline, narrow, acuminate; awn straight, 16-22 mm, somewhat longer than lemma, arising ¼-⅓ down from apex of lemma. Stamens 1-3. 2n=14. Fl. March–May.

Hab.: Ruderal, on walls, roofs and in other arid waste places. Common. Sharon Plain, Philistean Plain; Upper and Lower Galilee, Mt. Carmel, Samaria, Shefela, Judean Mts., Judean Desert, N. and C. Negev; Dan Valley, Upper and Lower Jordan Valley, Dead Sea area, Arava Valley; Golan, Gilead, Ammon, Moav, Edom.

Area: Mediterranean, Irano-Turanian and Saharo-Arabian, extending into the Euro-Siberian.

"In Palestine the two varieties differ in their ecology and form separate populations in Judean Mts., Carmel, Negev, Ammon and Edom, var. *hirsutus* in drier habitats than var. *tectorum*" (Yavin, *op. cit.* 198).

10. Bromus fasciculatus C. Presl, Cyper. Gram. Sic. 39 (1820); Bor, Fl. Iran. 70:127; Yavin, *op. cit.* 203; Boiss., Fl. 5:650; Post, Fl. 2:773. *B. rubens* L. var. *fasciculatus* (C. Presl) Trabut in Batt. & Trabut, Fl. Algérie Monocot. 226 (1895). [Plate 237]

Annual, 5-20(-30) cm. Culms erect, single or several, glabrous. Ligule 1-2 mm, hyaline. Panicle raceme-like, erect, cuneate, 3-6 cm (incl. awns), of several spikelets, sometimes of 1–2 spikelets only; branches or pedicels very short; pedicels hispidulous. Spikelets 30–40 mm (incl. awns), cuneate, strongly compressed, 6–10-flowered, green or purple; florets spaced on rachilla. Glumes linear-lanceolate, acuminate; lemma 1–1.5 mm broad, linear-subulate, keeled, narrowly hyaline-margined, involute; apical teeth 2 mm, hyaline, acuminate; awn 13–20 mm, about as long as to somewhat longer than lemma, at maturity recurved and widely spreading. Stamens 3. 2n=14. Fl. March–May.

Hab.: Batha and steppes. Common, especially in steppes. Coastal Galilee, Coast of Carmel, Sharon Plain, Philistean Plain; Upper and Lower Galilee, Mt. Carmel, Esdraelon Plain, Mt. Gilboa, Samaria, Shefela, Judean Mts., Judean Desert, N., C. and S. Negev; Upper and Lower Jordan Valley, Dead Sea area; Gilead, Ammon, Moav, Edom.

Var. **fasciculatus.** Glumes and lemma scabridulous, not hairy. Rare. Mt. Carmel.

Var. **fallax** Maire, Contr. Fl. Afr. Nord 3427 (1942). Glumes glabrous or nearly so; lemma hairy but not ciliate.

Var. **alexandrinus** Thell., Feddes Repert. 5:161 (1908). Glumes and lemma hairy; lemma long-ciliate at margin.

Var. *fallax* and var. *alexandrinus* are mostly found growing together.

Area of species: S. and E. Mediterranean, extending into the Saharo-Arabian and the W. Irano-Turanian.

11. Bromus rubens L., Cent. Pl. 1:5 (1755); Amoen. Acad. 4:265 (1759); Bor, Fl. Iraq 9:151; Yavin, *op. cit.* 203; Boiss., Fl. 5:650; Post, Fl. 2:773. [Plate 238]

Annual, 10–30 cm, erect. Culms erect, pubescent below panicle. Ligule up to 5 mm, lacerate, milky-white. Panicle 4–7 cm (incl. awns), very dense, brush-like, obovoid, erect, of very numerous spikelets, often reddish or purplish, with rachis-internodes usually not exceeding 2 mm; branches and pedicels erect, very short, hispidulous. Spikelets about 45–50 mm (incl. awns), cuneate, slightly compressed, 4–5(–8)-flowered. Glumes and lemmas densely appressed-hairy (var. *rubens*) or glabrous to scabridulous (var. *glabriglumis* Maire in Emberger & Maire, Cat. Pl. Maroc 4:943, 1941); lemma 2–2.5 mm broad, linear-lanceolate, acuminate, broadly membranous-margined, 5-veined; apical teeth hyaline, short, aristulate; awn 15–22 mm, arising at ¼ down from apex of lemma, as long as or longer than lemma, straight even at maturity, erect to somewhat spreading. Stamens 2 or 3. 2n=28. Fl. April–May.

Hab.: Usually on somewhat saline soils, steppes, batha and fallow fields. Acco Plain, Sharon Plain, Philistean Plain; Judean Mts., Judean Desert, N., W., C. and S. Negev; Lower Jordan Valley, Dead Sea area, Arava Valley; Ammon, Moav, Edom.

Area: Mediterranean, W. Irano-Turanian and Saharo-Arabian.

Var. *rubens* and var. *glabriglumis* grow together.

This species does not occur N. of the Tel-Aviv–Jerusalem line.

12. Bromus madritensis L., Cent. Pl. 1:5 (1755); Amoen. Acad. 4:265 (1759); Yavin, *op. cit.* 202; Boiss., Fl. 5:649; Post, Fl. 2:772. [Plates 239, 240]

Annual, 20–60 cm. Panicle 8–15 cm, interrupted; branches erect to erecto-patent, 2–3-spiculate. Spikelets 35–50 mm (incl. awns), green or purple, cuneate, 6–13-flowered; florets spaced on rachilla; callus at base of floret hemispheric. Glumes keeled; upper glume 12–15 mm, with prominent veins; lemma 12–20 mm (excl. awn), 1.5–2 mm broad, 7-veined; apical teeth short (1–2 mm), fine, hyaline; awn 10–20 mm, erect or spreading. Stamens 1–2. 2n=28. Fl. April–May.

Subsp. **madritensis**. [Plate 239]. At least part of panicle-branches 2–4 cm; internodes on rachis of panicle 1–3 cm; spikelets 25–35 mm (excl. awns). Glumes and lemmas papillose [var. *victorinii* (Sennen & Mauricio) Maire] or pilose (var. *ciliatus* Guss.). 2n=28.

Hab.: Batha and steppes. Common. Acco Plain, Coast of Carmel, Sharon Plain, Philistean Plain, Upper and Lower Galilee, Mt. Carmel, Esdraelon Plain, Mt. Gilboa, Samaria, Shefela, Judean Mts., Judean Desert, N., W. and C. Negev; Hula Plain, Upper Jordan Valley, Beit Shean Valley, Arava Valley; Gilead, Ammon, Edom.

Subsp. **delilei** (Boiss.) Maire & Weiller in Maire, Contr. Fl. Afr. Nord 2874 (1939). [Plate 240]. Branches of panicle not exceeding 1 cm; longest internodes on rachis of panicle 0.3–1 cm; spikelets 15–25 mm (excl. awns). Glumes and lemmas glabrous (var. *glabriglumis* Maire & Weiller) or hairy (var. *villiglumis* Maire & Weiller). 2n=28.

Hab.: Batha and steppes. Common. Coastal Galilee, Sharon Plain, Philistean Plain; Upper Galilee, Mt. Carmel, Esdraelon Plain, Shefela, Judean Mts., N. and C. Negev; Upper Jordan Valley, Beit Shean Valley, Lower Jordan Valley, Dead Sea area, Arava Valley; E. Gilead, Ammon, Edom.

Area of species: Mediterranean and Irano-Turanian, extending into the Euro-Siberian.

13. Bromus sterilis L., Sp. Pl. 77 (1753); Bor, Fl. Iraq 9:158, t.54; Yavin, *op. cit.* 198; Boiss., Fl. 5:648; Post, Fl. 2:771. [Plate 241]

Annual, 30–50 cm. Culms glabrous. Ligule membranous, lacerate. Panicle very lax, glabrous or scabrous; branches capillary, scabridulous or glabrous, spreading and curved upwards in their upper part, usually 1- to few-spiculate, the longest branches as long as or longer than spikelet. Spikelets 50–65(–70) mm (incl. awns), cuneate, 6–10-flowered, often purplish; callus at base of floret hemispheric, rounded at tip, not longer than broad. Upper glume 8–15(–17) mm; lemma 13–20 × 2.5 mm (excl. awn), oblong-lanceolate, narrowly membranous-margined, 7-veined, scabridulous to glabrous; apical teeth 1–3 mm, fine, hyaline; awn 20–35 mm. Stamens 3. 2n=28. Fl. April–May.

Hab.: Batha, fallow fields. Sharon Plain; Upper and Lower Galilee, Mt. Carmel, Esdraelon Plain, Mt. Gilboa, Samaria, Judean Mts.; Dan Valley, Hula Plain, Upper and Lower Jordan Valley; Golan, Gilead, Ammon.

Area: Mediterranean and Irano-Turanian, extending into the Euro-Siberian.

14. Bromus diandrus Roth, Bot. Abh. 44 (1787); Yavin, *op. cit.* 199; Bor, Fl. Iran. 70:125, t.19. *B. gussonei* Parl., Pl. Rar. Sic. 2:8 (1840) & Fl. Ital. 1:407 (1850). *B. rigidus* Roth subsp. *gussonei* (Parl.) Maire in Emberger & Maire, Cat. Pl. Maroc 4:942 (1941); Yavin, *op. cit.* 199. *B. rigidus* var. *gussonei* (Parl.) Boiss., Fl. 5:649 (1884). [Plate 242]

Annual, 35–80 cm. Ligule lacerate at apex. Panicle very lax, often broader than long; branches capillary, spreading at maturity, scabridulous, in clusters of 2–4 at nodes, 1(–2)-spiculate; the longest branches up to 10 cm. Spikelets 70–90 mm (incl. awns), cuneate, nodding at maturity, usually dark purple, 5–8-flowered; florets spaced on rachilla; callus at base of floret rounded, not longer than broad, obtuse at tip, glossy; callus-scar almost circular. Glumes broadly membranous-margined, lanceolate, acuminate, keeled, with prominent veins; upper glume about 20 mm; lemma 25–30 × 3–3.5 mm (excl. awn), oblong-lanceolate, acuminate, membranous-margined, scabridulous, keeled, 7-veined; apical teeth 2–3 mm, very fine, hyaline; awn 35–40 mm, straight or flexuose. Stamens usually 2. 2n=56. Fl. April–May.

Hab.: Disturbed ground, mostly on heavy soil, hillslopes. Coastal Galilee, Acco Plain, Coast of Carmel, Sharon Plain, Philistean Plain; Upper Galilee, Mt. Carmel, Mt. Gilboa, Samaria, Judean Mts.; Hula Plain, Upper Jordan Valley; Golan, Gilead, ?Ammon.

Area: Mediterranean, extending into the Irano-Turanian region (Palestine, Lebanon, Syria, Turkey, Cyprus, N. Iraq, Caucasus and Russian C. Asia).

15. Bromus rigidus Roth, Bot. Mag. (Zürich) 4(10):21 (1790); Boiss., Fl. 5:649; *B. villosus* Forssk., Fl. Aeg.-Arab. 23 (1775) non Scop. (1772); Post, Fl. 2:772. *B. maximus* Desf., Fl. Atl. 1:95, t.26 (1798). [Plate 243]

Annual, 30–50(–80) cm. Culms shortly tomentose below panicle. Panicle 8–15 cm, dense and erect, narrow, longer than broad; rachis, branches and pedicels antrorsely and densely hispidulous; branches in clusters, not exceeding 2.5 cm. Spikelets (50–)60–80 mm (incl. awns), cuneate, 4–6(–9)-flowered, scabridulous, often dark purple; callus at base of each floret cuneate, longer than broad, acute and sharp-pointed, compressed; callus-scar oblong. Glumes membranous, acuminate, aristulate, with prominent veins; lemma 20–30 mm, keeled, 3–5-veined, broadly membranous at margin; apical teeth 2–3 mm, hyaline, subulate; awn firm, straight, very scabrous, about twice as long as lemma. Stamens usually 2. 2n=42. Fl. March–May.

Hab.: Sands and sandy loam of the coastal plain. Coastal Galilee, Acco Plain, Coast of Carmel, Sharon Plain, Philistean Plain; W. and N. Negev; Arava.

Area: Mediterranean, extending into Europe and the Caspian Sea coast.

B. rigidus Roth "is a bad weed in the United States where it is known as Ripgut Grass or Ripgut Brome and considered an obnoxious weed on the ranges by the stockman. The sharp awns can cause festering wounds in livestock and may even pierce their intestines causing death" (Bor, Fl. Iraq 9:142).

Sect. 4. CERATOCHLOA (Beauv.) Griseb. ex Ledeb., Fl. Ross. 4:360 (1852). Perennials, tuft-forming. Spikelets ovate or ovate-lanceolate, strongly compressed. Lower glume 5-veined, the upper 7-9-veined; lemma strongly keeled on back. Awn usually much shorter than lemma, often absent.

16. Bromus catharticus Vahl, Symb. Bot. 2:22 (1791); P. Pinto-Escobar, Caldasia 9:9-16 (1976). *Ceratochloa cathartica* (Vahl) Herter, Rev. Sudamer. Bot. 6:144 (1940). *B. unioloides* Kunth in Humb., Bonpl. & Kunth, Nov. Gen. Spec. 1:151 (1816); Bor, Fl. Iran. 70:132. *Festuca unioloides* Willd., Hort. Berol. 3:t.3 (1803). *B. unioloides* (Willd.) Rasp., Ann. Sci. Nat. ser. 1, 5:439 (1825). *B. willdenowii* Kunth, Révis. Gram. 1:134 (1829); C.C. Maw, Kew Bull. 29:431-434 (1974). [Plate 244]

Perennial, rhizomatous, 100-150 cm, generally glabrous. Leaves long, 3-12 mm broad. Panicle very lax, 15-30(-40) cm; branches up to 10 cm, scabrous, spreading or nodding, mostly longer than spikelets. Spikelets 20-35(-40) mm, strongly compressed, lanceolate to ovate, acute, 5-10-flowered. Glumes lanceolate-acuminate, keeled; lemma 12-15 mm, ovate-lanceolate, acuminate, strongly keeled, scabridulous, membranous-margined, muticous, mucronate or with a weak 1-3 mm awn. Fl. April-August.

Hab.: Irrigated land. Philistean Plain; Judean Mts. (Jerusalem); Upper Jordan Valley, Dead Sea area, Arava Valley; Golan.

Area: Euro-Siberian, Mediterranean, Irano-Turanian regions, S. America. Described from Peru.

First collected in Palestine in 1939.
Sometimes cultivated for fodder and naturalized.

13. BOISSIERA Hochst. ex Steud.

Annuals. Leaf-blade flat or partly rolled; sheaths tubular for ⅔ of their length; ligules membranous. Inflorescence a dense obovate terminal panicle. Spikelets numerous, shortly pedicellate, terete, 3-10-flowered, 2-3 upper florets rudimentary; rachilla disarticulating above glumes and between florets. Glumes unequal, broadly hyaline-margined; the lower 1-3-veined, the upper 5-7-veined; lemma rounded on back, membranous, at maturity coriaceous, with a hyaline upper part, prominently 11-13-veined; part of the veins prolonged on back of lemma below hyaline tip into awns; awns erect, later recurved, often purple; palea as long as lemma. Lodicules 2, glabrous, cuneate, dentate or lobed. Stamens 3, minute. Caryopsis with a hairy apical appendage; embryo small; hilum linear.

A monotypic genus.

1. **Boissiera squarrosa** (Banks & Sol.) Nevski, Acta Univ. As. Med. ser. 8b, Bot. fasc. 17:30 (1934); Bor, Fl. Iraq 9:128. *Pappophorum squarrosum* Banks & Sol. in Russ., Nat. Hist. Aleppo ed. 2, 2:244 (1794). *B. bromoides* Hochst. ex Steud., Syn. Pl. Glum. 1:200 (1854) *nom. illeg.*; Boiss., Fl. 5:560. *B. pumilio* (Trin.) Hackel in Stapf, Denkschr. Akad. Wiss. Math.-Nat. Kl. (Wien) 50:9 (1885); Post, Fl. 2:743. [Plate 245]

Annual, 2–8(–15) cm, appressed-puberulent. Culms solitary or tufted, leafy up to base of inflorescence. Panicle 2–4 cm, fasciculate-capitate, obovate. Spikelets 20–30 mm (incl. awns), erect. Glumes lanceolate, shorter than lowest floret; lower glume 5–7 mm, the upper 6 9 mm, lemma 7–9-awned, pubescent or pilose; its hyaline tip lacerate, varying in length, about ¼–½ length of lemma proper; awns unequal; longer awns as long as to 2–3 times as long as lemma. Fl. April–May.

Hab.: Steppes. Judean Desert, N., C. and S. Negev; Dead Sea area; Gilead, Ammon, Edom.

Area: Irano-Turanian.

Tribe 5. AVENEAE Dumort. (1823). Annuals or perennials. Leaf-blades flat or rolled; sheaths usually split to base, without auricles; ligules membranous. Inflorescence a contracted or effuse panicle, sometimes a raceme, rarely a spike. Spikelets all alike, pedicellate, 2–7-flowered; florets all hermaphrodite or the upper reduced, or when florets 2, the lower staminate, the upper hermaphrodite; rachilla disarticulating above glumes and sometimes also between florets. Glumes 2, persistent, equal or unequal, often as long as or longer than spikelet; lemma 5–7-veined, entire or 2-lobed at apex, awned from below apex; lobes often produced as bristles; awn usually geniculate; palea hyaline, 2-veined, 2-keeled. Lodicules 2, lanceolate or 2-toothed, or absent. Stamens 3. Caryopsis with a small apical appendage, tightly or loosely enclosed by lemma and palea; embryo ⅕–½ length of caryopsis; hilum linear, as long as caryopsis, or punctiform. Chromosomes large; x=7.

14. AVENA L.

Annuals. Leaf-blades usually flat; ligule membranous. Inflorescence an effuse or more or less contracted panicle, often one-sided; branches and pedicels filiform. Spikelets often nodding, awned, with 1–6 hermaphrodite florets and 1–2 staminate or rudimentary florets; rachilla disarticulating above glumes and between florets, in some species above glumes only, in cultivated species not disarticulating at all. Glumes equal or distinctly unequal, herbaceous or chartaceous, rounded on back, persistent; lower glume 5–7-veined, the upper 7–9-veined; lemma 7-veined, indurate at maturity, 2-lobed or 2-aristulate and with a long dorsal awn; awn arising at about middle of lemma, usually geniculate, hygroscopic, usually with a thick twisted lower part (*column*) and a thinner straight upper part (*bristle*); callus mostly bearded; palea 2-fid, ciliate along keels. Lodicules 2. Stamens 3. Caryopsis cylindrical, grooved on ventral side; hilum linear. 2n=14, 28, 42. Dispersal unit: individual floret or a spikelet without glumes.

About 30 species, mainly in temperate and subtrop. regions.

Several species are important cereals, especially in temperate regions. All cultivated oats are characterized by non-shattering panicles. In Israel the large-grained hexaploid *A. sativa* L. and *A. byzantina* C. Koch and the diploid *A. strigosa* Schreb. are grown as irrigated forage crops. The latter, W. Mediterranean by origin, is now being preferred owing to its resistance to rust. The two main cultivated species, *A. sativa* and *A. byzantina*, probably originated from the wild hexaploid *A. sterilis*, were introduced from Europe. They differ from each other in the fracture of the rachilla during threshing; in *A. sativa* the rachilla breaks at the base of each floret, while a rachilla-segment remains attached to the upper floret in *A. byzantina*.

Literature: A. I. Malzev, Wild and cultivated oats, Section *Eu Avena* Griseb., *Bull. Appl. Bot. Pl.-Breed. (Leningrad), Suppl.* 38:1-522 (1930). G. Ladizinsky & D. Zohary, Genetic relationships between diploids and tetraploids in Series *Eubarbatae* of *Avena, Canad. Jour. Genetics and Cytology* 10:68-81 (1968); Notes on species delimitation, species relationships and polyploidy in *Avena* L., *Euphytica* 20:380-395 (1971). G. Ladizinsky, Biological flora of Israel. 2. *Avena* L., *Israel Jour. Bot.* 20:133-151 (1971). D. Zohary, Oats, in: P. H. Davis et al. (eds.), *Plant Life of South-West Asia*, 247-253, Bot. Soc. Edinburgh, 1971. T. Rajhathy & H. Thomas, *Cytogenetics of Oats (Avena L.)*, Ottawa, Canada, 1974. J. H. W. Holden, Oats, in: N. W. Simmonds (ed.), *Evolution of Crop Plants*, 86-90, London and New York, 1976. B. R. Baum, *Oats: Wild and Cultivated. A Monograph of the Genus Avena L. (Poaceae)*, Ottawa, Canada, 1977.

1. Glumes very unequal in length, the lower $\frac{1}{2}$-$\frac{2}{3}$ length of the upper   2
- Glumes equal or subequal   3
2. Rachilla disarticulating at maturity under each floret. Spikelets 3-5-flowered. Lemma glabrous.   **1. A. clauda**
- Rachilla disarticulating under lowermost floret only. Spikelets 2-3-flowered. Lemma pubescent in upper half.   **2. A. eriantha**
3(1). Rachilla disarticulating at maturity under each floret; dispersal unit: individual floret. Teeth of lemma ending in bristle-like 3-10 mm long aristulae, with tips reaching or overtopping knee of awn   4
- Rachilla disarticulating at maturity under lowermost floret only; dispersal unit: spikelet without glumes. Teeth of lemma hyaline not bristle-like, rarely exceeding 3 mm, with tips not reaching top of awn-column.   **6. A. sterilis**
4. Callus long, acute; callus-scar linear. Glumes (25-)30-45 mm.   **3. A. longiglumis**
- Callus short; callus-scar ovate to oblong-ovate. Glumes shorter than in above   5
5. Glumes 12-18(-20) mm. Aristulae of lemma mostly with a lateral setula on one side, rarely without a setula. Plants of steppe and desert districts (rarely somewhat extending along coastal plain).   **5. A. wiestii**
- Glumes 20-23(-25) mm. Aristulae of lemma without a lateral setula, rarely with a minute toothlet. Plants of Mediterranean districts.   **4. A. barbata**

1. Avena clauda Durieu in Duchartre, Rev. Bot. 1:360 (1845); Bor, Fl. Iraq

9:330, tt.122, 123; Baum, Monogr. 138, Map 63; Boiss., Fl. 5:542; Post, Fl. 2:738. [Plate 246]

Annual, 20–60 cm. Panicle 5–15 cm, lax; lower pedicels as long as spikelet. Spikelets 20–26(–30) mm (excl. awns), 3–5-flowered; rachilla disarticulating at maturity under each floret. Glumes very unequal; lower glume 3–5-veined, $\frac{1}{2}$–$\frac{2}{3}$ length of 5–7-veined upper glume; lemma 2-aristulate; aristulae often with lateral setulae; lemma of lower floret 20–25(–27) mm; dorsal awn 3–4.5 cm; callus very short, densely bearded; callus-scar linear. Dispersal unit: individual floret. 2n=14. Fl. March–May.

Hab.: Batha and steppes; calcareous soil. Rare. Upper and Lower Galilee, Mt. Carmel, E. Samaria, Judean Mts., Judean Desert, N. Negev; Upper Jordan Valley; Ammon.

Area: Mediterranean and Irano-Turanian.

2. Avena eriantha Durieu in Duchartre, Rev. Bot. 1:360 (1845); Bor, Fl. Iraq 9:334; Baum, Monogr. 145, Map 73, incl. var. *acuminata* Coss., Bull. Soc. Bot. Fr. 1:14 (1854). *A. pilosa* (Roem. & Schult.) Bieb., Fl. Taur.-Cauc. 3:84 (1819) non Scop. (1772); Boiss., Fl. 5:542; Post, Fl. 2:737. [Plate 247]

Annual, 30–60 cm. Spikelets 20–23 mm (excl. awns), 2–3-flowered; rachilla disarticulating under lower floret only. Glumes very unequal in length; the lower 3–5-veined, the upper 5–7-veined; lemma 2-aristulate, rarely not aristulate; lemma of lower floret shortly pubescent in upper $\frac{1}{2}$; callus short; callus-scar linear. Dispersal unit: spikelet without glumes. 2n=14. Fl. April–May.

Hab.: Very rare. Upper Galilee, Judean Mts. (Jerusalem).

Area: Mediterranean and Irano-Turanian.

Differs from *A. clauda* primarily in the type of dispersal unit.

3. Avena longiglumis Durieu in Duchartre, Rev. Bot. 1:359 (1845); Baum, Monogr. 221, Map 173; Post, Fl. 2:738. *A. barbata* Pott ex Link subsp. *longiglumis* (Durieu) Lindb. fil., Acta Soc. Sci. Fenn. nov. ser. B, 1:12 (1932). [Plate 248]

Annual, 30–80 cm. Panicle 10–30(–40) cm, lax, one-sided. Spikelets 20–35 mm (excl. awns), 2–3-flowered; rachilla disarticulating at maturity below each floret. Glumes equal or subequal, (25–)30–45 mm, 9–11-veined; lemma 2-aristulate, 20–25 mm (incl. aristulae), tough, long-villose in lower $\frac{1}{2}$ to $\frac{2}{3}$; aristulae often with a shorter lateral setula; dorsal awn much longer than glumes, inserted at about middle of lemma; callus acute and sharp-pointed, 3–4 mm; callus-scar linear, very narrow. Dispersal unit: individual floret. 2n=14. Fl. March–April.

Hab.: Batha and garigue, sandy loam. Rare. Sharon Plain; S. Judean Mts., Judean Desert, N. Negev; Dead Sea area; Edom.

Area: W., S. and E. Mediterranean and Saharo-Arabian.

Plants in steppe and desert districts with smaller spikelets (glumes 25 mm).

4. Avena barbata Pott ex Link in Schrad., Jour. für die Bot. 1799(2):315 (1800); Bor, Fl. Iraq 9:328, t.120; Baum, Monogr. 195, Map 140; Rocha Afonso, in Fl. Eur. 5:206; Boiss., Fl. 5:543; Post, Fl. 2:738. *A. alba* auctt. non Vahl (1791).* [Plate 249]

Annual, 40–80(–100) cm. Panicle 10–20(–50) cm. Spikelets 20–23(–25) mm (excl. awns), 2–3-flowered; rachilla disarticulating at maturity below each floret. Glumes subequal, 20–23(–25) mm, 7–9-veined; lemma 2-aristulate; aristulae very rarely with a minute lateral toothlet at base; lemma of lower floret 12–18 mm, villose up to insertion of dorsal awn; awn about twice as long as glumes; callus short; callus-scar ovate to elliptic. Dispersal unit: individual floret. 2n=28. Fl. March–April.

Subsp. **barbata**. Incl. *A. hirtula* Lag., Gen. Sp. Nov. 4 (1816). *A. barbata* subsp. *hirtula* (Lag.) Taborda Morais, Bol. Soc. Brot. ser. 2, 13:622 (1939). [Plate 249]. Aristulae at tip of lemma 3–4 mm. 2n=28.

Hab.: Batha, fallow fields, disturbed ground. Common. Coastal Galilee, Acco Plain, Sharon Plain, Philistean Plain; Upper and Lower Galilee, Mt. Carmel, Esdraelon Plain, Mt. Gilboa, Samaria, Shefela, Judean Mts., N. Negev; Hula Plain, Upper Jordan Valley, Beit Shean Valley; Golan, Gilead, Ammon, Moav, Edom.

Area of subspecies: Mediterranean, extending into the Irano-Turanian and the Euro-Siberian (C. Europe).

Subsp. **atherantha** (C. Presl) Rocha Afonso, Jour. Linn. Soc. London (Bot.) 76:358 (1978). *A. atherantha* C. Presl, Cyper. Gram. Sic. 30 (1820). Aristulae at tip of lemma 6–12 mm. 2n=14.

Hab.: Sandy loam and sandy soils, in association with *Desmostachya bipinnata*, *Centaurea procurrens* and *Lavandula stoechas*. Sharon Plain, Philistean Plain. Sporadic.

Area of subspecies: S. Mediterranean.

Ladizinsky (1971) treated this taxon under *A. hirtula* Lag. and showed on a map its area of distribution in Israel.

5. Avena wiestii Steud., Syn. Pl. Glum. 1:231 (1854); Bor, Fl. Iraq 9:340; Baum, Monogr. 239, Map 201; Boiss., Fl. 5:543; Post, Fl. 2:738. *A. barbata* subsp. *wiestii* (Steud.) Mansf., Kulturpfl. (Berlin), Beih. 2:479 (1959); Tsvelev, Nov. Syst. Pl. Vasc. (Leningrad) 11:70 (1974). [Plate 250]

Annual, 50–80 cm. Panicle 10–15(–25–50) cm. Spikelets 12–18(–20) mm (excl. awns), 2–3-flowered; rachilla disarticulating at maturity below each floret. Glumes subequal, 12–18(–20) mm, 5–7-veined; lemma 2-aristulate; aristulae with a lateral setula on one side; setulae much varying in length, rarely

---

* "*Avena alba* Vahl is an earlier name of *Arrhenatherum erianthum* Boiss. & Reut., now *Arrhenatherum album* (Vahl) W.D. Clayton (Kew Bull. 16:250, 1962)" (Bor, *loc. cit.*).

absent or appearing as minute teeth; lemma of lower floret 10–12 mm, hairy in lower ½; dorsal awn about twice as long as glumes; callus short; callus-scar narrowly ovate. Dispersal unit: individual floret. 2n=14. Fl. March–April.

Hab.: Steppes, loess, sandy or calcareous soils; often in the Artemisietum herbae-albae. Sharon Plain, Philistean Plain (occasional); Judean Desert, W., N. and C. Negev; Upper and Lower Jordan Valley, Dead Sea area, Arava Valley; Transjordan (Al-Eisawi).

Area: Irano-Turanian and Saharo-Arabian, slightly extending into the Mediterranean.

Plants intermediate between *A. wiestii* (diploid) and *A. barbata* (tetraploid) occur in the transition zone between the distribution areas of the two species (cf. also Ladizinsky & Zohary, 1968).

6. Avena sterilis L., Sp. Pl. ed. 2, 118 (1762); Bor, Fl. Iraq 9:340; Baum, Monogr. 334, Map 302; Boiss., Fl. 5:542; Post, Fl. 2:737. [Plate 251]

Annual, 40–100 cm. Panicle 12–30 cm, nearly one-sided. Spikelets large, 3–4-flowered; the lower 2–3 florets dorsally aristate and densely villose from base up to middle; the upper 1–2 florets awnless and glabrous; rachilla disarticulating below lowermost floret only. Glumes subequal, (27–)30–35(–40) mm, exceeding florets (excl. awns); lemma 20–26 mm, long-villose in lower ½, 2-dentate, teeth hyaline, not aristulate; dorsal awn much longer than glumes; callus short; callus-scar ovate. Dispersal unit: spikelet without glumes. 2n=42. Fl. March–May.

Subsp. **sterilis.** [Plate 251]. Ligule 5–6 mm. Florets 3–5. Glumes (27–)30–35(–40) mm; lemma of lower floret up to 26 mm.

Hab.: Batha, openings in maquis, fallow and cultivated ground, on various soils. Very common. Coastal Galilee, Coast of Carmel, Sharon Plain, Philistean Plain; Upper and Lower Galilee, Mt. Carmel, Esdraelon Plain, Mt. Gilboa, Samaria, Shefela, Judean Mts., Judean Desert, N. and C. Negev; Dan Valley, Hula Plain, Upper Jordan Valley, Beit Shean Valley, Dead Sea area, Arava Valley; Golan, Gilead, Ammon.

Subsp. **ludoviciana** (Durieu) Nyman, Consp. 810 (1882). *A. ludoviciana* Durieu, Bull. Soc. Linn. Bord. 20:41 (1855); Bor, Fl. Iraq 9:336, tt.125, 126. Ligule 2–4 mm. Florets 2, rarely 3. Glumes 18–25 mm; lemma of lower floret 20–22 mm.

Hab.: Rare and sporadic. Philistean Plain; Mt. Carmel, Esdraelon Plain, Judean Mts.; Gilead, Ammon, Moav.

Area of species: Mediterranean and Irano-Turanian.

Dubious Species

Avena carmeli Boiss., Diagn. ser. 1, 13:50 (1854); Fl. 5:548.

No *Avena* specimens which fit Boissier's description of *A. carmeli* have been collected in Palestine since the twenties of the current century. According to C.E. Hubbard (Blumea Suppl. 3:14, 1946) and Baum (Monogr. 390), Boissier's specimen under *A. carmeli* from Mt. Carmel is *Parapholis incurva* (L.) C.E. Hubbard.

15. ARRHENATHERUM Beauv.

Perennials. Culms erect, with or without globose tubers near base. Leaf-blades flat; ligule membranous. Inflorescence a more or less contracted panicle; branches filiform, semiverticillate, especially on lower nodes of main rachis; pedicels filiform. Spikelets awned, glossy, somewhat compressed laterally, 2-flowered; lower floret staminate (rarely hermaphrodite), sessile, the upper hermaphrodite (or pistillate), pedicellate; rachilla disarticulating above glumes and between florets, shortly produced beyond upper floret. Glumes unequal, hyaline, persistent; the lower about ½ length of the upper, 1-veined; the upper 3-veined; lemma 7-veined, veins elevated; lemma of lower floret with a long geniculate awn arising above base or nearly at base; callus hirsute; lemma of upper floret long-hirsute to a varying degree, shortly 2-fid at tip and with a straight weak bristle-like awn arising somewhat below fissure, rarely awn absent; palea ciliolate along keels. Lodicules 2. Stamens 3. Caryopsis oblong, hairy, enclosed in lemma and palea but free; embryo short; hilum linear.

Six to ten species in the Mediterranean and Irano-Turanian regions and in Europe.

1. Hairs on lemma of upper floret not exceeding insertion point of awn; lemma of lower floret sparsely hirtellous, sometimes also with a few long hairs near margins.
   1. A. palaestinum
- Hairs on lemma of upper floret reaching beyond insertion point of awn; lemma of lower floret with a row of long hairs along sides near margins. **2. A. kotschyi**

1. Arrhenatherum palaestinum Boiss., Diagn. ser. 1, 13:51 (1854); Bornm., Bot. Jahrb. 61, Beibl. 140:162 (1928); Jordanov, in Fl. Bulg. 1:324, t.17 f.7; Mout., Nouv. Fl. Liban Syrie 1:85. *A. elatius* (L.) Beauv. ex J. & C. Presl subsp. *palaestinum* (Boiss.) Bornm., Beih. Bot. Centralbl. 31 Abt. II:268 (1914); Opphr. & Evenari, Florula Cisiord., Bull. Soc. Bot. Genève ser. 2, 31:158 (1941); var. *palaestinum* (Boiss.) Boiss., Fl. 5:550 (1884; *"palestinum"*); Post, Fl. 2:739. [Plate 252]

Perennial, 40–80(–100) cm. Culm with (1–)2–3 superimposed globose tubers at base; tubers about 1 cm across. Leaves usually glabrous; ligule membranous, short, lacerate. Panicle 10–20(–30) cm, oblong, subinterrupted; branches erect or erecto-patent, unequal, the longer 2–3 cm. Spikelets (10–)11–13 mm (excl. awns). Glumes very unequal, glabrous, scabridulous at back, lanceolate, acute, somewhat exceeded by florets; lower glume shorter and much narrower than the upper; upper floret pedicellate; lemmas 2-fid at apex; lemma of the lower (staminate) floret hyaline at margins, hirsute at base, glabrous or more or less hirtellous on sides and sometimes with sparse long hairs near margins below; its awn 2.5–3.5 cm, arising a little above base of lemma, stout, brown and twisted below knee; lemma of upper floret long-hirsute up to its middle or at most to insertion of its awn; awn straight, setaceous, short, arising at ¼–⅓ down from apex. Anthers 4.5–5 mm. 2n=42=6x (D. Zohary, unpubl.). Fl. March–May

Hab.: Batha and rocky places. Upper and Lower Galilee, Mt. Gilboa, Samaria, Shefela, Judean Mts., Judean Desert; Golan, Gilead, Ammon.

Area: E. Mediterranean (Palestine, Lebanon, Syria, Turkey, Cyprus, Greece incl. Rhodes, S.E. Bulgaria).

A. palaestinum differs from European *A. elatius* in larger spikelets; its geniculate awn is stouter and longer and arises from near the base of the lower lemma; the lemma of the upper floret is long-hirsute at least up to its middle and not only to below the middle.

Eig (Schedae Fl. Exsicc. Pal., Cent. II:3, no. 108, 1932) wrote about *A. palaestinum* Boiss.: "It is a component of open Mediterranean associations (e.g., Poterietum spinosi). *A. palaestinum* is limited in Palestine to the Mediterranean mountain districts, where it is rather a rare plant". It is remarkable that it has not been found on Mt. Carmel.

2. Arrhenatherum kotschyi Boiss., Diagn. ser. 1, 7:122 (1846); Mout., Nouv. Fl. Liban Syrie 1:85; Bor, Fl. Iraq 9:325, t.119; Boiss., Fl. 5:550; Post, Fl. 2:740. [Plate 253]

Perennial, (20–)40–80 cm, with globose tubers at base of culms. It differs from *A. palaestinum* mainly in the lemma of the upper floret being long-hirsute all over and beyond insertion of its awn, and in the lemma of the lower floret being long-ciliate along margins. Fl. May–June.

Hab.: Batha on mountain slopes from 1,600 m upwards on Mt. Hermon. Rare in Golan and E. Galilee, at lower elevations. Recorded from Gilead and Ammon by Dinsmore.

Area: W. Irano-Turanian (Palestine, rare; Lebanon, Syria, Turkey, Iraq, Iran, Afghanistan).

Throughout its area (especially in Iran) *A. kotschyi* has been collected at high altitudes (1,600–3,000 m). On Mt. Hermon it replaces *A. palaestinum* from 1,600 m upwards. A population of dwarf plants (15–30 cm high), but not differing otherwise from *A. kotschyi*, has been found by A. Shmida.

A. kotschyi and *A. palaestinum* appear to hybridize. Plants transitional between the two species, especially in indumentum of lemmas, have been collected at elevations below 1,600 m on slopes of Mt. Hermon (1,000–1,500 m), in N. Golan and in E. Galilee, especially on Mt. Meiron (900–1,200 m). Some of these plants differ from typical *A. palaestinum* in that the lemma in the lower floret is characteristic of *A.*

kotschyi. Other plants from the same districts approach *A. kotschyi* even more. The hairs on the upper lemma are denser and reach somewhat beyond the insertion of the awn, while the lower lemma is hairy all over. Bor, who examined specimens from Mt. Meiron in 1970, noted on the label of one of the specimens (from the Herbarium of the Hebrew University): "This specimen needs investigation. Both lemmas are hairy, the upper excessively so. Spikelets also seem to be shorter and squatter".

16. GAUDINIA Beauv.

Annual. Leaf-blades flat; ligule membranous, glabrous. Inflorescence a long linear distichous spike; rachis fragile, disarticulating above insertion of spikelet. Spikelets laterally compressed, alternate and sessile in concavities of rachis and appressed to it, 4–11-flowered; florets distant from one another, hermaphrodite or the uppermost rudimentary. Glumes very unequal, much shorter than spikelet, oblong, with elevated veins, hyaline-margined; lemma and palea membranous; lemma lanceolate, broadly hyaline-margined, 2-dentate at tip, usually with a dorsal geniculate awn arising above its middle; palea 2-fid at apex. Lodicules 2. Stamens 3. Caryopsis fusiform, hairy at tip.

Three species in the Mediterranean and the Azores.

1. Gaudinia fragilis (L.) Beauv., Agrost. 95, 164 (1812); Boiss., Fl. 5:549; Post, Fl. 2:739. *Avena fragilis* L., Sp. Pl. 80 (1753). [Plate 254]

Annual, 20–50(–60) cm. Culms erect, glabrous. Leaf-sheaths and blades spreadingly pilose; ligule membranous, very short, lacerate. Spike 10–25 cm. Spikelets 1.5–2 cm, each falling with rachis-internode immediately below it. Lower glume acute, 3(–5)-veined, $\frac{1}{3}$–$\frac{1}{2}$ length and much narrower than the upper; upper glume 7–11-veined, 8–11 mm, obtuse; lemma with geniculate dorsal awn twisted above knee, about 1 cm long; florets spaced on rachilla. Fl. April–June.

Hab.: Mostly in plains and valleys, in somewhat humid habitats. Rather uncommon. Acco Plain, Sharon Plain, Philistean Plain; Upper and Lower Galilee, Mt. Carmel, Esdraelon Plain, Samaria, Shefela, Judean Mts.; Upper Jordan Valley; Golan, Gilead.

Area: Mediterranean, extending into the Euro-Siberian.

17. PILGEROCHLOA Eig

Annuals. Leaf-blades flat; ligule membranous, oblong, lacerate at tip. Inflorescence a lax panicle; branches capillary, erecto-patent; pedicels straight, club-shaped. Spikelets laterally compressed, 2–4-flowered; florets hermaphrodite, lower floret much longer than others, the uppermost rudimentary, all except lowermost stipitate; rachilla disarticulating above glumes and between florets. Glumes very unequal, persistent; the lower less than $\frac{1}{2}$ length of the upper, 1-veined; the upper 3-veined; lemma oblong-lanceolate, 2-dentate at tip and with dorsal awn arising at about its middle; awn geniculate and twisted below knee. Lodicules 2. Stamens 3. Caryopsis small, free.

A monotypic genus.

Literature: A. Eig, Zwei neue Gramineengattungen von der *Ventenata*-Gattung gesondert, *Feddes Repert.* 26:65–79 (1929).

1. **Pilgerochloa blanchei** (Boiss.) Eig, Feddes Repert. 26:74 (1929); Mout., Nouv. Fl. Liban Syrie 1:78, t.26 f.1; Post, Fl. 2:736. *Ventenata blanchei* Boiss., Fl. 5:539 (1884). [Plate 255]

Annual, 20–40 cm, glabrous. Leaves narrow; upper leaf-sheath partly enclosing panicle. Panicle 10–15(–20) cm, pyramidal, many-spiculate; rachis puberulent; branches semi-verticillate, scabridulous; pedicels generally not longer than spikelets. Spikelets 11–12 mm (excl. awns), 2–3 at ends of branches. Lower glume less than ½ length of the upper, 1-veined; upper glume 8–10 mm, lanceolate, long-acuminate, keeled, scabridulous along keel and elevated veins; lemma of lower floret 10–11 mm, 2-keeled, 2-aristulate at tip; dorsal awn 12–16 mm. Fl. April–May.

Hab.: Basalt-derived or calcareous soils which are flooded in winter and dry up in summer. Rare. Upper Galilee; Golan.

Area: E. Mediterranean (N. Palestine, Lebanon, Syria).

18. LOPHOCHLOA Reichenb.

Annuals. Leaf-blades flat or rolled; ligule membranous, short, irregularly dentate, truncate. Inflorescence a spike-like more or less dense panicle, often lobed; branches and pedicels mostly short. Spikelets compressed, (2–)3–5(–10)-flowered; florets hermaphrodite or the uppermost rudimentary; rachilla disarticulating above glumes and between florets; in some spikelets the upper 2 or 3 lemmas are smaller than others, much thickened, awnless or with very short reflexed awns; these lemmas are shed readily and may sprout as proliferating organs. Glumes subequal or unequal, keeled, mostly somewhat shorter than spikelet; the upper 3-veined, longer and broader than 1- or 3-veined lower glume; lemma boat-shaped, 5-veined, keeled, 2-lobed or emarginate, rarely entire; lobes obtuse or acute, with or without an awn in the sinus; awn straight or twisted at base; palea as long as lemma, 2-veined; each vein produced at tip into a short point or a longer bristle. Lodicules 2. Stamens 3; anthers minute. Caryopsis narrowly oblong; hilum punctiform.

Four species in the Mediterranean and Irano-Turanian regions.

Literature: K. Domin, Monographie der Gattung *Koeleria*, *Bibliotheca Botanica* (*Stuttgart*) 14, Heft 65:1–354, tt. 1–22, 2 maps (1907). J. Bornmüller, Zur Flora des Libanon und Antilibanon, *Beih. Bot. Centralbl.* 38 Abt. II:269–272 (1914).

1. Lower glume densely villose all over, slightly longer than upper glume. Terminal part of rachilla pilose.　　　　　　　　　　　　　　　　　　　　　　　　**4. L. pumila**

- Lower glume not as above, somewhat shorter than upper glume   2
2. Lobes at apex of lemma rounded, sometimes lemma emarginate; lemma devoid of awn, with or without a short mucro about as long as sinus between lobes.

3. L. obtusiflora
- Lobes at apex of lemma acute or subobtuse; midvein of lemma produced into an awn exserted from the sinus; awn sometimes very short   3
3. Palea ending in 2 bristle-like awns which are ⅕-½ length of palea proper. Spikelets (3-)4-6.5 mm (excl. awns).   **1. L. berythea**
- Palea ending in 2 very short points. Spikelets 3-4 mm (excl. awns).   **2. L. cristata**

1. **Lophochloa berythea** (Boiss. & Bl.) Bor, Taxon 16:68 (1967); Bor, Fl. Iraq 9:348, t.132. *Koeleria berythea* Boiss. & Bl. in Boiss., Diagn. ser. 2, 4:135 (1859). *K. phleoides* (Vill.) Pers. var. *grandiflora* Boiss., Fl. 5:573 (1884). *K. phleoides* subsp. *berythea* (Boiss. & Bl.) Domin, Biblioth. Bot. (Stuttgart) 14, Heft 65:270 (1907); Post, Fl. 2:749. [Plate 256]

Annual, 10-30(-40) cm. Culms glabrous. Leaf-blades and -sheaths with soft white hairs. Panicle 3-4(-8) × 0.6-1.2(-2) cm, dense, bristly, often lobed. Spikelets (3-)4-6.5 mm (excl. awns), 3-5-flowered. Glumes unequal, hyaline-margined, glabrous, sometimes puberulent; lower glume linear-lanceolate, acuminate; upper glume ovate, acute; lemma acute, keeled, broadly membranous-margined, glabrous or pilose; lobes of 2-lobed apex acute; midvein produced into a straight awn; awn varying in length (0.5-3 mm); palea ending in 2 long bristle-like awns, ⅕-½ length of palea proper. Fl. February–June.

Hab.: Fallow fields, roadsides, batha; various soils. Common. Acco Plain, Coast of Carmel, Sharon Plain, Philistean Plain; Upper and Lower Galilee, Mt. Carmel, Esdraelon Plain, Samaria, Judean Mts., Judean Desert, W., N. and C. Negev; Hula Plain, Upper and Lower Jordan Valley, Dead Sea area; Golan, Gilead, Ammon, Moav, Edom.

Area: E. Mediterranean and W. Irano-Turanian (Palestine, Lebanon, Syria, Cyprus, Iraq, Iran).

Post (*loc. cit.*) records three varieties under *K. phleoides* (Vill.) Pers. subsp. *berythea* (Boiss. & Bl.) Domin:

(1) var. *berythea*. *K. phleoides* var. *grandiflora* Boiss. (1884). Awns ⅓-½ length of lemma;
(2) var. *postiana* Domin. Awns as long or nearly as long as lemma;
(3) var. *biseta* (Steud.) Domin.

2. **Lophochloa cristata** (L.) Hyl., Bot. Not. 106:355 (1953). *Festuca cristata* L., Sp. Pl. 76 (1753). *Koeleria cristata* (L.) Bertol., Amoen. Ital. 67 (1819) non Pers. (1805). *F. phleoides* Vill., Fl. Delph. 7 (1786). *L. phleoides* (Vill.) Reichenb., Fl. Germ. Excurs. 42 (1830); Bor, Fl. Iraq 9:351. *K. phleoides* (Vill.) Pers., Syn. Pl. 1:97 (1805); Boiss., Fl. 5:572; Post, Fl. 2:748. [Plate 257]

Annual, 10-30(-40) cm. Culms glabrous. Leaf-blades and -sheaths hairy or glabrous. Panicle (2-)4-6(-10) × 0.5-0.8(-1.5) cm, spike-like, dense, sometimes lobed. Spikelets 3–4 mm, 3–5-flowered. Glumes unequal, hyaline-margined, glabrous or irregularly pubescent mainly along keel; lemma acute, keeled, membranous-margined, glabrous or pilose; lobes of 2-lobed apex acute; midvein produced into a short, straight awn; awn varying in length [0.2–1.5(–3) mm]; palea ending in 2 short points. Fl. March–June.

Hab.: Batha, fields, roadsides, on various soils. Coastal Galilee, Acco Plain, Sharon Plain, Philistean Plain, W. Negev; Upper Galilee, Mt. Carmel, Samaria, Judean Mts.; Upper and Lower Jordan Valley; Golan, E. Gilead, Ammon, Edom.

Area: Mediterranean and Irano Turanian, extending into the Euro-Siberian and the Sudanian.

Post (Fl. 2:749) records the following infraspecific taxa of *Koeleria phleoides* subsp. *phleoides*:

var. *typica* Domin (1907). Panicle cylindrical or oblong-cylindrical;
subvar. *genuina* Domin. Spikelets hairy;
subvar. *glabriflora* (Trautv.) Domin. Spikelets nearly glabrous;
var. *pseudolobulata* Degen & Domin. Panicle loosely lobed;
var. *vestita* Domin & Bornm. Spikelets long-pilose.

3. Lophochloa obtusiflora (Boiss.) Gontsch., in Fl. URSS 2:338 (1934); Bor, Fl. Iraq 9:348, t.133. *Koeleria obtusiflora* Boiss., Diagn. ser. 1, 7:121 (1846). *K. phleoides* (Vill.) Pers. var. *obtusiflora* (Boiss.) Boiss., Fl. 5:573 (1884). *K. phleoides* subsp. *obtusiflora* (Boiss.) Domin, Biblioth. Bot. (Stuttgart) 14, Heft 65:272 (1907); Post, Fl. 2:749. [Plate 258]

Annual, 15–30(–40) cm. Culms glabrous. Leaf-blades and -sheaths with soft white hairs. Panicle 3–10 × 0.6–1.2 cm, spike-like, dense, sometimes lobed. Spikelets (2–)3(–4) mm. Glumes unequal, glabrous or puberulent; lemma obtuse, keeled, often prominently veined, glabrous or pilose; apex emarginate or 2-lobed, with rounded lobes, with or without a short mucro in the sinus; palea produced above into 2 short points. Fl. April–May.

Hab.: Fallow fields, roadsides; various soils. Coast of Carmel, Sharon Plain, Philistean Plain; Judean Mts.; Dan Valley, Upper Jordan Valley, Dead Sea area; Golan, Gilead.

Area: E. Mediterranean and Irano-Turanian.

Post (*loc. cit.*) cites under *K. phleoides* subsp. *obtusiflora* (Boiss.) Domin the following infraspecific taxa:

var. *typica* Domin. Spikelets 4–5 mm;
var. *amblyantha* (Boiss.) Domin. Spikelets 2 mm;
var. *condensata* (Boiss.) Domin. Spikelets 3 mm. Glumes and lemmas hairy.

4. Lophochloa pumila (Desf.) Bor, Grasses Burma Ceylon India Pakistan 445 (1960); Bor, Fl. Iraq 9:354, t.134. *Avena pumila* Desf., Fl. Atl. 1:103 (1798). *Koeleria pumila* (Desf.) Domin, Feddes Repert. 2:31 (1906); Täckholm & Drar, Fl. Eg. 1:319 (1941); Post, Fl. 2:750. *K. sinaica* Boiss., Diagn. ser. 1, 13:53 (1854); Boiss., Fl. 5:573. *Trisetum pumilum* (Desf.) Kunth, Révis.

Gram. 1:102 (1829); Boiss., Fl. 5:534. *Trisetaria pumila* (Desf.) Maire, Fl. Afr. Nord 2:261 (1953). [Plate 259]

Annual, 5–12 cm. Leaves spreadingly hairy; ligule hairy. Panicle 2–3 cm; pedicels hispidulous. Spikelets 3–4 mm, oblong-lanceolate. Glumes somewhat shorter than spikelet; lower glume somewhat longer than the upper, densely felty-villose all over; upper glume nearly glabrous except along keel; lemma glabrous, scabridulous, entire at tip, awned below acute tip; awn straight, varying in length, at most as long as lemma; terminal part of rachilla pilose with hairs about 1 mm. Fl. March–April.

Hab.: Desert; saline sandy depressions. Dead Sea area.

Area: Saharo-Arabian and W. Irano-Turanian, extending into the W. Mediterranean (Spain).

19. AVELLINIA Parl.

Annuals. Leaves flat or rolled. Inflorescence a panicle, contracted at maturity; branches conspicuous, pedicels short. Spikelets laterally compressed, 2–4-flowered; florets distant on glabrous rachilla. Glumes very unequal, membranous; the lower 1-veined, the upper broader and longer, 3-veined, nearly as long as spikelet; lemma obscurely 3-veined, 2-fid at apex, with subulate teeth and a short awn arising from sinus; palea much shorter than lemma. Lodicules 2, glabrous. Stamens 3. Caryopsis linear, glabrous.

One or two Mediterranean species.

1. Avellinia michelii (Savi) Parl., Pl. Nov. 61 (1842). *Bromus michelii* Savi, Bot. Etrusc. 1:78 (1808). *Koeleria michelii* (Savi) Coss. & Durieu in Durieu, Expl. Sci. Algérie, Bot. II, 120 (1855); Boiss., Fl. 5:574. [Plate 260]

Annual, 7–30 cm. Culms and leaves densely pubescent; ligule less than 1 mm, truncate, lacerate, hairy on abaxial side. Panicle 2–7 cm, oblong-lanceolate, sometimes lobed; branches filiform; pedicels shorter than spikelets. Spikelets 3–5 mm (incl. awns). Lower glume about ¼ length of upper, linear-lanceolate, subulate at apex; upper glume lanceolate, mucronate; lemma linear-lanceolate; teeth very narrow; awn 1–2 mm. Fl. April.

Hab.: Calcareous sandstone hills and sandy soil. Very rare. Philistean Plain (collected only once in Gedera, in 1927).

Area: Mediterranean (Portugal, Spain, S. France, Italy, Jugoslavia, Greece, Crete, Cyprus, Algeria, Morocco).

Record new for the area of the Flora of Post.

20. TRISETARIA Forssk.

Annuals. Leaf-blades flat or rolled; ligule membranous, glabrous or short-hairy. Inflorescence a spike-like or open panicle, its base generally enclosed in uppermost leaf-sheath; branches and pedicels short; rachis and the branches of panicle scabridulous. Spikelets laterally compressed, 1-2(-5)-flowered, the upper floret borne on an elongated hairy rachilla-joint; rachilla disarticulating above glumes and between florets. Glumes subequal or unequal, acute or acuminate, persistent; lower glume smaller than upper, 1(-2)-veined; upper glume 3-veined; lemma membranous, 5-veined, concave, 2-dentate, ending in 2 apical white bristles and with a dorsal awn arising just below the sinus or towards middle; awn geniculate and twisted below knee; rarely bristles and awn absent; palea usually smaller than lemma. Lodicules 2. Stamens 3. Caryopsis narrowly oblong, grooved on ventral side; hilum punctiform.

Five Mediterranean, Irano-Turanian and Saharo-Arabian species.

Literature: S.A. Nevski, in *Schedae ad Herb. As. Med.* Ser. 8b, Fasc. 21, Tashkent, 1934. R. Maire, Contribution à l'étude de la Flore de l'Afrique du Nord, Fasc. 32, *Bull. Soc. Hist. Nat. Afr. Nord* 33:92-93 (1942). N.L. Bor, *Trisetaria,* in: *Flora of Iraq* 9:354-358, 1968.

1. Glumes distinctly unequal; lower glume ½-⅔ length of and much narrower than upper glume. Dorsal awn of lemma 1½-2 times as long as glume.
   4. T. macrochaeta
- Glumes slightly unequal　　　　　　　　　　　　　　　　　　　　　　　　2
2. Dorsal awn of lemma inserted at or below middle of lemma, not exserted or shortly exserted from glumes.　　　　　　　　　　　　　　　　**3. T. glumacea**
- Dorsal awn inserted just below the sinus between teeth or at ¼ below tip of lemma; awn sometimes absent　　　　　　　　　　　　　　　　　　　　3
3. Anthers 0.5-1 mm. Glumes acuminate. Apical bristles and dorsal awn of lemma much exserted from glumes; very rarely lemma nearly muticous.　　**1. T. linearis**
- Anthers 1.5-2 mm. Glumes usually acute. Lemma varying from muticous to short-mucronate or to ending in a short awn slightly exceeding glumes, without or with short apical bristles, or provided with both longer apical bristles and a longer dorsal awn.
   2. T. koelerioides

1. Trisetaria linearis Forssk., Fl. Aeg.-Arab. LX, 27-28 (1775); Mout., Nouv. Fl. Liban Syrie 1:79, t.24 f.4; Bor, Fl. Iraq 9:356, t.135. *Trisetum lineare* (Forssk.) Boiss., Diagn. ser. 1, 13:49 (1854) & Fl. 5:536; Post, Fl. 2:735. [Plate 261]

Annual, 15-40(-50) cm, glaucescent. Culms and leaves minutely appressed-hirtellous; leaf-blades flat or rolled; ligule 1-2 mm, truncate, lacerate. Panicle 5-15(-20) × 0.5-1(-1.5) cm (excl. awns), spike-like, oblong, dense, sometimes lobed; branches short, erect, bearing spikelets from their base; pedicels short, thickened at apex. Spikelets 5-6 mm (excl. awns), 2-flowered, with upper floret on an elongated hairy rachilla-joint, or 1-flowered mostly with upper floret reduced to an awn; rachilla produced beyond florets. Glumes slightly unequal, longer than florets (excl. awns and bristles), linear-lanceolate, acuminate, keeled, with unequal sides and with 1-3 elevated veins in lower part, scabridulous along keel; apical bristles of lemma 2 (rarely 3), straight, about as long or up to twice as long as lemma proper; dorsal awn scabrous, spreading, 2-3 times as long as lemma proper, inserted just below sinus or at ¼ below tip, pale brown, geniculate and twisted below knee; very rarely lemma nearly muticous [var. *submutica* Melderis in Rech. fil., Ark. Bot. ser. 2, 2(5):294, 1952]. Anthers 0.5–1 mm. Fl. March–May.

Hab.: Mostly on sandy soils, kurkar hills. Coastal Galilee, Sharon Plain, Philistean Plain; W., N. and C. Negev; Dead Sea area; Ammon, Edom.

Area: E. Mediterranean and Saharo-Arabian (Syrian Desert, Palmyra, Palestine, N. Egypt, Libya).

2. Trisetaria koelerioides (Bornm. & Hackel) Melderis in Rech. fil., Ark. Bot. ser. 2, 2(5):292 (1952), incl. var. *aristata* (Bornm. & Hackel) Melderis, *op. cit.* 292 & var. *longiaristata* Melderis, *op. cit.* 293 (1952); Mout., Nouv. Fl. Liban Syrie 1:80. *Trisetum (Trisetaria) koelerioides* Bornm. & Hackel in Bornm., Verh. Zool.-Bot. Ges. Wien 48:104 (1898) & var. *aristatum* Bornm. & Hackel & forma *intermedia* Bornm. & Hackel in Bornm., *op. cit.* 647 (1898); Post, Fl. 2:735. [Plate 262]

Annual, 15–30(–40) cm. Closely related to *Trisetaria linearis* and differing mainly in the following: Anthers 1.5–2 mm; glumes broader, usually acute (not acuminate). Lemma varying from muticous to short-mucronate (var. *koelerioides*) or shortly awned. Awn slightly exceeding glumes, without or with short apical bristles (var. *aristata*), or provided with both longer apical bristles and a longer dorsal awn (var. *longiaristata*). Fl. March–May.

Hab.: Sandy soils, mainly in the coastal plain. Together with *T. linearis* and less common. Acco Plain, Sharon Plain, Philistean Plain.

Area: E. Mediterranean (Palestine, Lebanon, Syria).

Melderis (*op. cit.*) remarks about *T. koelerioides* (Engl. translation): "This species is closely related to *T. linearis....* The collections from Palestine show that both species are highly polymorphic and contain parallel forms with different awn lengths, thus making the delimitation of species complicated in many cases.... After analyzing these two species Samuelsson came to the conclusion that one of the most important distinguishing characters between *T. koelerioides* and *T. linearis* is the length of the anthers".

The variability in *T. koelerioides* may be the result of hybridization with *T. linearis*. The two species were found growing together on various occasions.

Within the genus *Trisetaria*, *T. koelerioides* can be regarded as a terminal link in the process of speciation.

3. Trisetaria glumacea (Boiss.) Maire, Bull. Soc. Hist. Nat. Afr. Nord 33:93 (1942). *Trisetum glumaceum* Boiss., Diagn. ser. 1, 13:49 (1854); Täckholm & Drar, Fl. Eg. 1:317; Boiss., Fl. 5:536; Post, Fl. 2:734. [Plate 263]

Annual, 10–20 cm, with the aspect of *Triplachne nitens*, but finely pubescent. Leaf-blades hirtellous; sheaths retrorsely hairy; ligule membranous, fimbriate. Panicle 3–6 cm, oblong, glossy, sometimes lobed. Spikelets 4–5 mm (excl. awns), 2-flowered, with a rudiment of a 3rd floret often reduced to a hairy stalk; rachilla above lower floret long-hairy. Glumes nearly equal, longer than florets, lanceolate, acute, membranous, glossy, glabrous, scabridulous at keel; lemma proper about ½ length of glumes; apical bristles ⅓ length of lemma proper; dorsal awn pale brown in lower part, inserted at or below middle of lemma, not exserted or shortly exserted from glumes. Anthers 2 mm. Fl. April–May.

Hab.: Sandy soils in deserts. W. and N. Negev; Arava Valley; Edom.

Area: Saharo-Arabian (Palestine, Egypt).

4. Trisetaria macrochaeta (Boiss.) Maire, Bull. Soc. Hist. Nat. Afr. Nord 33:92 (1942). *T. macrochaeta* (Boiss.) Paunero Ruiz ex Melderis, in Rech. fil., Ark. Bot. ser. 2, 2(5):294 (1952). *Trisetum macrochaetum* Boiss., Diagn. ser. 1, 13:48 (1854) & Fl. 5:536; Post, Fl. 2:734. [Plate 264]

Annual, 6–20 cm. Leaves linear-setaceous, glabrescent, the lower rolled; ligule membranous, lacerate. Panicle 2.5–5(–7) cm, oblong-ovoid, more or less lax, contracted after flowering; branches capillary. Spikelets 4–5 mm (excl. awns), golden, glossy, 2-flowered, rachilla-joint above lower floret shortly hairy. Glumes unequal, longer than florets, acuminate and mucronate or aristulate, especially in lower part of panicle, scabridulous along keel; lower glume 3 mm, linear-lanceolate, 1-veined, ⅓ breadth of upper; upper glume 4 mm, lanceolate, 3-veined; apical bristles of lemma about ½ length of lemma proper and slightly exserted from glumes; dorsal awn much exserted from glumes, pale brown, inserted a little above middle of lemma and 1½–2 times as long as lemma. Anthers 1.5 mm. Fl. March–April.

Hab.: Steppes, calcareous and sandy ground. Philistean Plain; Shefela, Judean Mts., Judean Desert, W., N. and C. Negev; Upper and Lower Jordan Valley, Dead Sea area; Transjordan (Al-Eisawi).

Area: Saharo-Arabian (Palestine, Egypt, Libya, Spain).

21. Aira L.

Slender annuals. Leaf-blades flat or folded; ligule membranous. Inflorescence a lax panicle with capillary branches. Spikelets laterally compressed, 2-flowered, both flowers hermaphrodite; rachilla glabrous, disarticulating above glumes and between florets. Glumes subequal, acute, keeled, 1-veined, glabrous, usually longer than florets, persistent; lemma rounded on back, lanceolate, acute or acuminate, often 2-fid or ending in 2 apical bristles and with a dorsal awn arising below middle; awn geniculate, twisted below knee; palea 2-dentate, shorter than lemma. Lodicules 2. Stamens 3. Caryopsis oblong, grooved on ventral side, tightly enclosed by and adhering to lemma and palea; hilum minute.

Twelve species in temperate regions and in mountains of Trop. and S. Africa, as well as in Mauritania.

1. Aira elegantissima Schur, Verh. Mitt. Siebenb. Ver. Naturw. 4, App.: 85 (1853); Greuter in Greuter & Rech. fil., Boissiera 13:179 (1967). *A. elegans* Willd. ex Gaudin *nom. illeg. A. capillaris* Host, Gram. Austr. 4:20, t.35 (1809) non G. Savi (1792) nec Lag. (1805); Boiss., Fl. 5:529; Post, Fl. 2:732. [Plate 265]

Annual, 15–30 cm, glabrous. Leaves setaceous; ligule 2 mm, lanceolate, membranous. Panicle 4–10 × 2–6 cm, ovate in outline; branches in twos, erecto-patent, trichotomous; pedicels thickened at apex. Spikelets 1.5–2 mm. Glumes ovate-lanceolate, acute, membranous; lemma brown, long-attenuate and white above; lemma of lower floret often awnless; that of upper floret with a geniculate awn arising below middle, exserted from glumes and brown below knee. Fl. April.

Hab.: Batha. Rare. Sharon Plain, Philistean Plain; Upper and Lower Galilee, Mt. Gilboa; Golan.

Area: Mediterranean and Irano-Turanian, extending into the Euro-Siberian.

22. ANTINORIA Parl.

Slender annuals or perennials. Leaf-blades flat; ligule membranous. Inflorescence a lax panicle with capillary branches; pedicels thickened towards apex. Spikelets laterally compressed, 2-flowered; both florets hermaphrodite; lower floret sessile, the upper borne on elongated rachilla-internode; rachilla disarticulating above glumes and between florets. Glumes subequal, keeled, usually somewhat longer than florets; lemma shorter than glumes, rounded on back, muticous, 3-veined, broadest near apex, truncate and shallowly 3-lobed, awnless. Lodicules 2. Stamens 3. Caryopsis obovoid, flattened on ventral side, free.

Two species, mainly W. Mediterranean.

1. Antinoria insularis Parl., Fl. Palerm. 1:94 (1845); Mout., Nouv. Fl. Liban Syrie 1:88. *Aira insularis* (Parl.) Boiss., Fl. 5:528 (1884). *Airopsis insularis* (Parl.) Nyman, Syll. 411 (1854–1855). [Plate 266]

Annual, 10–20 cm, glabrous. Ligule oblong, about 1 mm. Panicle up to 10 cm, ovate in outline, lax. Spikelets 1–1.5 mm. Glumes obtuse, boat-shaped; lemma hyaline. Anthers 0.5 mm. Fl. May–July.

Hab.: On soils leached of chalk, which are flooded in winter and dry in summer. Rare. Golan.

Area: Mainly W. Mediterranean.

23. CORYNEPHORUS Beauv.

Annuals or perennials. Leaf-blades flat or setaceous; ligule membranous. Inflorescence an effuse panicle. Spikelets laterally compressed, 2-flowered; both florets hermaphrodite; the lower sessile, the upper pedicellate; rachilla produced beyond upper floret. Glumes subequal, boat-shaped, membranous, 1–3-veined, keeled, longer than florets; lemma acute, awned at back above base; callus bearded; awn articulated near middle, the lower part brownish, twisted, with a ring of short hairs at joint; the upper part pale, ending in a club-shaped tip; palea somewhat shorter than lemma. Lodicules 2. Stamens 3. Caryopsis oblong, adhéring to palea; hilum punctiform.

Six European and Mediterranean species.

1. **Corynephorus divaricatus** (Pourr.) Breistr., Procès-Verb. Soc. Dauph. Étud. Biol. (Grenoble) ser. 3, 17:3 (1950). *Aira divaricata* Pourr., Mém. Acad. Sci. Toulouse 3:307 (1788). *C. articulatus* (Desf.) Beauv., Agrost. 159 (1812); Boiss., Fl. 5:530. *Aira articulata* Desf., Fl. Atl. 1:70 (1798). *Weingaertneria articulata* (Desf.) F.W. Schultz, Arch. Fl. Jour. Bot. (1864); Aschers. & Graebn., Syn. Mitteleur. Fl. 2(1):302 (1899); Post, Fl. 2:732. [Plate 267]

Annual, 30–80 cm, often purplish. Culms slender. Ligule hyaline, acute, up to 8 mm. Panicle up to 20 cm; branches and pedicels capillary, widespreading during flowering. Spikelets 4–5 mm. Glumes subequal, acute to acuminate, scabridulous at keel, membranous to hyaline at tip and at margin, purplish or green-purplish at back; callus of florets bearded, silvery, the hairs up to ⅔ length of lemma on one side, shorter on other; awn 2.5–3.5 mm, not exceeding glumes. Fl. March–May.

Hab.: Sandy soils of the coastal plain and soils derived from the inland Nubian sandstone. Sharon Plain, Philistean Plain; W., N. and C. Negev; Upper Jordan Valley; Gilead, Ammon, Edom.

Area: Mediterranean, extending into the Irano-Turanian.

24. Holcus L.

Annuals or perennials. Leaf-blades flat; ligule membranous. Inflorescence a panicle. Spikelets laterally compressed, 2(–3)-flowered, at maturity falling entire from persistent pedicels. Glumes subequal in length, longer than florets, boat-shaped and keeled, the lower 1-veined, narrower than 3-veined upper glume; rachilla curved and elongated below lower floret; florets distant from each other; the lower hermaphrodite, the upper staminate or neuter; lemma of hermaphrodite floret muticous, of staminate floret awned from back below apex; palea nearly as long as lemma. Lodicules 2. Stamens 3. Styles very short. Caryopsis oblong, laterally flattened, grooved, glabrous; hilum punctiform.

Eight species in the Mediterranean and Irano-Turanian regions, 1 species in S. Africa.

1. **Holcus annuus** Salzm. ex C.A. Mey., Verz. Pfl. Cauc. 17 (1831); Griseb. in Ledeb., Fl. Ross. 4:410 (1852) non Salzm. ex Coss. & Durieu (1855); Tzvelev, Nov. Syst. Pl. Vasc. (Leningrad) 6:20 (1969); Boiss., Fl. 5:532; Post, Fl. 2:733, incl. var. *glabrescens* Eig, Bull. Inst. Agr. Nat. Hist. Tel-Aviv 6:64 (1927). *H. setosus* Trin., Mém. Acad. Sci. Pétersb. ser. 6, 5(7):87 (1840) *nom. illeg.*; Roshev., in Fl. URSS 2:238, t.18 f.1. [Plate 268]

Annual, 30–60 cm, soft-puberulent. Culms erect. Ligule membranous, lacerate, pubescent on back. Panicle 5–9 × 1–2 cm, lobed, contracted after flowering; branches and pedicels hispid. Spikelets oblong-ovoid, whitish, 2-flowered. Glumes ciliate along keel, shortly appressed-pubescent or glabrous, about 3.5–4 mm (excl. awns), acute and ending in a bristle-like awn, the lower lanceolate, tapering to an awn shorter than the glume proper, the upper twice as broad, ovate-lanceolate, with elevated lateral veins, ending in an awn about as long as glume proper; lemma of hermaphrodite floret much shorter than glumes, muticous, glabrous, glossy; staminate floret much narrower, its lemma glabrous, awned; awn sinuous when dry, included in glumes. Fl. March–May.

Hab.: Sandy, mostly non-calcareous soils. Coastal Galilee, Acco Plain, Sharon Plain, Philistean Plain; W. Negev.

Area: S. and E. Mediterranean, extending to shores of Caspian Sea.

25. MILIUM L.

Annuals or perennials. Leaf-blades flat; ligule membranous. Inflorescence an effuse or contracted panicle with whorled branches; pedicels capillary, thickened above. Spikelets ellipsoid, acute, 1-flowered, hermaphrodite. Glumes equal, membranous, 3-veined, longer than floret, persistent; lemma elliptic, coriaceous, muticous, faintly 5-veined, very smooth, indurate in fruit, clasping palea, devoid of callus; palea similar in texture. Lodicules 2, linear. Stamens 3. Caryopsis elliptic in outline, falling with lemma, but free; embryo about ⅕ length of caryopsis; hilum linear, short.

Four to six species in N. temperate regions.

1. Milium pedicellare (Bornm.) Roshev. ex Melderis in Rech. fil., Ark. Bot. ser. 2, 2(5):291 (1952); Bor, Fl. Iraq 9:278, t.98. *M. vernale* Bieb. var. *pedicellare* Bornm., Beih. Bot. Centralbl. 31 Abt. II:267 (1914); Post, Fl. 2:727. [Plate 269]

Annual, 30–60 cm, glabrous. Leaves flaccid; ligule 5 mm. Panicle 10–20(–25) × 5–15 cm, lax, ovate in outline; branches in pairs at nodes, spreading; pedicels 5–13 mm, spreading at a wide angle. Spikelets 3–3.5 mm, ovoid to ellipsoid, acute. Glumes membranous, ovate, acute, light green, later straw-coloured, minutely papillose; lower glume broadly clasping the upper; lemma elliptic, obtuse, glabrous, very smooth and glossy. Fl. March–April.

Hab.: Open spaces in maquis, garigue, limestone rocks. Upper and Lower Galilee, Mt. Tabor, ?Mt. Carmel, ?Samaria, Judean Mts.; Dan Valley; Golan, Gilead, Ammon.

Area: E. Mediterranean, extending into the W. Irano-Turanian (Palestine, Lebanon, Syria, Cyprus, Turkey, Iraq, Iran).

26. AMMOPHILA Host

Perennials, rhizomatous. Culms stout, erect. Leaf-blades long, stiff, usually folded; ligule long, scarious. Inflorescence a contracted and dense spike-like panicle. Spikelets 1-flowered, laterally compressed; floret hermaphrodite with a pedicel-like rudiment of a sterile floret; rachilla disarticulating above glumes. Glumes subequal, chartaceous; the lower 1-veined, the upper 3-veined; lemma similar in texture, keeled, 5-veined, 2-fid and with a short aristula in the sinus; callus short, long-bearded; palea nearly as long as lemma, with closely adjacent keels. Lodicules 2. Stamens 3. Caryopsis free, oblong-obovoid, grooved on ventral face; hilum linear, somewhat shorter than caryopsis.

Two species of sandy coasts in the Mediterranean and Euro-Siberian–Borealo-American regions.

1. Ammophila arenaria (L.) Link, Hort. Berol. 1:105 (1827); incl. subsp. *arundinacea* Lindb. fil., Acta Soc. Sci. Fenn. ser. nov. B, 1(2):10 (1932); Boiss., Fl. 5:526; Post, Fl. 2:732 as var. *arundinacea* (Host) Fiori. *Arundo arenaria* L., Sp. Pl. 82 (1753). *Ammophila arundinacea* Host, Gram. Austr. 4:24, t.41 ff.1, 2 (1809) *nom. illeg. Ammophila littoralis* (Beauv.) Rothm., Feddes Repert. 52:269 (1943). *Psamma littoralis* Beauv., Agrost. 143, t.6 f.1 (1812). [Plate 270]

Perennial, 50–120 cm. Rhizome creeping, stout, branched. Leaf-blades rigid, pungent, involute; sheaths ribbed; ligule membranous, 2-fid, 1–2.5 cm. Panicle 7–25 × 1.5–2.5 cm, cylindrical, straw-coloured; branches erect. Spikelets 10–12 mm. Glumes persistent, lanceolate, usually acute, about as long as lemma; lemma 8–12 mm, lanceolate, with a short rigid awn in the sinus; hairs of callus 3–5 mm. Fl. March–August.

Hab.: Sand dunes. Acco Plain, Sharon Plain, Philistean Plain; W. Negev.

Area: Mediterranean extending along the Atlantic coast of Europe.

Ammophila arenaria is treated here as a single species without further subdivision, as done also by Boissier (*loc. cit.*). This is at variance with the treatment in some recent Floras. Tutin in Flora europaea (5:236, 1980) accepted the subdivision of the species into the W. and N. European subsp. *arenaria* and the Mediterranean subsp. *arundinacea* Lindb. fil. (1932). Pignatti in Flora d'Italia (3:577, 1982) regarded the Mediterranean taxon as a species, *A. littoralis* (Beauv.) Rothm., and considered *A. arenaria* as limited to the Atlantic coast of Europe.

The distinction of the taxa is based on alleged differences in morphological characters, namely in glumes exceeding the lemma in the northern taxon vs glumes about equalling the lemma in the Mediterranean one, and in the hairs at the base of the lemma being longer in the Mediterranean taxon.

Comparison of herbarium specimens from Palestine and some Mediterranean countries with specimens from localities of the W. and N. Atlantic coasts of Europe showed that hairs of the lemma gradually lengthen in the southerly direction forming a cline (from 2–3 mm and 2–4 mm on the Atlantic coast to 3–5 mm in the Mediterranean). Moreover, there is no conspicuous difference in length of glumes compared to that of the lemma in the two areas.

It appears that it is neither possible to subdivide *A. arenaria* into two subspecies nor to accept *A. littoralis* as a distinct species. We prefer therefore to treat *A. arenaria* (L.) Link as a single taxon which is very characteristic of dunes along the Mediterranean and European Atlantic coasts.

Known as a sand-binder (Marram Grass) and planted for that purpose.

27. LAGURUS L.

Annual. Leaf-blades flat; ligule membranous. Inflorescence a dense panicle, head-like and soft-bristly; branches and pedicels very short. Spikelets laterally compressed, 1-flowered, hermaphrodite; lower spikelets of panicle mostly sterile. Glumes subequal, feathery, longer than floret (not incl. awn), persistent; rachilla breaking up at maturity above glumes; lemma shorter than glumes, with 2 aristulate teeth at apex and with a longer dorsal awn long-exserted from glumes; palea shorter than lemma. Lodicules 2. Stamens 3. Caryopsis oblong, without appendage, grooved, free, enclosed by hardened lemma and palea; hilum short.

A monotypic genus.

1. **Lagurus ovatus** L., Sp. Pl. 81 (1753); C. E. Hubbard, Grasses 289, t.288; Bor, Fl. Iran. 70:309; Boiss., Fl. 5:521; Post, Fl. 2:731. [Plate 271]

Annual, 10–40 cm. Leaf-sheaths long-villose. Panicle long-pedunculate, head-like, ovoid to globose, 0.5–6 × 0.5–2 cm (excl. awns), very dense, softly hairy and bristly, pale, rarely tinged with purple; branches very short. Spikelets 8–10 mm. Glumes narrowly lanceolate, each tapering into a fine bristle, membranous, 1-veined, densely covered with long spreading hairs; lemma 4–5 mm, elliptic, rounded on back, 5-veined, borne on short villose callus, 2-dentate at apex, each tooth ending in a fine bristle or aristula; awn arising from upper ⅓ of lemma, geniculate at about middle, twisted below, exserted from glumes. Fl. March–April.

Hab.: Sandy soils and batha on hillslopes. Coastal Galilee, Acco Plain, Coast of Carmel, Sharon Plain, Philistean Plain; Lower Galilee, Mt. Carmel, ?Samaria, Judean Mts.; Upper Jordan Valley; Golan, Gilead, Moav.

Area: Mediterranean, extending into the Euro-Siberian region.

This plant is used for ornament and sometimes dyed with bright colours for bouquets.

28. POLYPOGON Desf.

Annuals or perennials. Leaf-blades flat; ligule membranous, elongate, lacerate. Inflorescence a panicle with numerous very small, mostly awned spikelets; panicle very dense, contracted, spike-like, or lobed and interrupted, with semi-verticillate oblong branches. Spikelets 1-flowered, hermaphrodite, falling entire at maturity. Glumes equal, rounded on back, oblong, 1-veined, 2-fid, emarginate or entire at apex, with vein excurrent as a long scabrous awn, or glumes awnless; lemma about ⅓–½ length of glumes, hyaline, truncate, obscurely 5-veined, minutely denticulate at apex, awnless or midvein excurrent as a short awn. Lodicules 2. Stamens 3. Caryopsis ellipsoid; hilum punctiform.

Eight to ten species in the Mediterranean and Irano-Turanian regions.

1. Glumes awnless.   **3. P. viridis**

- Glumes long-awned   2
2. Lemma awnless. Glumes 2-fid down to ¼-⅓ of their length.   **1. P. maritimus**
- Lemma awned. Glumes emarginate at apex.   **2. P. monspeliensis**

1. Polypogon maritimus Willd., Ges. Naturf. Freunde Berlin Neue Schr. 3:442 (1801); Bor, Fl. Iraq 9:314, t.114; Boiss, Fl. 5:520; Post, Fl. 2:731. [Plate 272]

Annual, 5-30 cm. Panicle 1-5 × 0.8-2.5 cm (incl. awns), dense, cylindrical, sometimes tinged with purple; pedicels of spikelets jointed above middle. Spikelets 2 mm (excl. awns), falling at maturity with upper joint of pedicel (callus). Glumes membranous, obtuse, 2-fid down to ¼-⅓ of their length, hairy above, densely long-ciliate at margin and with short rigid points at back near base; awn about 4 times as long as glume, arising from sinus; lemma ⅓ length of glume, awnless, minutely denticulate at apex. Fl. February–May.

Hab.: Moist habitats. Acco Plain, Sharon Plain; Upper Galilee, Mt. Carmel, N. Negev; Lower Jordan Valley, Dead Sea area; Transjordan (Al-Eisawi).

Area: E. and S. Mediterranean and Irano-Turanian, extending into the Euro-Siberian.

2. Polypogon monspeliensis (L.) Desf., Fl. Atl. 1:67 (1798); Bor, Fl. Iraq 9:318, t.115; Boiss., Fl. 5:520; Post, Fl. 2:730, incl. forma *brevisetus* Opphr., Florula Transiord. 146, Bull. Soc. Bot. Genève ser. 2, 22:265 (1931). *Alopecurus monspeliensis* L., Sp. Pl. 61 (1753). [Plate 273]

Annual, (10-)20-80 cm, its height varying with soil humidity of habitat. Panicle 2-15 × 1.5-3 cm (incl. awns), dense, very soft-bristly, oblong to ovoid, sometimes lobed, much branched; pedicels of spikelets jointed at or above middle. Spikelets 2-3 mm (excl. awns), falling at maturity with thickened upper part of pedicel (callus). Glumes membranous, obtuse, emarginate at apex, ciliolate at margin, scabridulous; awn about 3 times as long as glume proper, arising from sinus just below apex; lemma ½ length of glume, short-awned; awn fragile. Fl. March–June.

Hab.: Moist habitats. Common. Acco Plain, Coast of Carmel, Sharon Plain, Philistean Plain; Upper and Lower Galilee, Esdraelon Plain, Samaria, Judean Mts., Judean Desert, W. and C. Negev; Dan Valley, Hula Plain, Upper Jordan Valley, Beit Shean Valley, Lower Jordan Valley, Dead Sea area; Golan, Gilead, Ammon, Edom.

Area: Mediterranean, Irano-Turanian and Saharo-Arabian regions, Trop. and S. Africa.

3. Polypogon viridis (Gouan) Breistr., Bull. Soc. Bot. Fr. 110 (Sess. Extr.):56 (1966). *Agrostis viridis* Gouan, Hort. Monsp. 546 (1762). *P. semiverticillatus* (Forssk.) Hyl., Nomencl. Syst. Nord. Gefässpfl., Uppsala Univ. Arsskr. no. 7:74 (1945); Bor, Fl. Iraq 9:318, t.116. *A. semiverticillata* (Forssk.) C. Christ., Dansk Bot. Ark. 4(3):12 (1922); *Phalaris semiverticillata* Forssk., Fl. Aeg.-Arab. 17 (1775). *A. verticillata* Vill., Prosp. Pl. Dauph. 16 (1779); Boiss., Fl. 5:513; Post, Fl. 2:729. [Plate 274]

Perennial, 20–50(–70) cm, stoloniferous, rooting at lower nodes. Panicle 4–15 × 0.6–2.5 cm, erect, oblong-pyramidal, lobed, branched; branches semiverticillate, erect, scabrous, with dense spikelets clustered right to base of branches; pedicels very short, jointed on branches. Spikelets 1.5–2.2 mm, falling at maturity with pedicel attached. Glumes awnless, lanceolate, subobtuse, appressed-puberulent; lemma about ½ length of glumes, awnless, broadly elliptic; palea about as long as lemma. Fl. April–August.

Hab.: Moist habitats. Coast of Carmel, Sharon Plain, Philistean Plain; Upper and Lower Galilee, Mt. Carmel, Esdraelon Plain, ?Samaria, Judean Mts., Judean Desert; Negev; Dan Valley, Hula Plain, Upper Jordan Valley, Beit Shean Valley, Lower Jordan Valley, Dead Sea area; Golan, Moav, Edom.

Area: Mediterranean and Irano-Turanian; naturalized in W. Europe.

Polypogon interruptus Humb., Bonpl. & Kunth, Nov. Gen. Sp. 1:134, 1816, has been collected once on the bank of Yarkon River in June 1923. This species is an American grass of wet places cited in the literature as adventive in Rotterdam (the Netherlands). It is a tall stoloniferous perennial, with a lobed panicle, awned glumes, the awn not exceeding the glume in length, and with a shortly awned lemma.*

29. GASTRIDIUM Beauv.

Annuals. Leaf-blades flat; ligule membranous, elongate, lanceolate. Inflorescence a spike-like panicle; branches short, erect. Spikelets numerous, shortly pedicellate, 1-flowered, hermaphrodite, laterally compressed; rachilla disarticulating above glumes, shortly overtopping floret. Glumes unequal, ventricose at base, keeled above, much longer than lemma, persistent; lower glume 1-veined, the upper 3-veined; lemma very short, broadly elliptic when spread out, truncate-dentate, 5-veined; median vein produced from dorsal surface into an awn twisted at base (rarely lemma awnless); palea hyaline. Lodicules 2. Stamens 3. Caryopsis obovoid to ellipsoid, tightly enclosed between lemma and palea; hilum punctiform.

Three species in the Mediterranean and Irano-Turanian regions.

1. Glumes lanceolate, acuminate, smooth, scabridulous along keel; lower glume subulate at tip; the upper ⅔ length of the lower; lemma usually hairy with dorsal awn, rarely nearly glabrous, awnless. Panicle lax during flowering.
                                                                    1. G. ventricosum
– Glumes ovate-lanceolate, acute, punctulate-scabridulous; upper glume ⅕ length of the lower; the lower not subulate; lemma usually not hairy, awnless, rarely hairy with dorsal awn. Panicle contracted even during flowering.   **2. G. scabrum**

1. Gastridium ventricosum (Gouan) Schinz & Thell., Viert. Naturf. Ges. Zürich 58:39 (1913); Bor, Fl. Iraq 9:304; Eig, Bull. Inst. Agr. Nat. Hist. Tel-

*Not illustrated.

Aviv 6:62 (1927); Post, Fl. 2:729. *Agrostis ventricosa* Gouan, Hort. Monsp. 39, t.1 f.2 (early 1762). *Milium lendigerum* L., Sp. Pl. ed. 2, 91 (late 1762). *G. lendigerum* (L.) Desv., Obs. Pl. Angers 48 (1818); Boiss., Fl. 5:519. [Plate 275]

Annual, (10-)15-40 cm. Panicle (3-)5-10 × 0.6-0.8 cm, silvery, lax during flowering. Glumes straight, lanceolate, scabrous along keel, glossy; lower glume 5-6 mm, acuminate, subulate at tip; upper glume 3-4 mm, acute to acuminate; lemma about 1 mm, hairy; awn longer than glumes, twisted below middle and geniculate at middle; occasionally lemma awnless and then glabrous or nearly so [forma *muticum* (Gaudin) Maire & Weiller] Fl. April-June.

Hab.: Heavy soils. Acco Plain, Coast of Carmel, Sharon Plain, Philistean Plain; Upper Galilee, Mt. Carmel, Esdraelon Plain, Judean Mts., C. Negev; Dan Valley, Hula Plain, Upper Jordan Valley; Golan.

Area: Mediterranean, extending into the W. Irano-Turanian.

Plants transitional to *G. scabrum* are sometimes found.

Bor (*loc. cit.*) cites *G. phleoides* (Nees & Meyen) C.E. Hubbard from Palestine. We could not find any specimens fitting *G. phleoides* in our material of *G. ventricosum*.

2. Gastridium scabrum C. Presl, Cyper. Gram. Sic. 21 (1820); Boiss., Fl. 5:519; Post, Fl. 2:730. *G. ventricosum* var. *scabrum* (C. Presl) Fiori, Nuova Fl. Anal. Italia 1:94 (1923). [Plate 276]

Annual, 10-20 cm. Panicle 4-5 × 0.3-0.4 cm, contracted even during flowering; branches very short. Glumes acute, scabridulous, somewhat shorter and more convergent than in *G. ventricosum*; lemma not hairy, awnless; rarely awn present and then lemma hairy. Fl. April-May.

Hab.: Mainly on sandy soils. Rare. Sharon Plain; Upper Galilee, Mt. Carmel, Judean Mts.

Area: Mediterranean.

Eig (Bull. Inst. Agr. Nat. Hist. Tel-Aviv 6:62, 1927) wrote: "In both species (*G. ventricosum* & *G. scabrum*) ... the pales (lemmas) are hairy when they are awned, and almost or quite glabrous when awnless. Hairiness of the pales can not serve as a diagnostic character for these two species, nor can the hairiness of the glumes which is not constant. On the other hand, the shape of the glumes, the density of the panicle (especially during flowering) and the length of the branches may be used.... Yet in these characters too there are transitions".

30. TRIPLACHNE Link

Annuals. Leaves flat; ligule membranous, lacerate. Inflorescence a spike-like panicle; branches and pedicels short. Spikelets 1-flowered, hermaphrodite, laterally compressed; rachilla prolonged beyond floret. Glumes unequal, not ventricose, straight, lanceolate, acute, much longer than lemma; lemma ⅓ length of glume, membranous, truncate, hairy, with shorter hairs in lower part and longer hairs in upper, obscurely 5-veined, bearing 2 bristle-like cusps

forming prolongation of 2 lateral veins and a dorsal longer awn; awn geniculate at middle and twisted below knee, arising from above base of lemma;
palea nearly as long as lemma. Lodicules 2. Stamens 3. Caryopsis ellipsoid.

A monotypic genus.

1. Triplachne nitens (Guss.) Link, Hort. Berol. 2:241 (1833); Täckholm &
Drar, Fl. Eg. 1:343; Post, Fl. 2:730. *Agrostis nitens* Guss., Ind. Sem. Horto
Boccad. (1825) & Fl. Sic. Prodr. 1:50 (1827). *Gastridium nitens* (Guss.) Coss. &
Durieu, Expl. Sci. Algérie, Bot. II, 68 (1855); Boiss., Fl. 5:519. [Plate 277]

Annual, 10–20 cm. Culms ascending. Upper leaf-sheaths inflated. Panicle
1.5–3 × 0.5–0.8 cm, dense, oblong, tapering at both ends. Glumes finely scabridulous in upper part; lower glume 4 mm, the upper 4.5 mm; lemma ⅓
length of glumes, short-villose, obtuse; terminal cusps white, about as long as
lemma; dorsal awn much longer, nearly as long as glumes, brown up to middle, white in the upper half. Fl. April–May.

Hab.: Sands near coast. Rare. Coastal Galilee, Acco Plain.

Area: S. Mediterranean (Palestine, Cyprus, S. Anatolia, N. Africa and the
Canary Isles, Sicily and S. Italian islands, Spain).

Bornmüller (Verh. Zool.-Bot. Ges. Wien 48:644, 1898) cited this rare species, under
Gastridium nitens, from "Haifa in den Sanden (Exsicc. Nr. 1653), zusammen mit
Aegialophila pumila".

Tribe 6. PHALARIDEAE Kunth (1829). Annuals or perennials. Leaf-blades flat;
sheaths split nearly to base, without auricles; ligule membranous. Inflorescence a dense spike-like panicle. Spikelets all alike, (2–)3-flowered; only upper
floret hermaphrodite, the (1–)2 lower staminate or reduced to scale-like lemmas; rachilla disarticulating above glumes (not between florets). Glumes 2,
persistent, equal, as long as spikelet; fertile lemma 5-veined, awnless, glossy,
indurate at maturity. Lodicules 2, sometimes absent. Stamens 3. Caryopsis
ellipsoid; hilum short. Chromosomes large; x=5, 6, 7.

31. PHALARIS L.

Annuals or perennials. Leaf-blades flat; ligule membranous, ovate or oblong,
dentate or lacerate. Inflorescence a dense spike-like panicle, oblong, ellipsoid
or ovoid. Spikelets strongly compressed laterally, (2–)3-flowered; the lower
(1–)2 florets reduced to minute scales (sterile lemmas), the upper hermaphrodite, falling at maturity from glumes together with sterile lemmas. Glumes
persistent, equal in size, longer than floret and enclosing it, membranous,
3-veined, compressed, boat-shaped, keeled; keel winged or wingless; sterile
lemma linear or lanceolate, appressed to base of fertile floret; lemma of fertile
floret coriaceous, compressed, ovate in outline, glabrous or hairy, obsoletely
5-veined; palea similar in texture and length. Lodicules 2. Stamens 3. Caryop-

sis tightly enclosed by lemma, somewhat compressed with a narrow ventral groove; hilum oblong.

About 15 species in N. and S. temperate regions.

Literature: D. E. Anderson, *Iowa State Jour. Sci.* 36, 1 (1961). A. Baytop, The genus *Phalaris* in Turkey, *Jour. Fac. Pharm. Istanbul* 5:9–26 (1969).

1. Spikelets in groups of 5–7, of two kinds; central spikelet in each group fertile, the others sterile; glumes of fertile spikelet attenuate at tip, each glume with a tooth-like wing on keel. Uppermost leaf-sheath inflated, usually enclosing base of panicle.
                                                                          4. P. paradoxa
 - Spikelets all alike
                                                                                  2
2. Perennial, with lower nodes of culms swollen. Panicle (2.5–)4–8(–10) cm; glumes not broader above than below middle; sterile lemmas 1, or 2 and distinctly unequal.
                                                                          5. P. tuberosa
 - Annual. Glumes broader above than below middle
                                                                                  3
3. Sterile lemma 1, or if 2, sterile lemmas very unequal; glumes irregularly dentate, very rarely entire.
                                                                          3. P. minor
 - Sterile lemmas 2, equal or subequal; glumes entire
                                                                                  4
4. Sterile lemmas minute (less than 1 mm), with tuft of hairs at base.
                                                                          2. P. brachystachys
 - Sterile lemmas 2–3.5 mm, about ½ length of fertile lemma, without a tuft of hairs.
                                                                          1. P. canariensis

1. Phalaris canariensis L., Sp. Pl. 54 (1753); Boiss., Fl. 5:471; Post, Fl. 2:713. [Plate 278]

Annual, 30–60 cm. Culms very leafy. Panicle 1.5–3.5 cm, ovoid or oblong. Spikelets shortly pedicellate, all alike. Glumes 7–8 mm, broader above than below middle, with a green band along wing; wing entire; fertile lemma 5–6 mm, appressed-hairy; sterile lemmas 2, subequal, 2–3.5 mm, lanceolate, nearly ½ length of fertile lemma. Fl. April–July.

Hab.: Fields and roadsides. Rare. Esdraelon Plain, Shefela, ?Judean Mts.; ?Lower Jordan Valley; Transjordan (Al-Eisawi).

Area: Mediterranean.

2. Phalaris brachystachys Link in Schrad., Neues Jour. Bot. 1(3):134 (1806); Bor, Fl. Iraq 9:362, t.138; Boiss., Fl. 5:471; Post, Fl. 2:713. [Plate 279]

Annual, 30–60 cm. Upper leaf-sheath inflated. Panicle 1.5–6 × 1–2 cm, dense, oblong or ovoid; axis scabrous. Spikelets shortly pedicellate, all alike. Glumes 6–8 mm, broader above than below middle, glabrous, winged; wing papery, entire, broad and oblique in upper part, tapering towards base; fertile lemma 5–6 mm, densely appressed-hairy; sterile lemmas 2, minute, equal, somewhat fleshy, brownish, glabrous, with a tuft of hairs at base. Fl. March–June.

Hab.: Fields. Acco Plain, Coast of Carmel, Sharon Plain, Philistean Plain; Upper and Lower Galilee, Mt. Carmel, Esdraelon Plain, Mt. Gilboa, Shefela,

Judean Mts., Judean Desert; Hula Plain, Upper Jordan Valley, Beit Shean Valley, Lower Jordan Valley, Dead Sea area; Ammon.

Area: Mediterranean.

3. Phalaris minor Retz., Obs. Bot. 3:8 (1783); Bor, Fl. Iraq 9:364, t.139; Boiss., Fl. 5:472; Post, Fl. 2:714. [Plate 280]

Annual, 30–60 cm. Upper leaf-sheath slightly or not inflated. Panicle (1.5–)2–5.5 × 1–1.5 cm, dense, oblong or ovoid, pale green; axis scabrous. Spikelets shortly pedicellate, all alike. Glumes 4.5–5.5 mm, winged, broader above than below middle; wing irregularly denticulate, rarely entire; fertile lemma 3–3.5 mm, glossy, appressed-hairy; sterile lemma single, 1.25 mm, sparsely appressed-hairy, subulate, ⅓–½ length of fertile lemma; rarely 2, and then sterile lemmas very unequal. Fl. March–June.

Var. **minor.** [Plate 280]. Glumes denticulate in upper part.

Hab.: Fields. Sharon Plain, Philistean Plain; Upper and Lower Galilee, Mt. Carmel, Esdraelon Plain, Samaria, Shefela, Judean Mts., N. and C. Negev; Upper Jordan Valley, Beit Shean Valley, Lower Jordan Valley, Arava Valley; Golan, Gilead, Ammon, Moav, Edom.

Var. **integra** Trabut in Batt. & Trabut, Fl. Algérie Monocot. 141 (1895). Glumes entire.

Hab.: Rare. N. Negev (Tel-Arad).

Area of species: Mediterranean and Irano-Turanian.

4. Phalaris paradoxa L., Sp. Pl. ed. 2, 1665 (1763); Bor, Fl. Iraq 9:366, t.140; Boiss., Fl. 5:472; Post, Fl. 2:714. [Plate 281]

Annual, 20–50 cm. Uppermost leaf-sheath usually inflated, enclosing base of panicle. Panicle 3–6(–10) cm, dense, cylindrical, narrowed towards base. Spikelets in groups of 5–7, only central spikelet fertile, the surrounding sterile ones pedicellate, shorter; sterile spikelets in lower part of panicle often deformed, hardened, clavate. Glumes of fertile spikelets 5–8 mm, lanceolate, produced into an attenuate long beak; each glume with a tooth-like wing on keel below beak; fertile lemma 2.5 mm, glossy, glabrous or sparsely pilose; sterile lemmas 2, minute, obtuse. Fl. March–August.

Hab.: Weed in crop fields. Common. Acco Plain, Coast of Carmel, Sharon Plain, Philistean Plain; Upper and Lower Galilee, Mt. Carmel, Esdraelon Plain, Mt. Gilboa, Shefela, Samaria, Judean Mts., Judean Desert, W., N. and C. Negev; Upper Jordan Valley, Beit Shean Valley, Dead Sea area; Golan, Gilead, Ammon, Moav.

Area of species: Mediterranean and Irano-Turanian, extending to the Atlantic coast of Europe.

Boissier, Post and Bor distinguish between the typical variety, var. *paradoxa*, and var. *praemorsa*, as follows:

Var. *paradoxa*. Sterile spikelets in all spikelet-groups similar.

Var. *praemorsa* (Lam.) Coss. & Durieu, Expl. Sci. Algérie, Bot. II, 25 (1855). Sterile spikelets in some spikelet-groups deformed, claviform; fertile spikelets in such groups often with shorter glumes.

In our material nearly all specimens have deformed sterile spikelets at least in the lower part of panicle. Boissier notes transition forms between the two varieties.

5. Phalaris tuberosa L., Mantissa Alt. 557 (1771); Bor, Fl. Iraq 9:368, t.141; Bor, Fl. Iran. 70:348; Post, Fl. 2:714. *P. aquatica* L., Cent. Plant. 1:4 (1755) & L., Amoen. Acad. 4:264 (1759) *nom. confus. P. nodosa* L. ex Murr., Syst. Veg. ed. 13:88 (1774) *nom. superfl.* cf. Bor, Fl. Iran *loc. cit.*; Boiss., Fl. 5:473. *P. bulbosa* auct. non L. [Plate 282]

Perennial, tufted, 40–150 cm; lowermost nodes of culms often swollen. Panicle (2.5–)4–8(–10) cm, oblong. Spikelets shortly pedicellate, all alike. Glumes 5–6 mm, narrowly winged, not broader above middle; wings entire; fertile lemma 4 mm, appressed-hairy; sterile lemmas hairy, 1(–2), if 2, unequal. Fl. April–August.

Hab.: Fields. Acco Plain; Philistean Plain: Upper and Lower Galilee, Mt. Carmel, Esdraelon Plain, Samaria, Judean Mts., N. Negev; Hula Plain; Golan, Gilead, Ammon, Moav.

Area: Mediterranean, extending into the Irano-Turanian.

Bor (Fl. Iran. 70:348) points out that *P. aquatica* L. is a *nomen confusum* according to Art. 69 of the International Code of Nomenclature.

Tribe 7. PHLEEAE Dumort. (1823). Annuals or perennials. Leaf-blades flat; sheaths split nearly to base, not auriculate; ligule membranous. Inflorescence a very dense spike-like panicle, usually cylindrical, more rarely head-like or consisting of a number of dense spikes borne along central rachis. Spikelets on very short branchlets, laterally compressed, 1-flowered; floret hermaphrodite (in *Beckmannia* 2-flowered with upper floret staminate); rachilla articulated above glumes, rarely not articulated, and then spikelets breaking off together with part of pedicel. Glumes 2, about as long as spikelet, equal or subequal in shape and length; lemma (3–)5(–9)-veined, awned or awnless; palea (1–)2-veined, 2-keeled, often absent. Lodicules 2 or absent. Stamens 3, rarely 2. Caryopsis ellipsoid and often slightly laterally compressed, usually with a beak at apex; embryo ¼–½ length of caryopsis; hilum small. Chromosomes large; x=7, rarely 5.

Six genera in extratropical countries of both hemispheres, and also in trop. mountains.

32. BECKMANNIA Host

Tall perennials resembling *Paspalidium geminatum* and *Echinochloa eruciformis*. Leaves flat. Inflorescence compound, of several appressed one-sided spikes that are racemosely arranged in 2 rows on one side of central rachis;

rachis of spikes 3-quetrous. Spikelets in 2 rows, all alike, 1-2-flowered, somewhat compressed laterally and inflated above, subcircular in outline, nearly sessile and densely imbricate, disarticulating below glumes and falling entire. Glumes equal, somewhat shorter than spikelet, subcoriaceous, boat-shaped, keeled, rounded above, 3-veined, green towards keel, hyaline at margins; lemma membranous, lanceolate, 5-veined, in lower floret mucronate or apiculate and exceeding glumes, in upper floret muticous; palea nearly as long as lemma. Lodicules 2. Stamens 3. Caryopsis elliptic, grooved on ventral side; embryo much shorter than caryopsis; hilum punctiform. $x=7$.

Two species in Eurasia and America.

1. Beckmannia eruciformis (L.) Host, Gram. Austr. 3:5 (1805); Fl. URSS 2:288, t.22 f.5; Mout., Nouv. Fl. Liban Syrie 1:61; Boiss., Fl. 5:451. *Phalaris erucaeformis* L., Sp. Pl. 55 (1753). [Plate 283]

Perennial, 50-100 cm, glabrous, rhizomatous. Culms erect, tuberous at base. Leaf-sheaths ribbed; ligule hyaline, acute, up to 1 cm. Inflorescence 10-30 cm; spikes 10-30, 1-2.5 cm, subsessile, the lower more or less distant. Spikelets 2.5-3 mm. Lemma puberulent, keeled, apiculate, with apex exserted from glumes. Fl. May-June.

Hab.: Marshes. Rare. Golan.

Area: Mediterranean, Euro-Siberian and Boreo-American.

33. PHLEUM L.

Annuals or perennials. Leaves flat; ligule hyaline. Inflorescence a dense spike-like panicle, cylindrical or ovoid; branches erect and very short; pedicels very short. Spikelets numerous, strongly compressed laterally, 1-flowered, hermaphrodite, deciduous from pedicels at maturity; rachilla disarticulating above glumes, sometimes prolonged above floret as a pedicel-like stalk. Glumes equal, as long as spikelet, free, boat-shaped, keeled, 3-5-veined, abruptly awned or mucronate at apex or gradually acuminate, often coarsely ciliate on keels; lateral veins close to dorsal one; lemma shorter than glumes, membranous, truncate and short-dentate, awnless, (1-)3-5(-9)-veined; palea slightly shorter than lemma. Lodicules 2, rarely absent. Stamens 3. Caryopsis obovoid or ellipsoid, enclosed by indurated lemma and palea; hilum punctiform.

Twelve species in the temperate regions.

Literature: A. Eig, A second contribution to the knowledge of the Flora of Palestine, *Bull. Inst. Agr. Nat. Hist. Tel-Aviv* 6:60-61 (1927).

1. Spikelets 2-3 mm, broadly elliptic. Glumes ending in a very short mucro, connivent at tip; keel strongly curved, very rarely ciliate. **1. P. subulatum**
- Spikelets 3-4 mm, lanceolate to oblong. Glumes not connivent, ending in a 0.5 mm awn; awns straight, parallel; keel slightly curved, ciliate.
2. P. graecum subsp. **aegaeum**

1. Phleum subulatum (Savi) Aschers. & Graebn., Syn. Mitteleur. Fl. 2(1):154 (1899); Bor, Fl. Iran. 70:287; Post, Fl. 2:718. *Phalaris subulata* Savi, Fl. Pis. 1:57 (1798). *Phleum tenue* (Host) Schrad., Fl. Germ. 1:191 (1806); Boiss., Fl. 5:480. [Plate 284]

Annual, 30–40 cm. Panicle (2–)3–10 cm, long-pedunculate, narrowly cylindrical, slender, rounded at both ends. Spikelets 2–3 mm, broadly elliptic, strongly compressed. Glumes semi-elliptic, short-mucronate, connivent at tips; keel strongly curved, not ciliate [very rarely ciliate, in var. *ciliatum* (Boiss.) Halácsy, 1904]; lemma up to ½ length of glumes, 5-veined, denticulate. Fl. April–August.

Hab.: Batha and fallow fields, on various soils and rocks. Coast of Carmel, Sharon Plain, Philistean Plain, W. Negev; Upper and Lower Galilee, Mt. Carmel, Esdraelon Plain, Mt. Gilboa, Samaria, Shefela, Judean Mts.; Upper Jordan Valley, Beit Shean Valley; Golan, Gilead, Ammon.

Area: Mediterranean.

"A widely distributed, but uncommon grass" (Eig, Schedae Fl. Exsicc. Pal., Cent. I:2, no. 5, 1930).

2. Phleum graecum Boiss. & Heldr. in Boiss., Diagn. ser. 1, 13:42 (1854) subsp. **aegaeum** (Vierh.) Greuter in Greuter & Rech. fil., Boissiera 13:180 (1967); Boiss., Fl. 5:481 & Post, Fl. 2:718 as *P. arenarium* auct. non L. (1753). *P. arenarium* L. subsp. *aegaeum* Vierh., Verh. Zool.-Bot. Ges. Wien 69:304 (1919). [Plate 285]

Annual, 10–30(–40) cm. Panicle 1–5 cm, long-pedunculate, very dense, narrowly cylindrical. Spikelets 3–4 mm, lanceolate to oblong, strongly compressed. Glumes gradually narrowed at both ends, abruptly short-awned, not connivent at tips; awns straight, parallel; keel slightly curved, spreadingly ciliate (often from middle upwards); lemma ⅓ length of glumes, puberulent, 3–5–7-veined. Anthers 1.5–2 mm. Fl. March–May.

Hab.: Sandy soil. Acco Plain, Coastal Carmel, Sharon Plain, Philistean Plain.

Area of subsp.: E. Mediterranean (Aegean Islands, ?W. Turkey, Lebanon, Palestine).

34. RHIZOCEPHALUS Boiss.

Annuals, nearly stemless. Culms very short. Leaves basal; blades flat or folded, acuminate-subulate above; sheaths scarious; ligule short, lacerate. Inflorescences head-like panicles, borne on very short peduncles, crowded and nearly concealed between basal leaf-sheaths; branches and pedicels very short. Spikelets 1-flowered, hermaphrodite, falling entire at maturity with part of pedicel. Glumes subequal, very shortly connate at base, urn-shaped in outline, abruptly contracted at tip to a very short mucro, subcoriaceous, 3-veined, glabrous at centre; lateral veins much thickened, hirtellous, with clavate hairs in lower part; lemma considerably longer than glumes, oblong-lanceolate and ending in a cusp, convolute, coriaceous, with 5 elevated veins, hirtellous and covered with clavate hairs in lower half; palea lanceolate, hyaline. Lodicules absent. Stamens 2, with minute anthers. Caryopsis ending in short beak; embryo ¼–⅓ length of caryopsis; hilum punctiform.

A monotypic genus.

1. Rhizocephalus orientalis Boiss., Diagn. ser. 1, 5:68 (1844); Bor, Fl. Iraq 9:322, t.177; Boiss., Fl. 5:478. *Heleochloa orientalis* (Boiss.) Dinsmore in Post, Fl. 2:117 (1933). [Plate 286]

Annual, 3–8 cm, forming a more or less hemispherical tuft, nearly glabrous. Leaf-sheaths elongate, veined. Panicles 1–1.5 × 0.5–1 cm, ovoid. Glumes 3 mm, somewhat recurved; lemma 5.5–6 mm, enclosing palea. Fl. April.

Hab.: Steppes. Artemisietum herbae-albae. C. Negev; Golan, Edom.
Area: Irano-Turanian.

35. ALOPECURUS L.

Annuals or perennials. Leaf-blades flat; ligule hyaline. Inflorescence a dense spike-like panicle; pedicels short, with cupuliform tips. Spikelets strongly compressed laterally, 1-flowered, hermaphrodite, falling entire at maturity. Glumes equal or subequal, as long as spikelet, connate by margins in lower ¼–½, 3-veined, keeled and ciliate on keels; lemma membranous, 3–5-veined, with a dorsal awn arising from its middle or near base; palea usually absent. Lodicules absent. Stamens 3. Caryopsis laterally compressed, loosely enclosed in lemma; embryo ⅓ length of caryopsis; hilum punctiform.

About 20 species in temperate regions.

1. Perennial, with a creeping rhizome. Glumes long-ciliate along keels from base to tip. **1. A. arundinaceus**
- Annuals. Glumes shortly ciliate along keels or nearly glabrous in upper part 2
2. Panicle narrowly cylindrical, 7–20 times as long as broad. **2. A. myosuroides**
- Panicle ovate or lanceolate in outline, 2–3(–5) times as long as broad. **3. A. utriculatus**

1. Alopecurus arundinaceus Poir. in Lam., Encycl. Méth. Bot. 8:776 (1808); Bor, Fl. Iraq 9:287; Boiss., Fl. 5:487. *A. ventricosus* Pers., Syn. Pl. 1:80 (1805) non Huds. (1778); Post, Fl. 2:721. [Plate 287]

Perennial, up to 1 m, glabrous, with a creeping rhizome. Leaf-blades 12–25 cm; upper leaf-sheaths slightly inflated. Panicle 3–6 × 1 cm, cylindrical, obtuse, very dense, silvery hairy; branches 3–4-spiculate. Spikelets 3.5–4.5 mm. Glumes equal, connate in lower ¼, long-ciliate along keels from base to tip, silky along lateral veins; lemma as long as glumes, awned from below middle; awn included or somewhat exserted from glumes. Fl. March–April.

Hab.: Alluvial soil. Rare. Philistean Plain (Wadi Rubin); Upper Galilee; Golan, Gilead, Ammon.
Area: Euro-Siberian, Mediterranean and Irano-Turanian.

According to Flora URSS 2:149 (1934), *A. arundinaceus* is an excellent fodder grass, especially suitable for cultivation on humid saline soils under arid steppe conditions.

2. Alopecurus myosuroides Huds., Fl. Angl. 23 (early 1762); Bor, Fl. Iraq 9:288, t.102. *A. agrestis* L., Sp. Pl. ed. 2, 89 (late 1762); Boiss., Fl. 5:485; Post, Fl. 2:719, incl. var. *minimus* Post, Fl. ed. 1, 857 (1883–1896). [Plate 288]

Annual, 20–50(–70) cm. Panicle narrowly cylindrical, 3–7(–10) × (0.2–) 0.5–0.7 cm (excl. awns), green or purplish. Spikelets 4–6(–8) mm. Glumes equal, connate in lower ⅓ or ½, winged in upper ½ (especially in upper part of panicle), shortly ciliate along keels, hairy in lower part; lemma about as long as glumes to slightly exceeding them, long-awned from base; awn much longer than glumes, erecto-patent, rarely short and included. Fl. March–June.

Hab.: Damp habitats, fields. Fairly common. Coastal Galilee, Acco Plain, Coast of Carmel, Sharon Plain, Philistean Plain; Upper and Lower Galilee, Mt. Carmel, Esdraelon Plain, Samaria, Shefela, Judean Mts.; Dan Valley, Hula Plain, Upper and Lower Jordan Valley, Arava Valley; Golan, Gilead.

Area: Euro-Siberian, Mediterranean and Irano-Turanian.

A. myosuroides is variable in size of panicles and spikelets and especially in width of the wings on its glumes.

3. Alopecurus utriculatus Banks & Sol. in Russ., Nat. Hist. Aleppo ed. 2, 2:243 (1794); Bor, Fl. Iraq 9:292, t.104. *A. anthoxanthoides* Boiss., Diagn. ser. 1, 13:42 (1854); Fl. 5:486; Post, Fl. 2:720. [Plate 289]

Annual, 20–40(–50) cm. Upper leaf-blades somewhat inflated. Panicle 2–4.5 × 0.6–1.2 cm, lax, green or purplish; branches very short. Spikelets 6–6.5 mm. Glumes subequal, acute, connate to just below middle, narrowly winged from lower ⅓ upwards, shortly ciliate along keels, long-hairy in lower ½, otherwise glabrous; lemma long-awned from base; awn 2–3 times as long as lemma, erecto-patent. Fl. February–June.

Hab.: Maquis, garigue, batha and other habitats; calcareous, sandy and other soils. Fairly common. Acco Plain, Coast of Carmel, Sharon Plain; Upper and Lower Galilee, Mt. Carmel, Esdraelon Plain, Mt. Gilboa, Samaria, Shefela, Judean Mts., Judean Desert; Upper Jordan Valley, Beit Shean Valley; Golan, Gilead, Moav, Edom.

Area: E. Mediterranean extending into the Irano-Turanian (Palestine, Lebanon, Cyprus, Syria, Turkey, Iraq, Iran).

In parallel to *A. myosuroides* Huds., *A. utriculatus* varies in size of panicles and spikelets and in width of wings on the glumes. Several varieties have been described in this species under *A. anthoxanthoides*: var. *alatus* Post, Bull. Herb. Boiss. 3:166 (1895), var. *bornmülleri* (Domin) Bornm. & Domin, var. *confusus* Bornm. & Domin and var. *pseudo-alatus* Bornm. & Domin in Bornm., Beih. Bot. Centralbl. 30 Abt. II: 266 (1913). However, these varieties are connected by transition forms and hardly deserve taxonomic status.

36. Cornucopiae L.

Annuals. Leaf-blades flat; leaf-sheaths inflated, partly enclosing peduncles; ligule membranous. Inflorescence a short dense spike-like panicle enclosed in its lower part in a cup-shaped coriaceous involucre and borne on a more or less thickened long peduncle; peduncle breaks off at base at maturity. Spikelets 1-flowered, hermaphrodite, strongly compressed laterally. Glumes equal, as long as spikelet, narrowly elliptic to oblong, keeled, 3-veined, connate by their margins in lower part, ciliate on keels; lemma connate by its margins in lower part, 5-veined; the midvein produced from dorsal surface into a short or long awn; palea absent. Lodicules absent. Stamens 3; filaments long. Caryopsis compressed; hilum punctiform.

Two Mediterranean species, one of them slightly extending into the W. Irano-Turanian.

Literature: A. Eig, A second contribution to the knowledge of the Flora of Palestine, *Bull. Inst. Agr. Nat. Hist. Tel-Aviv* 6:56–58 (1927); On the synonymy of *Cornucopiae alopecuroides* L., *Proc. Linn. Soc. London* 145, 2:69–73 (1932–1933).

1. Peduncle curved. Involucre distinctly dentate. Lemma awnless or with a short awn included in glumes; glumes truncate.   **1. C. cucullatum**
- Peduncle straight. Involucre crenate to entire. Lemma with an awn much exceeding glumes, at least in upper spikelets; glumes acute.   **2. C. alopecuroides**

1. Cornucopiae cucullatum L., Sp. Pl. 54 (1753); Bor, Fl. Iraq 9:300, t.109; Boiss., Fl. 5:474; Post, Fl. 2:715. [Plate 290]

Annual, 10–45 cm. Peduncles of panicles curved, several at end of branches, arising from axil of inflated leaf-sheath. Panicle about 1 cm, ovoid or oblong; involucre cup- or funnel-shaped, irregularly dentate. Glumes truncate or obtuse, connate nearly to middle; lemma about as long as glumes, awnless or with a short awn arising from somewhat below its middle. Fl. February–May.

Hab.: Fields and roadsides. Acco Plain, Coast of Carmel, Sharon Plain; Upper Galilee, Mt. Carmel, Esdraelon Plain, ?Judean Desert (Mar-Saba, Boiss.); Dan Valley, Upper and Lower Jordan Valley; Golan, Gilead.

Area: E. Mediterranean, somewhat extending into the W. Irano-Turanian (Palestine, Lebanon, Syria, Turkey, Aegean Islands, rare in N. Iraq).

"After flowering the retrorsely scabrid peduncle curves into a half circle and thus aids distribution by animals or man" (Bor, *loc. cit.*).

2. Cornucopiae alopecuroides L., Mantissa 29 (1767); Eig, Proc. Linn. Soc. London 145, 2:69–73 (1932–1933); incl. var. *heteraristatum* (Eig) Eig, *op. cit.* 72 (1932–1933); Mout., Nouv. Fl. Liban Syrie 1:58; Post, Fl. 2:715. *Alopecurus involucratus* Post, Jour. Linn. Soc. London (Bot.) 24:440 (1888); Post, Fl. ed. 1, 858 (1896). *C. involucratum* (Post) Mez, Feddes Repert. 17:293 (1921) and var. *heteraristatum* Eig, *op. cit.* 56 (1927). [Plate 291]

Annual, 6–20 cm. Peduncles of panicles straight, clavate, somewhat constricted under involucre. Panicle 0.8–2 × 0.5–1 cm (excl. awn), ovoid to oblong; involucre cup- or funnel-shaped, shallowly crenate to entire, parallel-veined. Glumes scarious, connate nearly to middle, oblong, triangular-ovate at apex, acute, long-ciliate along keel in lower part; lemma somewhat shorter than glumes, subacute, with a basal awn up to twice its length at least in upper spikelets of panicle. Fl. March–June.

Hab.: Marshes and other moist habitats; alluvial soils inundated in winter. Coast of Carmel, Sharon Plain, Philistean Plain; Upper and Lower Galilee; Golan, Gilead.

Area: E. Mediterranean (Palestine, Syria).

Endemic to Palestine and Syria; not known from Italy despite the indication by Linnaeus: "Habitat in Italia".

"This species... is strictly limited to the localities inundated in winter in the maritime plain and in Galilee" (Eig, Feinbrun & Zohary, Schedae Fl. Exsicc. Pal., Cent. III:3 no. 208, 1934).

Eig (*loc. cit.*, 1927) described var. *heteraristatum* of *C. involucratum* (Post) Mez as follows: "Spikelets of the lower ½–⅔ part of the spike with very short awns, only 0.5–3 mm long, included in glumes, sometimes muticous". Though he assumed that the lower spikelets of the panicle were shortly awned or awnless also in the authentic specimen, he described his new var. *heteraristatum*. On p. 57 he says: "one may thus suppose that Post simply was not exact in his description, and there is in general no *C. involucratum* whose awns in all the spikelets correspond to Post's description". Indeed Post's description (Fl. ed. 1, 858 and ed. 2, 2:715) does not mention absence of exserted awns in the lower spikelets, whereas in his corresponding illustration of the panicle the lower spikelets lack exserted awns. Thus one is bound to conclude that var. *heteraristatum* cannot be maintained as a distinct taxon.

Tribe 8. POEAE R. Br. (1814). Annuals or perennials. Leaf-blades flat to setaceous; sheaths split to a various degree, with or without auricles; ligule membranous. Inflorescence a contracted or more or less effuse panicle, rarely a raceme or terminal spike. Spikelets 1–10(–15)-flowered, all hermaphrodite, or in some genera some upper spikelets sterile and reduced to empty glumes and lemmas; rachilla disarticulating above glumes and between florets. Glumes (1–)2, usually persistent, coriaceous to membranous, mostly shorter than spikelets and than lemma of lower floret, rarely absent; lemma similar to glumes in texture, (1–)3–7(–13)-veined, usually keeled, awnless or with a terminal awn; palea 2-veined, 2-keeled. Lodicules 2, usually lanceolate-acute, sometimes lobed, rarely absent. Stamens 3, rarely 2 or 1. Caryopsis free or slightly adherent to lemma and palea, grooved; hilum linear, as long as caryopsis or punctiform. Chromosomes large; x=7, rarely 5, 6 or 11.

About 50 genera in extratropical countries of both hemispheres, or in trop. mountains.

37. Festuca L.

Tufted perennials. Leaf-blades flat, folded or rolled; sheaths split to base or tubular, auriculate in some species; ligule short. Inflorescence a long panicle. Spikelets pedicellate, laterally compressed, several-flowered; rachilla disarticulating above glumes and between florets. Glumes narrow, shorter than lowest floret, mostly keeled, membranous-margined; lower glume 1-veined, smaller than 3-veined upper glume; lemma rounded on' back, 5-veined, awned, mucronate or muticous. Lodicules 2, usually 2-lobed. Stamens 3. Caryopsis fusiform, grooved; hilum linear.

A cosmopolitan genus of about 80 species.

1. Festuca arundinacea Schreb., Spicil. Fl. Lips. 57 (1771); Bor, Fl. Iraq 9:82, t.29; Post, Fl. 2:764. *F. elatior* L. subsp. *arundinacea* (Schreb.) Hackel, Monogr. Festuca 153 (1882); Boiss., Fl. 5:622. [Plate 292]

Perennial, 1 m or more. Leaves rigid, long, provided with falcate ciliate auricles; ligule very short, truncate, sometimes a mere rim. Panicle more or less contracted, many-spiculate; branches of panicle short, lowest branch much shorter than panicle. Spikelets 8–12 mm, 3–5(–8)-flowered, elliptic to oblong. Glumes equal to slightly unequal; lemma muticous, mucronate, or with a 1–2 mm awn. Fl. April–August.

Hab.: Marshes, banks of rivers, moist grassy places. Acco Plain, Sharon Plain, Philistean Plain; Upper and Lower Galilee, Mt. Carmel, Esdraelon Plain, Samaria, Shefela, Judean Mts.; Hula Plain; Golan, Gilead, Ammon, Moav.

Area: Euro-Siberian, Mediterranean and Irano-Turanian.

38. Lolium L.

Annuals or perennials. Culms erect to subprostrate and ascending. Leaf-blades flat or more or less folded, in young shoots either rolled or folded, generally with auricles; ligule short, usually truncate. Inflorescence a terminal spike of solitary spikelets. Spikelets laterally compressed, lying edgewise to concavities of more or less flattened rachis, sometimes sunk in them; florets 2–22 (0–2 apical florets rudimentary), imbricate, hermaphrodite, except the uppermost; rachilla disarticulating above glume and between florets, leaving a short segment attached at base of floret. Glume single (upper glume) in lateral spikelets, 2 in terminal spikelet and subequal, rounded on back, 3–9-veined; lemma rounded on back, hyaline at apex, with or without a subterminal awn; awn, when present, straight, slender, up to 2.5 cm; palea similar to lemma in size and shape, usually minutely scabridulous along keels. Lodicules 2. Stamens 3. Caryopsis tightly adhering to lemma and palea or sometimes partly free; hilum linear.

Eight species in the Mediterranean, Irano-Turanian and Euro-Siberian regions.

Valuable pasture and fodder plants.

Literature: E. E. Terrell, A taxonomic revision of the genus *Lolium, U.S. Dept. Agric., Techn. Bull.* 1392:1-65 (1968).

According to Terrell (*op. cit.*), all *Lolium* species studied for chromosome numbers have 2n=14.

1. Perennial of Mediterranean hill districts, tufted, with non-flowering shoots at anthesis and with brown leaf-sheaths of preceding years at base; leaf-blades flat or folded along midrib in young shoots.   **1. L. perenne**
- Annuals (rarely biennials or short-lived perennials), without non-flowering shoots at anthesis; leaf-blades mostly rolled in young shoots   2
2. Lemma elliptic to ovate, very turgid at maturity. Mature caryopsis 2-3 times as long as broad. Glume usually 1.5-3 cm, as long as or longer than florets. Tall weed in crop fields.   **5. L. temulentum**
- Lemma oblong to oblong-lanceolate, not turgid at maturity. Mature caryopsis more than 3 times as long as broad. Glume mostly shorter   3
3. Spikes 20 cm or longer. Spikelets 11-22-flowered. Glume ¼-½ length of spikelet; lemma awned, especially in middle and upper spikelets. Very rare.
                                                              2. L. multiflorum
- Spikes usually shorter. Spikelets 2-10(-11)-flowered   4
4. Lemma long-awned; awn (8-)10-15 mm, as long as or longer than lemma proper. Rare plants of northern districts (Upper Galilee, Golan).   **6. L. persicum**
- Lemma awnless; rarely upper florets with awns not exceeding 3(-5) mm   5
5. Glume in lower and middle spikelets 20 mm or longer, longer than florets. Spikelets 2-4-flowered. Not common.   **4. L. subulatum**
- Glume not reaching 20 mm, longer to shorter than florets. Spikelets 3-8(-11)-flowered. Very common.   **3. L. rigidum**

1. Lolium perenne L., Sp. Pl. 83 (1753); Terrell, *op. cit.* 7 & 46-49, f.1; Bor, Fl. Iraq 9:93; Boiss., Fl. 5:679; Post, Fl. 2:789. [Plate 293]

Perennial, 10-25(-40) cm, with numerous sterile shoots and with brown leaf-sheaths of preceding years at base. Culms leafy. Leaf-blades 2-4 mm broad, flat, folded in young shoots. Spikes straight or slightly curved, (3-)5-15 cm; rachis smooth. Spikelets 8-15 mm, 5-9(-10)-flowered. Glume 7-10 mm, ½-⅘ length of spikelet, erecto-patent; lemma 4-5 times longer than broad; awns usually absent. Fl. April-June.

Hab.: Calcareous and basalt soils. Rather rare. Usually forming pure stands. Upper Galilee, Mt. Carmel, Samaria, Judean Mts.; Golan, Gilead, Ammon.

Area: Euro-Siberian, Mediterranean and Irano-Turanian.

According to Terrell (*op. cit.* 7), the chromosome number is 2n=14.

"As far as we have observed this species in the East-Mediter. countries, it seems to be here of lesser phytosociological importance than in the Eurosib.-Boreoam. region. In Palestine, a rare plant limited in Cisjordania almost exclusively to the higher parts of the Med. mountain range. In Mediterranean Transjordan with its more continental conditions it seems to be somewhat more common. In Palestine, as in Europe, it often is of a ruderal character. If not disturbed by man, *L. perenne* generally forms nearly pure stands by itself" (Eig, Feinbrun & Zohary, Schedae Fl. Exsicc. Pal., Cent. III:6 no. 214, 1934).

2. Lolium multiflorum Lam., Fl. Fr. 3:621 (1779); Terrell, *op. cit.* 10 & 49–52, f.2; Boiss., Fl. 5:679; incl. var. *pumilum* Boiss., *loc. cit.* (1884) & var. *aristatum* Gaudin, Fl. Helv. 1:354 (1828). *L. italicum* A. Braun, Flora (Regensb.) 17:259 (1834). *L. gaudinii* Parl., Fl. Ital. 1:532 (1850); Post, Fl. 2:789, incl. var. *pumilum* (Boiss.) Dinsmore & var. *aristatum* (Post) Dinsmore in Post, *op. cit.* 790 (1933). [Plate 294]

Annual, biennial or short-lived perennial, 30–100 cm. Leaf-blades 2–8 mm broad, rolled in young shoots. Spikes 15–30 cm, straight or slightly curved; rachis usually scabridulous. Spikelets 10–30 mm (excl. awns), 11–22-flowered. Glume 5–14(–18) mm, ¼–½ length of spikelet; lemma glabrous or glabrescent, typically awned, rarely awnless; awn up to 15 mm, more or less straight, arising below apex. Fl. April–August.

Hab.: Very rare. Acco Plain (marshes), Sharon Plain; Upper Jordan Valley (Lissan marl); Golan.

Area: Mediterranean, extending into the Euro-Siberian and the W. Irano-Turanian.

Probably introduced with seeds of lawn grasses.
According to Terrell (*op. cit.* 12), the chromosome number is 2n=14.

3. Lolium rigidum Gaudin, Agrost. Helv. 1:334 (1811); Terrell, *op. cit.* 15 & 52–54, ff.4, 5, 6, 8; Boiss., Fl. 5:680; Post, Fl. 2:790. [Plates 295, 296, 297]

Annual, 15–40(–70) cm. Leaf-blades folded in bud. Spikes straight or curved, cylindrical or flattened; rachis glabrous to scabrous. Spikelets 8–25 mm, 3–8(–11)-flowered, lying edgewise to concavities of rachis or more or less sunk in rachis. Glume lanceolate-oblong, slightly longer than spikelet to about ½ its length, obtuse or acute, 3–7(–9)-veined; lemma usually awnless; rarely short awns present in upper florets. Fl. March–May.

1. Glumes at most ½ length of spikelet. Plants of sandy soils in the Negev.
                                                                  subsp. **negevense**
– Glume ⅔ length of spikelet to slightly exceeding florets   2
2. Lower rachis-internodes (1.5–)2–3.5 mm broad, often distinctly flattened at right angles to side on which spikelets are inserted. Glume usually not shorter than florets. Culms up to 20 cm, often incurved, decumbent to subprostrate, rarely erect.
                                                                  subsp. **lepturoides**
– Rachis narrower. Glume much varying in length. Culms 20–40(–70) cm, usually erect.   subsp. **rigidum**

Subsp. **rigidum.** Incl. var. *longiglume* Melderis in Rech. fil., Ark. Bot. ser. 2, 2(5):299 (1952). Var. *rigidum*; Terrell, *op. cit.* 20 & 52, f.6. *L. rigidum* Gaudin, *loc. cit.*; Bor, Fl. Iraq 9:94, t.33. *L. strictum* C. Presl, Cyper. Gram. Sic. 49 (1820). [Plate 295]. Culms usually 25–40(–70) cm, mostly erect. Spikes straight, 9–30 cm; rachis cylindrical, slender to somewhat thickened, 1–1.5(–2) mm broad at lower internodes. Spikelets usually about 10 mm, 5–6-flowered. Glume varying in length, generally about as long as spikelet to slightly longer, or rarely only ⅔–⅚ its length, coriaceous, tip membranous; florets mostly almost concealed by glume; sometimes glumes spreading and shorter than florets, the spike resembling that of *L. perenne*. Lemma 4.5–8.5 mm.

Hab.: Batha, fields, roadsides; various soils. Common. Coastal Galilee, Acco Plain, Coast of Carmel, Sharon Plain, Philistean Plain; Upper and Lower Galilee, Mt. Carmel, Esdraelon Plain, Mt. Gilboa, Samaria, Shefela, Judean Mts., Judean Desert, W., N. and C. Negev; Dan Valley, Hula Plain, Upper and Lower Jordan Valley, Dead Sea area; Golan, Gilead, Ammon, Moav.

Area of subsp.: Mediterranean and Irano-Turanian. Introduced in many parts of the world.

According to Terrell (*op. cit.* 20), 2n=14.

Subsp. **lepturoides** (Boiss.) Sennen & Mauricio, Cat. Fl. Rif Or. 135 (1933). *Lolium lepturoides* Boiss., Diagn. ser. 1, 13:67 (1854). *L. rigidum* Gaudin var. *rottbollioides* Heldr. ex Boiss., Fl. 5:680 (1884); Terrell, *op. cit.* 23 & 54, f.8. *L. loliaceum* (Bory & Chaub.) Hand.-Mazz., Ann. Naturh. Mus. (Wien) 28:32 (1914). *L. rigidum* var. *loliaceum* (Bory & Chaub.) Halácsy, Consp. Fl. Graec. 3:446 (1904); Post, Fl. 2:790. [Plate 296]. Culms usually less than 30 cm, mostly decumbent to subprostrate. Spikes usually incurved, flattened, 3–7(–20) cm, generally shorter than in subsp. *rigidum*; rachis rigid, often distinctly flattened, 1.5–3.5 mm broad at lower internodes. Spikelets 7–9 mm, generally 4–5-flowered; florets often concealed by glume. Glumes more indurate than in subsp. *rigidum*, coriaceous to tip, acute, (6–)9–10 mm, slightly longer to somewhat shorter than florets; lemma 3.2–5.5(–7) mm.

Hab.: Sandy soil near coast and alluvial soils. Rare. Coastal Galilee, Acco Plain, Coast of Carmel; Upper Galilee, Mt. Carmel, Judean Mts.; Hula Plain.

Area of subsp.: Mainly E. Mediterranean (Palestine, Lebanon, Syria, S. and N. Turkey, Crimea, Crete, Greece, Jugoslavia, Italy, Cyrenaica, Canary Islands and Madeira).

According to Terrell (*op. cit.* 23), 2n=14.

Subsp. **negevense** Feinbr. subsp. nov.* [Plate 297]. Culms erect, 30–50 cm. Spikes straight, (5–)10–20 cm; internodes of rachis ⅔ length of spikelet or longer; rachis as in subsp. *rigidum*. Spikelets 14–25 mm, 6–8(–10)-flowered, pale green; florets conspicuously spaced on rachilla; internodes of rachilla 1.5–2 mm. Glume 5–9(–11) mm, at most ½ length of spikelet, often shorter; tip and margins of glume membranous; lemma 5–7.5 mm, hyaline in its upper part.

Hab.: Sandy and sandy-loess soils. Forming extensive steppe-like stands. W. and N. Negev, Philistean Plain. Rare in Sharon Plain in the *Ceratonia-Pistacia lentiscus* association.

Area of subsp.: Endemic, extending south to El Arish (Sinai).

Area of species: Mediterranean and Irano-Turanian.

L. rigidum Gaudin is a very polymorphic species. Terrell (*op. cit.*) shows the extent of variation in spike traits in figure 5. Spikes range from very slender, cylindrical,

* See Appendix, p. 397.

with spikelets sunk in the rachis and completely covered by glume, to laterally compressed and broad with diverging spikelets and florets that overtop glume considerably.

The entire polymorphic range shown by Terrell is well represented in Palestine. Subsp. *negevense* Feinbr. is a variant shown on the extreme left of Terrell's figure 5. Its spikelets are at least twice as long as the glumes and generally longer than in subsp. *rigidum*; florets are spaced on the elongated rachilla; the rachis, too, is elongated, its internodes sometimes exceeding the spikelets in length. Terrell (*op cit.* 9) refers to this variant as possibly meriting taxonomic recognition. Indeed, subsp. *negevense* is distinct also in its ecology. It grows in nearly pure steppe-like populations on sandy and sandy-loess soils in the Negev, extending further south. Thus it occupies the southernmost part in the distribution area of *L. rigidum* in the E. Mediterranean. Occasionally we found specimens of subsp. *rigidum* at the side of a road that had been cut through a vast stand of subsp. *negevense*. No compatibility data are as yet available for *L. rigidum* in general (Terrell, *op. cit.* 4), but the occurrence of the two subspecies in close proximity to one another suggests a compatibility barrier between them.

In its laterally compressed spikes subsp. *negevense* resembles *L. perenne*, as well as a Portuguese annual of maritime habitats described as *L. parabolicae* Sennen ex Samp. (Bol. Soc. Brot. ser. 2, 1:125, 1922. Cf. also Kerguélen, Lejeunia, n.s. 65, 1972 & Bull. Soc. Bot. Fr. 125, 7-8:395, 1978).

Various aberrant forms are recorded in the literature for *L. perenne, L. multiflorum, L. rigidum* and *L. temulentum*. In our material the following were found in *L. rigidum* subsp. *rigidum*:

(1) Plants with short awns on lemmas in upper florets of spikelets. The validly published name cited by Terrell for such plants is *L. rigidum* var. *aristatum* Hackel ex Stuck., Anal. Mus. Nac. Buenos Aires ser. 3, 6:530 (1906).

(2) Several plants in our Herbarium are characterized by broad and rather short flattened spikes with closely approximated spikelets. These fit the description of *L. rigidum* var. *compressum* (Boiss. & Heldr.) Boiss. as cited by Boissier in Fl. 5:680 (1884). It seems clear that the spikes in these specimens have developed by an abnormal shortening of the rachis. On one sheet the same specimen bears culms with normal spikes as well as abnormal ones.

(3) The most common of the aberrant *L. rigidum* plants in our Herbarium display different degrees of branching at various points of the spike. In one plant the inflorescence is a broad raceme with spreading short secondary spikes. These branched variants can be designated by the validly published name *L. rigidum* forma *ramosum* Bouly de Lesd., Publ. Soc. Dunk. 1934:91 (1934). They are clearly cases of teratological development.

4. Lolium subulatum Vis., Fl. Dalm. 1:90, t.3 (1842) emend. Terrell, *op. cit.* 26 & 54, f.10 (1968). *L. loliaceum* auct. p.p. non (Bory & Chaub.) Hand.-Mazz. (1914); Bor, Fl. Iraq 9:92, t.31. *L. rigidum* var. *subulatum* (Vis.) Charrel, Österr. Bot. Zeitschr. 42:411 (1892). [Plate 298]

Annual, 20–40 cm. Culms erect, stout. Spikes 10–20 cm, straight to somewhat curved, cylindrical. Spikelets 2–4-flowered, sunk in concavities of rachis and entirely covered by glumes. Glumes acute, appressed to rachis; glume in lower and middle spikelets exceeding 2 cm, usually longer than florets and

2-2.8 mm broad; lemma of lower and middle spikelets about 1 cm; rachilla segments long; awns usually absent; in upper florets awns sometimes up to 3 mm, rarely up to 5 mm [forma *aristatum* Lindb. fil., Öfvers. Finska Vet.-Soc. Forhandl. 48(13):12, 1905–1906]. Fl. April–May.

Hab.: Fields and batha, mainly calcareous and basalt soils. Fairly rare. Upper and Lower Galilee, Mt. Carmel, Samaria, Judean Mts.; Upper Jordan Valley; Gilead, Ammon.

Area: E. Mediterranean (Palestine, Lebanon, Syria, Cyprus, S. Turkey, Greece, Jugoslavia).

L. subulatum Vis. deserves the status of a species, though Boissier and Post cited it as a synonym of *L. rigidum* var. *rottbollioides* (Boiss., Fl. 5:680) or of *L. rigidum* var. *loliaceum* (Post, Fl. 2:790).

5. Lolium temulentum L., Sp. Pl. 83 (1753); Terrell, *op. cit.* 35 & 55, f.15; Bor, Fl. Iraq 9:96; Boiss., Fl. 5:681; Post, Fl. 2:790. [Plate 299]

Annual, 40–80(–100) cm. Leaf-blades rolled in bud. Spikes straight. Spikelets 8–28 mm (excl. awns), 5–10-flowered, lying edgewise to concavities of rachis. Glume of lower and middle spikelets 7–30 × 1.8–3 mm, ³⁄₄–1½ times as long as spikelet (mostly longer than florets); lemma of lower and middle florets 5–8.5 × 1.5–3 mm, elliptic to ovate, very turgid at maturity, usually 2–3½ times as long as broad; awns present or absent; when present usually less than 17 mm, arising 0.5–2 mm below apex. Mature caryopsis oblong, often partly exposed, 4–7 mm, about 2–3 times as long as broad. Fl. March–June.

Forma **temulentum**. Forma *macrochaeton* (A. Braun) Junge, Jahrb. Hamb. Wiss. Anst., Beih. 3, 30:314 (1913). Var. *macrochaeton* A. Braun, Flora (Regensb.) 17:252 (1834). [Plate 299]. Lemma long-awned; awn slender, rigid.

Hab.: Weeds, mainly in fields of cereal crops. Sharon Plain, Philistean Plain; Upper Galilee, Mt. Carmel, Esdraelon Plain, Shefela, Samaria, Judean Mts.; Hula Plain, Upper Jordan Valley, Beit Shean Valley.

Forma **arvense** (With.) Junge, Jahrb. Hamb. Wiss. Anst., Beih. 3, 30:314 (1913). *L. arvense* With., Arr. Brit. Pl. ed. 3, 2:168 (1796). *L. temulentum* var. *muticum* Boiss., Fl. 5:681 (1884). Forma *muticum* (Boiss.) Rech. fil., Fl. Aeg. 787 (1943). Lemma awnless or with awn not exceeding 3 mm.

Hab.: As above. Coastal Galilee, Sharon Plain, Philistean Plain; Upper Galilee, Shefela, Judean Mts.; Hula Plain, Upper and Lower Jordan Valley, Dead Sea area; Golan, Gilead, Ammon.

Area of species: Euro-Siberian, Mediterranean and Irano-Turanian.

We agree with Terrell that the rank of *forma* suits the awned and the awnless variants of the species.

According to Terrell (*op. cit.* 4), *L. temulentum* is self-pollinated and self-compatible. Its chromosome number is 2n=14.

The poisonous properties of *L. temulentum* are known since ancient times. The caryopses contain the alkaloid *temulin*. The flour of cereals that had not been cleaned properly and contained grains of *L. temulentum* is reputed to have caused sometimes fatal cases of food poisoning, especially in wet years. The poisonous properties of the grain of *L. temulentum* are known to be connected with the presence of fungal mycelium between the seed coat and aleurone layer.

6. Lolium persicum Boiss. & Hohen. in Boiss., Diagn. ser. 1, 13:66 (1854); Bor, Fl. Iraq 9:94, t.32; Terrell, *op. cit.* 41, f.17; Boiss., Fl. 5:680. [Plate 300]

Annual, 30–80 cm. Spikes straight, up to 25 cm. Spikelets 12–20 mm (excl. awns), 4–9-flowered, lying edgewise to concavities of rachis. Glume usually equalling spikelet, sometimes ½–¾ its length; lemma awned, lanceolate, acute or tapering above, apex 2-fid; awn (8–)10–15 mm, arising below apex of lemma. Caryopsis about 3½–5 times as long as broad. Fl. April–May.

Hab.: Fields, basalt rocks and soil. Very rare. Upper Galilee, Golan.

Area: Irano-Turanian.

39. VULPIA C.C. Gmel.

Annuals, rarely perennials. Leaf-blades narrow, flat or rolled; sheaths split nearly to base; ligule membranous. Inflorescence a slightly open or contracted panicle, or rarely a spike-like raceme with short pedicels; a pedicellate spikelet and a branch bearing pedicellate spikelets are usually present at lower nodes of panicle. Spikelets laterally compressed, broadening towards apex, 3–12-flowered, disarticulating above glumes and between florets, sometimes falling entire together with pedicel; florets spaced on rachilla, cleistogamous or with anthers barely exserted, or chasmogamous; apical florets mostly sterile or staminate. Glumes markedly unequal, acute to acuminate, awned in some species; rarely glumes subequal; lower glume often minute, 1-veined or veinless, rudimentary in some spikelets; the upper 1–3-veined; lemma chartaceous, hyaline-margined, lanceolate, rounded on back, (3–)5-veined, tapering into a long straight scabrous awn; palea about as long as lemma. Lodicules 2. Stamen 1, infrequently 3; filaments short. Caryopsis narrowly ellipsoid, adherent to palea; hilum linear, somewhat shorter than caryopsis.

About 25 species, mainly Mediterranean; widely naturalized in temperate regions and along the Pacific coast of N. and S. America.

The length of the lower glume compared to that of the upper is an important diagnostic character.

Literature: J. T. Henrard, A study of the genus *Vulpia, Blumea* 2:299–326 (1937). R. Cotton & C.A. Stace, Taxonomy of the genus *Vulpia* (Gramineae) I. Chromosome numbers and geographical distribution of the Old World species, *Genetica* 46:235–255 (1976). C.A. Stace & R. Cotton, Nomenclature and distribution of *Vulpia membranacea* (L.) Dumort. and *V. fasciculata* (Forsskal) Samp., *Watsonia* 11:117–123 (1976). R. Cotton & C.A. Stace, Morphological and anatomical variation of *Vulpia* (Gramineae), *Bot. Not.* 130:173–187 (1977). P. Auquier, Le genre *Vulpia* C.C. Gmel. (Poaceae) en Belgique, *Bull. Jard. Bot. Nat. Belg.* 47:117–137 (1977); *Vulpia*, in: *Nouvelle Flore de la Belgique*, etc., ed. 2, 747–748, 1978. C.A. Stace, Changing concepts of the genus *Nardurus* Reichenb., *Jour. Linn. Soc. London (Bot.)* 76:344–350 (1978). C.A. Stace & P. Auquier, Taxonomy and variation of *Vulpia ciliata* Dumort., *Jour. Linn. Soc. London (Bot.)* 77:107–112 (1978). C.A. Stace & R. Cotton, *Vulpia*, in: T.G. Tutin et al. (eds.), Flora europaea 5:154–156 (1980). C.A. Stace, Generic and infrageneric nomenclature of annual Poaceae: Poeae related to *Vulpia* and *Desmazeria, Nord. Jour. Bot.* 1:17–26 (1981).

Measurements of glumes should be taken in several spikelets from the middle of inflorescence; those of the lemma (excl. awn) from the basal floret of the spikelet.

1. Glumes subequal in length, long-awned; lower glume subulate. Dwarf desert annual.   **5. V. brevis**
- Glumes very unequal; sometimes lower glume minute or obsolete   2
2. Lemma densely long-ciliate at margin and mostly long-hairy on back.   **4. V. ciliata**
- Lemma not ciliate, glabrous or scabrous, rarely short-hairy on back   3
3. Upper glume awned; awn 8–12 mm; lower glume minute.   **1. V. fasciculata**
- Upper glume awnless or, if awned, awn at most 2 mm   4
4. Inflorescence a narrow raceme with pedicels 1–1.5 mm and spikelets 4–8 mm (excl. awns), appressed sidewise to axis. Anthers exserted at anthesis.   **6. V. unilateralis**
- Inflorescence a contracted panicle; spikelets larger. Anthers usually included at anthesis   5
5. Upper glume nearly as long as adjacent lemma (excl. awn). Base of panicle generally well exserted from uppermost leaf-sheath.   **3. V. muralis**
- Upper glume about ½ length of adjacent lemma. Panicle partly enclosed by uppermost leaf-sheath.   **2. V. myuros**

Sect. 1. MONACHNE Dumort., Obs. Gram. Belg. (1824).

1. Vulpia fasciculata (Forssk.) Samp., Lista Esp. Herb. Port. 24 (1913); Stace & Cotton, Watsonia 11:119 (1976). *Festuca fasciculata* Forssk., Fl. Aeg.-Arab. 22 (1775). *V. uniglumis* (Ait.) Dumort., Obs. Gram. Belg. 101 (1824); Boiss., Fl. 5:629. *F. uniglumis* Sol. in Ait., Hort. Kew. 108 (1789). *V. membranacea* auct. non (L.) Dumort.; Post, Fl. 2:765 as *V. membranacea* (L.) Link. [Plate 301]

Annual, 10–30 cm. Inflorescence a rigid raceme or a contracted sparingly branched panicle, 5–10(–15) cm (incl. awns), usually well exserted from uppermost leaf-sheath; pedicels 3–7 mm, dilated distally. Spikelets 9–15 mm (excl. awns; 30–40 mm incl. awns), cuneate, 4–6-flowered; florets diminishing in size towards apex, the apical 1–2 florets small, sterile. Glumes very unequal; lower glume minute, triangular; the upper about as long as adjacent lemma, long-acuminate and awned; awn 8–12 mm; lemma ending in a scabridulous awn about as long to twice as long as lemma proper (lemma of lower floret about 1 cm; awn about 2 cm); callus elongate, pointed. Stamens 3; anthers 1–2 mm, included at anthesis. Ovary and caryopsis pubescent at apex. Fl. March–May.

Hab.: Mostly sandy soils of the coastal plain. Coastal Galilee, Acco Plain, Sharon Plain, Philistean Plain.

Area: Mediterranean, extending into the Euro-Siberian.

According to Stace & Cotton (Watsonia 11:117-123, 1976), the omni-Mediterranean *V. fasciculata* is tetraploid (2n=28), while the W. Mediterranean *V. membranacea* (L.) Dumort. is diploid (2n=14).

As pointed out by Auquier (1977), the binomial *V. membranacea* (L.) Dumort. is a *nomen dubium*. The correct name for the species is *V. pyramidata* (Link) Rothm. (Feddes Repert. 52:270, 1943).

Sect. 2. VULPIA. Sect. *Euvulpia* Boiss., Fl. 5:628 (1884).

2. Vulpia myuros (L.) C.C. Gmel., Fl. Bad. 1:8 (1806); Bor, Fl. Iran. 70:90; Stace & Cotton, in Fl. Eur. 5:156; Boiss., Fl. 5:628; Post, Fl. 2:764. *Festuca myuros* L., Sp. Pl. 74 (1753). [Plate 302]

Annual, 15-45 cm. Inflorescence a sparingly branched panicle, 5-15 cm, partly enclosed by uppermost leaf-sheath, or a raceme; branches rather short, appressed to axis; pedicels 1-2 mm. Spikelets 6-10 mm (excl. awns), cuneate, 4-6-flowered, the majority of florets fertile. Lower glume $\frac{1}{10}$-$\frac{2}{5}$ length of the upper, in some spikelets minute or absent; upper glume 4-5 mm, acuminate, sometimes ending in an awn 1-2 mm long; adjacent lemma 6-7 mm (excl. awn); awn 2-3 times as long as lemma proper; callus very short. Stamens 1(-3); anthers usually included at anthesis. Fl. February-May.

Hab.: Batha, walls, roofs and other arid habitats. Acco Plain, Sharon Plain, Philistean Plain, W. Negev; Upper and Lower Galilee, Mt. Carmel, Esdraelon Plain, Mt. Gilboa, Samaria, Judean Mts.; Hula Plain, Upper and Lower Jordan Valley; Golan, Gilead, Ammon.

Area: Euro-Siberian and Irano-Turanian regions, N. and S. America.

Stace & Cotton (*loc. cit.*) report *V. myuros* to be hexaploid with 2n=42.

3. Vulpia muralis (Kunth) Nees, Linnaea 19:694 (1847); Stace & Cotton, in Fl. Eur. 5:156. *Festuca muralis* Kunth, Syn. Pl. 1:218 (1822). *V. broteri* Boiss. & Reuter, Pugillus 128 (1852). *V. dertonensis* var. *broteri* (Boiss. & Reuter) Hegi, Ill. Fl. Mitteleur. 1:327 (1908). *V. dertonensis* var. *longaristata* (Willk.) Aznavour, Magyar Bot. Lapok 10:17 (1911). [Plate 303]

Annual, 10-50 cm. Inflorescence a sparingly branched panicle or rarely a raceme, 3-15 cm, usually well exserted from uppermost leaf-sheath; pedicels 0.5-3.5 mm. Spikelets 8-10 mm (excl. awns), 6-7-flowered; the majority of florets fertile. Lower glume $\frac{1}{4}$-$\frac{1}{2}$ length of the upper; upper glume 4-6 mm, acuminate, sometimes with a 1 mm long awn; adjacent lemma 4-8 mm; awn usually 2-3 times as long as lemma proper; callus very short. Stamens 1(-3); anthers usually included at anthesis. Fl. February-May.

Hab.: Batha, rocks, walls, roofs and other arid habitats. Sharon Plain; Upper Galilee, Esdraelon Plain, Judean Mts.; Upper Jordan Valley; Golan, Gilead, Ammon.

Area: Mediterranean.

According to Stace & Cotton (*loc. cit.*), *V. muralis* is diploid with 2n=14.

Records on *V. bromoides* and *V. dertonensis* from Palestine should be referred to *V. muralis*.

4. Vulpia ciliata Dumort., Obs. Gram. Belg. 100 (1824); Stace & Auquier, Jour. Linn. Soc. London (Bot.) 77:107–112 (1978); Stace & Cotton, in Fl. Eur. 5:156; Bor, Fl. Iran. 70:88; Boiss., Fl. 5:628; Post, Fl. 2:765. Incl. var. *plumosa* Boiss., *op. cit.* 629 (1884). *Festuca ciliata* Danth. ex DC. in Lam. & DC., Fl. Fr. ed. 3, 3:55 (1805) non Gouan (1762) nec Link (1799). *V. aetnensis* Tineo, Pl. Rar. Sic. fasc. 2:22 (1846). [Plate 304]

Annual, 10–30(–40) cm. Inflorescence a sparingly branched contracted panicle or raceme, 3–20 cm, partly enclosed in uppermost leaf-sheath; branches erect, appressed to axis; pedicels 0.2–2 mm. Spikelets 6–8 mm (excl. awns), 2–7-flowered, as a rule only 1–2 lowermost florets fertile; upper florets reduced to empty lemmas. Glumes membranous, markedly unequal, glabrous; lower glume minute; upper glume ¼–⅓ length of adjoining lemma; lemma of fertile florets 5–6.5 mm, long-ciliate at margins and villose on back; empty lemmas long-ciliate, villose at back or glabrous; awn as long to twice as long as lemma proper. Stamens 1(–3); anthers usually included at anthesis. Fl. March–April.

Hab.: Batha, stone walls and similar habitats, sandy and basalt soils. Coastal Galilee, Sharon Plain, Philistean Plain; Upper and Lower Galilee, Esdraelon Plain, Mt. Gilboa, E. Samaria, Shefela, Judean Mts., Judean Desert, N. Negev; Upper and Lower Jordan Valley, Dead Sea area; Golan, Gilead, Ammon.

Area: Mediterranean and Irano-Turanian, extending into the Euro-Siberian.

Reported to be tetraploid (2n=28) by Cotton & Stace (Genetica 46:242, 1976).

Sect. 3. SPIRACHNE (Hackel) Boiss., Fl. 5:630 (1884).

5. Vulpia brevis Boiss. & Ky. in Boiss., Diagn. ser. 2, 4:139 (1859); Post, Fl. 2:765. *V. inops* Hackel, Flora (Regensb.) 63:476 (1880); Boiss., Fl. 5:630. *Festuca inops* Del., Fl. Aeg. Ill. 52 no. 110 (1813) *nom. nud.*; Del. in Steud., Nomencl. Bot. 1:337 (1821) *nom. nud*; Del., Fl. Eg. Suppl. t.63 f.1 in C. & W. Barbey (1882) non De Not. (1848). [Plate 305]

Densely tufted dwarf annual, 3–8(–15) cm. Leaf-sheaths broad, striate; blades very narrow, folded. Inflorescence a short panicle of a few spikelets; branches short, upright; pedicels compressed, dilated distally, falling off with spikelet at maturity. Spikelets 5–6 mm (excl. awns), single or in threes, compressed, cuneate, glabrous or hairy, 4–9-flowered with only lowermost floret fertile. Glumes subequal in length, ending in a capillary awn about as long as glume proper; lower glume subulate, the upper linear-lanceolate; lemma of fertile floret similar to upper glume, long-awned; sterile florets reduced to small awned lemmas overtopped by awns of glumes and of fertile lemma. Stamens 3; anthers short. Fl. March–April.

Hab.: Steppes. Very rare. W., N. and C. Negev.

Area: Saharo-Arabian and W. Irano-Turanian (Syria, Palestine, Sinai, Egypt, Libya).

Sect. 4. APALOCHLOA (Dumort.) Stace, Nord. Jour. Bot. 1:24 (1981). Sect. *Nardurus* (Reichenb.) Stace, Jour. Linn. Soc. London (Bot.) 76:350 (1978).

6. Vulpia unilateralis (L.) Stace, Jour. Linn. Soc. London (Bot.) 76:350 (1978). *Triticum unilaterale* L., Mantissa 35 (1767). *Nardurus maritimus* (L.) Murb., Contrib. Fl. Nord-Ouest Afr. ser. 1, 4:25 (1900); Bor, Fl. Iraq 9:99; Post, Fl. 2:766 non *V. maritima* S.F. Gray. *N. tenuiflorus* (Schrad.) Boiss., Fl. 5:632 (1884) non Boiss., Voy. Bot. Midi Esp. 2:667 (1844). [Plate 306]

Annual, 5–20 cm. Inflorescence a spike-like one-sided narrow raceme, 3–6(–8) cm, rarely with a few branches near base. Spikelets 4–8 mm (excl. awns), appressed with their narrow side to axis, 4–5-flowered, most florets fertile; pedicels very short (1–1.5 mm), thick. Glumes markedly unequal, lanceolate-linear, acuminate, keeled; the lower 1-veined, about ⅓ length of 3-veined upper glume; the latter shorter than adjacent floret; lemma of lower floret 3–5 mm, lanceolate-linear, very acute, awned, glabrous or hairy; awn ½ length of lemma to 3 times its length. Stamens 3; anthers 0.7–2 mm, exserted at anthesis. Fl. March–April.

Hab.: Steppes, ploughed and fallow fields. Fairly rare. Samaria, Judean Mts., Judean Desert; Lower Jordan Valley; Gilead, Ammon, Moav, Edom.

Area: Mediterranean and Irano-Turanian, extending into the Euro-Siberian.

40. LOLIOLUM Krecz. & Bobrov

Dwarf annuals. Inflorescence a spike. Spikelets laterally compressed, arranged in 2 rows with their narrow side in concavities on one face of flattened rachis, on thick, 0.2–0.3 mm pedicels. Glumes subequal, with 1 thick vein and membranous margins towards base; lemma coriaceous, ovate, obtuse, rounded on back, obscurely 5-veined, ending in a short scabrous awn, rarely awnless; palea 2-dentate. Lodicules 2, 2-dentate. Stamens 3. Caryopsis slightly adherent to palea; hilum short, linear.

A monotypic genus.

Literature: C.A. Stace, Changing concepts of the genus *Nardurus* Reichenb. (Gramineae), *Jour. Linn. Soc. London (Bot.)* 76:344–350 (1978).

1. Loliolum subulatum (Banks & Sol.) Eig, Jour. Bot. (London) 75:189 (1937); Stace, Jour. Linn. Soc. London (Bot.) 76:348 (1978). *Triticum subulatum* Banks & Sol. in Russ., Nat. Hist. Aleppo, ed. 2, 2:244 (1794). *Nardurus subulatus* (Banks & Sol.) Bor, Kong. Danske Vid. Selsk. Biol. Skr. 14, 4:67 (1965); Bor, Fl. Iraq 9:100, t.34. *L. orientale* (Boiss.) Krecz. & Bobrov, in Fl. URSS 2:544 (1934). *N. orientalis* Boiss., Diagn. ser. 1, 7:127 (1846); Boiss., Fl. 5:633; Post, Fl. 2:767. [Plate 307]

Annual, 3-10 cm. Leaf-blades short, folded; uppermost leaf-sheath enveloping base of spike; ligule short, dentate. Spike 3–6 cm, flattened, slightly
curved; rachis stiff. Spikelets 2–3-flowered. Glumes linear-lanceolate, about as
long as spikelet; upper glume slightly longer than the lower; lemma pubescent or glabrous, with a short awn slightly below tip; palea ciliate along
keels. Fl. March–May.

Hab.: Steppes, mainly Artemisietum herbae-albae; stony ground. Fairly
rare. Judean Desert, W., N. and C. Negev; Lower Jordan Valley, Dead Sea
area; E. Gilead, Moav, Edom.

Area: Irano-Turanian.

Described from Syria.

41. CATAPODIUM Link

Annuals. Leaves glabrous; ligule truncate, irregularly lacerate at tip. Inflorescence a panicle, raceme or spike, rather stiff, narrow, linear to ovate; branches
and pedicels 3-angled. Spikelets 3- to many-flowered, laterally compressed,
appressed to one side of rachis or of branches. Glumes slightly unequal,
keeled, shorter than lowest floret; lower glume 1–3-veined; upper 3-veined;
lemma firm, rounded on back or keeled above, obtuse, muticous, obscurely
5-veined; palea as long as lemma. Stamens 3. Caryopsis ellipsoid-oblong,
strongly adherent to palea; hilum punctiform. x=7.

Two species, Mediterranean, extending into the W. Irano-Turanian and
into W. Europe.

1. Inflorescence a panicle, branched especially in its lower part; lemma rounded on
   back; spikelets about 1.5 mm broad.   **1. C. rigidum**
- Inflorescence a spike-like raceme or spike; pedicels very short (1 mm); lemma
   keeled in upper ½; spikelets 2–2.5 mm broad.   **2. C. marinum**

1. Catapodium rigidum (L.) C.E. Hubbard in Dony, Fl. Bedfordshire 427
(1953). *Poa rigida* L., Cent. Pl. 1:5 (1755). *Desmazeria rigida* (L.) Tutin in
Clapham, Tutin & E.F. Warburg, Fl. Brit. Isles, 1434 (1952). *Scleropoa rigida*
(L.) Griseb., Spicil. Fl. Rumel. 2:431 (1846); Boiss., Fl. 5:638; Post, Fl. 2:769.
[Plate 308]

Annual, 10–25 cm. Culms erect or ascending. Inflorescence a rigid narrow
panicle, with ascending branches and pedicels, very rarely unbranched;
branches 1–4-spiculate; pedicels 0.5–2.5 mm. Spikelets about 1.5 mm broad,
narrowly oblong, green or purplish, 5–10-flowered. Lemma rounded on back.
Fl. March–May.

Hab.: Roadsides and near habitations. Common. Acco Plain, Coast of Car-
mel, Sharon Plain, Philistean Plain; Mt. Carmel, Upper and Lower Galilee,
Esdraelon Plain, Mt. Gilboa, Samaria, Shefela, Judean Mts., Judean Desert,
N. Negev; Dan Valley, Upper and Lower Jordan Valley, Dead Sea area;
Golan, Gilead, Ammon, Moav.

Area: Mediterranean, extending into the W. Irano-Turanian and W. Europe.

2. Catapodium marinum (L.) C.E. Hubbard, Kew Bull. 1954:375 (1954). *Festuca marina* L., Amoen. Acad. 4:96 (1759). *C. loliaceum* (Huds.) Link, Hort. Berol. 1:45 (1827); Boiss., Fl. 5:634; Post, Fl. 2:767. *Desmazeria marina* (L.) Druce, Scott. Bot. Rev. 1:156 (1912). *Poa loliacea* Huds., Fl. Angl. 35 (1762). *D. loliacea* (Huds.) Nyman, Syll. 426 (1854–1855). [Plate 309]

Annual, 10–15 cm. Culms erect or ascending. Inflorescence a spike-like raceme or spike or sometimes a panicle in which branching has been reduced; lower 1–3 spikelets often borne on about 1 cm erect branches, while other spikelets almost sessile. Spikelets 2–2.5 mm broad, oblong, 7–11-flowered. Lemma keeled in upper half. Fl. April–May.

Hab.: Sandy soil. Rare. Coast of Carmel, Sharon Plain.

Area: S. Mediterranean, extending along the Atlantic coast of Europe.

42. Ctenopsis De Not.

Annuals. Leaf-blades setaceous, rolled. Inflorescence a spike-like unilateral raceme. Spikelets several-flowered, arranged in 2 rows at right angles to the flattened continuous rachis; rachilla disarticulating below each floret. Glumes very unequal, the lower minute, squamiform, the upper about as long as lowest lemma; lemma 5-veined, keeled towards tip, mucronate or awned; palea nearly as long as lemma. Lodicules 2. Stamens 3. Caryopsis narrowly ellipsoid, more or less free; hilum less than ½ length of caryopsis.

Three species, of which two are W. Mediterranean and one is Saharo-Arabian.

1. Ctenopsis pectinella (Del.) De Not., Ann. Sci. Nat. Bot. ser. 3, 9:325 (1848); Bor, Fl. Iraq 9:68, t.23. *Festuca pectinella* Del., Ind. Sem. Hort. Monsp. 24 (1836). *Vulpia patens* Boiss., Diagn. ser. 1, 13:62 (1854). *V. pectinella* (Del.) Boiss., Fl. 5:631 (1884); Post, Fl. 2:766. *C. patens* (Boiss.) Melderis in Rech. fil., Ark. Bot. ser. 2, 2(5):297 (1952). [Plate 310]

Annual, 5–30 cm. Culms capillary, smooth and glabrous. Inflorescence spike-like, 2–3 cm, 9–15-spiculate, comb-like after dispersal of florets. Spikelets sessile or subsessile, 3–5(–7)-flowered. Glumes persistent, coriaceous, acute; lemma mucronate. Fl. March–May.

Hab.: Sandy soil. Philistean Plain; W., N. and C. Negev.

Area: Saharo-Arabian (Syria, S. Palestine, N. Africa).

43. Cutandia Willk.*

Annuals. Leaf-blades flat or more or less rolled; ligule hyaline. Inflorescence a panicle; branches 3-angled, divaricate at maturity. Spikelets laterally com-

* Dr C.A. Stace (Leicester) kindly helped with elucidating literature citations.

pressed, 2- to many-flowered, disarticulating under each floret, at base of pedicel or of branch. Glumes markedly unequal, 1-3-veined, more or less keeled, awnless or shortly awned, shorter than lowest floret; lemma keeled, conspicuously 3-veined, muticous or ending in a mucro or a short awn; palea nearly as long as lemma. Lodicules 2. Stamens 3. Caryopsis narrowly ellipsoid, more or less free; hilum punctiform.

Six Mediterranean species.

Literature: C.A. Stace, Notes on *Cutandia* and related genera, *Jour. Linn. Soc. London (Bot.)* 76:350-352 (1978); *Cutandia*, in: *Flora europaea* 5:158 (1980).

1. Spikelets 2-3(-4)-flowered, 1-1.5(-2) mm broad   2
- Spikelets 5-15-flowered, 3-5 mm broad   3
2. Lemma acuminate, ending in a sharp mucro.   **3. C. memphitica**
- Lemma subacute, muticous.   **4. C. dichotoma**
3(1). Lemma glabrous, subacute, ending in a mucro. Inflorescence glaucous.
                                                                                      1. C. maritima
- Lemma obtuse, muticous, with short capitate hairs below. Inflorescence usually purplish at maturity.   **2. C. philistaea**

1. Cutandia maritima (L.) W. Barbey, Fl. Sard. Comp. 72 (1885). *Triticum maritimum* L., Sp. Pl. ed. 2, 128 (1762). *Scleropoa maritima* (L.) Parl., Fl. Ital. 1:468 (1850); Boiss., Fl. 5:637; Post, Fl. 2:768. [Plate 311]

Annual, 10-40 cm, glaucous. Uppermost leaf-blade reaching and overtopping base of panicle. Panicle 3-12 cm, oblong-ovoid. Spikelets elliptic to oblong, 8-15 mm, 5-9(-12)-flowered. Lower glume 4.5-5.5 mm, the upper 5.5-6.5 mm; lemma subacute, produced in a mucro, glabrous. Fl. March-May.

Hab.: Sands and sandy soils on the coast. Coastal Galilee, Acco Plain, Coast of Carmel, Sharon Plain, Philistean Plain.

Area: Mediterranean.

2. Cutandia philistaea (Boiss.) Jackson, Index Kewensis 675 (1893). *Scleropoa philistaea* Boiss., Diagn. ser. 1, 13:60 (1854); Boiss., Fl. 5:636; Post, Fl. 2:768. *Desmazeria philistaea* (Boiss.) H. Scholz, Willdenowia 6:291 (1971). [Plate 312]

Annual, 10-30(-45) cm, purplish. Uppermost leaf remote from panicle; blades short. Panicle 5-10 cm, ovate-oblong. Spikelets elliptic to oblong, 7-10 mm, 5-9(-12)-flowered, greenish or purplish. Lower glume 3-3.5 mm, the upper 4 mm; lemma white-margined, obtuse and muticous, sometimes emarginate and obsoletely mucronulate, with minute capitate hairs below. Fl. March-May.

Hab.: Calcareous sandstone, sandy loam, sandy soil. Coastal Galilee, Acco Plain, Coast of Carmel, Sharon Plain, Philistean Plain; W. Negev.

Area: E. Mediterranean (Palestine, Lebanon, Egypt, Libya, Tunisia).

3. Cutandia memphitica (Spreng.) K. Richter, Pl. Eur. 1:77 (1890); Stace, in Fl. Eur. 5:159; Bor, Fl. Iraq 9:70, t.24 f.10. *Dactylis memphitica* Spreng., Bot. Gart. Halle Nachtr. 1:20 (1801); Roth, Catalecta Bot. 3:18 (1806). *Scleropoa memphitica* (Spreng.) Parl., Fl. Ital. 1:471 (1850); Boiss., Fl. 5:639; Post, Fl. 2:769. [Plate 313]

Annual, 10-30(-40) cm. Culms numerous, often bearing panicles at each node. Leaf-sheaths glabrous, conspicuously dilated, partly enclosing panicles. Panicle 5-10 cm, lax, much branched; branches scabridulous, strongly divaricate at maturity. Spikelets 7-10 mm, narrowly linear, 2-3(-4)-flowered, glabrous or sometimes hirtellous. Glumes lanceolate, acute and mucronate, the lower 3-4 mm, the upper 4-5 mm; lemma acuminate, produced in a long mucro. Fl. March-April.

Hab.: Maritime and desert sand dunes, sandy soil. Coastal Galilee, Acco Plain, Sharon Plain, Philistean Plain; W., N. and C. Negev; Dead Sea area, Arava Valley; E. Gilead, Edom.

Area: Saharo-Arabian and Irano-Turanian, extending into the Mediterranean along the coast.

Eig (1927) recorded this species (under *Scleropoa*) from the En-Harod railway station (Esdraelon Plain) and remarked: "It was undoubtedly introduced here with sea sand from the environs of Haifa brought for building purposes".

4. Cutandia dichotoma (Forssk.) Trabut in Batt. & Trabut, Fl. Algérie Monocot. 237 (1895); Bor, Fl. Iraq 9:70, t.24 ff.1-9. *Festuca dichotoma* Forssk., Fl. Aeg.-Arab. 22 (1775). *Scleropoa dichotoma* (Forssk.) Parl., Fl. Ital. 1:471 (1850); Boiss., Fl. 5:639. *S. memphitica* var. *dichotoma* (Forssk.) Bonnet in Bonnet & Barratte, Cat. Pl. Tunisie 483 (1896); Post, Fl. 2:770. [Plate 314]

Annual. Differs from the closely related *C. memphitica* as follows: Plants generally smaller (5-25 cm). Spikelets often somewhat smaller (6-8 mm). Glumes subacute; lemma subacute, muticous. Fl. March-April.

Hab.: Sands, sandy loess and Nubian sandstone. W., N. and C. Negev; Edom.

Area: Saharo-Arabian and Irano-Turanian.

44. SPHENOPUS Trin.

Annual. Culms slender. Leaf-blades flat or filiform; ligule membranous, long. Inflorescence a lax panicle; branches capillary, widely spreading. Spikelets minute, 2-5-flowered, awnless, laterally compressed, borne on slender club-shaped pedicels about as long as spikelets; upper floret often sterile; rachilla disarticulating above glumes and below each floret. Glumes markedly unequal, membranous, mostly veinless, much shorter than lowest floret; lemma muticous, hyaline above, 3-veined, keeled; palea shorter than lemma. Lodicules 2. Stamens 3. Caryopsis oblong; hilum punctiform.

Two or three Mediterranean species.

1. Sphenopus divaricatus (Gouan) Reichenb., Fl. Germ. Excurs. 45 (1830); Bor, Fl. Iraq 9:108, t.38; Boiss., Fl. 5:575; Post, Fl. 2:750. *Poa divaricata* Gouan, Obs. Bot. 4, t.2 f.1 (1773). [Plate 315]

Dainty annual, up to 30 cm. Leaf-blades short. Culms smooth and glabrous, with dark nodes. Panicle 6 × 4 cm; branches in pairs at each node of rachis, di- and trichotomously branched. Spikelets 2(-2.5) mm. Lower glume minute, veinless; the upper longer, veinless or with 1-3 short veins; lemma obtuse. Fl. March–May.

Hab.: Arid soils, sandy, gravelly and silty, often saline. Acco Plain, Coast of Carmel, Sharon Plain, Philistean Plain; Judean Desert, N. Negev; Beit Shean Valley, Lower Jordan Valley, Dead Sea area; E. Ammon, Moav, ?Edom.

Area: Mediterranean, Irano-Turanian and Saharo-Arabian.

45. PSILURUS Trin.

Annuals. Leaf-blades filiform. Inflorescence a terminal filiform spike of spaced distichous spikelets borne in cavities of rachis; rachis disarticulating tardily. Spikelets 1-2-flowered; lower floret hermaphrodite, the upper rudimentary, borne on a long rachilla. Lower glume absent in lateral spikelets, in terminal spikelet very small; upper glume triangular, also small, several times shorter than spikelet, inserted somewhat laterally; lemma linear-lanceolate, membranous, with an awn arising between 2 short teeth; palea as long as lemma; rudimentary floret sometimes awned. Lodicules 2. Stamen 1. Caryopsis slender; hilum linear.

A monotypic genus.

1. Psilurus incurvus (Gouan) Schinz & Thell., Viert. Naturf. Ges. Zürich 58:40 (1913); Bor, Fl. Iraq 9:102, t.35; Post, Fl. 2:791. *Nardus incurvus* Gouan, Hort. Monsp. 33 (I. 1762). *N. aristatus* L., Sp. Pl. ed. 2, 78 (IX. 1762). *P. aristatus* (L.) Lange, Vid. Meddel. Dansk Naturh. Foren. Kjøbenhavn 1860:59 (1860). *P. nardoides* Trin., Fund. Agrost. 93 (1820); Boiss., Fl. 5:682. [Plate 316]

Annual, (10-)15-30(-40) cm. Culms slender. Spike 10-20 cm. Spikelets (excl. awn) subulate, shorter than internodes of rachis. Lemma 4 mm; awn straight, about as long as lemma. Fl. March–May.

Hab.: Batha and fallow fields. Coastal Galilee, Acco Plain, Coast of Carmel, Sharon Plain, Philistean Plain; Upper and Lower Galilee, Mt. Carmel, Esdraelon Plain, Samaria, Shefela, Judean Mts., ?Judean Desert, W. and N. Negev; Upper and Lower Jordan Valley; Golan, Gilead, Ammon.

Area: Mediterranean and Irano-Turanian.

46. POA L.

Annuals or perennials. Leaf-blades flat to folded; sheaths tubular or split; ligule membranous. Inflorescence an effuse or contracted panicle; branches filiform. Spikelets laterally compressed, 2–10-flowered; florets hermaphrodite, the upper rudimentary; rachilla disarticulating above glumes and between florets; spikelets sometimes proliferating (viviparous). Glumes somewhat unequal, keeled, the lower 1-veined, the upper 3-veined; lemma keeled, 5-veined, hyaline at tip and margins, awnless, glabrous or hairy along keel and veins; palea aculeolate or ciliate along keels; callus short, glabrous or with a bundle of long wool. Lodicules 2, acuminate, often with an additional tooth. Stamens 3. Caryopsis elliptic in outline, flattened on ventral side; hilum punctiform.

About 300 species in temperate regions and in mountains in the Trop.

Literature: N. Feinbrun, Poa series Bulbosae Roshev. of Palestine and Syria, *Kew Bull.* 1940:277–285 (1940). C.C. Heyn, Studies of bulbous *Poa* in Palestine. I. The agamic complex of *Poa bulbosa*, *Bull. Res. Counc. Israel* 11D:117–126 (1962); Biosystematic approaches to the solution of taxonomic problems in Israel, in: P.H. Davis et al. (eds.), *Plant Life of South-West Asia*, 181–193, Bot. Soc. Edinburgh, 1971.

1. Perennial; base of culms bulbous. Lemma obscurely veined   2
- Annual or perennial; base of culms not bulbous. Lemma mostly with conspicuously prominent veins   5
2. Lemma of lowermost floret with bundle of long wavy wool at base.*
1. P. bulbosa var. **bulbosa**
- Lemma without wool at base   3
3. Lemma of lower floret 3.5–4 mm (in Moav 2.75 mm). Basal leaves with up to 2–4 mm long hyaline ligules; sheaths of basal leaves broadly hyaline-margined; hyaline tips of old leaf-sheaths protruding above tufts of plant. Plants of steppes in C. Negev, Ammon, Moav and Edom.   **3. P. sinaica**
- Lemma of lower floret 2–2.5(–3) mm. Plant without hyaline tips protruding above tufts   4
4. Panicle 2.5–5(–7) cm; height of plant 10–25 cm; tufts 1–1.5(–2) cm high; sheaths of basal leaves not hyaline-margined. Plants of Judean Desert and N. and C. Negev with dense compact tufts forming large continuous patches of vegetation.
2. P. eigii
- Panicle 5–10 cm; height of plant 30–50 cm; tufts 2–3 cm high. Plants mostly of non-desert districts of country, forming small patches.   **1. P. bulbosa** var. **hackelii**
5(1). Perennial, 30–70 cm   6
- Annual, 10–20 cm   7
6. Ligule 3.5–10 mm, acute. Lemma prominently veined. Plants of wet habitats.
4. P. trivialis
- Ligule 0.5 mm, truncate or absent. Lemma obscurely veined. Plants of maquis in Golan.   **5. P. nemoralis**
7(5). Anthers 0.2–0.5 mm, hardly longer than broad. Lemma of lower floret at most 2.5 mm; florets widely spaced on rachilla; uppermost floret somewhat longer than joint of rachilla below it. Common.   **7. P. infirma**

---

* Viviparous specimens (var. *vivipara*) which often grow in association with plants bearing regular non-proliferating spikelets cannot be identified by characters of the lemma.

– Anthers 0.7–1.3 mm, 2–3 times as long as broad. Lemma of lower floret exceeding 2.5 mm; florets crowded; uppermost floret at least twice as long as joint of rachilla below it. Fairly rare.   **6. P. annua**

1. Poa bulbosa L., Sp. Pl. 70 (1753); Feinbr., Kew Bull. 1940:278, f.1; Bor, Fl. Iraq 9:114; Boiss., Fl. 5:605; Post, Fl. 2:759. [Plate 317]

Perennial, 15–50 cm, tufted, forming small patches. Culms smooth, glabrous, bulbous at base and with residues of old leaf-sheaths. Leaves mainly basal; leaf-blades narrow, folded or rolled; basal sheaths broadened. Panicle (3–)4–8(–10) cm, contracted except during anthesis. Spikelets 1 5 mm, ovate-elliptic in outline, 3–5-flowered, often proliferating. Glumes acute; lemma of lower floret 2–2.5(–3) mm; lemma obsoletely veined, ciliate on lower ½ of keel and margins, often with copious wool on callus. Seed set generally low. 2n=40–58 (Heyn, *op. cit.*, 1962). Fl. March–April.

Hab.: Degraded batha, disturbed habitats, on terra rossa or greyish calcareous soil, mainly in the hill districts.

Var. **bulbosa**. Callus of lower florets with a long strand of wavy wool. Height of plant 15–30 cm.

Hab.: Common. Sharon Plain, Philistean Plain; Upper and Lower Galilee, Mt. Carmel, Samaria, Judean Mts., Judean Desert, N. and C. Negev (rare); Golan, Gilead, Ammon, Moav, Edom.

Area: As of species.

Var. **vivipara** Koeler, Descr. Gram. 189 (1802). Subsp. *vivipara* (Koeler) Arcangeli, Comp. Fl. Ital. 785 (1882). Proliferating; all or a part of florets transformed into leafy buds.

Hab.: Often together with var. *bulbosa*. Sharon Plain; Upper and Lower Galilee, Esdraelon Plain, Mt. Gilboa, Shefela, Judean Mts., C. Negev; Golan, Gilead, Ammon, Moav, Edom.

Area: As of species.

Var. **hackelii** (Post) Feinbr. comb. nov. *P. hackelii* Post in Post & Autran, Bull. Herb. Boiss. 5:760 (1897). Callus of florets without a strand of wool. Bulbous base of culm generally longer than in var. *bulbosa*. Height of plant usually 30–50 cm.

Hab.: Fairly common. Coastal Galilee, Acco Plain, Sharon Plain; Upper and Lower Galilee, Esdraelon Plain, Judean Mts., N. Negev; Gilead, Ammon, Edom.

Plants transitional between var. *hackelii* and var. *bulbosa* occur, with florets bearing a few long or short hairs at base of lemma. Post distinguished his *P. hackelii* from *P. bulbosa* mainly by the lack of a strand of long wool on the callus of florets characteristic of *P. bulbosa*.

Area: E. Mediterranean (Palestine, Syria, S. Anatolia).

Area of species: Mediterranean, Irano-Turanian and Euro-Siberian. In Palestine and Syria *P. bulbosa* is confined to Mediterranean territories.

It is remarkable that the main differential character of *P. bulbosa* within the group of bulbous species, namely presence of wool at base of lemma, which is a stable trait throughout the area of the species, breaks down in Palestine, Syria and S. Anatolia where *P. bulbosa* meets *P. sinaica*. Could this be the result of introgression from *P. sinaica*?

2. Poa eigii Feinbr., Kew Bull. 1940:280, t.7 f.3 (1940). [Plate 318]

Perennial, 10–25 mm, densely tufted; tufts short, compact, forming large patches. Base of culms bulbous, with a dense cover of residues of leaf-sheaths of previous years. Leaves mostly basal; leaf-blades very narrow, rolled; basal leaf-sheaths not hyaline-margined; ligule short (1–1.5 mm). Panicle 2.5–5(–7) cm, narrowly oblong-lanceolate to oblong; branches short, drawn together. Spikelets 3–4 mm, 3–5-flowered, ovate in outline, sometimes proliferating. Glumes ovate-oblong, acuminate, 2–2.5 mm, the upper broader than the lower; lemma acutish, obsoletely veined, densely long-ciliate on lower half of keel and margins, rarely glabrous (forma *glabrescens*); lemma of lower floret 2–2.5 mm; wool on callus lacking. 2n=44, 47 (Heyn, *op. cit.*, 1962). Fl. February–March.

Hab.: Steppes. Important component and prominent in the spring aspect of the following plant associations: Artemisietum herbae-albae, Noaeetum mucronatae, Phlomidetum brachyodontis, Salsoletum villosae, Suaedetum asphalticae. Judean Desert, N. and C. Negev; Lower Jordan Valley, Dead Sea area; E. Ammon (forma *vivipara*).

Area: W. Irano-Turanian. Endemic.

3. Poa sinaica Steud., Syn. Pl. Glum. 1:256 (1854); Feinbr., Kew Bull. 1940:280, f.2; Bor, Fl. Iraq 9:118, t.40; Boiss., Fl. 5:606; Post, Fl. 2:760. [Plate 319]

Perennial, 15–40(–50) cm, tufted; tufts very dense. Culms bulbous at base and clothed with residues of old leaf-sheaths. Leaf-blades of basal leaves very narrow, rolled; ligule of basal leaves 2–4 mm, hyaline; leaf-sheaths broadly hyaline-margined; hyaline tips of old leaf-sheaths protrude above tuft. Panicle (4–)6–8(–13) cm, contracted or effuse. Spikelets 3–6 mm, 4–5(–7)-flowered, sometimes proliferating (forma *vivipara*). Glumes acute; lower glume elliptic, 2.5 mm, the upper broadly elliptic, 3 mm; lemma obscurely veined; lemma of lower floret 3.5–4 mm (rarely shorter); wool on callus lacking. 2n=~ 40–50 (Heyn, *op. cit.*, 1962). Fl. March–April.

Var. **sinaica**. Var. *aegyptiaca* Schweinf. in Sickenberger, Contrib. Fl. Eg. 310 (1901). Lemma ciliate on lower half of keel and margins.

Hab.: Steppes; very common and stand-forming, often in Artemisietum herbae-albae. C. Negev; E. Ammon, Moav, Edom.

Var. **glabrescens** Feinbr., *op. cit.* 283 (1940), incl. subsp. *moabitica* Feinbr., *loc. cit.* Lemma glabrous on keel and margins.

Hab.: Steppes. Moav, Edom.

Area of species: Irano-Turanian (Egypt, Palestine, Syria, Iraq, C. Anatolia, Transcaucasia, Iran, Afghanistan, W. Pakistan).

4. Poa trivialis L., Sp. Pl. 67 (1753); Bor, Fl. Iraq 9:113, t.39; Boiss., Fl. 5:602; Post, Fl. 2:759. [Plate 320]

Perennial, 30–60 cm, shortly stoloniferous. Leaf-blades flat; ligule long, pointed. Panicle 10–20 cm, ovoid-pyramidal; branches usually 5 on lower nodes; pedicels up to 2 mm. Spikelets 3–4 mm, 2–4-flowered. Lemma distinctly 5-veined, with short hairs on keel up to its middle, not conspicuously hairy along veins, with long wool on callus (very rarely callus glabrescent). Fl. March–May.

Subsp. **trivialis.** Stolons terete, slender.

Hab: Marshes. Coastal Galilee, Acco Plain, Coast of Carmel, Sharon Plain, Philistean Plain; Upper and Lower Galilee, Mt. Carmel, Esdraelon Plain, Samaria, Judean Mts., Dan Valley, Hula Plain, Upper Jordan Valley, Dead Sea area; Gilead.

Subsp. **sylvicola** (Guss.) Lindb. fil., Öfvers. Finska Vet.-Soc. Förhandl. 48(13):9 (1906). Stolons moniliform, fleshy.

Hab.: Marshes. Upper Galilee; Sharon Plain.

Area of species: Euro-Siberian, extending into the Mediterranean and Irano-Turanian.

5. Poa nemoralis L., Sp. Pl. 69 (1753); Mout., Nuov. Fl. Liban Syrie 1:106; Boiss., Fl. 5:607; Post, Fl. 2:760. [Plate 321]

Perennial, 50–70 cm, devoid of rhizome. Root fibrous. Culms slender, leafy up to inflorescence. Leaf-blades flat; ligule up to 0.5 mm or absent. Panicle oblong, 7–10(–20) cm, contracted except during anthesis; branches capillary, semiverticillate, 2–5 at lower node; pedicels shorter than spikelet. Spikelets 3–4 mm, ovate-lanceolate in outline, 2–5-flowered. Glumes shorter than spikelet, lanceolate, acute; lower glume narrower; lemma lanceolate, subacute, keeled, rather obscurely veined, without or with a few long hairs on callus. Fl. July.

Hab.: *Quercus calliprinos* maquis, basalt. Rare. Golan.

Area: Euro-Siberian–Borealo-American, extending into the Mediterranean and the forest-covered mountains of the Irano-Turanian.

Very variable species reaching in the Golan the southern limit of its area.

6. Poa annua L., Sp. Pl. 68 (1753); Bor, Fl. Iraq 9:122; Boiss., Fl. 5:601; Post, Fl. 2:758. [Plate 322]

Annual or short-lived perennial, 10–20 cm. Panicle 3–5 cm, more or less pyramidal; lower branches in pairs, spreading or deflexed after anthesis. Spikelets few, not crowded towards ends of panicle-branches, 3–5-flowered. Lemma broadly hyaline, silky-hairy on keel and lateral veins. Anthers 0.7–1.3 mm, 2–3 times as long as broad. Fl. February–April.

Hab.: Irrigated sites. Rare. Acco Plain; Philistean Plain; Upper Galilee, Mt. Carmel, N. Negev; Hula Plain; Ammon.

Area: Euro-Siberian, Mediterranean and Irano-Turanian.

7. Poa infirma Kunth in Humb., Bonpl. & Kunth, Nov. Gen. Sp. 1:158 (1816); Bor, Fl. Iraq 9:124, t.42. *P. annua* L. subsp. *exilis* (Tommasini) Murb. in Aschers. & Graebn., Syn. Mitteleur. Fl. 2(1):390 (1900). [Plate 323]

Annual, 10-20 cm, Panicle 3-8 cm, narrowly ovoid; branches erecto-patent after anthesis. Spikelets few, appressed to rachis, 5-6-flowered. Lemma broadly hyaline, hairy on keel and lateral veins. Anthers minute (0.2-0.5 mm), scarcely longer than broad, about ½ length of those of *P. annua*. Fl. December–June.

Hab.: Near habitations, irrigated sites, roadsides. Fairly common. Sharon Plain, Philistean Plain; Upper Galilee, Mt. Carmel, Esdraelon Plain, Samaria, Judean Mts., Judean Desert; Upper Jordan Valley, Dead Sea area; Golan, Gilead.

Area: Mediterranean.

47. CATABROSA Beauv.

Perennials of damp or marshy habitats. Leaf-sheaths tubular for ½–⅔ of their length. Inflorescence a panicle with spreading or erect branches and short pedicels. Spikelets numerous, small, usually 2-flowered, awnless; rachilla disarticulating under each floret. Glumes unequal, much shorter than lowest floret, lower glume obovate or ovate, veinless, the upper broadly elliptic, 1-3-veined or veinless; lemma elliptic, truncate, eroded at tip, prominently 3-veined, 3-keeled; palea as long as lemma. Lodicules 2, minute. Stamens 3. Caryopsis ellipsoid, loosely enclosed between lemma and palea; hilum short.

Four temperate species.

1. Panicle effuse, with widely spreading branches.   **1. C. aquatica**
- Panicle narrow, with short ascending branches.   **2. C. capusii**

1. Catabrosa aquatica (L.) Beauv., Agrost. 97, 157 (1812); Bor, Fl. Iraq 9:58, t.18; Boiss., Fl. 5:576; Post, Fl. 2:750. *Aira aquatica* L., Sp. Pl. 64 (1753). [Plate 324]

Perennial, 30-50(-60) cm, with a creeping rhizome. Panicle oblong-pyramidal, 10-25 × 5-7 cm; branches whorled, first ascending, then spreading or deflexed. Spikelets 2.5-3 mm, 2(-3-4)-flowered; some spikelets sessile, the others short-pedicellate. Glumes obtuse, membranous at tip. Fl. March–May.

Hab.: Streams, marshes. Sharon Plain, Philistean Plain; Upper Galilee, Mt. Carmel, Esdraelon Plain, Judean Mts., Judean Desert; Hula Plain, Lower Jordan Valley; Golan, Gilead, Ammon.

Area: Euro-Siberian, Mediterranean and Irano-Turanian.

2. Catabrosa capusii Franchet, Ann. Sci. Nat. ser. 6, 18:272 (1884); Nevski in Fl. URSS 2:446, t.33 f.7; Bor, Fl. Iraq 9:60, t.19.*

Perennial, up to 20 cm, stoloniferous, smooth and glabrous. Panicle 4–5 (–12) × 0.5–1.5 cm, contracted, more or less interrupted; branches very short, ascending. Spikelets 3.5–4 mm, nearly always 2-flowered. Glumes membranous, eroded at apex; lower glume ovate, the upper broadly elliptic. Fl. April.

Hab.: Marsh, 520 m. E. Ammon.

Area: Irano-Turanian.

48. PUCCINELLIA Parl.

Perennials, rarely annuals or biennials. Leaf-sheaths split nearly to base. Inflorescence an effuse or contracted panicle. Spikelets (2–)3–10-flowered, somewhat laterally compressed; rachilla disarticulating above glumes and between florets. Glumes usually unequal, concave, obtuse, muticous, shorter than adjacent lemmas; lower glume 1(–3)-veined, the upper 3–7-veined; lemma 5-veined, rounded on back, rarely keeled, concave, obtuse or truncate, sometimes eroded at apex; palea nearly as long as lemma, ciliolate along keels. Lodicules 2, oblong, oblique, lobed, glabrous. Stamens 3. Ovary glabrous; styles arising on side of ovary. Caryopsis oblong, glabrous, free; hilum punctiform.

Four species in the Euro-Siberian, Mediterranean and Irano-Turanian regions.

1. Puccinellia distans (L.) Parl., Fl. Ital. 1:367 (1850). *Poa distans* L., Mantissa 1:32 (1767). *Glyceria distans* (L.) Wahlenb., Fl. Upsal. 36 (1828). *Atropis distans* (L.) Griseb. in Ledeb., Fl. Ross. 4:388 (1852); Boiss., Fl. 5:615; Post, Fl. 2:762. [Plate 325]

Perennial, tufted, up to 80 cm. Leaf-blades folded; ligule membranous, ovate. Panicle 15–30 cm, broadly pyramidal, erect; branches 5–6 at lower nodes of rachis, spreading at anthesis, naked in their lower part. Spikelets 4–7 × 1.5 mm, oblong, 4–7-flowered, shortly pedicellate. Glumes very unequal, broadly membranous-margined, ovate, obtuse to truncate; lemma 2.5–3 mm, glabrous or slightly hairy near base, obsoletely veined. Anthers about 2 mm. Fl. April–June.

Hab.: Salty marshes. Hula Plain; Transjordan (Al-Eisawi).

Area: Euro-Siberian, Mediterranean and Irano-Turanian.

Our plants differ from *P. distans* from Europe by taller stature, a larger panicle and longer spikelets.

* Not illustrated.

49. Sclerochloa Beauv.

Annuals. Leaf-blades flat or folded; sheaths tubular in lower $\frac{1}{4}$-$\frac{1}{2}$; ligule membranous. Inflorescence a spike-like panicle; branches and pedicels short and thick; branches arranged on one side of rachis, sometimes bearing single spikelets in upper part of panicle. Spikelets 3–8-flowered; lower florets hermaphrodite, the uppermost sterile; rachilla thick, tough, disarticulating below each floret or pedicel. Glumes very unequal, much shorter than spikelet, broadly hyaline-margined, prominently veined, keeled, obtuse; lower glume 3–5-veined, about $\frac{1}{2}$ length of the upper; upper glume 5–9-veined; lemma membranous to coriaceous with broad hyaline margins, prominently 5–7-veined, keeled, glabrous; palea nearly as long as lemma. Lodicules 2. Stamens 3. Caryopsis lanceolate-oblong in outline, flattened on one side, with a prominent 2-toothed beak (stylopodium); hilum punctiform.

Three Mediterranean and Irano-Turanian species, extending into C. and Atlantic Europe.

1. Sclerochloa dura (L.) Beauv., Agrost. 98, 177, t.19 f.4 (1812); Bor, Fl. Iraq 9:104, t.36; Boiss., Fl. 5:635; Post, Fl. 2:768. *Cynosurus durus* L., Sp. Pl. 72 (1753). [Plate 326]

Annual, 3–10(–15) cm. Culms numerous, leafy to base of inflorescences. Panicle 1.5–3.5 × 0.7–1.5 cm, dense, ovate to ovate-oblong in outline. Spikelets 6–10(–12) mm, 3–5-flowered; rachilla disarticulating tardily above glumes and between lower florets. Glumes obtuse to emarginate; lower glume ovate, the upper oblong; lemma oblong-lanceolate, obtuse, sometimes mucronate. Fl. March–May.

Hab.: Alluvial soils wet in winter. Not common. Sharon Plain, Philistean Plain; Upper and Lower Galilee, Mt. Carmel, Esdraelon Plain, Judean Mts.; Golan.

Area: Mediterranean and Irano-Turanian, extending into the Euro-Siberian.

50. Dactylis L.

Perennials. Leaves flat or rolled inwards; ligule hyaline. Inflorescence a spike-like one-sided dense panicle; branches erect and appressed to rachis, but widely spreading at anthesis, bearing spikelets in dense strongly compressed clusters, each cluster of several spikelets. Spikelets 2–5-flowered, laterally compressed; florets hermaphrodite, the uppermost sterile; rachilla disarticulating above glumes and between florets. Glumes unequal to subequal, shorter than lower floret, keeled, 1–3-veined, acute and mucronate, inequilateral, broadly hyaline-margined, persistent; lemma 5-veined, keeled, mostly ciliate on keel; palea about as long as lemma. Lodicules 2. Stamens 3. Caryopsis ellipsoid, flattened on ventral side; embryo small; hilum punctiform.

One very polymorphic widely distributed species.

Introduced as a valuable pasture grass into temperate regions throughout the world.

Literature: G. L. Stebbins & D. Zohary, Cytogenetic and evolutionary studies in the genus *Dactylis, Univ. Calif. Publ. Bot.* 31:1–40 (1959). U. Nur & D. Zohary, Distribution patterns of diploid and tetraploid forms of *Dactylis glomerata* L. in Israel, *Bull. Res. Counc. Israel* 7D:13–22 (1959).

1. Dactylis glomerata L., Sp. Pl. 71 (1753); Bor, Fl. Iraq 9:76, t.26; Boiss., Fl. 5:596; Post, Fl 2:756. [Plate 327]

Perennial, 50–80 cm, tufted. Leaves glaucous; leaf sheaths tubular up to ½–¾ of their length; ligule 4–6 mm. Panicle 3–12 cm. Spikelets 5–9 mm. Lower glume 3–3.5 mm, the upper 4–4.5 mm; lemma 4–7 mm, mostly emarginate and mucronate to shortly awned. $2n = 14, 28, 28 + 1$ to 4B. Fl. April–June.

Hab.: Batha dominated by *Sarcopoterium spinosum* on various soils with 550 mm annual rainfall and semi-steppe plant associations where annual rainfall does not exceed 300 mm. Coastal Galilee, Coast of Carmel, Sharon Plain, Philistean Plain; Upper and Lower Galilee, Mt. Carmel, Esdraelon Plain, Mt. Gilboa, Samaria, Shefela, Judean Mts., Judean Desert, N. Negev; Dead Sea area; Golan, Gilead, Ammon, Moav.

Area: Mediterranean, Irano-Turanian and Euro-Siberian.

According to Stebbins & D. Zohary (*op. cit.*), *Dactylis glomerata* is a "single large polyploid complex". It comprises eleven diploid taxa (subspecies), endemic to more or less restricted geographical areas, and a continuous tetraploid superstructure spread over Europe, W. Asia and N. Africa. The diploid taxa can be distinguished by external morphology, ecological preferences and geographical distribution. The tetraploid part of the complex is of hybrid origin. Tetraploids have not been studied sufficiently with a view to classification. "Their variability falls almost entirely within the limits set by the diploids, except for the greater size of the reproductive structures... which can be ascribed directly to the polyploid conditions". (Stebbins & D. Zohary, *op. cit.*)

The diploid taxon found in Israel and the neighbouring countries is subsp. *judaica* Stebbins & D. Zohary (Nur & D. Zohary, *op. cit.*). The diploid populations occupy only relatively humid true Mediterranean habitats in the hill districts and are restricted in extent in comparison with the tetraploids. The tetraploids occupy a much wider range of habitats and are spread over most of the Mediterranean territories of Israel and some of the more arid semi-steppe habitats in the border zone. Nur & Zohary assume that two diploid forms, the local subsp. *judaica* and the Iranian subsp. *woronowii* (Ovczinn.) Stebbins & D. Zohary, contributed their chromosomes to the formation of the E. Mediterranean tetraploid level.

51. CYNOSURUS L.

Annuals or perennials. Leaves flat; ligule membranous, long. Inflorescence a one-sided, capitate or spike-like panicle. Spikelets dimorphic, in clusters;

clusters containing sterile and fertile spikelets. Sterile spikelets distichous, often pectinate, consisting of aristate glumes and numerous aristate empty lemmas borne on a continuous rachilla. Fertile spikelets few, enclosed and concealed by sterile ones, 1–5-flowered, laterally compressed, with a rachilla disarticulating above glumes and between florets. Glumes of fertile spikelets subequal, membranous to hyaline, acute; lemma concave, 3–5-veined, awned or mucronate from a 2-dentate apex; palea about as long as lemma, 2-fid. Lodicules 2. Stamens 3. Caryopsis elliptic in outline; hilum punctiform.

Five to six species in the N. temperate regions of the Old World.

1. Panicle narrow, with longish branches and a whitish transverse band at each joint; peduncle curved under the panicle. Spikelets hairy.   **3. C. elegans**
- Panicle with very short branches and devoid of whitish bands as above; peduncle not curved as above. Spikelets scabrous or glabrous, not hairy   2
2. Fertile spikelets 1-flowered. Glumes 5–7 mm (incl. awns), subulate and long-aristate from a lanceolate base; awn of lemma 5–8 times as long as lemma proper.
                                                                                   2. C. callitrichus
- Fertile spikelets 2–3-flowered. Glumes 9–10 mm, lanceolate, short-aristate; awn of lemma 2–3 times as long as lemma proper.   **1. C. echinatus**

1. Cynosurus echinatus L., Sp. Pl. 72 (1753); Bor, Fl. Iraq 9:74; Boiss., Fl. 5:571; Post, Fl. 2:748. [Plate 328]

Annual, (10–)20–60 cm. Culms erect. Panicle ovoid to oblong, 2–4(–5) cm (incl. awns). Sterile spikelets pectinate, containing about 15–20 aristate lemmas; awns purple. Fertile spikelets 2–3-flowered; glumes hyaline, 9–10 mm, lanceolate, attenuate above and short-aristate, 1-veined; lemma scabridulous in its upper part, that of lowest floret ending in an awn longer than lemma proper. Fl. March–June.

Hab.: Batha. Coast of Carmel, Sharon Plain, Philistean Plain; Upper and Lower Galilee, Mt. Carmel, Esdraelon Plain, Samaria; Shefela, Judean Mts.; Golan.

Area: Mediterranean and Irano-Turanian.

2. Cynosurus callitrichus W. Barbey in C. & W. Barbey, Herb. Levant 165, t.10 (1882); Boiss., Fl. 5:571; incl. var. *viridis* Bornm., Verh. Zool.-Bot. Ges. Wien 1898:106 (1898). *C. echinatus* L. var. *callitrichus* (W. Barbey) Bornm., Beih. Bot. Centr. 31 Abt. II:269 (1914); Post, Fl. 2:748. *C. coloratus* Lehm. ex Steud., Nomencl. Bot. ed. 2, 1:465 (1840) *pro syn.* [Plate 329]

Annual, 5–15(–20) cm. Culms erect, leafy. Panicle dense, long-aristate, ovoid, 2–4 cm (incl. awns), mostly purple to violet, usually subtended by upper leaf-sheath. Lemmas of sterile spikelets more or less spirally arranged, narrow, ending in long purple awns. Fertile spikelets 1-flowered; glumes very narrow, long-aristate, much longer than floret; lemma ending in an awn 5–8(–10) times as long as lemma proper. Fl. March–May.

Hab.: Batha and grassy places. Sharon Plain; Upper and Lower Galilee, Mt.

Carmel, Samaria, Shefela, Judean Mts., N. and C. Negev; Golan, Gilead, Ammon, Moav.

Area: E. Mediterranean (Syria, Palestine, Egypt, Libya, Crete). Introduced in S. Africa.

3. Cynosurus elegans Desf., Fl. Atl. 1:82, t.17 (1798); Bor, Fl. Iraq 9:73, t.25; Boiss., Fl. 5:571; Post, Fl. 2:748. [Plate 330]

Annual, 15–30(–40) cm. Panicle narrow, semi-oblong, 1.5–4.5(–5) cm (incl. awns), with longish branches; peduncle curved under panicle. Sterile spikelets pectinate; lemmas rather shortly aristate. Fertile spikelets 1–3-flowered; glumes very narrow, about as long as spikelet, lanceolate at base and attenuate into a long subulate cusp; lemma often hairy, ending in an awn many times as long as lemma proper. Fl. March–May.

Hab.: Rocky batha. Upper and Lower Galilee, Mt. Gilboa, Judean Mts.; Hula Plain; Golan, Gilead, Ammon, Moav.

Area: Mediterranean and Irano-Turanian.

52. Lamarckia Moench

Annual. Leaf-blades flat; ligule membranous. Inflorescence a one-sided dense panicle. Spikelets dimorphic, flattened, in clusters falling at maturity as a unit; each cluster containing 1 fertile (upper) spikelet and several sterile (lower) ones. Fertile spikelet with 1 hermaphrodite pedicellate floret and often with an awned rudiment of a second floret; glumes subequal, narrow, 1-veined, acute, as long as floret; lemma broader than glumes, 2-fid, long-awned; palea 2-keeled. Sterile spikelets linear, reduced to several awnless distichous imbricate glumes and lemmas.

A monotypic genus.

1. Lamarckia aurea (L.) Moench, Meth. 201 (1794); Bor, Fl. Iran. 70:17; Boiss., Fl. 5:570; Post, Fl. 2:747. *Cynosurus aureus* L., Sp. Pl. 73 (1753). [Plate 331]

Annual, 7–20(–40) cm, pale green. Leaves glabrous, soft; leaf-sheaths compressed, keeled, lax; ligule oblong, reaching 1 cm. Panicle 3–7 × 1.5–2.5 cm, pale yellow, sometimes purple-tinged, with nodding hispid branchlets bearing deciduous clusters of spikelets. Glumes membranous, tapering at both ends, acute. Lemma of sterile spikelets 3-veined, obtuse, rounded and denticulate at apex. Fl. January–May.

Hab.: Arid sites. Coast of Carmel, Sharon Plain, Philistean Plain; Upper and Lower Galilee, Mt. Carmel, Shefela, Judean Mts., Judean Desert, N. and C. Negev; Hula Plain, Upper and Lower Jordan Valley; Golan, Gilead, Ammon.

Area: Mediterranean and Irano-Turanian, extending into the Sudanian.

53. BRIZA L.

Annuals or perennials. Leaf-blades flat; ligule membranous. Inflorescence a lax panicle with filiform branches and often pendulous spikelets. Spikelets 4-20-flowered, laterally compressed, broadly ovate to triangular in outline, with imbricate hermaphrodite florets. Glumes subequal, ovate-orbicular, cordate at base, concave, 3-9-veined, muticous, rounded on back, shorter than lowest floret; glumes and lemma membranous at margins; lemma suborbicular to broadly ovate, cordate, muticous, 5-9-veined; palea much smaller, obtuse. Lodicules 2. Stamens 3. Caryopsis obovate in outline, flattened on adaxial side, enclosed in lemma and palea.

About 20 species in the N. temperate zone and S. America.

1. Spikelets 8-20 mm.   **3. B. maxima**
- Spikelets not exceeding 6 mm   2
2. Pedicels longer than spikelets. Spikelets rounded at apex, broader than long to as broad as long, nodding.   **1. B. minor**
- Pedicels shorter than spikelets. Spikelets acutish, longer than broad, not nodding.
                                                                                  2. B. humilis

1. Briza minor L., Sp. Pl. 70 (1753); Bor, Fl. Iraq 9:56; Boiss., Fl. 5:593; Post, Fl. 2:755. [Plate 332]

Annual, 10-40 cm. Ligule elongate. Panicle erect, effuse; branches spreading; pedicels longer than spikelets. Spikelets 3-5 mm, as broad as long or broader, 5-7-flowered, nodding. Lemma orbicular-cordate, broader than long, very obtuse. Caryopsis free; hilum punctiform. Fl. February–April.

Hab.: Grassy and moist places. Coastal Galilee, Acco Plain, Coast of Carmel, Sharon Plain, Philistean Plain; Upper and Lower Galilee, Mt. Carmel, Esdraelon Plain, Samaria, Shefela, Judean Mts.; Dan Valley, Hula Plain, Upper Jordan Valley, Beit Shean Valley; Gilead.

Area: Mediterranean, Irano-Turanian and Euro-Siberian; introduced into various parts of the world.

2. Briza humilis Bieb., Fl. Taur.-Cauc. 1:66 (1808); Bor, Fl. Iraq 9:56, t.17. *B. spicata* Sm. in Sibth. & Sm., Fl. Graeca 1:60, t.77 (1808) non Burm. fil. (1768) nec Lam. (1783); Boiss., Fl. 5:593; Post, Fl. 2:755. [Plate 333]

Annual, 20-40 cm. Ligule oblong, lacerate. Panicle raceme-like, often one-sided; upper branchlets often bearing a single spikelet on a pedicel shorter than spikelet. Spikelets 4-5.5 mm, ovate or oblong, subacute, 5-9-flowered. Glumes obovate, apiculate; lemma subacute; glumes and lemmas finely scabridulous. Caryopsis free; hilum oblong. Fl. February–April.

Hab.: Batha. Gilead.

Area: E. Mediterranean and Irano-Turanian, extending into S. Russia.

3. Briza maxima L., Sp. Pl. 70 (1753); Boiss., Fl. 5:593; Post, Fl. 2:755. [Plate 334]

Showy annual, 30–40(–50) cm. Ligule oblong. Panicle one-sided, with slender branches. Spikelets 8–20 × 5–10 mm, ovate-oblong to oblong, 7–20-flowered, nodding, often pubescent, straw-coloured or purplish at base. Glumes often purplish; lemma rounded on back, ovate, apiculate, broadly membranous-margined. Caryopsis adnate to palea; hilum linear. Fl. February–May.

Hab.: Batha and grassy places. Coastal Galilee, Coast of Carmel, Sharon Plain, Philistean Plain; Upper and Lower Galilee, Mt. Carmel, Mt. Gilboa, Samaria, Shefela, Judean Mts.; Hula Plain, Upper Jordan Valley; Golan, Gilead, Moav.

Area: Mediterranean.

Tribe 9. MONERMEAE C.E. Hubbard (1948). Annuals. Leaf-blades flat or rolled; sheaths split nearly to base; ligule membranous. Inflorescence a true spike, with spikelets sunk in cavities of rachis-internodes. Spikelets solitary, alternate, 1(–2)-flowered. Glumes 2, seated side by side, or 1, the lower glume suppressed (in lateral spikelets), coriaceous, as long as spikelet; lemma hyaline with 3 faint veins, awnless; palea 2-veined. Lodicules 2, acuminate. Stamens 3. Caryopsis free, ellipsoid, with a short and broad appendage; hilum short, linear to elliptic. Chromosomes large; x=7, 13.

54. PARAPHOLIS C.E. Hubbard

Annuals. Leaves flat or rolled; ligule membranous. Inflorescence a cylindrical spike, disarticulating at maturity into spikelets, each spikelet with adjoining rachis-internode; internodes with strong prominent veins. Spikelets 1-flowered, sessile and sunk in cavities in rachis. Glumes 2, subequal, glabrous, coriaceous, veined, seated side by side, closing cavity in rachis and concealing floret, lanceolate, acute to acuminate, asymmetric, the keel dividing each glume into 2 very unequal parts, the inner part (next to adjacent glume) coriaceous, the outer sharply reflexed inwards, very narrow, hyaline; keel winged in some species; lemma somewhat shorter than glumes, hyaline, turned with its side towards rachis; palea nearly as long as lemma, 2-veined. Lodicules 2. Stamens 3. Caryopsis oblong with an apical appendage; embryo small. Chromosomes large; x=7, 9, 19.

Six species in the Mediterranean and Irano-Turanian regions and along the Atlantic coast of Europe.

Literature: H. Runemark, A revision of *Parapholis* and *Monerma* in the Mediterranean, *Bot. Not.* 115:1–17 (1962).

1. Spike thick, strongly curved. Anthers about 1 mm, included. Keel of glume not winged.   **1. P. incurva**
- Spikes thinner, usually straight. Anthers 2–3.5 mm, exserted during flowering. Keel of glume winged.   **2. P. filiformis**

Boulos & Al-Eisawi (1977) reported *P. marginata* Runemark from Ras en-Naqb in Edom (no. 6220). *P. marginata* differs from *P. incurva* in winged keel of glumes, from *P. filiformis* in anthers less than 1 mm.

1. Parapholis incurva (L.) C. E. Hubbard, Blumea, Suppl. 3:14 (1946); Bor, Fl. Iraq 9:266, t.93. *Aegilops incurva* L., Sp. Pl. 1051 (1753). *A. incurvata* L., Sp. Pl. ed. 2, 1490 (1763). *Lepturus incurvatus* (L.) Trin., Fund. Agrost. 123 (1820); Boiss., Fl. 5:684; Post, Fl. 2:792. *Pholiurus incurvatus* (L.) Hitchcock, U.S. Dept. Agric. Bull. 772:106 (1920). *Pholiurus incurvus* (L.) Schinz & Thell., Viert. Naturf. Ges. Zürich 66:265 (1921). [Plate 335]

Annual, 5-25 cm, glabrous. Culms decumbent or ascending, usually numerous; nodes of culms usually covered by leaf-sheaths. Leaves rolled or flat; uppermost sheath strongly inflated, partly enclosing spike even in mature plants. Spikes numerous, 5-10 cm × 1.25-1.75 mm, 10-20-spiculate, usually strongly curved. Spikelets 4.5-6 mm. Glumes lanceolate, acuminate, about 1½ times as long as lemma and up to 1½ times as long as rachis-internode; keel not winged. Anthers 1-1.25 mm, included. Cleistogamous. Fl. April-May.

Hab.: Wet places. Coastal Galilee, Acco Plain, Coast of Carmel, Sharon Plain, Philistean Plain; Mt. Carmel, Judean Mts., W., N. and C. Negev; Lower Jordan Valley, Dead Sea area; Edom.

Area: Mediterranean and Irano-Turanian, extending into the Euro-Siberian.

Chromosome number 2n=38 is recorded from Greece (Runemark, *op. cit.* 10).

2. Parapholis filiformis (Roth) C. E. Hubbard, Blumea, Suppl. 3:14 (1946). *Rottboellia filiformis* Roth, Catalecta Bot. 1:21 (1797). *Lepturus filiformis* (Roth) Trin., Fund. Agrost. 123 (1820); Boiss., Fl. 5:684; Post, Fl. 2:792. *Pholiurus filiformis* (Roth) Schinz & Thell., Viert. Naturf. Ges. Zürich 66:265 (1921). [Plate 336]

Annual, 10-40(-60) cm, glabrous. Culms erect or ascending. Leaves rolled or flat. Spikes 3-15 cm × 1.0-1.25 mm, 10-20-spiculate, usually straight. Spikelets 4-6 mm. Glumes with a winged keel, linear-lanceolate, acuminate, somewhat longer than rachis-internode. Anthers (2-)2.5-3.5 mm. Chasmogamous. Fl. May-July.

Hab.: Saline marshes; Juncetum acuti. Acco Plain, Coast of Carmel.

Area: Mediterranean, extending into the Euro-Siberian.

Chromosome number 2n=14 is recorded from Portugal (Runemark, *op. cit.* 10).

55. MONERMA Beauv.
(*Hainardia* Greuter)

Annuals. Leaves flat, later rolled; ligule a narrow membranous rim. Inflorescence a cylindrical spike; rachis tough, disarticulating tardily, each spikelet falling with a rachis-internode; internodes with thick prominent veins. Spikelets 1-flowered, sessile and sunk in cavities in rachis. Glumes 2 in terminal spikelet, 1 (the upper) in lateral spikelets, indurate, strongly veined, entirely concealing floret; lemma hyaline, somewhat shorter than glume, turned with its back to rachis; palea hyaline, 2-veined. Lodicules 2, fleshy. Stamens 3. Caryopsis oblong; embryo small; hilum linear, short. x=13.

A monotypic genus.

Greuter (1967) rejected the name *Monerma* and named the genus *Hainardia* Greuter on the contention that, since the type species of *Lepturus* R. Br. is included by P. Beauvois in his *Monerma*, the latter name is illegitimate. In fact, P. Beauvois cited, following the description of *Monerma*, 3 species as follows: "Spec. *Rottb. repens*, *subulata* Savi; *monandra* Lin."

Tzvelev (1976, p. 530) does not accept the illegitimacy of *Monerma* and reasons that "in the description of the genus *Monerma* not the species *Rottboellia repens* Forst. (the type species of *Lepturus* R. Br.) is included, but Mediterranean specimens of *Monerma*, erroneously identified by some authors as *Rottboellia repens*".

Indeed, the name *repens* in the above citation is not followed by an author's name. Furthermore, *Rottboellia repens* Forst. fil. is an Australian perennial, while the other two species, *subulata* Savi and *monandra* L., are Mediterranean annuals. In addition, Plate XX f.X by P. Beauvois clearly shows that the spikelets of *Monerma* have a single glume; spikelets of *Lepturus* R. Br. have, however, 2 glumes.

1. Monerma cylindrica (Willd.) Coss. & Durieu, Expl. Sci. Algérie, Bot. II, 214 (1855); Bor, Fl. Iraq 9:268, t.94; Boiss., Fl. 5:683; Post, Fl. 2:791. *Rottboellia cylindrica* Willd., Sp. Pl. ed. 4, 1:464 (1797). *Lepturus cylindricus* (Willd.) Trin., Fund. Agrost. 123 (1820). *Hainardia cylindrica* (Willd.) Greuter in Greuter & Rech. fil., Boissiera 13:178 (1967) *nom. superfl.* [Plate 337]

Annual, 5–20 cm, glabrous. Uppermost leaf-sheath enclosing spike at beginning of flowering. Spike 5–10(–15) cm × 1.5–1.75 mm, straight, sometimes curved, glabrous. Glumes 5–7 mm, lanceolate, acute, rigid, 5–7-veined, much longer than adjacent rachis-internode, spreading at first, then appressed to rachis. Anthers 2.5–3.5 mm. Fl. April–May.

Hab.: Wet places. Acco Plain, Sharon Plain, Philistean Plain; Judean Mts.
Area: Mediterranean, extending into the Irano-Turanian.

Tribe 10. SESLERIEAE Koch (1837). Annuals or perennials. Leaf-blades flat or folded; sheaths tubular in their lower ⅓ or split to just above base; ligule membranous, generally short, often shortly ciliate. Inflorescence a very dense spike-like or head-like panicle. Spikelets 2–5- or rarely 6–14-flowered, the uppermost 1–2 florets rudimentary; rachilla disarticulating under each fertile floret. Glumes 2, shorter than spikelet, membranous or coriaceous-membranous; lemma (3–)5(–7)-veined, usually produced at apex into a mucro or a short awn, more rarely lemma divided into (3–)5(–7) lanceolate-subulate lobes; palea 2-keeled. Lodicules 2 or absent. Stamens 3. Ovary glabrous or short-hairy; stigmas sparsely short-hairy. Caryopsis broadly elliptic or obovate in outline, flattened, free. Chromosomes large; x=7 or 9.

Six genera, mainly Mediterranean, extending also into Europe and the Middle East.

56. ECHINARIA Desf.

Annuals. Leaf-blade flat; sheaths tubular; ligule membranous, ciliolate. Inflorescence a dense head-like panicle, prickly, globose or ovoid, long-pedunculate. Spikelets 2–4-flowered, shortly pedicellate to subsessile. Glumes subequal, membranous, slightly compressed, obtuse, shorter than spikelet; lower glume with 2 veins, each produced into a mucro; the upper with 1 median vein produced into a mucro; lemma coriaceous, broad, with (3-)5(-7) flattened lobes or awns prickly and reflexed at maturity; palea rigid, as long as lemma, 2-awned. Lodicules 2. Stamens 3. Caryopsis obovate-truncate in outline, pubescent at apex; hilum punctiform.

A monotypic genus.

1. Echinaria capitata (L.) Desf., Fl. Atl. 2:385 (1799); Bor, Fl. Iraq 9:78, t.27; Boiss., Fl. 5:565; Post, Fl. 2:746. *Cenchrus capitatus* L., Sp. Pl. 1049 (1753). [Plate 338]

Annual, 4–25(-30) cm. Culms smooth and glabrous or pubescent under nodes. Leaf-blades narrow, strongly veined, often spinescent. Panicle up to 2 cm, borne on a peduncle longer than leaves; branches of panicle short, 1–2-spiculate. Lower glume shorter than the upper, 2-keeled. Fl. March–June.

Hab.: Batha, rocky hillsides or sandy places. Philistean Plain, W. Negev; Upper and Lower Galilee, Mt. Carmel, Mt. Gilboa, Samaria, Judean Mts., Judean Desert; Golan, Gilead, Ammon, Moav, Edom.

Area: Mediterranean, extending into the W. Irano-Turanian.

57. AMMOCHLOA Boiss.

Annuals. Leaves flat; sheaths split to just above base; ligule membranous, lacerate. Inflorescence a dense ovoid or globose head of nearly sessile spikelets. Spikelets laterally compressed, many-flowered, hermaphrodite; rachilla disarticulating below each floret; lemma falling with rachilla-joint immediately above it. Glumes subequal, shorter than spikelet, ovate, oblique, acute, chartaceous at centre, broadly membranous-margined, 1-veined, keeled, inequilateral; lemma broadly ovate, oblique, 5-veined, becoming coriaceous and ending in a firm mucro; palea narrower. Lodicules absent. Stamens 3. Caryopsis elliptic-oblong in outline, 3-angular in section, rostrate; hilum punctiform.

Two to three species.

1. Ammochloa palaestina Boiss., Diagn. ser. 1, 13:52 (1854); Bor, Fl. Iran. 70:9; Post, Fl. 2:746. *A. subacaulis* Balansa ex Coss. & Durieu, Bull. Soc. Bot. Fr. 1:317 (1854); Boiss., Fl. 5:566. [Plate 339]

Dwarf annual, 3–10 cm. Culms numerous, short, unequal, striate and glabrous. Leaf-blades longer than culms, acuminate, glabrous; sheaths usually inflated. Heads 2–6-spiculate, overtopped by leaves. Spikelets about 1 cm, subsessile, strongly compressed, oblong, 8–14-flowered. Glumes 1-veined, acute, forming a kind of involucre at base of head. Fl. March–May.

Hab.: Sandy soil, semistabilized dunes. Sharon Plain, Philistean Plain; W., N. and C. Negev; E. Gilead, Edom.

Area: E. Mediterranean, Saharo-Arabian and W. Irano-Turanian.

Tribe 11. MELICEAE Reichenb. (1828). Perennials. Leaf-blades flat or rolled; sheaths tubular, devoid of auricles; ligules membranous or absent. Inflorescence a panicle, either effuse or contracted, with filiform branches, or racemelike. Spikelets (1–)2- to several-flowered, hermaphrodite; rachilla produced above uppermost floret and crowned with several sterile lemmas wrapped together to form a compact knob; rarely spikelets falling entire. Glumes 2, more or less unequal, usually shorter than spikelet; lemma rounded on back, 5- to many-veined, awnless, or rarely with a straight or curved awn arising from near its tip; palea 2-veined, 2-keeled. Lodicules 2, small, truncate, sometimes connate. Stamens 3, rarely 2. Caryopsis ellipsoid, grooved on adaxial side; embryo about $\frac{1}{5}$ length of caryopsis; hilum linear, short. Chromosomes large; x=8, 9 or 10.

Seven to nine genera mainly in subtrop. and temperate areas of both hemispheres; some taxa arctic or of mountains in the Tropics.

58. MELICA L.

Perennials, tufted or with creeping rhizomes. Leaf-blades flat or rolled, narrow; leaf-sheaths tubular; ligule membranous, rarely absent. Inflorescence a more or less contracted panicle with filiform branches, or raceme-like with shortly pedicellate or subsessile spikelets. Spikelets slightly compressed laterally, containing 1–3(–5) hermaphrodite florets and above them 1–3 sterile lemmas forming a compact stipitate, clavate or ellipsoid apical structure. Glumes more or less unequal, concave, membranous or coriaceous-membranous, 3–5-veined; lemma of hermaphrodite florets membranous, rounded on back, coriaceous in fruit, 5- to many-veined, glabrous or more or less pilose. Lodicules 2, laterally fused. Stamens 3. Caryopsis oblong-ellipsoid; embryo small; hilum linear, short. Chromosomes fairly large; x=9.

Seventy-five to ninety species in temperate regions, mainly of the N. hemisphere, also in the Capensis and S. America.

Some species grown in Europe as ornamentals.

Literature: C. Papp, Monographie der europäischen Arten der Gattung *Melica* L., *Bot. Jahrb.* 65:275–384 (1932); Monographie der asiatischen Arten der Gattung *Melica* L., *Acad. Rom. Mem. Sect. Ştiint.* 3, 12:187–267 (1938). N.L. Bor, *Melica*, in: *Fl.*

iranica 70:249-260, 1970. W. Hempel, Taxonomische und chorologische Untersuchungen an Arten von *Melica* L. Subgen. *Melica, Feddes Repert.* 81:131-145 (1970); Vorarbeiten zu einer Revision der Gattung *Melica* L. I., *ibid.* 81:657-686 (1971); II., *ibid.* 84:533-568 (1973).

1. Lemma long-villose.   **2. M. cupani**
- Lemma glabrous.   **1. M. minuta**

1. Melica minuta L., Mantissa 32 (1767); Boiss., Fl. 5:585 *pro syn.*; Post, Fl. 2:753. Incl. var. *parviflora* (Boiss.) Bornm. & forma *planifolia* Bornm., Beih. Bot. Centralbl. Abt. II, 31:272 (1914). *M. angustifolia* Boiss. & Bl. in Boiss., Diagn. ser. 2, 4:131 (1859). [Plate 340]

Perennial, 20-60(-80) cm, tufted, smooth. Leaves numerous, glabrous; blades flat or rolled and filiform, stiff; ligule truncate or lacerate, rarely absent. Panicle 5-15(-20) cm, more or less lax, with spikelets often turned to one side; branches mostly erect, rarely spreading. Spikelets 5-9 mm, sometimes purplish, deciduous at maturity as a unit with thickened hispidulous tip of pedicel at base; hermaphrodite florets 1-2; sterile florets 1-2. Glumes unequal, acute; lower glume 5-6 mm, the upper 7-8 mm, slightly shorter or longer than lemma; lemma ovate, prominently 9-11-veined, glabrous, more or less coriaceous; lemma of lower fertile floret acute, of second floret (when present) obtuse. Fl. April-June.

Hab.: Cliffs and rocky ground in the shade of maquis and garigue. Upper Galilee, Mt. Carmel, Samaria, Judean Mts.; Edom.

Area: Mediterranean.

Boissier and Blanche in Boissier (1859), who described *M. angustifolia* from littoral Lebanon, doubted that it differed from *M. minuta*.

2. Melica cupani Guss., Fl. Sic. Prodr., Suppl. 17 (1832); Boiss., Fl. 5:590; Post, Fl. 2:754. *M. schischkinii* Iljinskaja, Not. Syst. (Leningrad) 12:29 (1950); Bor, Fl. Iran. 70:259. [Plate 341]

Perennial, 30-50 cm. Culms erect. Leaves rolled or flat, glabrous; upper leaf-blades usually erect, rigid; ligule truncate, rarely absent. Panicle 5-10(-15) cm, spike-like, cylindrical or unilateral, dense or lax, with short appressed branches. Spikelets 7-10 mm, shortly pedicellate, deciduous at maturity as a unit with hispidulous tip of pedicel at base; hermaphrodite florets 1(-2); sterile florets 1-2, turbinate or clavate, scabrous or glabrous, rarely hairy. Glumes unequal, acute, straw-coloured or purplish; lower glume 4-6 mm, ovate, acute; the upper 7-9 mm, lanceolate, acuminate; lemma densely long-villose along all veins from base to apex. Fl. April-May.

Hab.: Calcareous and basalt rocks and stony ground, batha. Upper and Lower Galilee, Samaria, Judean Mts.; Golan.

Area: S. Mediterranean and Irano-Turanian.

Hempel (Feddes Repert. 81:137, 1970) described Sect. *Cupani* and showed its distribution area on a map (f.9). He distinguished several species within this section, among them *M. schischkinii* Iljinskaja. The latter species is quoted by Bor (Fl. Iran. 70:259) as found in Palestine. Bor remarked, however, as follows: "*M. schischkinii* is being maintained for the present but it may disappear into synonymy on revision". He was aware of the fact that "the distinctions drawn between certain species are too fine to be maintained in a key" (*loc. cit.*).

Tribe 12. STIPEAE Dumort. (1823). Annuals or perennials. Leaf-blades mostly rolled or folded; sheaths split nearly to base, without auricles; ligules membranous, often ciliate at margin. Inflorescence an effuse or contracted panicle, sometimes raceme-like. Spikelets all alike, hermaphrodite, 1-flowered; rachilla short, disarticulating above glumes. Glumes 2, persistent, as long as spikelet, more delicate in texture than lemma and palea, 1–3-veined; lemma coriaceous, becoming indurated at maturity, with convolute or involute margins, 3–5-veined, entire or 2-dentate at apex, with a terminal awn and a bearded, curved, often sharp callus; awn often twisted in lower part, often geniculate, plumose or glabrous; palea 2-veined. Lodicules mostly 3, rarely 2, glabrous. Stamens 3. Stigmas plumose. Caryopsis ellipsoid to narrowly cylindrical, often with a short beak; embryo about $\frac{1}{5}$ length of caryopsis; hilum linear, nearly as long as caryopsis. Chromosomes small; x=7, 9, 11, 12, 13, 17; aneuploidy frequent.

Ten to twelve genera, mainly in extratropical regions of both hemispheres.

59. STIPA L.

Perennials, rarely annuals. Leaf-blades rolled, sometimes flat; ligule membranous. Inflorescence an effuse or contracted panicle. Spikelets somewhat dorsally compressed, 1-flowered; rachilla disarticulating above glumes, not produced above floret. Glumes persistent, subequal or unequal, hyaline or membranous, lanceolate, acuminate, longer than lemma (excl. awn); lemma very firm, with a bearded, mostly pungent callus, convolute, awned at tip or between 2 lobes; awn articulated to the lemma, consisting of a twisted *column* and a generally long, glabrous, hairy or plumose *bristle*; palea hyaline. Lodicules 3, rarely 2. Stamens 3, very rarely 2; anthers often with a tuft of hairs at apex. Caryopsis fusiform, tightly enclosed by lemma and palea; hilum linear, nearly as long as caryopsis.

About 300 species widely distributed in temperate and warm-temperate regions of the world, many of them xerophytic.

Literature: R. Roshevits, *Stipa*, in: *Flora URSS* 2:79–112, 1934. B. de Winter, The South African Stipeae and Aristideae (Gramineae), *Bothalia* 8:199–404 (1965). N. L. Bor, *Stipa*, in: *Flora of Iraq* 9:395–410, 1968.

1. Bristle of awn plumose	2
– Bristle of awn glabrous, scabrous or hairy	3
2. Column hairy.	**3. S. barbata**
– Column scabrous or glabrous.	**4. S. hohenackeriana**

3(1). Upper glume 7–8(–10) mm; the lower 1½–2 times as long as the upper.
1. S. parviflora

– Glumes subequal or somewhat unequal in length, 10 mm or longer   4

4. Awn 1.5–2 cm. Plant of maquis, garigue and batha.   **6. S. bromoides**

– Awn 4 cm or longer. Plants of steppes and batha   5

5. Annual. Lemma 4–5 mm. Awn 5–10 cm (very rarely longer).   **2. S. capensis**

– Perennial. Lemma 15–20 mm. Awn 18–25 cm.   **5. S. lagascae**

1. Stipa parviflora Desf., Fl. Atl. 1:98, t.29 (1798); Bor, Fl. Iraq 9:408, t.153; Boiss., Fl. 5:499; Post, Fl. 2:724. [Plate 342]

Perennial, 30–60(–80) cm, growing in small tufts. Culms erect, glabrous. Upper leaf-sheath often enclosing panicle at base. Panicle 10–20 cm (excl. awns), lax, nodding; branches capillary. Glumes very unequal, gaping, membranous-hyaline, 3-veined; the lower 10–16 mm, acuminate-aristate, the upper 7–8(–10) mm, acute; lemma covered with silky hairs; callus hairy except at its tip; awn 6–8(–12) cm, more or less geniculate above lower ¼–⅓, twisted below knee, then arcuate. Fl. April–June.

Hab.: Steppes. Judean Desert, N., C. and S. Negev; Gilead, Ammon, Moav, Edom.

Area: Irano-Turanian, extending to a degree into the Mediterranean (Spain, Crete, Balkans).

2. Stipa capensis Thunb., Prodr. Fl. Cap. 19 (1794); Bor, Fl. Iraq 9:402, t.151. *S. tortilis* Desf., Fl. Atl. 1:99, t.31 (1798); Boiss., Fl. 5:500; Post, Fl. 2:725. [Plate 343]

Annual, 15–30(–40) cm, tufted. Upper leaf-sheath inflated, enclosing young panicle; ligule a membranous ciliolate rim; auricles bearded. Panicle 3–10 cm, contracted, narrow; branches short, unequal; pedicels shorter than spikelets. Spikelets silvery. Glumes 12–20 mm, subequal, hyaline, narrowly linear-lanceolate tapering to a fine cusp; lemma 4–5 mm, hairy; callus appressed-hairy except at its sharp naked tip; awn 5–10 cm; column geniculate twice, twisted, hairy; bristle scabridulous; awns of a panicle often twisted together. Fl. March–May.

Hab.: Steppes and batha. Very common and abundant, often forming extensive stands. Philistean Plain, W. Negev; Upper and Lower Galilee, Mt. Carmel, Mt. Gilboa, Samaria, Shefela, Judean Mts., Judean Desert, N., C. and S. Negev; Upper Jordan Valley, Beit Shean Valley, Lower Jordan Valley, Dead Sea area, Arava Valley; Golan, Gilead, Ammon, Moav, Edom.

Area: Irano-Turanian and Saharo-Arabian, extending into the Mediterranean.

"... grazing of pastures abounding in *S. capensis* can be injurious to livestock during the period when the grass is fruiting, though it is innocuous before heading. The spear-like grain is armed with a very sharp callus and a long slender awn, spirally coiled like a corkscrew, which twists or untwists as the humidity of the air varies. When these awns are caught up in the vegetation, the twisting action helps to drive the grass seed into the ground, and they are prevented from being easily dragged out of the ground by retrorse hairs of the callus. As mentioned by Guest (1933) the same action tends to drive the pointed seeds into the eyes, mouth and other tender parts of sheep—causing inflammation and sores, and damaging the fleece" (Bor, Fl. Iraq 9:404).

3. **Stipa barbata** Desf., Fl. Atl. 1:97, t.27 (1798); Bor, Fl. Iraq 9:399, t.150 & Fl. Iran. 70:391; Boiss., Fl. 5:503; Post, Fl. 2:725. *S. arabica* Trin. & Rupr., Sp. Gram. Stip. 77 (1842). *S. ehrenbergiana* Trin. & Rupr., Sp. Gram. Stip. 75 (1842). *S. szovitsiana* Trin. in Hohenacker, Bull. Soc. Nat. Moscou 11:943 (1838) *nom. nud. S. szovitsiana* (Trin.) Griseb. in Ledeb., Fl. Ross. 4:450 (1852) *descr.*; Roshevits, in Fl. URSS 2:91; Bor, Grass. Burma Ceylon India Pakist. 647. *S. damascena* Boiss., Diagn. ser. 1, 13:45 (1854). [Plate 344]

Perennial, 30–70 cm, tufted. Culms covered with a few leaf-sheaths up to panicle. Leaves mainly basal, rolled, filiform, scabrous on outside and with dense short silky hairs on inside; ligule short, lacerate. Panicle elongate, narrow, enveloped at base by uppermost leaf-sheath; branches short, upright. Glumes 25–35 mm, subequal, hyaline, narrowly lanceolate, attenuate to a very slender awn-like cusp longer than glume proper; lemma (8–)10–11 mm, indurate, with 2 longitudinal rows of hairs mainly in its lower ½ and partly hairy between rows and with a crown of hairs below articulation with awn; callus appressed-hairy with a naked tip; awn 10–15 cm, 1–2-geniculate above lower ⅓ or ¼; column beneath knee twisted and covered with 0.6–1.2 mm appressed or spreading hairs; bristle plumose with 1.5–2.5 mm erecto-patent hairs. Fl. March–May.

Hab.: Steppes. S. Judean Mts., Judean Desert, N. and C. Negev; E. Ammon, Moav, Edom.

Area: Irano-Turanian (Morocco, Algeria, Tunisia, Libya, Spain, Sicily, Sinai, Palestine, Syria, Turkey, Transcaucasia, Iraq, Iran, W. Pakistan, Afghanistan, N.W. India, C. Asia).

S. barbata is treated here in a wide sense, as a plant of steppes of the Irano-Turanian region. Its area extends from Morocco in the West to C. Asia in the East. *S. arabica* Trin. & Rupr. described from the Sinai Peninsula, *S. ehrenbergiana* Trin. & Rupr. from Syria and *S. damascena* Boiss. from Damascus are regarded as synonymous with *S. barbata*. Their differential characters are within the limits of the intraspecific variation of *S. barbata*. This conception of *S. barbata* is in conformity with the treatment by Bor (Flora of Iraq 9, 1968 & Flora iranica 70, 1970) and by Täckholm & Drar (Flora of Egypt 1, 1950).

Bor (Fl. Iraq 9:400) stresses that grazing during the fruiting period may be injurious to livestock.

4. **Stipa hohenackeriana** Trin. & Rupr., Sp. Gram. Stip. 80 (1842); Roshevits, in Fl. URSS 2:92; Bor, Fl. Iraq 9:405. *S. barbata* var. *pabotii* Mout., Nouv. Fl. Liban Syrie 1:54 (1966). *S. assyriaca* Hand.-Mazz., Ann. Naturh. Mus. (Wien) 28:14, f.5 (1914). [Plate 345]

Perennial, 30–50(–80) cm, tufted. Culms with pubescent internodes, covered with a few leaf-sheaths up to panicle. Leaf-blades glabrous on outside, densely short-hairy on inside. Panicle contracted, enclosed at base by uppermost leaf-sheath. Glumes subequal, 3–4 cm, attenuate to a long slender and fragile awn-like cusp; lemma 9–12 mm, diffusely silky-hairy in lower ½, with a crown of hairs below articulation of awn; awn 10–18 cm, geniculate; column glabrous or scabrous; bristle plumose with hairs 2–3 mm long; callus appressed-hairy except at naked tip. Fl. March–May.

Hab.: Steppes. ?Judean Desert, C. Negev; E. Ammon, Edom.

Area: Irano-Turanian.

5. Stipa lagascae Roem. & Schult., Syst. Veg. ed. 15, 2:333 (1817); Bor, Fl. Iraq 9:398 & 406; Boiss., Fl. 5:500. *S. holosericea* Trin. & Rupr., Sp. Gram. Stip. 71 (1842). *S. fontanesii* auct. non Parl., Fl. Ital. 1:167 (1850); Post, Fl. 2:725. [Plate 346]

Perennial, 50–100 cm, tufted. Culms covered with a few long leaf-sheaths up to panicle. Leaves mainly basal, filiform; ligule membranous, (2–)5 mm. Panicle elongate, contracted, enveloped at base by uppermost leaf-sheath; branches short, upright. Spikelets few. Glumes 2.5–4(–6) cm, much varying in length, subequal, hyaline, lanceolate, attenuate to a slender awn-like cusp; cusp longer than glume proper, fragile; lemma 15–20 mm, indurate, with silky hairs in 2 rows mainly on lower ½ and partly between rows, and with a crown of hairs below articulation of awn; callus obconical, silky-hairy, tip naked; awn 18–25 cm, obscurely 1–2-geniculate, especially at maturity; column of awn shortly and densely hairy; bristle scabridulous or minutely hairy. Anthers glabrous at apex. Fl. April–June.

Hab.: Sands or stony ground. Philistean Plain; Judean Mts., Judean Desert, W., N. and C. Negev; E. Gilead, Ammon, Moav, Edom.

Area: Mediterranean and W. Irano-Turanian.

6. Stipa bromoides (L.) Dörfler, Herb. Norm. Cent. 34, no. 129 (1897); Bor, Fl. Iraq 9:400; Post, Fl. 2:726. *Agrostis bromoides* L., Mantissa 30 (1767). *Aristella bromoides* (L.) Bertol., Fl. Ital. 1:690 (1835); Boiss., Fl. 5:504. *S. aristella* L., Syst. Nat. ed. 12, 3 App.: 229 (1768). [Plate 347]

Perennial, 50–70(–100) cm, tufted. Culms slender, rigid, glabrous. Leaf-blades glabrous; ligule none or very short. Panicle 10–30 cm, contracted, linear, interrupted at base; branches short, appressed to rachis, in threes on nodes of rachis, scabridulous. Spikelets erect, 8–10 mm. Glumes 11–12 mm, subequal, prominently 3-veined, long-acuminate; lemma 6–10 mm, covered with appressed silky hairs; callus very short, rounded; awn 1.5–2 cm, straight, not geniculate, glabrous to scabridulous. Fl. March–June.

Hab.: Maquis, garigue and batha in hills. Sharon Plain; Upper and Lower Galilee, Mt. Tabor, Mt. Carmel, Mt. Gilboa, Samaria, Judean Mts.; Gilead.

Area: Mediterranean, extending into the Irano-Turanian.

60. PIPTATHERUM Beauv.

Perennials. Leaf-blades flat or rolled; ligule membranous. Inflorescence an effuse panicle. Spikelets dorsally compressed, 1-flowered; rachilla disarticulating above glumes, not prolonged above floret. Glumes persistent, subequal, membranous, longer than lemma (excl. awn); lemma elliptic or ovate, dorsally flattened, ending in a straight articulated deciduous awn; lemma and palea often coriaceous, at maturity tightly enclosing caryopsis. Lodicules 3, the 2 lower ovate, the upper narrower. Stamens 3.

Twenty five species in the Old World, mainly in the Irano-Turanian and Mediterranean regions.

Literature: H. Freitag, The genus *Piptatherum* (Gramineae) in Southwest Asia, *Notes Roy. Bot. Gard. Edinb.* 33:341–408 (1975).

1. Spikelets 3–4 mm (excl. awn).   **1. P. miliaceum**
- Spikelets 7–14 mm (excl. awn)   2
2. Lemma 4–5 mm at maturity, obtuse at apex, mostly glabrous; awn (3–)4–5(–6) mm.
                                                  2. P. blancheanum
- Lemma 6.5–9 mm at maturity, acute at apex, mostly hairy; awn 11–16(–18) mm.
                                        3. P. holciforme subsp. **longiglume**

1. Piptatherum miliaceum (L.) Coss., Not. Pl. Crit. 129 (1851); Freitag, *op. cit.* 362; Boiss., Fl. 5:506. [Plate 348]

Perennial, 60–120 cm. Culms remaining green after leaves have dried off. Panicle lax, 15–30(–40) cm; branches slender, in whorls of 3 to many. Spikelets ovate to lanceolate, 3–4 mm (excl. awn). Both glumes 3-veined; lemma glabrous; awn 3–5 mm; mature lemma obovate. Fl. February–July.

1. Panicle loosely verticillate; lower whorls with 3–5 branches, all spikelet-bearing.
                                                  var. **miliaceum**
- Panicle densely verticillate; lower whorls with 15–30 branches, mostly sterile.
                                                  var. **thomasii**

Var. **miliaceum.** *Agrostis miliacea* L., Sp. Pl. 61 (1753). *Oryzopsis miliacea* (L.) Aschers. & Schweinf., Mém. Inst. Egypt 2:169 (1887); Bor, Fl. Iraq 9:414, t.156; Post, Fl. 2:726. [Plate 348]

Hab.: Maquis, garigue, batha, hedges. Common. Coast of Carmel, Sharon Plain, Philistean Plain; Upper and Lower Galilee, Mt. Carmel, Esdraelon Plain, Mt. Gilboa, Samaria, Judean Mts., Judean Desert, N. and C. Negev; Hula Plain, Upper and Lower Jordan Valley, Dead Sea area; Golan, Gilead, Ammon, Moav, Edom.

Var. **thomasii** (Duby) Boiss., Fl. 5:507 (1884); Freitag, *op. cit.* 363 as subsp. *Milium thomasii* Duby, Bot. Gall. 1:505 (1828). *Piptatherum thomasii* (Duby) Kunth, Enum. Pl. 1:177 (1833).

Hab.: As above. Less common. Sharon Plain, Philistean Plain; Upper Galilee, Mt. Carmel, Esdraelon Plain, Judean Mts.; Dan Valley, Upper Jordan Valley.

Area of species: Mediterranean, extending into the Saharo-Arabian and W. Irano-Turanian.

2. Piptatherum blancheanum Desv. ex Boiss., Diagn. ser. 2, 4:127 (1859); Freitag, *op. cit.* 365. *P. holciforme* (Bieb.) Roem. & Schult. var. *blancheanum* (Desv. ex Boiss.) Boiss., Fl. 5:508 (1884). *Oryzopsis holciformis* (Bieb.) Hackel var. *blancheana* (Desv. ex Boiss.) Bornm., Beih. Bot. Centralbl. Abt. II, 31:267 (1914); Post, Fl. 2:727. [Plate 349]

Perennial, 50–100 cm, loosely tufted with a few vegetative shoots. Leaf-sheaths smooth or scabridulous. Panicle lax, 10–25 cm; branches ascending to spreading, the longest about ½ length of panicle. Spikelets lanceolate, 7–9 mm (excl. awn). Glumes purple at back, yellowish at margins, the lower 7–9-veined, the upper 5–7-veined; mature lemma narrowly elliptic, (4–)4.5–5 mm, obtuse at apex, glabrous except 2 tufts of hairs near base; awn (3–)4–5(–6) mm, straight, caducous. Fl. April–June.

Hab.: Batha and garigue. Sharon Plain, Philistean Plain; Upper and Lower Galilee, Mt. Carmel, Mt. Gilboa, Samaria, Judean Mts.; E. Gilead.

Area: E. Mediterranean.

Eig (*op. cit.* 62, 1927) erroneously reported this plant as *Oryzopsis coerulescens* from several localities in the hill districts of Palestine; these localities were later cited in the second edition of Post's Flora (p. 726). In fact, *P. coerulescens* (Desf.) Beauv. does not occur in Palestine.

3. Piptatherum holciforme (Bieb.) Roem. & Schult., Syst. Veg. ed. 15, 2:328 (1817) subsp. **longiglume** (Hausskn.) Freitag, Notes Roy. Bot. Gard. Edinb. 33:372 (1975). *P. holciforme* var. *longiglume* Hausskn., Mitt. Thür. Bot. Ver. Nov. ser. 13, 14:42 (1899); Boiss., Fl. 5:508 as *P. holciforme*. Bor, Fl. Iraq 9:412, t. 155; Fl. Iran. 70:408 & Post, Fl. 2:726 as *Oryzopsis holciformis* (Bieb.) Hackel, Denkschr. Akad. Wiss. Math.-Nat. Kl. (Wien) 50(2):8 (1885). [Plate 350]

Perennial, 50–80 cm, in tufts, with several vegetative shoots. Panicle lax, 15–30 cm; branches erect or ascending to spreading, lower branches 2–4 in a whorl, the longer ones at least ½ length of panicle. Spikelets ovate to lanceolate, 10–14 mm (excl. awn). Glumes green at back, hyaline to yellowish at margins and tip; mature lemma ovate to broadly lanceolate, 6.5–9 mm, covered with ascending appressed short hairs, or rarely glabrous; awn 11–16(–18) mm, slightly bent, caducous. Fl. April–June.

Var. **longiglume;** Freitag, *op. cit.* 372. *Oryzopsis holciformis* var. *longiglumis* (Hausskn.) Halácsy, Consp. Fl. Graec. 3:354 (1904); Bor, Fl. Iran. 70:409. Lemma hairy.

Hab.: Batha and garigue, rocky places. Sharon Plain; Upper and Lower Galilee, Mt. Carmel, Mt. Gilboa, Samaria, Judean Mts.; Golan, Gilead, Ammon, Moav, Edom.

Var. **epilosum** Freitag, Notes Roy. Bot. Gard. Edinb. 33:372 (1975). Lemma completely glabrous.

Hab.: Batha. Upper Galilee, Judean Mts. (Jerusalem, in tree-shaded yard).
Area of subsp.: E. Mediterranean and W. Irano-Turanian.

Tribe 13. ARUNDINEAE Dumort. (1823). Perennials, reed-like. Leaf-blades broad, flat, auriculate or rounded at base; sheaths split nearly to base; ligule a short, densely ciliolate membranous rim. Inflorescence a large plumose, more or less effuse or contracted panicle. Spikelets all alike, laterally compressed, hermaphrodite, 1-10-flowered; 1-3 upper florets usually reduced; rachilla disarticulating above glumes and between florets. Glumes 2, usually shorter than spikelet, hyaline or membranous, acuminate, persistent; lemma 3-veined (sometimes with 2-4 additional veins), glabrous or hairy, often subulately acuminate, awnless or with a short awn (up to 2.5 mm); palea 2-veined, 2-keeled. Lodicules 2 or 3, obtuse, free. Stamens 3. Caryopsis rarely developed, ellipsoid, free; embryo about ½ length of caryopsis; hilum linear, oblong or punctiform. Chromosomes small; x=12.

Two or three genera, nearly cosmopolitan.

61. ARUNDO L.

Tall stout reed-like perennials, rhizomatous. Culms hollow. Leaf-blades flat, broad; ligule a rim of very short ciliolate hairs. Inflorescence a large plumose panicle, much branched; lower branches in fascicles; rachis scabridulous. Spikelets laterally compressed, few-flowered; florets hermaphrodite, the uppermost rudimentary; rachilla scabridulous, disarticulating above glumes and between florets. Glumes subequal, nearly as long as spikelet, membranous, lanceolate, acute, 3-5-veined, persistent; lemma with long dense silky hairs on lower ½ of back, 3-5-veined; callus short, obtuse, nearly glabrous; palea about ½ length of lemma, densely ciliolate along keels. Lodicules 2. Stamens (2-)3. Caryopsis free; hilum oblong.

Twelve species of trop. and temperate regions.

Literature: C.E. Hubbard, *Arundo,* in: A. W. Hill (ed.), *Flora of Tropical Africa* 10(1):158-159, 1937.

1. Hairs of lemma 7-8 mm, arising from base; lemma 2-fid at apex. Spikelets 3-4-flowered. Plant 3-6 m, with upper leaves nodding during flowering.   **1. A. donax**
- Hairs of lemma not reaching 5 mm, arising from above base; apex of lemma entire. Spikelets 1-2-flowered. Plant 1.5-2.5 m, with upper leaves erect during flowering.

2. A. plinii

1. Arundo donax L., Sp. Pl. 81 (1753); Bor, Fl. Iraq 9:372, t.142; Boiss., Fl. 5:564; Post, Fl. 2:745. [Plate 351]

Perennial, 3-6 m. Rhizome creeping, with tuber-like swellings. Culms erect, woody, 2 cm or more in thickness at base, smooth and glabrous. Leaf-blades broad (up to 5-6 cm), up to 1 m long, smooth or scabridulous at margins, with broad folded auricles; upper leaves nodding during flowering; sheaths tight, glabrous, or hairy at mouth. Panicle 40–60 × 5–8 cm, contracted and dense; branches numerous, mostly erect, spreading during flowering. Spikelets 10–13(–15) mm, 3–4-flowered; pedicels as long as or shorter than spikelet. Glumes oblong-lanceolate, tapering to a fine point, glabrous, purplish or brownish; lemma slightly exceeding glumes, with long hairs from its base, 2-fid at apex and with a short awn from sinus. Fl. July–December.

Hab.: Near water on banks of rivers. Sharon Plain, Philistean Plain; Upper Galilee, Mt. Carmel, Samaria, Judean Desert, N. Negev; Dan Valley, Hula Plain, Upper and Lower Jordan Valley, Dead Sea area; Arava Valley; Golan, ?Gilead, ?Ammon, Moav.

Area: Mediterranean, Irano-Turanian and Euro-Siberian.

2. **Arundo plinii** Turra, Farsetia Nov. Gen. 11 (1765); Mout., Nouv. Fl. Liban Syrie 1:74; Post, Fl. 2:746. *Donax plinii* (Turra) C. Koch, Dendrol. 2(2):352 (1873); Aschers. & Graebn., Syn. Mitteleur. Fl. 2(1):334. *A. pliniana* Turra, Fl. Ital. Prodr. 1:63 (1780); Boiss., Fl. 5:564. *A. mauritanica* Desf., Fl. Atl. 1:106 (1798) non Poir. (1789). [Plate 352]

Perennial, 1.5–2.5 m. Differs from *A. donax* L. in smaller stature, more slender culms, upper leaves not nodding during flowering, a smaller panicle and especially smaller spikelets with 1–2 florets only; lemma hairy from above base; apex entire. Fl. June–December.

Hab.: Banks of rivers. Rare. Sharon Plain, Philistean Plain; Esdraelon Plain.

Area: Mediterranean.

62. PHRAGMITES Adans.

Tall stout reed-like perennials, with creeping rhizomes. Culms hollow. Leaf-blades flat. Inflorescence a very large plumose panicle, much branched. Spikelets conspicuously silky with long hairs on rachilla, 2–10-flowered, slightly compressed laterally; lower floret staminate or neuter, the others hermaphrodite; rachilla disarticulating above glumes and between florets. Glumes unequal, shorter than spikelet, membranous, 3-veined, persistent; lemma glabrous, longer than glumes, membranous, lanceolate, long-acuminate, 3-veined; palea ½–⅖ length of lemma. Lodicules 2. Stamens 3. Caryopsis enclosed by lemma and palea; hilum oblong.

Three cosmopolitan species.

Literature: C.E. Hubbard, *Phragmites*, in: A.W. Hill (ed.), *Flora of Tropical Africa* 10(1):152–158, 1937. W. Greuter & K.H. Rechinger, *Boissiera* 13:174 (1967). W.D. Clayton, Studies in the Gramineae: XIV, *Kew Bull.* 21:113–117 (1967); The correct name of the common reed, *Taxon* 17:168–169 (1968); *Phragmites*, in: F.N. Hepper (ed.), *Flora of West Tropical Africa*, ed. 2, 3:374, 1972.

1. **Phragmites australis** (Cav.) Trin. ex Steud., Nomencl. Bot. ed. 2, 2:324 (1841) subsp. **altissimus** (Benth.) Clayton, Taxon 17:169 (1968); Bor, Fl. Iraq 9:374, t.143 as *P. australis*. *Arundo altissima* Benth., Cat. Pl. Pyr. 62 (1826). *P. communis* Trin., Fund. Agrost. 134 (1820) subsp. *maximus* (Forssk.) Clayton, Kew Bull. 21:116 (1967) *nom. dub.* *A. maxima* Forssk., Fl. Aeg.-Arab. (1775) *nom. dub.* *P. communis* var. *isiacus* (Del.) Coss. & Durieu, Expl. Sci. Algérie, Bot. II, 125 (1855); Boiss., Fl. 5:563; Post, Fl. 2:745. *A. isiaca* Del., Fl. Aeg. Ill. 52 (1813) *nom. illeg.* [Plate 353]

Perennial, 3–6 m. Rhizome long. Culms leafy, smooth and glabrous, 1 cm thick at base or thicker, woody. Leaf-blades generally broad and long, glabrous, auriculate; sheaths glabrous, appressed to culm; ligule long-hairy. Panicle 20–40(–60) cm, dense, brownish before flowering, silky-hairy and yellowish during flowering, erect or somewhat nodding; branches and pedicels capillary; pedicels shorter than spikelet. Spikelets 10–15 mm, widely gaping during flowering, 2–7-flowered; rachilla silky with long hairs. Glumes very unequal; lower glume about ½ length of the upper; upper glume oblong, obtuse to 3-denticulate; lemma smooth and glabrous; lemma of lower (sterile) floret lanceolate, acute, about twice as long as upper glume, devoid of bearded callus; lemma of upper (fertile) florets lanceolate, acuminate, with a long hairy callus. Fl. mainly October–January.

Hab.: Banks of rivers and lakes. Common. Acco Plain, Sharon Plain, Philistean Plain; Upper and Lower Galilee, Esdraelon Plain, Samaria, Judean Mts., Judean Desert; Hula Plain, Upper and Lower Jordan Valley, Dead Sea area, Arava Valley; Golan, Ammon, Moav, Edom.

Area of subsp.: Mediterranean, Irano-Turanian, Saharo-Arabian and Trop.

Var. *stenophyllus* (Boiss.) Bor, Fl. Iran. 70:353, 1970 (based on *P. communis* var. *stenophylla* Boiss., Fl. Or. 5:563, 1884) is an ecotype of maritime sands and salines, characterized by short rigid culms and abbreviated, usually pungent and convolute leaves. Ascherson & Graebner cite a similar ecotype of dry habitats in the typical subspecies as var. *pumila* [Syn. Mitteleur. Fl. 2(1):330].

Clayton (1967) points out that subsp. *altissimus* (cited as subsp. *maximus* Clayton) differs from subsp. *australis* in taller stature and a larger panicle, as well as in the shape of its upper glume which is oblong, obtuse to 3-denticulate, and not lanceolate, acute to apiculate.

P. australis (Cav.) Trin. ex Steud. subsp. *australis* is a distinctly shorter plant with a smaller panicle, and is a Borealo-Trop. taxon with a very wide distribution area.

The synonymy of *P. australis* subsp. *australis* is: *Arundo australis* Cav., Anal. Ci. Nat. 1:100 (1799); *P. communis* Trin., Fund. Agrost. 134 (1820), based on *A. phragmites* L.

Tribe 14. DANTHONIEAE (G. Beck) C.E. Hubbard. Annuals or perennials. Leaf-blades narrow, linear or setaceous, rolled or flat; sheaths split nearly to base, devoid of auricles; ligule a fringe of long hairs. Inflorescence a panicle, contracted or more or less effuse, sometimes raceme-like. Spikelets all alike, hermaphrodite, 2–10-flowered (upper floret often rudimentary); rachilla disarticulating above glumes and between florets. Glumes 2, equal, usually equalling spikelet in length, often hyaline-margined and 3–11-veined; lemma coriaceous or membranous, (3–)5–11-veined, 2-lobed, awned in sinus, rarely devoid of awn; palea 2-keeled, hyaline. Lodicules 2, glabrous, small, cuneate, truncate. Stamens 3. Stigma plumose. Caryopsis free, ellipsoid, often with 2 small apical beaks; hilum small, elliptic. Chromosomes small; x=6, 9, 12.

About 30 genera, mainly in extratrop. countries of both hemispheres.

63. ASTHENATHERUM Nevski

Perennials, tufted, glaucous. Leaf-blades pungent; ligule reduced to a fringe of long hairs. Inflorescence a dense spike-like panicle enclosed at base in the uppermost leaf–sheath. Spikelets 2–3-flowered, laterally compressed; rachilla disarticulating above glumes and between florets. Glumes enclosing all florets, persistent, subequal, chartaceous, 7–11-veined; veins elevated; callus slender, acute; lemma coriaceous, rounded on back, many-veined, pilose, 2-fid at apex into acuminate hyaline lobes, with an awn in sinus; palea membranous, oblong, 2-keeled, smaller than lemma. Lodicules 2. Stamens 3. Caryopsis oblong-oblanceolate, concave on one side; embryo about ⅓ length of caryopsis; hilum punctiform.

Two species in the Saharo-Arabian and Irano-Turanian regions.

Literature: S.A. Nevski, *Asthenatherum, Acta Univ. As. Med.* Ser. 8b, Bot. fasc. 17:8 (1934). C.E. Hubbard, *Danthonia*, in: A.W. Hill (ed.), *Flora of Tropical Africa* 10(1):134–143, 1937.

1. Asthenatherum forsskalii (Vahl) Nevski, Acta Univ. As. Med. ser. 8b, Bot. fasc. 17:8 (1934); Bor, Fl. Iraq 9:382, t.145. *Avena forsskalii* Vahl, Symb. Bot. 2:25 (1791). *Danthonia forsskalii* (Vahl) R. Br. in Denh. & Clapp., Narr. Trav. North & Centr. Afr., App. 244 (1826); Boiss., Fl. 5:551; Post, Fl. 2:740. [Plate 354]

Perennial, 40–80(–100) cm, rhizomatous. Rhizome indurated; roots very thick, sand-covered. Culms velvety. Leaves crowded at base; leaf-blades acuminate, pungent, usually shortly and densely pubescent on both faces. Panicle 5–20 cm, narrow, partly enclosed in dilated uppermost leaf-sheath; branches short. Spikelets 7–9 mm, 2–3-flowered, pale yellow or sometimes tinged with purple. Glumes oblong-lanceolate, more or less acuminate, upper glume narrower than the lower; lemma short-hairy along veins, with each vein ending in a tuft of long white hairs, equalling lobes of lemma in length; awn brown, somewhat shorter than lemma proper; callus elongated, pungent, short-hairy. Fl. March–May.

Hab.: Dunes. Acco Plain, Sharon Plain, Philistean Plain, W. Negev; Arava Valley.

Area: Saharo-Arabian and Irano-Turanian.

64. SCHISMUS Beauv.

Annuals. Leaf-blades setaceous, rolled; ligule reduced to a fringe of long hairs. Inflorescence a usually dense raceme-like panicle; pedicels short, articulated to branches. Spikelets laterally compressed, 5–10-flowered; florets distant on rachilla, hermaphrodite or the uppermost reduced; rachilla disarticulating above glumes and between florets. Glumes persistent, subequal, oblong-lanceolate, with elevated veins and scarious margins, much longer than the lowest lemma, often about as long as spikelet; lower glume 5–7-veined, broader than 3 veined upper glume; lemma obovate to elliptical, rounded on back, 7–9-veined, 2-lobed, hyaline on margins and tip, awnless, sometimes with a mucro in fissure between lobes; callus minute; palea nearly as long as lemma or shorter, entire, 2-keeled below. Lodicules 2. Stamens 3. Caryopsis obovate, obtuse, orange-coloured, transparent, glossy, loosely enveloped by lemma and palea; embryo ⅓–½ length of caryopsis; hilum elliptic.

Five species in the Mediterranean and Saharo-Sindian regions and S. Africa.

1. Lobes of lemma of lowest floret much longer than broad; tip of palea reaching base of fissure between lobes of lemma; palea considerably shorter than lemma.

1. S. arabicus

– Lobes of lemma of lowest floret nearly as long as broad; tip of palea exceeding base of fissure, or palea nearly as long as lemma.   **2. S. barbatus**

Bor (1968) remarks: "To determine whether a specimen belongs to one or the other of these species a spikelet must be dissected. The lowest floret in the spikelet must be taken and the shape of the apical lobes, the length of the fissure and the relative length of the palea to the lemma determined. It will be observed that in the succeeding florets the lobes tend to become shorter and less acute or less acuminate. The fissure, too, may also disappear. Hence if the key characters are applied to the uppermost florets the observer may easily fall into error. When measuring the length of the glumes the terminal spikelet of the panicle should be chosen. As a general rule glumes over 5 mm long are found in *S. arabicus* and glumes under 5 mm long in *S. barbatus*; the remaining key characters are usually associated. It must be pointed out, however, that certain authorities (Täckholm, Maire, etc.) do not maintain both these species as distinct and include *S. arabicus* in *S. barbatus* as a subordinate taxon. It is proposed in this work to retain them as distinct, mainly on the authority of Haines [in Agnew, Bull. Coll. Sci. Baghdad 6 (Suppl.):87 (1962)] who has had very considerable experience of this genus in the field. It is extremely likely that hybridization occurs since both these species grow together, and this may be the real reason for much of the difficulty" (Bor, Fl. Iraq 9:377).

1. Schismus arabicus Nees, Fl. Afr. Austr. 422 (1841); Bor, Fl. Iraq 9:378; Boiss., Fl. 5:597, incl. var. *minutus* Boiss., *op. cit.* 598 (1884); Post, Fl. 2:757. [Plate 355]

Annual, 7–20 cm. Nodes of culms brown. Mouth of leaf-sheaths long-bearded. Panicle 1–5 cm, often slightly tinged with purple; pedicels short. Spikelets 5–7 mm. Glumes glabrous, acuminate; lemma hairy in lower ½; lobes of lemma of lowest floret much longer than broad; fissure between lobes

⅓–½ length of lemma; palea considerably shorter than lemma, its tip reaching base of fissure. Fl. March–April.

Hab.: Arid sites, steppes; extending further northwards within the country than *S. barbatus*. Sharon Plain, Philistean Plain; Esdraelon Plain, Judean Mts., Judean Desert, W., N. and C. Negev; Upper and Lower Jordan Valley, Dead Sea area, Arava Valley; Gilead, Ammon, Moav, Edom.

Area: Irano-Turanian and Saharo-Arabian, extending into the Mediterranean.

2. Schismus barbatus (L.) Thell., Bull. Herb. Boiss. ser. 2, 7:391 (1907) in obs.; Bor, Fl. Iraq 9:378, t.144. *Festuca barbata* L., Demonstr. Pl. 3 (1753). *S. calycinus* (L.) C. Koch, Linnaea 21:397 (1848); Boiss., Fl. 5:597. *Festuca calycina* L., Amoen. Acad. 3:400 (1756) & Sp. Pl. ed. 2, 110 (1762). [Plate 356]

Annual, 7–20 cm. Nodes of culms brown. Mouth of leaf-sheath long-bearded. Panicle 1–4 cm, often tinged with purple; pedicels short. Spikelets 5–7 mm. Glumes glabrous, acuminate; lemma hairy in lower ½ to ⅔; lobes of lemma of lowest floret as long as to slightly longer than broad; fissure between lobes short, cutting lemma to only ⅙–¼ of its length; sometimes mucro present in fissure; tip of palea exceeding base of fissure or palea as long as lemma. Fl. March–April.

Hab.: Arid places and steppes. Philistean Plain; Judean Desert, W., N., C. and S. Negev; Upper and Lower Jordan Valley; Arava Valley; Edom.

Area: Irano-Turanian and Saharo-Arabian, extending into the Mediterranean.

Tribe 15. ARISTIDEAE C.E. Hubbard ex Bor (1960). Perennials, rarely annuals. Leaf-blades mostly rolled, pungent, rarely flat; sheaths split nearly to base, without auricles or with very short auricles; ligule a fringe of hairs. Inflorescence a contracted or effuse panicle. Spikelets all alike, hermaphrodite, 1-flowered; rachilla disarticulating above glumes. Glumes 2, membranous, usually unequal, acute, mucronate or shortly awned; lemma convolute at maturity, indurated, cylindric, 3-veined, awned; awn 3-partite from base, or at tip of a straight or twisted column; branches of awn all glabrous or all or only central one plumose; palea hyaline, at most ½ length of lemma, mostly 2-veined, 2-keeled. Lodicules 2–3, elongate. Stamens 3 or 1. Ovary glabrous; stigmas plumose. Caryopsis very narrowly ellipsoid or cylindrical; embryo ⅓–⅔ length of caryopsis; hilum linear, nearly as long as caryopsis. Chromosomes small; x=11.

Three genera, mainly in trop. and subtrop. regions of both hemispheres.

65. ARISTIDA L.

Perennials, rarely annuals. Leaf-blade narrow, often rolled; ligule a dense fringe of hairs. Inflorescence an effuse or contracted panicle. Spikelets pedicellate, awned, 1-flowered, hermaphrodite; rachilla disarticulating above glumes, not prolonged beyond floret. Glumes persistent, equal or unequal, lanceolate, acuminate, rarely the lower longer than the upper; lemma chartaceous to coriaceous, lanceolate, 3-veined, narrowed above into an awn and with a bearded, pungent, obtuse or rarely forked callus; awn persistent or deciduous, divided into 3 branches; awn-branches glabrous or scabrous, not plumose; median awn-branch longer than lateral ones; palea 2-veined or veinless. Lodicules 2 or absent. Stamens 3 or 1. Caryopsis tightly enclosed by lemma; hilum linear. Chromosomes small; x=11. The majority of species diploid.

About 260 species in the Tropics and Subtropics of both hemispheres.

Literature: J.T. Henrard, A critical revision of the genus *Aristida, Mededeel. Rijks Herb. Leiden,* 54 & 54A,B,C (1926-1928, 1933); A monograph of the genus *Aristida, ibid.* 58 & 58A (1929, 1932, 1933). B. de Winter, The South African Stipeae and Aristideae (Gramineae), *Bothalia* 8:199-404 (1965).

1. Awn 4-9 cm, articulated to lemma, deciduous.   **3. A. sieberiana**
- Awn shorter, not articulated to lemma, persistent   2
2. Perennial. In majority of spikelets upper glume slightly shorter than lemma.
                                                                          1. A. coerulescens
- Annual. In majority of spikelets upper glume distinctly shorter than lemma.
                                                                          2. A. adscensionis

1. **Aristida coerulescens** Desf., Fl. Atl. 1:109, t.21 f.2 (1798); Bor, Fl. Iraq 9:384, t.146; Boiss., Fl. 5:491; Post, Fl. 2:721. *A. adscensionis* L. var. *coerulescens* (Desf.) Hackel in Stuck., Anal. Nac. Buenos Aires 11:90 (1904); Nábělek, Publ. Fac. Sci. Univ. Masaryk 111:6 (1929). [Plate 357]

Perennial, 40-60 cm, tufted. Leaf-blades folded or rolled. Panicle 15-20 × 2 cm, contracted, often purplish. Glumes unequal, lanceolate-subulate; the lower 7-8 mm, the upper 8-10 mm; lemma 8-9 mm; callus short, bearded towards tip; awn not articulated to lemma, persistent, devoid of column; awn-branches 15-20 mm, scabrous; central awn-branch at least twice as long as lemma. Fl. January-June.

Hab.: Arid hillslopes and steppes. Acco Plain, Sharon Plain, Philistean Plain; Upper and Lower Galilee, Mt. Carmel, Esdraelon Plain, Judean Mts., Judean Desert; Upper Jordan Valley, Beit Shean Valley, Dead Sea area, Arava Valley; Golan, Gilead.

Area: Mediterranean, Irano-Turanian, Saharo-Arabian, Trop.

2. **Aristida adscensionis** L., Sp. Pl. 82 (1753); Henrard, Monogr. 322, t.158; Bor, Fl. Iraq 9:384; Clayton, in Fl. W. Trop. Afr. 3, 2:378-379 (1972). *A. pumila* Decne., Ann. Sci. Nat. ser. 2:85 (1835); Boiss., Fl. 5:491. *A. adscensionis* L. var. *pumila* (Decne.) Coss. & Durieu, Expl. Sci. Algérie, Bot. II, 83 (1855); Post, Fl. 2:722. [Plate 358]

Annual, 8-30(-40) cm. Leaf-blades often rolled. Panicle 6-15 cm, contracted, narrow, one-sided, interrupted. Spikelets often purplish. Glumes unequal,

lanceolate-subulate; the lower 4.5–6 mm, the upper 6–10 mm, ending in a short mucro; lemma 10–15 mm, mostly longer than glumes, glabrous or scabrous in upper part, very rarely tuberculate (var. *ehrenbergii* Henrard); callus short, bearded towards tip; awn not articulated to lemma, persistent, devoid of column; awn-branches scabrous; central awn-branch 17–22 mm. Fl. January–May.

Hab.: Mainly steppe and desert slopes. E. Samaria, N. and S. Negev; Upper and Lower Jordan Valley, Dead Sea area, Arava Valley; Ammon, Moav, Edom.

Area: Saharo-Arabian.

Henrard (1932) divided the species into several varieties. In general our plants fit var. *pumila* characterized by a relatively long lemma and awns. Two specimens which display tuberculate lemmas (En Gedi 1936; Es-Salt 1929 Eig & Zohary), can be regarded as belonging to var. *ehrenbergii* Henrard. Clayton (*op. cit.* 379, 1972), considering the species as a whole, remarks: "Very variable in shape of inflorescence and length of glumes and lemmas; a number of varieties have been proposed (see Henrard, *op. cit.* 322), but they intergrade one with another and their significance is doubtful".

3. Aristida sieberiana Trin. in Spreng., Neue Entdeck. 2:61 (1820–1821); Clayton, Kew Bull. 23:210 (1969); Boiss., Fl. 5:492; Post, Fl. 2:722. [Plate 359]

Perennial, 30–100 cm, loosely tufted. Culms erect, stiff, slender, dark-coloured or purplish, glabrous, branched from lower and upper nodes. Leaf-blades rolled, 10–20 cm; basal sheaths broad, papery, straw-coloured, hyaline at margins. Panicle 10–15 cm, erect, contracted, purplish; branches short, upright. Glumes very unequal; the lower 8–15 mm, setaceous-acuminate, ending in a (1–)2–5 mm awn; upper glume 15–20 mm, 2-fid at apex and with a (1–)3–9 mm awn, arising between 2 terminal lobes; lemma 10–30 mm (incl. callus); callus bearded towards top; awn 4–9 cm, articulated to lemma, deciduous and consisting of a twisted column and of 3 awn-branches; awn-branches nearly equal, (4–)6–9 cm, scabridulous; column 1–3 cm. Fl. October–May.

Hab.: Sands and sandy loam. Sharon Plain, Philistean Plain.

Area: Sudanian, extending into the E. Mediterranean.

Type of species described from Palestine.

66. STIPAGROSTIS Nees

Perennials, densely tufted, or annuals. Leaf-blades rolled or folded, rarely flat; ligule a dense fringe of short hairs; mouth of leaf-sheaths bearded in some species. Inflorescence an effuse or contracted panicle. Spikelets pedicellate, awned, 1-flowered, hermaphrodite; rachilla disarticulating above glumes, not prolonged beyond floret. Glumes persistent, equal or unequal, lanceolate, mostly 3-veined; lemma coriaceous, lanceolate, 3-veined, narrowed above into a deciduous awn which consists of a short *column* divided into 3 branches; all awn-branches or the median one only plumose; median awn-branch longer than lateral ones; palea short; callus well developed, pungent or minutely 2-fid, usually oblique, glabrous or bearded. Lodicules usually 2, or absent. Stamens 3. Ovary glabrous. Caryopsis terete, tightly enclosed by lemma; hilum linear, almost as long as caryopsis; embryo $\frac{1}{3}$–$\frac{1}{2}$ its length. x=11. The majority of species tetraploid.

About 50 species in desert and semi-desert areas of Africa and in the Irano-Turanian and Saharo-Arabian regions.

Literature: J.T. Henrard, A monograph of the genus *Aristida*, *Mededeel. Rijks Herb. Leiden* 58 & 58A (1929, 1932, 1933). B. de Winter, Notes on the genus *Aristida* L. (Gramineae), *Kirkia* 3:132–137 (1963); The South African Stipeae and Aristideae (Gramineae), *Bothalia* 8:199–404 (1965). H. Scholz, *Stipagrostis scoparia* (Trin. & Rupr.) de Winter auch in Libyen gefunden, *Willdenowia* 6:163 (1970); Der *Stipagrostis-plumosa*-Komplex (Gramineae) in Nord-Afrika, *Willdenowia* 6:519–552 (1972); Verbreitungskarte von *Stipagrostis scoparia* (Gramineae), *Willdenowia* 8:507–509 (1979).

1. Lateral awn-branches naked; median branch plumose   2
- All three awn-branches plumose, rarely one lateral branch naked or hairy   6
2. Nodes of culms (incl. upper ones) with a conspicuous ring of dense long hairs.
                                                                              1. S. ciliata
- Nodes of culms without a ring of long hairs   3
3. Plumose awn-branch obtuse; tip not naked. Densely tufted plants with congested short basal leaves.
                                                                              2. S. obtusa
- Plumose awn-branch acute; tip naked. Plants not as above   4
4. Median awn-branch naked in its lower $\frac{1}{3}$ or $\frac{1}{2}$. Lower leaf-sheaths covered with dense fugacious wool.
                                                                              3. S. plumosa
- Median awn-branch plumose or hairy above branching point   5
5. Glumes hirtellous, usually purple at base; upper glume 10–12 mm, longer than lower.
                                                                              6. S. hirtigluma
- Glumes glabrous, often pale-purple-tinged all over; lower glume 15–18 mm, longer than upper.
                                                                              4. S. raddiana
6(1). Awn-branches distinctly unequal in length; median awn-branch 3 cm or longer; one of lateral branches sometimes naked or nearly so. Lower internodes densely woolly.
                                                                              5. S. lanata
- Awn-branches equal or subequal in length, not exceeding 2 cm. Lower culm-internodes not covered with wool.
                                                                              7. S. scoparia

1. Stipagrostis ciliata (Desf.) de Winter, Kirkia 3:133 (1963); Bor, Fl. Iraq 9:386, t.147. *Aristida ciliata* Desf. in Schrad., Neues Jour. Bot. 3:255 (1809); Boiss., Fl. 5:494; Post, Fl. 2:722. [Plate 360]

Densely tufted perennial, 30–60 cm. Culms erect or geniculately ascending, with a conspicuous ring of long hairs at nodes; internodes glabrous. Leaves crowded at base; basal leaf-sheaths glabrous, pale yellow, long-bearded at mouth; leaf-blades rolled, filiform, recurved. Panicle 10–15 cm, upright. Glumes subequal, 9–12.5 mm, pale, sometimes purplish at base; lemma tapering into the column, articulated at about middle, its upper $\frac{1}{2}$ deciduous to-

gether with awn; median awn-branch plumose from its lower ⅓ or ¼ upwards and naked at tip; lateral branches naked, less than ½ length of median one; callus bearded. Fl. March–June.

Hab.: Loess soil and gravel in steppes. Philistean Plain; W., N. and C. Negev; Arava Valley; S. Golan, Edom.

Area: Saharo-Arabian region and S. Africa.

2. Stipagrostis obtusa (Del.) Nees, Linnaea 7:293 (1832); Bor, Fl. Iraq 9:388. *Aristida obtusa* Del., Fl. Eg. 13 (1813–1814); t.13 f.3 (1826); Boiss., Fl. 5:494; Post, Fl. 2:722. [Plate 361]

Densely tufted perennial, 10–30 cm. Culms erect or geniculately ascending, 1-noded, usually bent at node; node purple. Leaf-sheaths short, persistent, crowded, long-bearded at mouth; leaf-blades short, rolled, filiform, usually recurved. Panicle up to 15 cm. Glumes subequal, 10–12 mm, acute; lemma about 3 mm (incl. callus), its tip articulated to column; median awn-branch plumose in upper ⅔, obtuse, tip not naked;; lateral awn-branches naked, less than ½ length of median one; callus of lemma long-bearded. Fl. March–May.

Hab.: Deserts, hammada on loess or sandy soil overlaying compact soil. W., N., C. and S. Negev; Arava Valley; Edom.

Area: Saharo-Arabian and Sudanian, also Trop. and S. Africa.

3. Stipagrostis plumosa (L.) Munro ex T. Anders., Jour. Linn. Soc. London (Bot.) 5, Suppl. 1:40 (1860); Bor, Fl. Iraq 9:389, t.148. *Aristida plumosa* L., Sp. Pl. ed. 2, 1666 (1763); Boiss., Fl. 5:495; Post, Fl. 2:723. [Plate 362]

Tufted perennial, 10–30(–40) cm. Culms erect or shortly geniculate; lower internodes and lower leaf-sheaths covered with fugacious wool; leaf-blades rigid, rolled, filiform, often recurved. Panicle 10–15 cm. Glumes unequal; the lower 15 mm, the upper 18 mm; lemma 5 mm (incl. callus), articulated to column; lateral awn-branches naked, median branch 2.5–6 cm, plumose in upper ⅔, acute; tip often naked; callus 2 mm, bearded. Fl. April–May.

Hab.: Rocky deserts, loess and sand derived from Nubian sandstone. W., N., C. and S. Negev; Arava Valley; Edom.

Area: Saharo-Arabian and Irano-Turanian, extending into the Sudanian.

Our plants belong to var. *plumosa* characterized by woolly basal leaf-sheaths.

4. Stipagrostis raddiana (Savi) de Winter, Kirkia 3:135 (1963); Bor, Fl. Iraq 9:392, t.149. *Aristida raddiana* Savi, Mem. Modena 21:198 (1837); Täckholm & Drar, Fl. Eg. 1:364 (1941); Boiss., Fl. 5:497 & Post, Fl. 2:723 as *A. caloptila*. [Plate 363]

Perennial, 20–40 cm. Culms erect, glabrous; nodes glabrous; the lowermost internodes sometimes tomentose. Leaf-blades rolled, setaceous; sheaths long-bearded at mouth. Panicle up to 20 cm; spikelets erect, often tinged with purple. Glumes glabrous; lower glume 15–18 mm, the upper slightly shorter; lemma 4 mm, articulated to column at tip; median awn-branch 5–7 cm, acute

278 GRAMINEAE

in outline, plumose from branching point of awn; tip naked; lateral awn-branches 1.5–1.7 cm, naked; callus of lemma bearded. Fl. January–May.

Hab.: Rocks in deserts, granite and Nubian sandstone. W., N., C. and S. Negev; Dead Sea area; Arava Valley; Moav, Edom.

Area: Saharo-Arabian.

Täckholm in Täckholm & Drar (*op. cit.* 365) remarks: "This species should not be confused with *A. caloptila* (Jaub. & Spach) Schweinf. ex Boiss. (1884) which has glabrous (in *A. raddiana* pilose) culms. It only occurs in Southern Arabia and Somaliland".

5. Stipagrostis lanata (Forssk.) de Winter, Kirkia 3:135 (1963); Bor, Fl. Iraq 9:394. *Aristida lanata* Forssk., Fl. Aeg.-Arab. 25 (1775); Täckholm & Drar, Fl. Eg. 1:360. *A. forsskahlei* Tausch, Flora (Regensb.) 19:506 (1836); Boiss., Fl. 5:496; Post, Fl. 2:723. [Plate 364]

Perennial, 20–50(–60) cm, with a woody rhizome and thick roots. Culms erect or decumbent; the lower internodes densely woolly; the upper pilose or glabrous. Leaf-sheaths glabrous; leaf-blades narrow, rolled, up to 10 cm. Panicle 10–15 cm, contracted, V-shaped in outline, more or less enclosed in upper leaf-sheath. Glumes unequal, the lower 11–12 mm, the upper 13–14 mm, acuminate. Column of awn 4–5 mm; all 3 awn-branches more or less plumose, rarely one lateral branch naked or hairy; median awn-branch 3–3.5 cm, longer than lateral ones, plumose except near its base; tip appressed-hairy or naked; callus long, bearded in its upper part. Fl. April–August.

Hab.: Maritime and desert sand dunes. Sharon Plain, Philistean Plain, W. Negev; Transjordan (Al-Eisawi).

Area: Saharo-Arabian (Egypt, S. Iraq, Sinai, Palestine, S. Lebanon).

6. Stipagrostis hirtigluma (Steud. ex Trin. & Rupr.) de Winter, Kirkia 3:134 (1963). *Aristida hirtigluma* Steud., Nomencl. Bot. ed. 2, 1:131 (1840) *nom. nud.*; Steud. ex Trin. & Rupr., Sp. Gram. Stip. 171 (1842); Boiss., Fl. 5:496; Post, Fl. 2:723. [Plate 365]

Perennial or annual, 20–40 cm. Culms glabrous. Leaf-blades rolled, filiform; sheaths with a tuft of long hairs at mouth. Panicle 15–20 cm, upright, contracted; spikelets numerous. Glumes hirtellous; the upper long-acuminate, 10–12 mm, longer and narrower than obtuse lower glume; lemma 4 mm, papillose-muricate, articulated to awn above middle; median awn-branch 4–7 mm, plumose from base or middle; tip naked; lateral awn-branches naked, ¼–⅓ length of median one; column 7–8 mm long, hairy below insertion of lateral awn-branches, twisted; callus with 2 collars of hairs, the upper longer. Fl. March–April.

Hab.: Sands and rocks in deserts. W. and S. Negev; Dead Sea area; Arava Valley.

Area: Saharo-Arabian, Sudanian (incl. Punjab), Trop. and S. Africa.

7. Stipagrostis scoparia (Trin. & Rupr.) de Winter, Kirkia 3:136 (1963). *Aristida scoparia* Trin. & Rupr., Sp. Gram. Stip. 176 (1842); Henrard, Monogr. Gen. Aristida 1:55-56 (1929); Post, Fl. 2:724. *A. pungens* Desf. var. *scoparia* (Trin. & Rupr.) Boiss., Fl. 5:498 (1884). [Plate 366]

Perennial, 60-100 cm, tufted, nearly suffrutescent, glabrous, with creeping rhizome. Culms indurate, branched also above base. Leaves numerous; leaf-blades 15-30 cm, rolled, rigid. Panicle 10-25 cm, very lax, effuse; branches erecto-patent, repeatedly branched; pedicels capillary. Glumes unequal, the lower 18-20 mm, the upper 15-17 mm; lemma articulated at apex to awn; the 3 awn-branches more or less equal in length, about 1.5 cm, densely plumose from base, naked at tip, acute; column 1 mm; callus nearly glabrous. Fl. April-June.

Hab.: Sands, mainly in deserts. Philistean Plain; W., N. and C. Negev; Arava Valley.

Area: Saharo-Arabian.

Tribe 16. AELUROPODEAE Nevski ex Bor (1965). Perennials. Leaf-blades often coriaceous, folded or rolled, acuminate, pungent; sheaths split nearly to base, without auricles; ligule a very narrow membranous rim with a fringe of short hairs. Inflorescence compound, of spikes gathered into dense oblong or globose terminal heads or spaced along a vertical rachis. Spikelets all alike, hermaphrodite, laterally compressed, 2- to many-flowered; rachilla irregularly disarticulating above glumes and between florets, prolonged beyond uppermost floret. Glumes 2, persistent, shorter than spikelet, coriaceous or membranous; lower glume smaller than the upper; lemma longer than glumes, coriaceous, keeled, awnless, apiculate or mucronate; palea hyaline, 2-keeled. Lodicules 2, free, glabrous or ciliate. Stamens 3. Ovary glabrous, with very long styles; stigmas plumose. Caryopsis free, with 2 horns (remnants of styles); embryo ½ length of caryopsis; hilum punctiform, dark-coloured. Chromosomes small; $x=10$.

67. AELUROPUS Trin.

Perennials, rhizomatous or stoloniferous; stolons slender, branched, producing sterile leafy shoots and numerous erect or ascending flowering culms. Inflorescence a terminal panicle, elongate and spike-like, or dense and head-like, with spikelets in spikes borne in 2 rows on short branchlets of main rachis; spikes spaced or crowded. Spikelets subsessile, imbricate, laterally compressed, 4-14(-20)-flowered; florets hermaphrodite, the uppermost reduced; rachilla disarticulating irregularly at maturity. Glumes unequal or subequal, shorter than spikelet, ovate-oblong, muticous or mucronate; the lower 1-3-veined, the upper 5-7-veined; lemma broadly lanceolate or ovate, keeled, 7-11-veined, apiculate or mucronate, glabrous or hairy; palea hyaline. Lodicules 2. Stamens 3. Caryopsis broadly elliptic in outline, dorsally compressed. $x=10$.

Five species, Mediterranean, Saharo-Arabian and Irano-Turanian.

Literature: N.L. Bor, The strange case of the genus *Aeluropus* Trin., *Webbia* 24: 401–409 (1969).

1. Spikelets hairy. Panicle globose, ellipsoid or oblong, about 1 cm (rarely up to 2 cm), of densely crowded fascicles of spikelets.   **2. A. lagopoides**
- Spikelets glabrous. Panicle elongate, spike-like, 3–10 cm, of several spike-like fascicles of spikelets; fascicles spaced or crowded in the upper part of panicle.
                                                                                    1. A. littoralis

Plants intermediate between *A. littoralis* and *A. lagopoides* are apparently the result of hybridization between the two species, which often grow together in the same habitats in the Jordan Valley. Cf. also Bor, Fl. Iran. 70:420 (1970).

1. Aeluropus littoralis (Gouan) Parl., Fl. Ital. 1:461 (1850); Bor, Fl. Iraq 9:422, t.159; Boiss., Fl. 5:594; Post, Fl. 2:756. *Poa littoralis* Gouan, Fl. Monsp. 470 (1764). [Plate 367]
Perennial, with flowering culms up to 30 cm. Basal leaf-sheaths glabrous; leaf-blades of flowering culms up to 5 cm, usually glabrous, lanceolate-acuminate, subulate at apex, stiff, more or less spreading. Panicle 3–10 cm, spike-like, consisting of several spikelet-bearing branchlets, the upper often crowded, the lower usually spaced. Spikelets 3–4(–5) mm, glabrous, 5–9(–11)-flowered, ovate-oblong. Glumes subequal, glabrous; lemma 2–3 mm, oblong-lanceolate, abruptly apiculate, membranous-margined, glabrous or very sparsely pilose. Fl. April–July.
Hab.: Saline marshes, inundated in winter. Acco Plain, Coast of Carmel, Sharon Plain, Philistean Plain; Esdraelon Plain; Upper Jordan Valley, Beit Shean Valley, Lower Jordan Valley, Dead Sea area, Arava Valley; E. Ammon.
Area: Mainly Mediterranean and Irano-Turanian.

"This species is one of characteristic plants of saline soils of the Jordan Valley and of the coastal plain. It is resistant to high salinity, and one of its adaptations is exudation of salt onto the leaf surface. In spite of its high salt content the plant is grazed by animals" (N. Feinbrun, M. Zohary & R. Koppel, Flora of the Land of Israel, Iconography, Plates 101–151, 1958).

2. Aeluropus lagopoides (L.) Trin. ex Thwaites, Enum. Pl. Zeyl. 374 (1864) excl. syn.; Bor, Fl. Iraq 9:423, t.160; Bor, Fl. Iran. 70:419; Bor, Webbia 24:401–409 (1969). *Dactylis lagopoides* L., Mantissa 33 (1767). *A. littoralis* var. *repens* (Desf.) Coss. & Durieu, Expl. Sci. Algérie, Bot. II, 155 (1855); Boiss., Fl. 5:594. *A. repens* (Desf.) Parl., Fl. Ital. 1:462 (1850); Post, Fl. 2:756. [Plate 368]
Perennial, with flowering culms up to 15 cm. Basal leaf-sheaths hairy; leaf-blades of flowering shoots short (up to 3 cm), hairy or glabrous, stiff, pointed, more or less spreading. Inflorescence a dense head, globose, ellipsoid or oblong, 0.7–1.5 cm (sometimes up to 2–2.5 cm). Spikelets 4–6 mm, densely aggregated on short branchlets, 4–7(–11–20)-flowered, villose. Glumes slightly unequal, villose; lemma broadly ovate, apiculate, rounded at back, villose, membranous-margined. Fl. summer.

Hab.: Salines that are inundated in winter; sometimes together with *A. littoralis*; often in deserts. ?Acco Plain; Upper Jordan Valley, Beit Shean Valley, Lower Jordan Valley, Dead Sea area; Transjordan (Al-Eisawi).

Area: Saharo-Arabian and Irano-Turanian, extending into the Mediterranean and Sudanian.

Tribe 17. PAPPOPHOREAE Kunth (1829). Annuals or perennials. Leaf-blades often rolled; sheaths split nearly to base, without auricles; ligule a row of hairs. Inflorescence a dense spike-like panicle. Spikelets all alike, 2- to many-flowered; the lower florets hermaphrodite, the others staminate or neuter and often much reduced; rachilla disarticulating above glumes, continuous between florets. Glumes 2, more or less unequal, 3–7-veined, as long as spikelet (excl. awns); the lower 3–7-veined, the upper 5–9-veined; lemma rounded on back, 9-veined, with veins produced into awns that are longer than lemma proper; palea 2-veined, 2-keeled. Lodicules 2, minute, cuneate. Stamens 3. Ovary glabrous; stigmas plumose. Caryopsis broadly elliptic in outline, flattened at back, free; embryo $\frac{3}{4}$–$\frac{5}{6}$ length of caryopsis; hilum punctiform. Chromosomes small; x=10.

Four genera, mainly in trop. and subtrop. regions of both hemispheres.

68. ENNEAPOGON Desv. ex Beauv.

Annuals or perennials. Leaves flat or rolled. Inflorescence a spike-like, somewhat lobed panicle of numerous shortly pedicellate spikelets borne on pubescent branches. Spikelets 2–6-flowered, the lower 1(–3) florets hermaphrodite, the others sterile; rachilla breaking up above glumes, but not between florets; florets falling at maturity as a unit. Glumes unequal, persistent; lemma of fertile florets much smaller than glumes, broadly elliptic, truncate when flattened, 9-veined; veins more or less elevated, produced above into 9 mostly feathery awns of which 5 are longer than the alternate 4; awns straight, longer than lemma proper; palea ciliate on keels. Lodicules 2. Stamens 3. Caryopsis elliptic in outline; hilum punctiform.

About 30 species in warm parts of Australia, Africa and Asia; 1 species in N. and C. America.

Literature: S.A. Nevski, *Enneapogon*, in: *Schedae Herb. Fl. Asiae Med.*, fasc. 21 no. 524, Tashkent (1934). S.A. Renvoize, The Afro-Asian species of *Enneapogon* Beauv., *Kew Bull.* 22:393–401 (1968). H. Freitag, Über den Fund von *Enneapogon persicus* in Spanien, *Collect. Bot. (Barcelona)* 7:483–493 (1968).

1. Panicle 2–3(–4) cm. Spikelet 5–6 mm. Awns 2–2½ times as long as lemma proper.
                                                          2. E. brachystachyus

– Panicle 5–8(–12) cm. Spikelet 9–11 mm. Awns 3–5 times as long as lemma proper.
                                                          1. E. persicus

1. Ennneapogon persicus Boiss., Diagn. ser. 1, 5:71 (1844); Bor, Fl. Iraq 9:418, t.158. *Pappophorum persicum* (Boiss.) Steud., Syn. Pl. Glum. 1:200 (1854); Boiss., Fl. 5:558; Post, Fl. 2:742. [Plate 369]

Perennial, 15–40(–50) cm, finely glandular-pubescent, densely tufted, with many leafy vegetative shoots. Culms erect. Leaves short, spreading, rolled. Panicle 5–8(–12) cm, spike-like; branches and pedicels rather short. Spikelets 9–11 mm (incl. awns), 2–3-flowered, tinged with purple. Glumes lanceolate, acute to acuminate, velvety, many-veined, somewhat shorter than spikelet; veins elevated; lower glume 6 mm, broader than upper, the upper 7–8 mm; lemma velvety; awns plumose up to ⅔, scabrous above, 3–5 times as long as lemma proper. Fl. March–April.

Hab.: Rocks (limestone, conglomerate, granite). Judean Desert, S. Negev; Dead Sea area; Transjordan (Al-Eisawi).

Area: Irano-Turanian.

2. Enneapogon brachystachyus (Jaub. & Spach) Stapf in Dyer, Fl. Cap. 7:654 (1900); Bor, Fl. Iran. 70:415. *Pappophorum brachystachyum* Jaub. & Spach, Ann. Sci. Nat. ser. 3, 14:365 (1850) & Ill. Pl. Or. 4:34, t.324 (1851); Täckholm & Drar, Fl. Eg. 1:220; Boiss., Fl. 5:558. *P. foxii* Post, Fl. Syria Palest. Sinai ed. 1, 876 (1883–1896); Post, Fl. ed. 2, 2:742. [Plate 370]

Perennial, 5–15(–20) cm, finely glandular-pubescent, rarely subglabrous. Culms geniculately ascending, with 2–4 nodes; nodes densely villose. Leaf-blades very narrow, usually rolled; mouth of sheaths shortly bearded. Panicle 2–3(–4) × 0.6–1.0 cm, spike-like, dense, light to dark grey. Spikelets 4–5 mm, 2(–3)-flowered; only lower floret fertile. Glumes exceeded by awns of florets, oblong, obtuse or emarginate, hyaline, sparsely pubescent, (3–)5(–7)-veined, lateral veins evanescent above; lower glume 3.5 mm, broader than upper; upper glume 4.5 mm; lemma white, hirsute in lower part; awns about 3 mm, feathery in lower ⅓–½, scabridulous above, 2–2½ times as long as lemma proper. Fl. January–April.

Hab.: Sandy soil, rock fissures. S. Negev; Dead Sea area, Arava Valley.

Area: Saharo-Arabian, Irano-Turanian and Sudanian.

Tribe 18. CYNODONTEAE Dumort. (1823). Perennials or annuals. Leaf-blades flat or rolled; sheaths split nearly to base, without auricles; ligules membranous, but usually reduced almost from base to a row of dense hairs, more rarely ligules longer and short-ciliate on margin. Inflorescence an effuse or compact panicle, sometimes spike-like, or consisting of variously arranged spikes or racemes. Spikelets 1- to many-flowered; lower floret always hermaphrodite; rachilla disarticulating under each floret, or under lower floret only, or not disarticulating. Glumes 2, usually shorter than spikelet; lemma (1–)3(–5)-veined, awned or awnless; palea 2(–1)-veined, 2-keeled. Lodicules 2, glabrous, more or less cuneate, truncate, rarely absent. Stamens 3, rarely 2. Caryopsis

ellipsoid to nearly globose, free; embryo $\frac{1}{3}$-$\frac{3}{4}$ length of caryopsis; hilum punctiform. Chromosomes small; x=9, 10.

About 100 genera, mainly in trop. and subtrop. regions of both hemispheres.

69. ERAGROSTIS N.M. Wolf

Annuals or perennials. Leaf-blades flat, with or without glands along margins, more rarely rolled; ligule generally reduced to a fringe of hairs. Inflorescence a panicle, effuse or contracted, rarely spike-like. Spikelets several- to many-flowered, laterally compressed, muticous, oblong, ovate or ovate-lanceolate, long- or short-pedicellate, disarticulating either from below upwards, with lemma and caryopsis falling off while palea remains on persistent rachilla for some time, or rachilla disarticulating from above downwards, with lemma and palea falling off together; florets usually hermaphrodite. Glumes unequal or nearly equal, narrow, acute, 1(-3)-veined, shorter than lowest floret; lemma concave, 3-veined, keeled or rounded on back, membranous to coriaceous; palea hyaline. Lodicules 2. Stamens 3 or 2. Caryopsis globose, ellipsoid or obovoid; hilum punctiform. x=10.

About 350 species, mostly in warm countries.

Literature: O. Stapf, *Eragrostis*, in: J.D. Hooker, *Flora of British India* 7:312-327, 1896; *Eragrostis*, in: Dyer (ed.), *Flora capensis* 7:594-631, 1900. A.S. Hitchcock, *Manual of the Grasses of the United States* ed. 2, 140-169, 1950. W.D. Clayton, *Eragrostis*, in: F.N. Hepper (ed.), *Flora of West Tropical Africa* ed. 2, 3(2):383-393, 1972. Le Roy Harvey, *Eragrostis*, in: F.W. Gould, *The Grasses of Texas*: 177-201, 1975.

1. Mouth of leaf-sheaths not bearded. Spikelets 2-2.5 mm.   **1. E. japonica**
- Mouth of leaf-sheaths bearded. Spikelets longer than above   2
2. Spikelets 4-5 × 3.4 mm, less than twice as long as broad, ovate to nearly elliptic in
· outline.   **2. E. echinochloidea**
- Spikelets at least twice as long as broad, oblong to lanceolate in outline   3
3. Panicle with a whorl of long hairs (about 5 mm) at insertion of panicle-branches, especially of the lower ones (not to be confused with the beard at mouth of leaf-sheaths!).   **6. E. pilosa**
- Panicle without a whorl of hairs at insertion of panicle-branches   4
4. Panicle contracted, with branches erect and close to panicle-axis. Spikelets 1-1.5 mm broad. Perennials   5
- Panicle with branches spreading, or if branches close to panicle-axis, spikelets 2-3.5 mm broad. Mostly annuals   6
5. Panicle-branches solitary, undivided for some distance from base. Pedicels very short.   **3. E. sarmentosa**
- Panicle-branches in fascicles, repeatedly branched. Pedicels 2-3 mm or longer.   **4. E. prolifera**
6(4). Culms bearing, apart from terminal panicles, small axillary panicles arising from lower leaf-sheaths.   **7. E. barrelieri**
- All panicles terminal   7

7. Glumes subequal in length. Leaf-blades with tubercle-like glands along margins, rarely without glands. Annual weeds rarely exceeding 50 cm in height, common   8
- Lower glume much shorter than the upper. Leaf-blades without glands along margins. Tall plant (about 50 cm or taller), recently introduced, fairly rare.
                                                                          5. E. palmeri

8. Spikelets 2–3.5 mm broad; lemma 2–2.8 mm.   **9. E. cilianensis**
- Spikelets 1–2 mm broad; lemma 1.8–2 mm.   **8. E. minor**

1. Eragrostis japonica (Thunb.) Trin., Mém. Acad. Sci. Pétersb., ser. 6, 1:405 (1831); Cope, Poaceae, in Nasir & Ali, Flora of Pakistan 143:88 (1982). *Poa japonica* Thunb., Fl. Japon. 51 (1784). *E. diarrhena* (Schult.) Steud., Syn. Pl. Glum. 1:266 (1854); Bor, Fl. Iraq 9:444, t.169. *E. namaquensis* Nees var. *diplachnoides* (Steud.) Clayton, Kew Bull. 25:251 (1971) & in: F.N. Hepper (ed.), Fl. W. Trop. Afr. ed. 2, 3(2):387 (1972). *E. diplachnoides* Steud., Syn. Pl. Glum. 1:268 (1854); Bor, Fl. Iraq 9:444, t.170. Boiss., Fl. 5:583 as *E. nutans* Retz. [Plate 371]

Annual, up to 50 cm. Leaf-blades flat; mouth of leaf-sheaths not bearded. Panicle 15–40 × 1–1.5 cm, contracted, oblong, repeatedly branched; branches nearly erect, in semi-whorls on nodes of rachis; longer branches up to 12 cm; pedicels about as long as spikelets. Spikelets erect, numerous, 2–2.5 × 1–1.25 mm, 5–9-flowered, pale, sometimes marked with purple, breaking up from above downwards. Glumes somewhat unequal; lemma keeled, about 1 mm, obtuse, with conspicuous lateral veins. Caryopsis light brown, seen at maturity through translucent lemma. Fl. September–November.

Hab.: Banks of lakes and rivers. Rare. Philistean Plain (Yarkon River); Upper Jordan Valley (Jordan River).

Area of species: Trop. and subtrop. regions of the Old World (Egypt, Iraq, India, S.E. Asia).

Known since 1929 from the banks of Wadi Musrara (tributary to the Yarkon River) near Tel Aviv. In recent years collected several times on the banks of the Jordan River and of Lake Kinneret.

Probably introduced by migratory birds.

2. Eragrostis echinochloidea Stapf,* in Dyer (ed.), Fl. Cap. 7:627 (1900). [Plate 372]

Annual or biennial, 50–80 cm, glaucous. Leaves flat or rolled, rigid, long-hairy on upper surface, especially when young; sheaths bearded at mouth. Panicle contracted to effuse; pedicels much shorter than spikelets. Spikelets 4–5 × 3–4 mm, ovate to nearly elliptic in outline, subacute, strongly compressed, whitish, 7–13-flowered, breaking up from above downwards. Glumes somewhat unequal, acute to acuminate, boat-shaped, strongly keeled; lemma broadly ovate, acute, boat-shaped, strongly keeled; side-veins prominent. Fl. November.

_____
* Kindly identified by T.A. Cope (Kew).

Hab.: Weed in irrigated gardens. Arava Valley.
Area: S. Africa.

This species was collected for the first time in 1980 in this country. It represents another case of an adventitious plant of irrigated habitats in a desert area along the Rift Valley, and was presumably introduced by migratory birds.

3. Eragrostis sarmentosa (Thunb.) Trin.,* Gram. Gen. 398 & Mém. Acad. Sci. Pétersb., ser. 6, 1:398 (1831); Stapf, in Dyer (ed.), Fl. Cap. 7:618. *Poa sarmentosa* Thunb., Prodr. Pl. Cap. 21 (1794). *E. kneuckeri* Hackel & Bornm. in Bornm., Feddes Repert. 10:472 (1912); Täckholm & Drar, Fl. Eg. 1:191; Post, Fl. 2:752. *E. hackeliana* Bornm. & Kneucker, Feddes Repert. 10:381 *descr.* (1912) non *E. hackelii* Hassler, Feddes Repert. 8:47 (1911). [Plate 373]

Perennial, 15–40 cm, tufted, stoloniferous. Leaf-blades flat, acuminate; mouth of leaf-sheaths bearded. Panicle 4–10(–15) cm × 7–12 mm, contracted, spike-like; branches solitary, up to 6 cm, erect, appressed to rachis, undivided for some distance. Spikelets short-pedicellate, 4–7 × 1.25–1.5 mm, 7–17-flowered, narrowly lanceolate, acute, greyish-brown to purplish. Glumes unequal; lemma keeled, acute. Fl. April–December.

Hab.: Marshes and banks of rivers, sandy soil. Sharon Plain, Philistean Plain, W. Negev; Arava Valley.

Area: Trop. Asia, S. Africa.

Casual; apparently introduced by migratory birds.

4. Eragrostis prolifera (Swartz) Steud.,** Syn. Pl. Glum. 1:177 (1854). *Poa prolifera* Swartz, Nov. Gen. Sp. Pl. 27 (1788). *E. fascicularis* Trin., Mém. Acad. Sci. Pétersb., ser. 6, 1:403 (1831); Clayton, in Fl. W. Trop. Afr. ed. 2, 3(2):391 in obs. (1972). [Plate 374]

Perennial, 80–100 cm, tufted, glaucous. Culms rigid, erect, branched above base, with erect branches. Leaves fasciculate; leaf-blades rolled; mouth of leaf-sheaths bearded. Panicle 12–20 × 1.5–2.5 cm, narrowly oblong, contracted; branches in fascicles, repeatedly branched, erect, close to rachis; pedicels shorter than spikelet. Spikelets 5–9 × 1–1.25 mm, 7–12-flowered, oblong, pale, breaking up from below upwards; rachilla persistent; florets loosely imbricate. Glumes unequal, membranous; lemma 1 mm, keeled, obtuse, with elevated lateral veins. Caryopsis ellipsoid, light brown, seen through translucent lemma. Fl. July–August.

Hab.: Sandy loam. Casual. Acco Plain, Sharon Plain.

Area: Trop. Africa and West Indies.

---

* Kindly identified by A. Melderis (British Museum) and W.D. Clayton (Kew).
** Kindly identified by Le Roy Harvey (Smithsonian Institution, Washington D.C.).

5. Eragrostis palmeri S. Wats., Proc. Amer. Acad. Sci. 18:182 (1883); Hitch-cock, Manual Grasses U.S. ed. 2, 160, f.219 (1950); Gould, Grasses Texas 184 (1975). [Plate 375]

Perennial, 50–100 cm, tufted. Mouth of leaf-sheaths bearded. Panicle 15–30 × 10–20 cm, very lax, oblong, glabrous at insertion of branches; branches spreading, repeatedly branched; branches and pedicels capillary; pedicels mostly shorter than spikelets. Spikelets 5–8 × 1.25–1.5 mm, 7–9-flowered, oblong, grey, breaking up from below upwards. Glumes unequal; lemma 2 mm, lightly keeled. Fl. June–August.

Hab.: Somewhat saline soil. Rare. Sharon Plain.
Area: Trop. C. America, Mexico, S. USA.

Recently introduced (collected in Israel in 1970 for the first time).

6. Eragrostis pilosa (L.) Beauv., Agrost. 71, 162 (1812); Bor, Fl. Iraq 9:447, t.171; Boiss., Fl. 5:581; Post, Fl. 2:751. *Poa pilosa* L., Sp. Pl. 68 (1753). [Plate 376]

Annual, 20–40(–60) cm. Leaf-blades flat, narrow, eglandular on margins; mouth of leaf-sheaths bearded. Panicle 15–20 × 2–10 cm, very lax, ovate in outline, narrowly oblong when young, long-pilose at insertion of branches, especially of lower ones; branches erecto-patent, repeatedly branched; lower branches often in whorls of 3–5 or more; branches and pedicels capillary; pedicels as long as or longer than spikelets. Spikelets 3–5 × 0.75–1.25 mm, 5–10-flowered, oblong, breaking up from below upwards. Glumes unequal; lemma 1.25–1.5 mm, subacute. Fl. September–October.

Hab.: Irrigated ground. Sharon Plain, Philistean Plain; Judean Mts. (Jeru-salem), W., N. and C. Negev; Upper Jordan Valley, Dead Sea area; Golan; Transjordan (Al-Eisawi).
Area: Borealo-Trop.

7. Eragrostis barrelieri Daveau, Bull. Herb. Boiss. 2:651–660, t.32 f.A (1894) & in Morot, Jour. Bot. (Paris) 8:289 (1894); Bor, Fl. Iraq 9:438, t.165. [Plate 377]

Annual, 15–30(–40) cm. Leaf-blades flat, without tubercle-like glands on margins; mouth of leaf-sheaths bearded. Apart from terminal panicles, there are small axillary panicles arising from lower nodes and emerging from tops of sheaths. Panicle lax, oblong-lanceolate in outline; branches erecto-patent, repeatedly branched, capillary; pedicels usually shorter than spikelets, with 1–2 crateriform glands under spikelet. Spikelets 6–12(–25) × 1.5–2 mm, 10–20-flowered, oblong, breaking up from below upwards. Glumes subequal; lemma elliptic, obtuse. Caryopsis oblong. Fl. March–October.

Hab.: Irrigated ground. Acco Plain; N. Negev; Dead Sea area; Gilead.
Area: Mediterranean and Saharo-Arabian.

Resembling *E. minor* but differing mainly in the axillary panicles in lower and basal nodes, and in the lack of tubercle-like glands along leaf-margins. The chromo-

some number cited in the literature for *E. barrelieri* is 2n=60, while it is given as 2n=30, 40 for *E. minor* and 2n=20 for *E. cilianensis*.

8. Eragrostis minor Host, Gram. Austr. 4:15 (1809); Daveau, Bull. Herb. Boiss. 2:656, t.32 f.B; Post, Fl. 2:751. *Poa eragrostis* L., Sp. Pl. 68 (1753). *E. poaeoides* Beauv., Agrost. 162 (1812) *nom. nud.*; Beauv. ex Roem. & Schult., Syst. Veg. ed. 15, 2:316 (1817) *descr.*; Hitchcock, *op. cit.* 869; Bor, Fl. Iraq 9:447; Boiss., Fl. 5:580. [Plate 378]

Annual, 20–40(–70) cm. Culms erect, with terminal panicles only. Leaf-blades flat, mostly with tubercle-like glands along margins; mouth of leaf-sheaths bearded; blade and especially sheath with scattered long hairs. Panicle up to 15 × 6 cm, lax, ovate-oblong in outline (rarely dense, 3 cm broad); branches usually spreading, repeatedly branched, capillary; pedicels usually shorter than spikelet, with 1–2 crateriform glands under spikelet. Spikelets 5–9 × 1–2.2 mm, 8–15(–20)-flowered, breaking up from below upwards; rachilla persistent. Glumes subequal; lemma keeled, without or rarely with few tubercle-like glands on keel. Fl. July–November.

Hab.: Irrigated ground. Upper Galilee, N. Negev; Dan Valley, Hula Plain, Upper Jordan Valley, Dead Sea area; Golan, Gilead, Ammon.

Area: Euro-Siberian, Mediterranean and Irano-Turanian, also Trop. Africa.

9. Eragrostis cilianensis (All.) F.T. Hubbard, Philippine Jour. Sci. (Bot.) 8: 159 (1913). *E. cilianensis* (All.) Vignolo-Lutati, Malpighia 18:386 (1904) *comb. illeg.*; Bor, Fl. Iraq 9:440, t.167. *Poa cilianensis* All., Fl. Pedem. 2:246, t.91 f.2 (1785). *E. major* Host, Gram. Austr. 4:14, t.24 (1809). *P. megastachya* Koeler, Descr. Gram. 181 (1802). *E. megastachya* (Koeler) Link, Hort. Berol. 1:187 (1827); Boiss., Fl. 5:580; Post, Fl. 2:751. [Plates 379, 380]

Annual, 15–60 cm. Culms erect, with terminal panicles only. Leaf-blades flat, with tubercle-like glands along margins; mouth of leaf-sheaths bearded; blade and sheath glabrous. Panicle up to 15 × 6 cm, oblong to ovate in outline, steel-grey to purplish, rarely dense; branches erecto-patent, capillary; pedicels shorter than spikelets. Spikelets 6–15 × 2–3.5 mm, many-flowered, breaking up from below upwards. Glumes subequal; lemma broadly ovate, often mucronulate, keeled, sometimes with tubercle-like glands on keel. Fl. July–November.

Subsp. **cilianensis.** [Plate 379]. Panicle lax, 4–6 cm broad, with fairly long branches; pedicels of lateral spikelets up to 4 mm.

Hab.: Irrigated ground. Fairly common. Sharon Plain, Philistean Plain; Upper and Lower Galilee, Esdraelon Plain, Samaria, Judean Mts., Judean Desert; Dan Valley, Hula Plain, Upper and Lower Jordan Valley, Dead Sea area; Golan; Transjordan (Al-Eisawi).

Subsp. **starosselskyi** (Grossh.) Tzvelev, Poaceae URSS, 634 (1976). *E. starosselskyi* Grossh., Not. Syst. (Leningrad) 4:18 (1923). [Plate 380]. Panicle dense, oblong, with very short branches; pedicels of lateral spikelets 0.5–2 mm.

Hab.: Rare. Dead Sea area (En Gedi).
Area of species: Borealo-Trop.

70. ELEUSINE Gaertn.

Annuals or perennials. Leaf-blades flat or folded; ligule reduced to a fringe of
hairs. Inflorescence of several spikes digitately arranged at apex of culm.
Spikelets 2- to several-flowered, laterally compressed, awnless, sessile and
densely imbricate, in 2 rows on one side of spike-rachis; rachilla disarticulating above glumes and between florets; florets usually hermaphrodite. Glumes
persistent, unequal, acute, membranous-margined, shorter than spikelet;
lower glume 1-2-veined, the upper 3-7-veined; lemma acute, 1-3-veined.
Lodicules 2. Stamens 3. Caryopsis 3-gonous, furrowed on ventral side; hilum
punctiform; pericarp membranous, not adnate to seed; seed dark brown, rugulose, adapted to zoochory. x=9.
Nine trop. and subtrop. species of both hemispheres.

1. **Eleusine indica** (L.) Gaertn., Fruct. Sem. Pl. 1:8 (1788); Täckholm & Drar,
Fl. Eg. 1:389; Bor, Fl. Iran. 70:437. *Cynosurus indicus* L., Sp. Pl. 72 (1753).
[Plate 381]
Annual, 15-90 cm, tufted. Culms erect. Leaf-sheaths compressed. Spikes
2-12, 3-8(-12) cm, in a terminal whorl, sometimes with 1 or 2 spikes below
whorl. Spikelets 5-6 mm, 3-6-flowered, elliptic. Glumes oblong-lanceolate,
acute, with elevated veins; lower glume shorter and narrower than upper;
lemma lanceolate, keeled; keel broadened above middle; palea shorter than
lemma, 2-dentate, 2-keeled and narrowly winged. Fl. June–November.
Hab.: Marshy or saline soil, irrigated fields and gardens, loess soil. Acco
Plain, Sharon Plain, Philistean Plain; Mt. Carmel, Esdraelon Plain, Judean
Mts., N. Negev; Upper Jordan Valley; Golan.
Area: Trop. and Subtrop.

Adventitious. First collected here in the thirties of the current century.
Forage grass in India. Bor (Grasses Burma Ceylon India Pakist., 493, 1960) says:
"This is one of the species which are said to possess cyanogenetic glycosides in the
leaves and is therefore dangerous to stock, especially when wilted... Owing to the
rapidity with which it occupies disturbed ground, it is especially useful, even though
it is an annual, as a soil-binder". Seeds used as food in times of scarcity in India.

71. DACTYLOCTENIUM Willd.

Annuals or perennials. Leaf-blades flat; ligule a row of hairs. Inflorescence of
several spikes digitately arranged at apex of culm. Spikelets 3-5-flowered, laterally compressed, sessile, crowded and imbricate, in 2 rows on one side of
spike-rachis; rachis ending in an acuminate tip; florets hermaphrodite;
rachilla disarticulating above lower glume. Glumes awned or mucronate,
subequal (excl. awn), 1-veined, keeled and usually scabrous along keel; the

lower persistent, the upper deciduous with lemmas; awns curved; lemma similar to upper glume, 3-veined, scabridulous along keel, ending in an awn or mucro; palea acute, hyaline. Lodicules 2. Stamens 3. Caryopsis angular, coarsely ridged or sculptured; hilum punctiform.

Ten species in warm parts of the Old World.

1. Dactyloctenium aegyptium (L.) Willd., Enum. Pl. Horti Berol. 1029 (1809; *"aegyptiacum"*); corr. K. Richter, Pl. Eur. 1:68 (1890); Bor, Fl. Iraq 9:426, t.161; Boiss., Fl. 5:556; Post, Fl. 2:741. *Cynosurus aegyptius* L., Sp. Pl. 72 (1753). [Plate 382]

Annual, 40–60 cm. Culms prostrate below, rooting at nodes. Leaves sparsely pilose, hairs with tubercle at base. Spikes 3–5, widely spreading, 2–3 cm, flattened; rachis hairy at base of spike. Glumes scabridulous at keel; the lower ovate-oblong, acute; the upper broadly ovate, obtuse or retuse, abruptly contracted at apex into a robust awn which is ½ length of to as long as glume proper. Fl. June–October.

Hab.: Irrigated ground; weed. Sharon Plain, Philistean Plain; Upper Galilee; Dead Sea area.

Area: Trop. extending into S. Mediterranean; introduced to America.

72. DESMOSTACHYA (Stapf) Stapf

Perennials. Leaves long, stiff, resembling those of *Imperata*; ligule a fringe of dense hairs. Inflorescence narrow, compound, consisting of numerous short spikes borne singly or in clusters on an elongate central rachis; spikes sessile, spreading or ascending, short, dense. Spikelets 6–18-flowered, laterally compressed, sessile or subsessile, crowded along one side of the short spike-rachis, falling entire. Glumes very unequal, lanceolate, membranous, 1-veined; lemma ovate, acute, keeled, faintly 3-veined, shiny, often suffused with purple; palea shorter than lemma. Lodicules 2. Stamens 3. Stigmas plumose. Caryopsis ovoid, 3-quetrous, with a small apical appendage; hilum punctiform.

A monotypic genus.

1. Desmostachya bipinnata (L.) Stapf in Dyer (ed.), Fl. Cap. 7:632 (1900); Täckholm & Drar, Fl. Eg. 1:177; Bor, Fl. Iraq 9:429, t.162. *Briza bipinnata* L., Syst. Nat. ed. 10, 2:875 (1759). *Eragrostis bipinnata* (L.) K. Schum. in Engler, Pflanzenwelt Ost-Afr. C, 113 (1895). *E. bipinnata* (L.) Muschler, Verhandl. Bot. Ver. Prov. Brandenb. 49:74 (1907); Post, Fl. 2:752. *E. cynosuroides* (Retz.) Beauv., Agrost. 162 (1812); Boiss., Fl. 5:583. [Plate 383]

Perennial, about 1 m or more, tufted, coarse, with widely spreading rhizomes. Culms terete, leafy. Basal leaves in a compact rosette surrounded by glossy yellow empty sheaths; leaf-blades up to 60 cm, flat, stiff, becoming rolled when dry, scabrous on margins, long-tapering towards tip and often spinescent; ligule a fringe of short hairs. Inflorescence 20–40 × 1.5–2.5 cm, cylindrical, dense; rachis finely villose; spikes short (1–2 cm), bearing spikelets from base to tip. Spikelets about 5 mm, 6–18-flowered, linear, imbricate, pale straw- or leaden-purplish-coloured. Fl. June–October.

Hab.: Sandy and sandy-clay soils along the coastal plain and on compact sandy soils in desert areas where the water table is not far below the ground level. Abundant locally. Acco Plain, Sharon Plain, Philistean Plain, W. Negev; Lower Jordan Valley, Dead Sea area, Arava Valley; Transjordan (Al-Eisawi).

Area: Sudanian, extending into the Saharo-Arabian, Mediterranean and Irano-Turanian.

When not in flower, resembles *Imperata cylindrica* from which it can be distinguished by the ligule.

73. DINEBRA Jacq.

Annuals. Leaf-blades flat. Inflorescence a narrow pyramidal raceme of subsessile spikes borne on a central rachis. Spikelets 2–3-flowered, laterally compressed, sessile, in 2 rows on one side of flattened spike-rachis; florets hermaphrodite; rachilla disarticulating above glumes. Glumes subequal, coriaceous, narrow, lanceolate, acuminate-aristate, keeled, 1-veined, much exceeding the florets; lemma membranous, ovate, acute, keeled, obsoletely veined; palea as long as lemma, retuse. Lodicules 2. Stamens 3. Caryopsis elliptic in outline, 3-gonous, with a small apical appendage; embryo about ½ length of caryopsis; hilum punctiform.

A monotypic genus.

1. Dinebra retroflexa (Vahl) Panz., Denkschr. Akad. Wiss. München 1813:270, t.12 (1814); Boiss., Fl. 5:557; Post, Fl. 2:742. *Cynosurus retroflexus* Vahl, Symb. Bot. 2:20 (1791). [Plate 384]

Annual, 15–60 cm. Raceme 8–25(–40) cm, erect; spikes 2–5 cm, numerous, alternate, opposite or whorled, erecto-patent, then reflexed. Spikelets 5–6.5 mm, steel-grey. Midvein of glumes well pronounced; lemma with midvein prolonged as a mucro from emarginate tip. Fl. July–December.

Hab.: Irrigated places, ditches. Rare. Coastal Galilee, Sharon Plain, Philistean Plain; Esdraelon Plain, Upper Jordan Valley; Golan.

Area: Trop. (Asia and Africa).

74. DIPLACHNE Beauv.

Perennials or annuals. Leaves flat or rolled; ligule membranous. Inflorescence consisting of slender spike-like racemes attached on all sides of a main rachis and bearing spaced spikelets from base to tip. Spikelets 3–10-flowered, hermaphrodite, compressed, linear, awnless, very shortly pedicellate; rachilla dis-

articulating above glumes and between florets. Glumes equal or unequal, 1-veined; lemma 1-veined, keeled, compressed, chartaceous, oblong, tapering to a 2-dentate or emarginate tip; midvein excurrent into a straight mucro; palea similar in texture, 2-fid, ciliolate along keels. Lodicules 2, truncate, dentate. Stamens 3. Caryopsis elliptic in outline, acute below, dorsally compressed, free; embryo about ½ length of caryopsis; hilum punctiform.

About 15 species in the trop. and subtrop. regions.

1. **Diplachne fusca** (L.) Roem. & Schult., Syst. Veg. ed. 15, 2:615 (1817);* Bor, Fl. Iraq 9:434, t.164 under *D. fusca* (L.) Beauv.; Boiss., Fl. 5:561. *D. fusca* (L.) Beauv. ex Stapf in Dyer, Fl. Cap. 7:591 (1900) fide Clayton, Fl. W. Trop. Afr. 3(2):398 (1972). *Festuca fusca* L., Syst. Nat. ed. 10, 2:876 (1759); Sp. Pl. ed. 2, 109 (1762). *D. reptatrix* (L.) Druce, Brit. Pl. List ed. 2, 129 (1928); Post, Fl. 2:744. [Plate 385]

Perennial, 60–100 cm, glabrous with a creeping rhizome. Culms leafy, with coriaceous leaf-sheaths at base; ligule long, lacerate. Inflorescence 10–30 cm, contracted; racemes 10 or more, up to 15 cm, ascending. Spikelets 5–15, 5–10 mm, 5–7-flowered, narrowly linear, olive-green, appressed to rachis. Glumes scabridulous at keel; lower glume linear-lanceolate, nearly ½ length of upper; lemma oblong, obtuse, rounded or emarginate at apex and shortly mucronate, with a prominent midvein, ciliate at margins in lower half. Fl. late summer.

Hab.: Sewage ditches, irrigated ground. Rare. Esdraelon Plain, N. Negev; Lower Jordan Valley; Transjordan (Al-Eisawi).

Area: Trop.

75. TETRAPOGON Desf.

Perennials or annuals. Leaf-blades folded; ligules membranous, short-hairy on back, short-ciliate on margin. Inflorescence a solitary 4-ranked spike formed by 2 spikes united at their inner faces by means of long hairs; sometimes inflorescence split upwards into two 2-ranked spikes. Spikelets 3–4-flowered, sessile, in 2 rows on outer face of flattened rachis; 2 lower florets hermaphrodite, the others neuter, reduced to awned empty lemmas; rachilla disarticulating above glume, continuous between florets. Glumes persistent, unequal, 1-veined, hyaline, keeled, acuminate; lemma 3-veined, keeled, obtuse or retuse, long-awned just below apex; palea shorter than lemma, retuse. Lodicules 2. Stamens 3. Caryopsis oblong; hilum punctiform. x=10.

About 5 species in the Saharo-Sindian region, Trop. and S. Africa.

1. **Tetrapogon villosus** Desf., Fl. Atl. 2:389, t.255 (1799); Boiss., Fl. 5:555; Post, Fl. 2:741. [Plate 386]

* Thanks are due to Dr W.D. Clayton who kindly sent me the correct reference for this species.

Perennial, 15–40 cm, tufted. Rhizome thickish, with a rosette of basal leaves. Culms sparingly leafy. Leaves flat or folded; ligule short, membranous, shortly ciliate. Spike 4–6 × 1–1.5 cm (incl. awns), erect, often purplish, sometimes split into two. Glumes lanceolate, about as long as lemma (excl. awn); the lower narrow, acute, mucronate, the upper much broader, mucronate-aristulate from a rounded apex; lemma coriaceous, obovate, obtuse, keeled, densely silky, long-hairy at back and with spreading hairs along marginal vein; apex obtuse or retuse, with a straight awn arising below apex, 2–4 times as long as lemma; lemmas of neuter florets glabrescent and shortly awned. Fl. February–May.

Hab.: Limestone, Nubian sandstone and granite rocks. Judean Desert, N., C. and S. Negev; Lower Jordan Valley, Dead Sea area, Arava Valley; Moav, Edom.

Area: Saharo-Sindian and Sudanian.

76. CHLORIS Swartz

Annuals or perennials, sometimes stoloniferous. Leaf-blades flat or rolled; ligule membranous, short-hairy at back, short-ciliate at margin. Inflorescence of spikes digitately arranged at apex of culm; spikes few or many. Spikelets 2–4-flowered, laterally compressed, sessile, crowded, in 2 rows on one side of spike-rachis; lower floret hermaphrodite, the others neuter, empty; rachilla disarticulating above glumes, continuous between florets. Glumes persistent, usually unequal, hyaline, lanceolate, keeled; lemma of hermaphrodite floret 3-veined, often ciliate on marginal veins, awned; awn straight, arising below emarginate tip; callus short, hairy; palea as long as lemma; lemmas of neuter florets smaller, glabrous. Lodicules 2. Stamens 3. Caryopsis oblanceolate in outline, 3-gonous; embryo ¾ length of caryopsis; hilum punctiform. x=10.

About 40 trop. and warm-temperate species. Several species known as pasture-grasses.

1. Awn of lemma of lower floret as long as or shorter than lemma proper; cilia on marginal veins of lemma somewhat exceeding tip of lemma. Perennial.
                                                                1. C. gayana
- Awn of lemma of lower floret 2–4 times as long as lemma proper; cilia on marginal veins much exceeding tip of lemma. Annual.   **2. C. virgata**

1. Chloris gayana Kunth, Révis. Gram. 89 (1829) *nom. nud.*; 293, t.58 (1830); Bor, Fl. Iraq 9:452, t.173. [Plate 387]

Perennial, 50–150 cm, stoloniferous. Culm-nodes brown. Leaf-blades flat; ligule short, membranous, long-ciliate. Spikes 4–20, 7–11 cm, not widely spreading. Spikelets with 2 awns, 6–7 mm (incl. awns). Glumes sparsely appressed-hairy; lower glume shorter, acute, the upper emarginate, mucronate; lemma of lower (hermaphrodite) floret oblong-lanceolate, ciliate along margins in upper half and provided with a straight awn shorter than or as long as lemma proper, arising just below emarginate tip; cilia slightly exceed-ing tip of lemma; lemma of second floret shorter than that of lower floret, with a shorter awn; upper rudiments minute, truncate. Fl. summer.

Hab.: Weed on irrigated ground and escapee from cultivation. Samaria; Hula Plain, Upper Jordan Valley.

Adventive. Orig. from Trop. Africa. Important fodder grass (Rhodes Grass).

2. Chloris virgata Swartz, Fl. Ind. Occ. 1:203 (1797); Täckholm & Drar, Fl. Eg. 1:385; Bor, Fl. Iraq 9:450, t.172. [Plate 388]

Annual, 30–50 cm, glabrous. Leaf-blades flat; upper sheaths often inflated; ligule short. Spikes 4–15, 3–8 cm, soft, silky, not widely spreading, sometimes partly enclosed in uppermost leaf-sheath, often purplish. Spikelets with 2 awns, 12–13 mm (incl. awns). Glumes glabrous, narrow; lower glume shorter, acute, the upper emarginate, aristulate; lemma of lower (hermaphrodite) floret ovate, long-ciliate along margins in its upper half, short-ciliate in lower half and provided with a straight awn 2–4 times as long as lemma; cilia much exceeding tip of lemma; callus hairy; lemma of second (neuter) floret smaller, glabrous, truncate, awned. Fl. summer.

Hab.: Irrigated land. Sharon Plain; W. Galilee, Judean Mts., W. and N. Negev.

Area: Trop. and Subtrop.

Adventive.

77. Cynodon L.C.M. Richard

Perennials. Leaves flat or rolled. Inflorescence of several spikes digitately arranged at apex of culm. Spikelets 1-flowered, laterally compressed, awnless, sessile, in 2 rows on one side of spike-rachis and more or less appressed to it; floret hermaphrodite; rachilla prolonged beyond floret and bearing a rudi-ment of floret at tip, at maturity disarticulating above glumes. Glumes per-sistent, subequal, membranous, 1-veined, keeled; lemma longer and broader than glumes, boat-shaped, compressed, keeled, obscurely 3-veined; palea nearly as long as lemma. Lodicules 2. Stamens 3. Caryopsis oblong, laterally compressed, free; embryo ⅓ length of caryopsis; hilum punctiform. x=9.

About 10 species in Trop. and Subtrop.

1. Cynodon dactylon (L.) Pers., Syn. Pl. 1:85 (1805); Bor, Fl. Iraq 9:454, t.174; Boiss., Fl. 5:553; Post, Fl. 2:740. *Panicum dactylon* L., Sp. Pl. 58 (1753). [Plate 389]

Perennial, up to 30 cm, with stolons and much-branched scaly rhizomes. Culms glabrous. Leaves short, acuminate, rigid, spreading, glabrous or pilose; mouth of leaf-sheaths pilose; ligule a rim of short hairs. Spikes 3–6(-7), 2–6 × 0.12–0.2 cm, straight, often purplish; rachis triquetrous. Spikelets about 2 mm. Glumes lanceolate, acute to subulate-mucronate;

lemma semi-ovate in profile, subobtuse or minutely apiculate; keel puberulent-ciliolate. Fl. summer.

Hab.: Irrigated land. Very common throughout the country.

Area: Borealo-Trop.

Used as a lawn grass (Bermuda Grass). Tends to be a troublesome weed.

78. SPOROBOLUS R. Br.

Perennials, mostly with creeping rhizomes. Leaves flat or rolled; ligule a fringe of short hairs. Inflorescence a lax or contracted panicle. Spikelets 1-flowered, muticous; rachilla disarticulating above glumes. Glumes usually unequal, 1-veined, persistent or singly caducous; the upper about as long as or shorter than lemma; lemma 1-veined, muticous; palea nearly as long as lemma, 2-fid at apex. Lodicules 2. Stamens 2-3. Caryopsis with a gelatinous pericarp, usually red, free within lemma and palea, adapted to zoochory. x=9.

One hundred and fifty trop. and warm-temperate species.

Literature: W.D. Clayton, Studies in the Gramineae:VI, The *Sporobolus indicus* complex, *Kew Bull.* 19:287-296 (1965).

N.L. Bor (Grasses Burma Ceylon India Pakist., 625, 1960) says about caryopses in the genus *Sporobolus*: "In this genus the grains are loosely or firmly enclosed in a gelatinous pericarp. When wetted the pericarp swells and, if it be loosely attached, turns inside out and ejects the seed, which remains attached to the lemma. In this condition the seeds are sticky and adhere to the feet of passing animals and so are distributed".

1. Panicle oblong-ovate to ovate-lanceolate in outline, 3-6 cm. Leaves numerous, distichously arranged, crowded. Plant of maritime sandy soils.   **1. S. pungens**
- Panicle cylindrical, slender, spike-like, 10-40 cm. Leaves few, long, not distichously arranged. Plant of wet sites, very rare.   **2. S. indicus**

1. Sporobolus pungens (Schreb.) Kunth, Révis. Gram. 68 (1829); Boiss., Fl. 5:512. *Agrostis pungens* Schreb., Beschr. Gräser 2:46, t.27 (1772). *S. arenarius* (Gouan) Duv.-Jouve, Bull. Soc. Bot. Fr. 16:294 (1869); Post, Fl. 2:728. *A. arenaria* Gouan, Obs. Bot. 3 (1773). *S. virginicus* (L.) Kunth, *op. cit.* 67 var. *arenarius* (Gouan) Maire in Jahandiez & Maire, Cat. Pl. Maroc 928 (1941). [Plate 390]

Perennial, 20-50 cm, with creeping rhizome. Culms numerous, ascending, simple or branched from lower nodes, densely distichously leafy. Leaves rigid; leaf-sheaths imbricate, bearded at mouth; blades rolled when dry, hairy on adaxial surface, ending in a pungent point. Panicle 3-5(-6) × 1-1.5 cm, oblong-ovate in outline, partly enclosed by uppermost leaf. Spikelets 2.5-3 mm, glabrous, shortly pedicellate. Glumes unequal, lanceolate, subacute; lower glume shorter and narrower than the upper; upper glume nearly as long as lemma; lemma and palea similar in texture and length. Fl. June-October.

Hab.: Maritime sandy soils, often exposed to sea-spray. Common. Coastal Galilee, Acco Plain, Coast of Carmel, Sharon Plain, Philistean Plain.

Area: Mediterranean, extending along the Atlantic coast of Europe.

2. Sporobolus indicus (L.) R. Br., Prodr. Fl. Nov. Holl. 170 (1810); Clayton, Kew Bull. 19:291. *Agrostis indica* L., Sp. Pl. 63 (1753). *S. poiretii* (Roem. & Schult.) Hitchcock, Bartonia 14:32 (1932); Hitchcock, Manual Grasses U.S. ed. 2, 418, f.597. [Plate 391]

Perennial, 60–100 cm, tufted, generally glabrous. Culms unbranched, erect. Leaves mostly basal; blades glabrous, rolled, long, tapering to a long filiform tip; sheaths ciliate-margined, densely hairy at mouth. Panicle 10–40 × (0.5–)1–2 cm, greyish, contracted, cylindrical, often interrupted; branches erect. Spikelets 1.8–2.5 mm, shortly pedicellate. Glumes unequal, translucent, the upper about ½ length of spikelet; lower glume shorter and narrower than the upper; lemma and palea similar in texture to glumes, gaping open beyond top of caryopsis. Caryopsis red-brown. Fl. late summer.

Hab.: Heavy soil. Very rare. Sharon Plain, W. Negev.

Area: Trop. Naturalized in S. Europe.

79. CRYPSIS Ait. emend. C.E. Hubbard
(incl. *Heleochloa* Host ex Roem.)

Annuals. Leaves flat or rolled; ligule a fringe of hairs. Inflorescence a variously shaped dense panicle, globose, ovoid or cylindrical, often partly enveloped at base by an involucre of uppermost subtending leaf-sheaths; branches and pedicels very short. Spikelets 1-flowered, strongly laterally compressed. Glumes somewhat unequal, 1-veined, compressed, keeled; lemma as long as glumes or exceeding them, acutely keeled, 1-veined, membranous, acute or mucronate; palea 1- or 2-veined. Lodicules absent. Stamens 2–3. Caryopsis oblong, with a free pericarp; embryo as long as or ¾ length of caryopsis; hilum punctiform. x=9.

Ten to twelve species in the Mediterranean, Irano-Turanian and Euro-Siberian regions; also in Trop. Africa.

Crypsis species show a striking preference for habitats that are swamped or inundated during winter.

Literature: C.E. Hubbard, in *Hooker's Icones Plantarum*, sub t. 3457 (1947). J. Lorch, A revision of *Crypsis* Ait. s.l. (Gramineae), *Bull. Res. Counc. Israel* 11D: 91–110, ff.1–13 & tt.1–5 (1962).

1. Panicle spike-like, cylindrical, at least 5 times as long as broad, terminal and single on generally unbranched culms; sheath of uppermost leaf hardly broadened   2
   - Panicle not as above; culms generally branched; sheaths of distal leaves distinctly broadened   3
2. Upper glume distinctly mucronate. Spikelets 2.5–3 mm. Anthers 2 mm.

   2. C. acuminata

- Upper glume not mucronate. Spikelets 2–2.5 mm. Anthers up to 1mm.

1. C. alopecuroides

3(1). Panicles head-like; heads very numerous and congested; primary (terminal) heads 1.5–3 mm broad, containing only 5–12 spikelets (rarely more).

4. C. minuartioides

- Heads broader than above, with numerous spikelets   4

4. Heads longer than broad, ovoid or oblong-ellipsoid, 5–8 mm broad, in greater part exserted from their involucre. Spikelets 3–3.5 mm.   **3. C. schoenoides**

- Heads broader than long, nearly wholly enclosed in involucre (only tips of spikelets exserted)   5

5. Spikelets 4–6 mm. Palea 2-veined. Stamens 3. Culms generally elongated (15 cm or longer).   **5. C. factorovskyi**

- Spikelets 3–4 mm. Palea 1-veined. Stamens 2. Culms short.   **6. C. aculeata**

1. Crypsis alopecuroides (Piller & Mitterp.) Schrad., Fl. Germ. 1:167 (1806);

Lorch, *op. cit.* 94, f.1 & t.1; Bor, Fl. Iraq 9:462, t.177. *Phleum alopecuroides* Piller & Mitterp., Iter Posegan. Slav. 147, t.16 (1783). *Heleochloa alopecuroides* (Piller & Mitterp.) Host ex Roem., Collect. 233 (1809); Host, Gram. Austr. 1:23, t.29 (1801) *nom. inval.*; Boiss., Fl. 5:476; Post, Fl. 2:717. [Plate 392]

Annual, 3–20(–40) cm, branched only at base. Distal internodes of culms longer than sheaths. Leaves flat, more or less hairy or glabrous; distal leaves not differing from others, with sheaths cylindrical and narrowly membranous-margined. Panicle 2–8 × 0.2–0.4 cm, terminal, spike-like, narrowly cylindrical, several times longer than broad, very dense, enveloped in uppermost sheath only at beginning of flowering. Spikelets 2–2.5 mm, numerous, sessile or short-pedicellate, those at base of panicle sterile. Glumes subacute, hispidulous at keel; upper glume somewhat longer and broader than the lower; lemma about as long as glumes and somewhat broader; palea 2-veined. Stamens 3. Fl. May–October.

Hab.: Sites that are inundated in winter. Rare. Jerusalem; Upper Jordan Valley; Golan.

Area: Mediterranean, W. Irano-Turanian and Euro-Siberian.

2. Crypsis acuminata Trin. in Spreng., Neue Entdeck. 2:57 (1820); Tzvelev,

Bot. Zhur. 51:1100 (1966). *C. borszczowii* Regel, Bull. Imp. Nat. Mosc. 41:306 (1868); Lorch, *op. cit.* 96, f.2 & t.1, f.2; Bor, Fl. Iraq 9:464. *Heleochloa acutiglumis* Boiss., Fl. 5:476 (1884); Post, Fl. 2:717. [Plate 393]

Annual, 5–30(–50) cm, branched only at base. Distal internodes longer than sheaths. Sheaths narrowly membranous-margined, sometimes fringed with hairs. Panicle (2–)4–8(–12) × 0.5–0.6 cm, terminal, spike-like, narrowly cylindrical, dense, bristly in profile, remote from uppermost leaf or partly enveloped by its sheath at beginning of flowering. Spikelets 2.5–3 mm (incl. mucro), numerous, short-pedicellate, sometimes sterile at base of panicle. Glumes membranous, slightly hairy along keel and margins; lower glume narrow, shortly mucronate; the upper broader, oblong-lanceolate, abruptly ending in a mucro ⅙-½ length of glume proper; lemma membranous, broader than glumes, obtuse and hairy at apex, abruptly ending in a mucro somewhat shorter than that of upper glume, but ending at same level; palea shorter, obtuse, 2-veined. Stamens 3. Caryopsis dark brown. Fl. May–August.

Hab.: Alluvial soil. Rare. Acco Plain; Esdraelon Plain.

Area: E. Mediterranean and W. Irano-Turanian.

This species is closely related to *C. alopecuroides*.

3. Crypsis schoenoides (L.) Lam., Tabl. Encycl. Méth. Bot. 1:166, t.42 f.1 (1791); Lorch, *op. cit.* 99, f.4 & t.3 f.1; Bor, Fl. Iraq 9:464, t.178. *Phleum schoenoides* L., Sp. Pl. 60 (1753). *Heleochloa schoenoides* (L.) Host ex Roem., Collect. 233 (1809); Host, Gram. Austr. 1:23 (1801) *nom. inval.*; Boiss., Fl. 5:476; Post, Fl. 2:717. [Plate 394]

Annual, 5–40 cm, often much branched at base, very variable in habit, compact or spreading. Culms procumbent or geniculately ascending. Panicle 1–4(-5) × 0.5–0.8 cm, ovoid or oblong-ellipsoid, longer than broad, enveloped in its lower ½ by an involucre of 1 or 2 distal leaves; sheaths of distal leaves often distinctly widened, broadly membranous-margined; leaf-blade abruptly tapering to a point; rachis of panicle short. Spikelets 3–3.5 mm, numerous, short-pedicellate. Glumes lanceolate, acute, membranous, hispidulous at keel and margins; upper glume broader than the lower and somewhat longer; lemma acute, exceeding glumes; palea ⅔ length of lemma, obtuse, 2-veined. Stamens 3; anthers 0.5–1 mm. Fl. May–September.

Hab.: Banks of rivers and other wet places. Coastal Galilee, Acco Plain, Coast of Carmel, Sharon Plain, Philistean Plain; Upper and Lower Galilee, Esdraelon Plain, Shefela, Judean Mts., W. and C. Negev; Hula Plain, Upper and Lower Jordan Valley, Dead Sea area; Golan, Gilead.

Since the use of herbicides along roads in Israel *C. schoenoides* has become abundant at roadsides.

Area: Mediterranean, Irano-Turanian and Euro-Siberian.

"*C. schoenoides* is highly variable in habit, being either procumbent or geniculately ascending, very compact or spreading" (Lorch, *op. cit.* 100). It is possible to distinguish it from *C. aculeata* by checking the apical leaves which show a clear discontinuity between sheath and blade in *C. schoenoides* (continuous in *C. aculeata*). Anthers do not exceed 1 mm in *C. schoenoides* (in *C. aculeata* 2 mm).

4. Crypsis minuartioides (Bornm.) Mez, Feddes Repert. 17:293 (1921); Lorch, *op. cit.* 102, ff.7, 8 & t.4 f.1; Post, Fl. 2:716. *Torgesia minuartioides* Bornm., Mitt. Thür. Bot. Ver. nov. ser. 30:83 (1913). *Heleochloa minuartioides* (Bornm.) Pilger in Engler & Prantl, Nat. Pflanzenfam. ed. 2, 14d:62 (1956). [Plate 395]

Annual, up to 25 cm, much branched at base. Culms prostrate, branched especially above; internodes short. Distal leaves numerous, clustered on upper internodes and forming involucres to panicles. Panicles very numerous, crowded, one at tip of each branch, 7–9 × 1.5–3 mm, narrowly oblong, longer than broad, compact, containing only 5–12 spikelets (rarely more); tips of spikelets protruding by 1–2 mm from sheaths of involucre; rachis 1–2 mm; involucre of each panicle narrow, of 2–4 short distal leaves; sheaths narrowly membranous-margined. Spikelets 4–5 mm. Glumes nearly equal in length, hyaline; lower glume linear, gradually subulate above, narrower and some-what shorter than the upper; upper glume mucronate; lemma nearly as long as upper glume, but broader; palea 2-veined. Stamens 3. Fl. May–July.

Hab.: Swampy soil. Sharon Plain, Philistean Plain.

Area: E. Mediterranean. Endemic to the coastal plain of Israel.

5. Crypsis factorovskyi Eig, Bull. Inst. Agr. Nat. Hist. Tel-Aviv 6:58 (1927); Lorch, *op. cit.* 101, ff.5, 6 & t.3 f.2; Post, Fl. 2:716. *Heleochloa factorovskyi* (Eig) Pilger in Engler & Prantl, Nat. Pflanzenfam. ed. 2, 14d:62 (1956). [Plate 396]

Annual, 30–90 cm, much-branched, sometimes purplish. Culms geniculate; internodes enveloped by leaf-sheaths up to ½ their length. Leaf-blades usually rolled, densely hairy especially on upper surface. Panicles head-like, com-pressed, of 2 kinds: terminal and axillary, the 2 sometimes combined in com-pound head. Terminal heads generally broader than long (5–8 × 5–11 mm), with numerous spikelets, enveloped by an involucre of 2 opposite coriaceous distal leaves, with much broadened many-veined membranous-margined sheaths. Axillary heads each situated singly at base of terminal ones, much narrower, with a few spikelets, enveloped on outer side by one long-bladed leaf, and on inner side by a shorter leaf which clasps culm and head. Spikelets 4–6 mm, on very short pedicels. Glumes shorter than spikelet, hyaline, lanceo-late, acuminate; the lower somewhat shorter and narrower than the upper; lemma obtuse, slightly broader and somewhat longer than glumes; palea shorter than lemma, 2-veined. Stamens 3. Fl. June–August.

Hab.: Marshes and wadis drying up in summer. Acco Plain, Coast of Car-mel, Sharon Plain, Philistean Plain, W. Negev; Mt. Carmel, Judean Mts.; Hula Plain, Upper Jordan Valley, Beit Shean Valley; Golan.

Area: E. Mediterranean, extending into the W. Irano-Turanian.

6. Crypsis aculeata (L.) Ait., Hort. Kew. 1:48 (1789); Lorch, *op. cit.* 105, f.9 & t.4 f.2; Bor, Fl. Iraq 9:460, t.176; Boiss., Fl. 5:475; Post, Fl. 2:716. *Schoenus aculeatus* L., Sp. Pl. 42 (1753). [Plate 397]

Annual, 5–20 cm. Culms often prostrate at base, frequently with numerous axillary shoots. Leaf-blades flat, sparingly pilose on both faces or only on upper surface. Panicle head-like, broader than long, up to 0.5 × 1 cm, envel-oped by an involucre formed by sheaths of 2 opposite distal leaves with short pointed blades. Spikelets 3–4 mm, numerous, slightly exserted from involucre. Glumes membranous; lower glume linear, acute; upper glume somewhat longer, much broader, obtuse; lemma slightly longer than glumes; palea 1-veined. Stamens 2. Fl. June–September.

Hab.: Marshes and water courses that are dry in summer. Very rare. Coast of Carmel, Philistean Plain, W. Negev.

Area: Mediterranean, Irano-Turanian and Euro-Siberian.

Tribe 19. PANICEAE R. Br. (1814). Annuals or perennials. Leaf-blades usually flat; sheaths split nearly to base, without auricles; ligule membranous, short, usually hairy at back and with a row of hairs at margin, rarely ligule absent. Inflorescences variously shaped panicles, sometimes consisting of spike-like racemes arranged along central rachis or digitate at or near its apex. Spikelets usually similar, hermaphrodite, rarely unisexual, usually deciduous from pedicels at maturity, 2-flowered; upper floret always hermaphrodite and fertile, the lower staminate or neuter; rachilla very short, not articulated. Glumes (1–)2, usually unequal, membranous or herbaceous; the lower smaller, rarely absent; the upper as long as or shorter than spikelet; lemma of lower floret (*lower lemma*) similar to glumes in texture; lower palea often absent; lemma of the upper floret (*upper lemma*) much firmer in texture, smooth or rugose, awnless or short-mucronate; rachilla not produced. Lodicules 2, usually small, cuneate, glabrous. Stamens 3. Stigmas plumose. Caryopsis firmly enclosed between lemma and palea; embryo about $\frac{1}{2}$–$\frac{3}{4}$ length of caryopsis; hilum punctiform. Chromosomes small; x=9, more rarely 7, 10, 15, 17, 19.

About 80 genera, the majority in trop. or subtrop. regions; in the temperate regions mainly weeds or cultivated.

80. Panicum L.

Annuals or perennials. Leaf-blades flat or rolled; ligule membranous, usually more or less hairy, densely ciliate at margin. Inflorescence an effuse or more rarely contracted panicle. Spikelets pedicellate, 2-flowered, awnless, falling entire from pedicels at maturity. Glumes unequal to subequal; the upper about as long as spikelet; lower floret staminate or neuter; its lemma (lower lemma) similar to upper glume; palea sometimes absent; upper floret hermaphrodite; its lemma (upper lemma) coriaceous; palea similar in texture. Lodicules 2. Stamens 3. Caryopsis elliptic in outline, enclosed by lemma and palea; hilum punctiform.

About 300 trop. and warm-temperate species.

1. Glumes similar in size and shape; lower glume nearly as long as spikelet. Culms branched from middle and upper nodes. Perennial plants of sands and deserts.
   6. P. turgidum
 – Glumes very unequal; lower glume $\frac{1}{4}$–$\frac{3}{4}$ length of upper   2
2. Lower glume obtuse or subacute, obscurely veined   3
 – Lower glume acuminate or acute, distinctly veined   4
3. Upper lemma transversely rugulose. Spikelets subacute, obscurely veined. Tall green perennials with a short rhizome.   **1. P. maximum**

- Upper lemma smooth. Spikelets acute, veined. Glaucous perennials with a long
  creeping rhizome.   **5. P. repens**
4(2). Spikelets 4–5 mm.   **4. P. miliaceum**
- Spikelets 2.5–3 mm   5
5. Leaf-blades and -sheaths long-hispid, with spreading tubercle-based hairs. Pedicels
   hispid, 1–3 cm or longer. Annuals.   **2. P. capillare**
- Leaves glabrous. Pedicels glabrous, shorter. Tall perennials.   **3. P. antidotale**

1. Panicum maximum Jacq., Icon. Pl. Rar. 1:2, t.13 (1781); Collect. Bot. 1:76
(1787); Hitchcock, Manual Grasses U.S. ed. 2, 695, f.1055; Bor, Fl. Iran.
70:473; Boiss., Fl. 5:439. [Plate 398]

Perennial, up to 3 m or more, densely tufted from a short stout rhizome.
Culms erect, with pubescent nodes. Leaf-blades flat, elongate, broadly linear,
long-pilose with tubercle-based hairs, or glabrescent; sheath pilose to glabrous; mouth of sheath densely long-pilose. Panicle 20–50 cm, lax, erect, with
capillary erecto-patent branches; lower branches in whorls. Spikelets 2.5–3
mm, erect, green or tinged with purple, narrowly elliptic, subacute, glabrous
or pubescent. Glumes very unequal; lower glume about ¼–⅓ length of spikelet, ovate, obtuse, membranous; upper glume and lower lemma similar in size
and shape, obtuse; upper lemma nearly as long, transversely rugulose, elliptic, more or less acute. Fl. summer.

Hab.; Cultivated ground, roadsides. Sharon Plain, Philistean Plain; Esdraelon Plain, Judean Mts.; Upper Jordan Valley.

Area: Trop.

Adventive. Recently introduced.
One of the best fodder grasses of the Tropics. In Africa brooms are made from the
inflorescence (Guinea Grass).

2. Panicum capillare L., Sp. Pl. 58 (1753); Hitchcock, Manual Grasses U.S.
ed. 2, 689, f.1042; Fiori, Nuov. Fl. Anal. Ital. 1:79; Eig, Bull. Inst. Agr. Nat.
Hist. Tel-Aviv 6:53; Post, Fl. 2:698. [Plate 399]

Annual, 10–50 cm. Culms papillose-hispid to nearly glabrous. Leaves flat,
broad; leaf-blades and -sheaths long-hispid, hairs spreading, with tubercle at
base; ligule reduced to a short ciliate rim. Panicle very diffuse, often ½ length
of entire plant, included in upper leaf-sheath until maturity; branches capillary, numerous, divaricately spreading at maturity; whole panicle breaking
away finally and tossed by wind as tumbleweed. Spikelets 2–3 mm, oblong-
ovate, cuspidate, long-pedicellate. Glumes very unequal; lower glume ovate,
⅓–⅖ length of upper, acute, 3-veined; upper glume oblong-lanceolate, cuspidate, 5-veined; upper lemma elliptic, smooth and glossy. Fl. summer.

Hab.: Cultivated ground, sometimes saline. Sharon Plain, Philistean Plain;
Esdraelon Plain, Judean Mts.; Beit Shean Valley.

Area: Native in S. and W. USA. Occasionally introduced.

Adventive. Eig (*loc. cit.*) recorded this species from Palestine in 1927. Subsequently
it was not collected again in the area until 1977.

3. Panicum antidotale Retz., Obs. Bot. 4:17 (1786–1787); Bor, Fl. Iran. 70:472; Boiss., Fl. 5:440. [Plate 400]

Tall perennial (up to 3 m), sometimes shrubby, with a stout creeping rhizome. Culms leafy; nodes brown. Leaves flat, long; sheaths smooth; ligule a rim of hairs. Panicle 15–30 cm, lax; branches slender, many-spiculate, ascending to spreading, the lower in dense fascicles. Spikelets 2.5–3 mm, glabrous, veined, ovate, acute, deciduous. Glumes membranous, hyaline-margined; lower glume ovate, acute, 3-veined, about ½ length of spikelet; upper glume 5–7-veined, slightly shorter than spikelet; lower lemma as long as spikelet; upper lemma chartaceous, glabrous, smooth and glossy. Filaments very short. Stigmas exserted, pale brown. Fl. April.

Hab.: Salines. Rare. Arava Valley.

Area: India, Afghanistan, Persia and Trop. of the Old World.

Adventive. Possibly an escapee from occasional cultivation.

4. Panicum miliaceum L., Sp. Pl. 59 (1753); Hitchcock, Manual Grasses U.S. ed. 2, 690, f.1047; Bor, Fl. Iraq 9:486; Boiss., Fl. 5:441; Post, Fl. 2:698. [Plate 401]

Annual, 30–100 cm. Leaves flat, pilose on both surfaces or glabrescent; hairs with tubercle at base; ligule a ciliate rim. Panicle 10–30 cm, effuse, nodding, often partly included in upper leaf-sheath; branches numerous, slender, ascending. Spikelets 4.5–5.5 mm, ovoid, acuminate, glabrous. Glumes very unequal, with elevated veins; lower glume broadly ovate, acuminate, 5-veined, about ½–⅔ length of upper; upper glume acuminate, 11-veined; lower lemma similar to upper glume; upper lemma shorter, turgid, smooth, glossy. Fl. spring.

Hab.: Casual. Near Tel-Aviv, irrigated ground; Jerusalem, near houses.

Area: Euro-Siberian and Irano-Turanian.

Formerly widely cultivated in Asia and E. Europe for its grain (Common Millet).

5. Panicum repens L., Sp. Pl. ed. 2, 87 (1762); Bor, Fl. Iraq 9:488, t.187; Boiss., Fl. 5:440; Post, Fl. 2:697. [Plate 402]

Perennial, 50–80 cm, glaucous, with a long creeping rhizome. Culms stiff. Leaf-sheaths ciliate; leaf-blades acuminate; ligule short, hairy. Panicle 6–10 cm, lax, much branched; branches erecto-patent, 1–3 on nodes of rachis, flexuose. Spikelets 2.5 mm, oblong, acute, white, often arranged in twos, upright. Glumes unequal; lower glume ¼–½ length of upper, rounded or truncate, broader than long, nearly cup-shaped; upper glume and lower lemma similar in size and shape, ovate, acute, 7–9-veined; hermaphrodite floret somewhat shorter, ovate, acute, white, smooth. Anthers orange-coloured. Stigmas violet. Fl. April–January.

Hab.: Banks of rivers and other moist habitats. Coastal Galilee, Acco Plain, Coast of Carmel, Sharon Plain, Philistean Plain, W. Negev; Esdraelon Plain,

?Samaria, ?Judean Mts. Dan Valley, Hula Plain, Upper and Lower Jordan Valley; Dead Sea area; Transjordan (Al-Eisawi).

Area: Mediterranean and Trop.

6. Panicum turgidum Forssk., Fl. Aeg.-Arab. 18 (1775); Bor, Fl. Iraq 9:490, t.188; Boiss., Fl. 5:441; Post, Fl. 2:698. [Plate 403]

Perennial, 60–150 cm, glabrous, growing in large tufts. Culms more or less lignified; nodes thickened and with clusters of branches. Leaf-blades short, ending in a sharp point. Panicle 10–15 cm, lax or more or less contracted; branches short, stiff. Spikelets 4 mm, white or purplish, turgid, ovoid, acute. Glumes nearly equal, similar in shape, broadly ovate, rounded on back, 7–9-veined, the upper at least ⅔ length of spikelet; lower lemma similar to upper glume and somewhat exceeding it, 9-veined; hermaphrodite floret somewhat shorter, ellipsoid, acute, glossy. Fl. March–January.

Hab.: Sands and deserts. Sharon Plain, Philistean Plain; W., N., C. and S. Negev; Lower Jordan Valley, Dead Sea area; Arava Valley; Edom.

Area: Saharo-Arabian and Sudanian.

Oppenheimer in Florula Transiordanica (Bull. Soc. Bot. Genève ser. 2, 22:264, 1931) remarks that in Aaronsohn's specimens from the southern Dead Sea area many inflorescences contain only sterile spikelets. We can confirm this observation; part of our specimens from the surroundings of the Dead Sea bear inflorescences with narrow, apparently sterile spikelets.

81. ECHINOCHLOA Beauv.

Annuals or perennials. Leaf-blades flat; ligule a fringe of hairs, absent in some species. Inflorescence a panicle of spike-like racemes arranged along a central rachis; racemes sometimes branched. Spikelets in 2–4 rows, densely packed on one side of flattened rachis, sessile or short-pedicellate, awned, or cuspidate or apiculate, ovate in outline, rounded on back, flat on ventral side, 2-flowered; lower floret staminate or neuter, the upper hermaphrodite. Glumes unequal, membranous; the lower much smaller, abaxial (turned away from rachis of raceme); upper glume as long as spikelet, apiculate, cuspidate, acuminate or shortly awned; lemma of lower floret (lower lemma) similar to upper glume; its palea hyaline; lemma of hermaphrodite floret (upper lemma) chartaceous to coriaceous, smooth and glossy, clasping palea of similar texture. Lodicules 2. Stamens 3. Caryopsis broadly ellipsoid, plano-convex; embryo ½–¾ length of caryopsis; hilum circular.

About 20 species mainly in the warmer regions of the globe.

1. Lower lemma cuspidate or awned; racemes 2.5–8 cm; rachis of racemes with long bristles at base of spikelets.   **2. E. crusgalli**
– Lower lemma acute or apiculate; racemes 1–2.5(–3) cm; rachis of racemes devoid of bristles at base of spikelets.   **1. E. colonum**

1. Echinochloa colonum (L.) Link, Hort. Berol. 2:209 (1833); Bor, Fl. Iraq 9:479, t.184. *Panicum colonum* L., Syst. Nat. ed. 10, 2:870 (1759); Boiss., Fl. 5:479; Post, Fl. 2:696. [Plate 404]

Annual, 30–70 cm. Leaf-blades flaccid, glabrous; ligule absent. Panicle 5–12 cm; racemes spaced, 1–2.5(–3) cm, appressed to main axis or ascending. Spikelets 2–3 mm, acute to apiculate, awnless, almost sessile, in about 3–4 rows, sometimes purple-tinged. Glumes and lemmas 5–7-veined; lower glume cordate-ovate, less than ½ length of upper. Fl. April–December.

Hab.: Irrigated places. Common, sometimes a noxious weed in alfalfa fields. Acco Plain, Sharon Plain, Philistean Plain; Upper and Lower Galilee, Mt. Carmel, Esdraelon Plain, Samaria, Shefela, Judean Mts., Judean Desert; Dan Valley, Hula Plain, Upper and Lower Jordan Valley, Dead Sea area; Golan, Gilead, Ammon.

Area: Mediterranean, Irano-Turanian and Trop.

Fodder plant in Trop. of Old World; the grains are eaten in times of shortage (Jungle Rice).

2. Echinochloa crusgalli (L.) Beauv., Agrost. 53, 161 (1812); Bor, Fl. Iraq 9:480, t.185. *Panicum crusgalli* L., Sp. Pl. 56 (1753); Boiss., Fl. 5:435; Post, Fl. 2:695. [Plate 405]

Annual, 50–100 cm, tufted. Leaf-blades broad, nearly glabrous; ligule absent. Panicle 7–15(–20) cm; rachis of panicle scabridulous and with tufts of long bristles at base of racemes; racemes spaced, 3–8 cm, solitary, sometimes 2–3 together, erect to ascending; rachis of raceme long-setose at insertion of spikelets. Spikelets 3–4 mm (excl. awns), acuminate to awned, tuberculate-hispid along veins, short-pedicellate, in about 4 rows, sometimes purplish-tinged. Lower glume cordate-ovate, ⅓ length of upper, 5-veined, cuspidate to short-awned; upper glume 5–7-veined, cuspidate to short-awned, with a cusp or awn shorter than that of lower lemma; lower lemma long- to short-awned or acuminate, or cuspidate. Fl. June–August.

Hab.: River banks and marshes (in Panicetum repentis association), fields, waste places. Acco Plain, Sharon Plain, Philistean Plain; Esdraelon Plain, Samaria, Judean Mts.; Dan Valley, Hula Plain, Upper and Lower Jordan Valley; Golan, Gilead.

Area: Mediterranean, Irano-Turanian, extending into the Euro-Siberian and Trop. Asia.

Fodder plant the grains of which are eaten in times of scarcity (Barnyard Millet).

E. crusgalli varies in appearance, owing to the variability of the lower lemma. This may be long- to short-awned or acuminate to cuspidate. The species has been accordingly subdivided into several varieties. Bor in Flora of Iraq (9:480–484, 1968) distinguished var. *crusgalli* ("Lower lemma long- to moderately long-awned") and var. *breviseta* (Doell) Neilr., 1859 ("Lower lemma acuminate or cuspidate"). But in Flora iranica (70:480–481, 1970), Bor subdivided the species into var. *crusgalli* ("Spiculae aristatae vel aristulatae") and var. *submutica* Neilr., 1859 ("Spiculae mucronatae").

Our herbarium specimens reflect the whole gamut of variation from the long-awned to the cuspidate spikelets. The majority of the long-awned specimens were collected in the Hula Plain and the Upper Jordan Valley; specimens with short-cuspidate and mucronate lemmas come from districts of the coastal plain. The difficulty in referring the specimens to one or the other variety stems not only from the numerous mixed collections made in the same locality, but also from the frequent occurrence of awned and cuspidate lemmas even in the same panicle.

82. BRACHIARIA (Trin.) Griseb.

Annuals or perennials. Leaf-blades flat; ligule a long-ciliate rim. Inflorescence a panicle of more or less spike-like one-sided racemes racemosely arranged on main rachis; racemes sometimes branched. Spikelets in 2 rows, subsessile or pedicellate, on one side of a 3-angled raceme-rachis, awnless, somewhat dorsally compressed, falling entire at maturity, 2-flowered; lower floret staminate; the upper hermaphrodite. Glumes unequal; the lower much smaller than the upper, adaxial (turned towards rachis of raceme); upper glume nearly as long as spikelet; lemma of the staminate floret (lower lemma) equal in shape and texture to upper glume; its palea hyaline; lemma of hermaphrodite floret (upper lemma) coriaceous to crustaceous. Lodicules 2. Stamens 3. Caryopsis oblong-elliptic in outline, flattened; hilum punctiform.

About 70 warm-temperate species.

1. Spikelets pubescent. Inflorescence linear, 0.5–1 cm broad, of several short erect racemes. Annual.　　　　　　　　　　　　　　　　　**2. B. eruciformis**
- Spikelets glabrous. Inflorescence oblong-ovate, much broader than in above, of long spreading compound racemes. Tall perennial.　　　　**1. B. mutica**

1. Brachiaria mutica (Forssk.) Stapf, in Fl. Trop. Afr. 9:526 (1919); Clayton, in Fl. W. Trop. Afr. ed. 2, 3:443. *Panicum muticum* Forssk., Fl. Aeg.-Arab. 20 (1775). *P. numidianum* Lam., Tabl. Encycl. Méth. Bot. 1:172 (1791); Boiss., Fl. 5:438; Post, Fl. 2:697. *B. purpurascens* (Raddi) Henrard, Blumea 3:434 (1940). [Plate 406]

Perennial, 60–180 cm, often rooting at lower nodes. Leaf-blades up to 1 cm broad. Panicle 10–20 cm, lax, of 5–20 racemes; racemes 2–10 cm, short-pedunculate, spreading, solitary or irregularly approximate, sometimes in whorls; lower racemes frequently compound; secondary racemes short, 6–3-spiculate; pedicels usually with a few slender bristles under spikelet. Spikelets 2.5–3.5 mm, oblong-lanceolate, acute, glabrous, often purplish, usually paired, one of the pair on a longer pedicel. Glumes membranous; lower glume broadly ovate, acute, ⅓–½ length of upper; upper glume and lower lemma similar in shape and size, 5-veined; upper lemma slightly shorter. Stigmas blackish-purple, conspicuous. Fl. April–December.

Hab.: Banks of rivers, hedges. Sharon Plain, Philistean Plain; Upper Jordan Valley, Beit Shean Valley, Lower Jordan Valley.

Area: Trop. Africa and S. America.

Eig (Bull. Inst. Agr. Nat. Hist. Tel-Aviv 6:55, 1927) discussed this species under *Panicum barbinode* Trin. forma *pilifera* Hackel, which had been recorded from Palestine for the first time by Bornmüller (Feddes Repert. 10:381, 1912).

It is doubtful whether this form deserves distinction. All specimens of *B. mutica* examined, both from Palestine and from S. America, display small bristles on the pedicels. It rather seems that these bristles are characteristic of *B. mutica*.

Contrary to Bornmüller, Eig did not doubt that this species is native in Palestine.

2. Brachiaria eruciformis (Sm.) Griseb. in Ledeb., Fl. Ross. 4:469 (1853); Bor, Fl. Iraq 9:472, t.181. *Panicum eruciforme* Sm. in Sibth. & Sm., Fl. Graec. Prodr. 1:46 (1806); Boiss., Fl. 5:437; Post, Fl. 2:696. [Plate 407]

Annual, 15–30(–40) cm. Leaf-blades and -sheaths soft-pubescent. Panicle 3–6 cm, erect, linear, of 5–10 racemes; racemes 1–2.5 cm, linear, erect, short-pedunculate, spaced on axis of panicle and appressed to it. Spikelets 2–2.5 mm, crowded, short-pedicellate, oblong-ovoid, subacute. Lower glume very small, nearly truncate, or absent; upper glume and lower lemma ovate, subacute, 5-veined, membranous, pubescent; upper lemma glabrous, glossy at maturity. Fl. July–November.

Hab.: Wet places. Rare. Samaria, Judean Mts.; Upper Jordan Valley.

Area: Mediterranean, Irano-Turanian, Trop.

83. PASPALUM L.

Annuals or perennials. Leaf-blades flat; ligule membranous, glabrous. Inflorescence of one-sided spike-like racemes arranged digitately (rarely single) or racemosely along a main rachis. Spikelets in 2 or 4 rows on one side of a strap-shaped raceme-rachis, dorsally compressed, plano-convex, orbicular or oblong, muticous, 2-flowered; lower floret reduced to an empty lemma, the upper hermaphrodite. Lower glume absent or minute; the upper membranous, covering dorsal surface of spikelet; lemma of lower floret (lower lemma) similar to upper glume in texture and shape; lemma of hermaphrodite floret (upper lemma) with involute margins, crustaceous. Lodicules 2. Stamens 3. Stigmas exserted. Caryopsis orbicular in outline, plano-convex; hilum punctiform.

About 250 species in trop. and subtrop. parts of both hemispheres, the majority in America.

Literature: A.R. Pinto da Silva, O género *Paspalum* em Portugal, *Agron. Lusit.* 2:5–23 (1940). F. Garbari, Il genere *Paspalum* L. (Gramineae) in Italia, *Atti Soc. Tosc. Sci. Nat. Mem.* ser. B, 79:52–65 (1972).

1. Inflorescence of several racemes spaced along main rachis; spikelets broadly ovate with a fringe of long white hairs on margins of upper glume.
   1. P. dilatatum

- Inflorescence a terminal pair of racemes (rarely 1 or 3 racemes); spikelets narrowly ovate or elliptic, devoid of a fringe of long hairs as in above.
   2. P. paspalodes

1. Paspalum dilatatum Poir. in Lam., Encycl. Méth. Bot. 5:35 (1804); Bor, Fl. Iraq 9:492, t.189. [Plate 408]

Perennial, up to 1 m, tufted, with short stout rhizomes. Culms and leaf-blades glabrous; leaf-sheaths bearded at mouth, upper part of sheath ciliate; ligule membranous, about 3mm. Racemes 5–8 cm, 4–8, alternate and distant from one another on main rachis of inflorescence, erecto-patent or spreading; rachis of raceme long-bearded at its insertion on main rachis. Spikelets 2.5–3.5 mm, broadly ovate, acute, shortly pedicellate, imbricate. Lower glume absent; upper glume broadly ovate and apiculate with long dense silky cilia along margins; lower lemma similar in shape and size to glume; upper lemma smaller, glabrous. Stigmas black-purple. Fl. June–December.

Hab.: In and near water; in the *Panicum repens–Typha* association. Acco Plain, Sharon Plain, Philistean Plain; Esdraelon Plain, Samaria, Shefela, Judean Desert; Dan Valley, Upper Jordan Valley.

Area: Trop. America. Now widely introduced as pasture grass (Dallis Grass).

2. Paspalum paspalodes (Michx.) Scribner, Mem. Torrey Bot. Club 5:29 (1894; "*paspaloides*"); Bor, Fl. Iraq 9:494, t.190. *Digitaria paspalodes* Michx., Fl. Bor.-Amer. 1:46 (1803). *P. distichum* auctt. non L. (1753). [Plate 409]

Perennial, 40–50(–80) cm, with widely spreading stolons and rhizomes. Culms glabrous, often with hairy nodes. Leaf-blades glabrous; leaf-sheaths bearded at mouth, upper part of sheath ciliate; ligule membranous, very short, lacerate. Inflorescence a terminal pair of spike-like racemes (rarely 1 or 3 racemes); racemes 3–5(–8) cm, linear; one of them subsessile, the other shortly pedunculate. Spikelets 3–3.5 mm, narrowly ovate to elliptic, acute, shortly pedicellate, muticous, imbricate. Lower glume minute, scale-like, or absent; the upper as long as spikelet, appressed-pubescent, 3–5-veined; upper lemma shorter. Stigmas black-purple. Fl. July–September.

Hab.: Marshes and ditches. Sharon Plain, Philistean Plain; Samaria; Hula Plain, Upper Jordan Valley, Beit Shean Valley; Golan; Transjordan (Al-Eisawi).

Area: Native in America; now widely naturalized.

84. PASPALIDIUM Stapf

Perennials, rhizomatous. Leaf-blades flat or rolled; ligule reduced to a ciliate rim. Inflorescence spike-like, consisting of spaced racemes appressed to main rachis. Spikelets in 2 rows on raceme-rachis, densely aggregated, glabrous, muticous, 2-flowered; lower floret staminate or reduced to a chartaceous lemma (lower lemma); the upper hermaphrodite. Glumes unequal; lower glume much shorter than upper, abaxial; lemma of upper floret (upper lemma) coriaceous, acute or apiculate, transversely very finely rugose. Lodicules 2. Stamens 3. Caryopsis broadly elliptic, nearly orbicular, strongly compressed dorsally; hilum punctiform.

Twenty species, mainly in warm regions of Old World, often aquatic.

1. **Paspalidium geminatum** (Forssk.) Stapf, in Fl. Trop. Afr. 9:583 (1920); Täckholm & Drar, Fl. Eg. 1:442; Bor, Fl. Iran. 70:475. *Panicum geminatum* Forssk., Fl. Aeg.-Arab. 18(1775); Eig, Bull. Inst. Agr. Nat. Hist. Tel-Aviv 6:54 (1927); Post, Fl. 2:696. *Panicum paspalodes* Pers., Syn. Pl. 1:81 (1805); Boiss., Fl. 5:436 (*"paspaloides"*). [Plate 410]

Perennial, 30–100 cm, glabrous; rhizomes long. Culms prostrate at base, rooting at nodes, then erect, often branched. Lower leaf-sheaths dilated; blades spreading. Panicle 10–30 cm, of several spike-like racemes; racemes 1–3 cm, linear, one-sided, alternate and approximate, subsessile. Spikelets 2–2.5 mm, ovoid, turgid, in 2 rows, shortly pedicellate, alternate on one side of strap-shaped rachis of raceme; rachis 3-quetrous, flat on back, ciliolate; pedicels setulose. Glumes membranous; lower glume scale-like, broadly obovate-truncate, about ⅓–½ length of upper, nearly veinless, turned away from rachis; upper glume shorter than spikelet, ovate-lanceolate, obtuse to sub-acute, 5-veined; lemma of staminate floret similar in shape, apiculate, somewhat longer. Stigmas brownish when dry. Fl. April–December.

Hab.: In water and moist soil. Acco Plain, Coast of Carmel, Sharon Plain, Philistean Plain; Lower Galilee, Esdraelon Plain, Samaria; Hula Plain, Upper Jordan Valley, Beit Shean Valley; Transjordan (Al-Eisawi).

Area: Trop. Asia and Africa.

Eig (*loc. cit.*) wrote: "A new species for the region of Post's Flora... In many places (banks of Wadi Musrara, shore of Lake Kinneret, En-Harod, etc.) this plant forms associations by itself without the intermixture of other plants, but it is frequently found in association with *Panicum repens* L."

85. DIGITARIA Haller

Annuals or perennials. Leaf-blades flat; ligules membranous, lacerate or dentate, sparsely pilose. Inflorescence of paired or (sub)digitately arranged racemes, often purplish; raceme-rachis flattened with a terete midrib or 3-quetrous, often more or less winged; pedicels unequal. Spikelets mostly all alike, awnless, 2-flowered, plano-convex, in 2–4 rows on one side of rachis and closely appressed to it, falling entire from pedicels; lower floret reduced to an empty lemma, the upper hermaphrodite. Glumes very unequal; lower glume abaxial, minute, hyaline, or absent; upper glume mostly shorter than spikelet, narrow, rarely absent; lemma of empty floret (lower lemma) as long as spikelet and forming its outline, 5–7-veined, membranous; palea absent or minute; lemma of hermaphrodite floret (upper lemma) chartaceous, faintly 3-veined; its palea similar in texture. Lodicules 2. Stamens 3. Caryopsis tightly enclosed by lemma and palea, oblong-elliptic; embryo ⅓ length of caryopsis; hilum elliptic.

Over 300 species in the warm parts of both hemispheres.

Literature: J.T. Henrard, *A Monograph of Digitaria*, Leiden, 1950. N.L. Bor, The genus *Digitaria* Heist. in India and Burma, *Webbia* 11:301–367 (1955).

1. **Digitaria sanguinalis** (L.) Scop., Fl. Carn. ed. 2, 1:52 (1771); Bor, Fl. Iraq 9:478, t.183; Bor, Fl. Iran. 70:491; Täckholm & Drar, Fl. Eg. 1:425–427. *Panicum sanguinale* L., Sp. Pl. 57 (1753); Boiss., Fl. 5:433, incl. var. *ciliare* Boiss., op. cit. 434 (1884); Post, Fl. 2:695. [Plate 411]

Annual, 30–80 cm. Culms glabrous except at bearded nodes. Leaf-sheaths keeled, more or less hairy; hairs with tubercle at base; ligules short. Panicle of several racemes varying in length; racemes 5–15 cm, subdigitately clustered on a rachis up to 2 cm, often purplish; rachis of racemes winged, slightly flex-uose. Spikelets in pairs, one sessile, the other pedicellate, 3–3.5 mm, oblong-lanceolate, acute. Lower glume triangular, minute or absent, the upper up to ½ length of spikelet, narrowly lanceolate, acute, 3-veined, shortly pubescent between veins or glabrous; lemma of empty floret lanceolate, acute, 5–7-veined, scabridulous on lateral veins at least above, often ciliate, with cilia varying in length, or lemma glabrous; lemma of hermaphrodite floret gla-brous, smooth. Fl. May–December.

Hab.: Weed on irrigated land. Acco Plain, Sharon Plain, Philistean Plain; Upper and Lower Galilee, Esdraelon Plain, Samaria, Judean Mts., Judean Desert; Hula Plain, Upper Jordan Valley; Edom.

Area: Borealo-Trop.

86. SETARIA Beauv.

Annuals or perennials. Leaf-blades flat or folded; ligule a membranous rim hairy at back and with a fringe of cilia at margin. Inflorescence a spike-like cylindrical panicle. Spikelets subsessile, awnless, 2-flowered, elliptic, dorsally compressed, supported by an involucre of one to several scabrous bristles (modified sterile branchlets); bristles longer than spikelets, persistent on rachis, while spikelets deciduous at maturity; lower floret staminate or neuter, the upper hermaphrodite, rounded on back, flat on ventral side. Glumes membranous, concave, unequal; the lower much shorter than spikelet, 3-veined; upper glume ½ length of spikelet or as long, 5–7-veined; lemma of lower floret membranous, 5–7-veined, subequal to lemma of upper floret; palea sometimes absent; lemma of upper floret coriaceous to crustaceous, muticous, enclosing palea; palea similar in texture and subequal in length. Lodicules 2. Stamens 3. Caryopsis broadly elliptic in outline, flattened, tightly enclosed in lemma and palea; hilum punctiform.

About 125 species in trop. and warm-temperate parts of the globe.

Literature: A. Baytop, The genus *Setaria* in Turkey, *Jour. Fac. Pharm. Istanbul* 5:34–45 (1969). E.E. Terrell, The correct names for Pearl Millet and Yellow Foxtail, *Taxon* 25, 2–3:294–304 (1976). P. Auquier, Le genre *Setaria* en Belgique et au Grand-Duché de Luxembourg, *Lejeunia* 97:1–13 (1979a); Les concepts de Dumortier dans le genre *Setaria* Beauv. (Poaceae), *Bull. Jard. Bot. Bruxelles* 49:427–433 (1979b).

1. Upper glume about ½ length of spikelet; upper lemma rugulose.   **1. S. glauca**
– Upper glume as long as spikelet; upper lemma smooth or finely punctulate   2

2. Bristles retrorsely scabrous.   **3. S. verticillata** var. **verticillata**
- Bristles antrorsely scabrous or scabridulous   3
3. Rachis of panicle antrorsely scabridulous and long-hairy; hairs about 1 mm; panicle dense, cylindrical; branches not verticillate.   **2. S. viridis**
- Rachis of panicle antrorsely scabrous, not long-hairy; panicle lobed, interrupted; branches verticillate.   **3. S. verticillata** var. **ambigua**

1. Setaria glauca (L.) Beauv., Agrost. 51, 178 (1812); Bor, Fl. Iraq 9:500, t.192; E.E. Terrell, Taxon 25:299, 303 (1976); Boiss., Fl. 5:442; Post, Fl. 2:699. *Panicum glaucum* L., Sp. Pl. ed. 2, 83 (1762). [Plate 412]

Annual, 10–50 cm. Panicle 2–6(–12) cm, dense, cylindrical to ovate-oblong; bristles of involucre 6–12, 5–8 mm, reddish-yellow, antrorsely scabridulous. Spikelets about 3 mm; lower floret staminate. Lower glume ⅓ length, the upper about ½ length of spikelet. Lemma of staminate floret ovate, as long as transversely rugulose lemma of hermaphrodite floret. Fl. April–September.

Hab.: Irrigated places. Rare. Philistean Plain; Judean Mts. (Jerusalem); ?Golan, Gilead.

Area: Borealo-Trop.

According to Hitchcock (Manual of Grasses of the United States, ed. 2, 933) *Setaria glauca* (L.) Beauv., Agrost. 51, 178 (1812) is based on *Panicum glaucum* L., but misapplied to *S. lutescens* (Weigel) F.T. Hubbard (1916).

2. Setaria viridis (L.) Beauv., Agrost. 51, 178 (1812); Eig, Bull. Inst. Agr. Nat. Hist. Tel-Aviv 6:55; Boiss., Fl. 5:443; Post, Fl. 2:699 excl. vars. *Panicum viride* L., Syst. Nat. ed. 10, 2:870 (1759). [Plate 413]

Annual, 10–60 cm. Panicle 2–6(–10) cm, cylindrical; rachis of panicle scabrous and hairy; hairs about 1 mm, tubercle-based; bristles of involucre 10–20, 4–10 mm, green, yellowish or purplish, antrorsely scabridulous. Spikelets 2–3 mm; lower floret neuter. Lower glume about ⅓ length of spikelet; upper glume as long as spikelet. Lemma of neuter floret as long as lemma of hermaphrodite floret; the latter marked with very fine dots and short transverse lines. Fl. April–September.

Hab.: Irrigated ground. Rare. Sharon Plain; Esdraelon Plain; Dan Valley; Golan.

Area: Euro-Siberian, Mediterranean and Irano-Turanian.

3. Setaria verticillata (L.) Beauv., Agrost. 51, 178 (1812); Bor, Fl. Iraq 9:503, t.193; Auquier, Lejeunia 97:4 (1979); Boiss., Fl. 5:443; Post, Fl. 2:699. *Panicum verticillatum* L., Sp. Pl. ed. 2, 82 (1762). [Plate 414]

Annual, 20–80 cm. Panicle 4–7(–10) cm, at first cylindrical, dense, later irregularly lobed; rachis of panicle without long hairs; bristles of involucre 6–10, 3–6 mm, green or purplish, retrorsely (rarely antrorsely) scabrous. Spikelets 2 mm; lower floret neuter. Lower glume ¼–½ length of spikelet; upper glume as long as spikelet. Lemma of neuter floret as long as lemma of hermaphrodite floret; the latter marked with very fine dots and short transverse lines. Fl. April–September.

Var. **verticillata.** [Plate 414]. Bristles of involucre retrorsely scabrous.
Hab.: Weed of irrigated places. Common. Sharon Plain, Philistean Plain; Upper and Lower Galilee, Samaria, Judean Mts., Judean Desert; Upper Jordan Valley, Dead Sea area; Transjordan (Al-Eisawi).
Area: Borealo-Trop.

Var. **ambigua** (Guss.) Parl., Fl. Palerm. 36 (1845). *Panicum verticillatum* var. *ambiguum* Guss., Fl. Sic. Prodr. 1.80 (1827). *S. verticillata* subsp. *ambigua* (Guss.) Trabut in Battandier & Trabut, Fl. Algér. Monocot. 135 (1895). *S. ambigua* (Guss.) Guss., Fl. Sic. Synops. 1:114 (1843); Baytop, *op. cit.* 43. *S. verticilliformis* Dumort., Florula Belg. 150 (1827); Auquier, *op. cit.* (1979 a & b). *S. verticillata* × *viridis* Beauv. in Lloyd, Fl. Ouest Fr. ed. 5, 391 (1898); Nevski, in Fl. URSS 4:44; Bor, Fl. Iraq 9:504; Fl. Iran. 70:499. Bristles of involucre antrorsely scabrous.
Regarded by some authors (Lloyd, Trabut and others) as natural hybrid between *S. verticillata* and *S. viridis.*
Hab.: Irrigated field edges. Rare. Sharon Plain.
Area: Mediterranean and Euro-Siberian.

Owing to the retrorsely scabrous bristles, mature panicles of *S. verticillata* var. *verticillata* adhere to clothes and to the fur of animals, thus becoming a nuisance.

87. PENNISETUM L.C.M. Richard ex Pers.

Perennials or annuals, tufted or with a creeping rhizome. Culms simple or branched. Leaves flat, often folded or rolled; ligule a membranous rim with a row of dense hairs. Inflorescence a spike-like cylindrical panicle. Spikelets awnless, (1–)2-flowered, dorsally compressed, solitary or in clusters of 2–5, surrounded by an involucre of numerous bristles (modified sterile branchlets); bristles of involucre unequal in length, simple or branched, scabridulous, the inner plumose or ciliate; involucre falling with spikelets at maturity. Glumes and lemmas hyaline; glumes shorter than spikelet, membranous, acute or acuminate, subequal or the lower smaller than the upper; lower floret of spikelet staminate or neuter, sometimes much reduced; upper floret hermaphrodite, with lemma as long as spikelet and palea nearly as long. Lodicules 2, minute or absent. Stamens 3. Stigmas long, exserted. Florets protogynous. Caryopsis obovoid or globose, free; hilum punctiform.
Over 100 species, trop. and subtrop., mainly in Africa.

P. americanum (L.) Schumann [*P. typhoides* (Burm.) Stapf & C.E. Hubbard], known as Duhn or Pearl Millet, is an economically important cereal grown for fodder and grain in Africa and India.

1. Involucre 3–4 times as long as spikelet; longest bristle 20–30 mm; rachis of spike

pilose. Lower floret of spikelet reduced to a narrow or bristle-like rudiment or absent altogether.   **3. P. asperifolium**
- Involucre at most twice as long as spikelet; longest bristle shorter than in above; rachis of spike not pilose. Lower floret of spikelet staminate   2
2. Culms 40-150 cm, branched at nodes up to middle. Longest bristle of involucre about twice as long as spikelet; bristles whitish to pale straw-coloured, rarely purplish.   **2. P. divisum**
- Culms 20-40 cm, not branched as above. Longest bristle of involucre about 3 times as long as spikelet; bristles purplish, rarely yellowish.   **1. P. ciliare**

1. Pennisetum ciliare (L.) Link, Hort. Berol. 1:213 (1827); Boiss., Fl. 5:445; Post, Fl. 2:701. *Cenchrus ciliaris* L., Mantissa Alt. 302 (1771); Bor, Fl. Iraq 9:474, t.182. [Plate 415]

Perennial, 20-40 cm, tufted. Culms herbaceous, leafy. Panicle (2-)3-5(-10) cm, purplish; rachis scabrous. Involucre sessile, 1½-2 times as long as spikelet; bristles purplish, rarely yellowish; inner bristles stout, flattened, spreadingly ciliate in lower ½, with long tubercle-based hairs; longest bristle about 15 mm, broader and thicker than others; outer bristles shorter, slender, scabridulous. Spikelets up to 5 mm, solitary or in clusters of 2-3, on short stout pedicels. Glumes subequal, acute. Lower floret staminate, its lemma as long as lemma of hermaphrodite floret, its palea shorter or absent. Fl. December-June.

Hab.: Rocks and stony places, mainly in deserts. Sharon Plain, Philistean Plain; Upper and Lower Galilee, Shefela, Judean Desert, N., C. and S. Negev; Upper and Lower Jordan Valley, Dead Sea area, Arava Valley; Golan, Moav, Edom.

Area: Saharo-Arabian, extending into the Sudanian and Mediterranean regions, also into S. Africa.

Lately *P. ciliare* has spread along roadsides in Jordan Valley and Golan.

Eig, Feinbrun & Zohary in Schedae Fl. Exsicc. Pal., Cent. III:3, no. 207 (1934) wrote on the habitats of *P. ciliare* in Palestine: "A common plant of rocky and stony habitats in the Saharo-Sindian Lower Jordan Valley, especially around the Dead Sea. Its northernmost stations are hot basaltic rocks in the northern part of the Sea of Galilee".

2. Pennisetum divisum (J.F. Gmel.) Henrard, Blumea 3:162 (1939) in obs.; Bor, Fl. Iraq 9:496, t.191; Täckholm, Stud. Fl. Eg. ed. 2, 754. *Panicum divisum* J.F. Gmel. in L., Syst. Nat. ed. 13, 2:156 (1791). *Pennisetum dichotomum* (Forssk.) Del., Fl. Eg. 159 (1813-1814), t.8 f.1 (1826) *comb. illeg.* Boiss., Fl. 5:444 (excl. syn. *Phalaris setacea* Forssk.); Post, Fl. 2:700. *Panicum dichotomum* Forssk., Fl. Aeg.-Arab. 20 (1775) non L. (1753). [Plate 416]

Perennial, 40-150 cm, glabrous, with a woody rhizome. Culms indurate, branched at nodes up to about middle, forming tufts resembling those of *Panicum turgidum*; branches often in clusters of 2-3. Leaf-sheaths of previous years persisting at branching nodes, scarious; uppermost sheath of branches often dilated, enveloping spike. Panicle 8–10 cm, dense, whitish; rachis scabridulous. Involucre sessile, 1½ times as long as spikelet; bristles whitish or pale straw-coloured, scabridulous, the inner plumose below; longest bristle 10–15(–18) mm. Spikelets 5–7 mm, solitary, rarely twin. Glumes unequal; upper glume acuminate, nearly as long as lemma of staminate and of hermaphrodite floret. Fl. April–June.

Hab.: Sandy soils, rocky slopes, mostly in deserts. N., C. and S. Negev; Lower Jordan Valley, Dead Sea area, Arava Valley; Transjordan (Al-Eisawi).

Penetrating northwards along the coastal plain to some extent and found in sandy habitats of the Philistean Plain.

Area: Saharo-Arabian, extending into Trop. Africa.

3. **Pennisetum asperifolium** (Desf.) Kunth, Révis. Gram. 49 (1829); Boiss., Fl. 5:446; Post, Fl. 2:701. *Cenchrus asperifolius* Desf., Fl. Atl. 2:388 (1799). *P. tiberiadis* Boiss., Diagn. ser. 1, 13:43 (1854). *P. setaceum* (Forssk.) Chiov. subsp. *asperifolium* (Desf.) Maire, Fl. Afr. Nord 1:333 (1952). [Plate 417]

Perennial, (25–)40–60(–100) cm, tufted. Culms erect, herbaceous, scabridulous under panicle, leafy. Leaves rigid, long, linear, acuminate. Panicle (7–)10–15(–20) cm, dense, usually purplish; rachis pilose. Involucre 3–4 times as long as spikelets, borne on a hispid 2–3 mm peduncle; bristles purplish or whitish, slender; inner bristles somewhat thicker, plumose below, with long hairs that are tuberculate at their base; longest bristle 25–30 mm, much exceeding the others. Spikelets solitary, rarely twin, 1-flowered. Glumes hyaline; the lower ovate-triangular, acute, ⅓–¼ length of upper or absent; upper glume oblong-lanceolate, mucronate-aristulate, 3–5-veined; lower floret absent or reduced to a narrow or bristle-like rudiment; lemma of hermaphrodite floret oblong-lanceolate, enclosing palea. Fl. nearly all year round, mainly in summer.

Hab.: Rocks. Coast of Carmel, Sharon Plain, Philistean Plain; Upper and Lower Galilee, Mt. Carmel, Esdraelon Plain, Samaria, Shefela, Judean Mts., Judean Desert; Hula Plain, Upper and Lower Jordan Valley, Dead Sea area; Golan, Gilead, Ammon, Moav.

Area: S. Mediterranean, Trop. and S. Africa.

This species has become established along roadsides in some localities, especially near springs along the Rift Valley since the seventies of the current century, following the elimination of roadside weeds through use of herbicides.

88. Cenchrus L.

Mostly annuals. Leaf-blades flat; ligule a membranous rim with a fringe of hairs at back and margin. Inflorescence a spike-like raceme of readily deciduous burrs (involucres enclosing a few spikelets); burr formed by partial fusion of bristles and/or rigid flattened spines (modified sterile branchlets), usually retrorsely barbed, especially towards apex. Spikelets few, in a cluster enclosed by burr, 2-flowered; lower floret staminate or neuter, the upper hermaphrodite. Glumes membranous, very unequal; the lower 1-veined, much smaller than the upper; the upper 3-veined, ovate; lemma of lower and upper florets ovate, acuminate, 5-veined. Lodicules absent. Stamens 3. Caryopsis oblong-ellipsoid or obovoid; embryo large; hilum punctiform.

About 25 trop. and temperate species.

1. Burrs with 1 or more whorls of slender outer bristles at base. Lemma of neuter floret nearly as long as lemma of fertile floret and distinctly longer than upper glume. **1. C. echinatus**
- Burrs with flattened spines only (devoid of slender bristles at base). Lemma of neuter floret distinctly shorter than lemma of fertile floret and somewhat longer than upper glume. **2. C. incertus**

1. Cenchrus echinatus L., Sp. Pl. 1050 (1753); Hitchcock, Manual Grasses U.S. ed. 2, 731. [Plate 418]

Annual, 20-40(-60) cm. Culms usually geniculate, rooting at lower nodes. Racemes 3-10 cm; burrs 15 mm broad (incl. spines), 8-9 mm high, often purplish, with one or more whorls of slender bristles at base; spines erect to spreading, ciliate to pubescent, partly barbed. Spikelets usually 4 to a burr. Glumes ovate, acute; lemma of neuter floret longer than upper glume and nearly as long as lemma of hermaphrodite floret. Fl. summer.

Hab.: Weed; heavy, generally irrigated soil. Sharon Plain, Philistean Plain.
Area: Native of Trop. America; adventive in most trop. countries.

Not recorded in the Floras of Boissier or Post.

2. Cenchrus incertus M.A. Curtis, Boston Jour. Nat. Hist. 1:135 (1837). *C. pauciflorus* Benth., Bot. Voy. Sulphur 56 (1844). Incl. *C. tribuloides* L., Sp. Pl. 1050 (1753); Hitchcock, Manual Grasses U.S. ed. 2, 735, ff.1120-1121.*

Annual or short-lived perennial, 30-40 cm, branched at base. Leaf-sheaths laterally compressed, much overlapping, glabrous or sparsely pilose; blades generally glabrous. Racemes 1.5-8 cm; burrs 10-15 mm broad (incl. spines), 10-12 mm high, densely woolly; spines erecto-patent or spreading. Spikelets usually 2 to a burr. Glumes ovate, acute; lemma of neuter floret slightly longer than upper glume and distinctly shorter than lemma of hermaphrodite floret. Fl. summer.

Hab.: Heavy soil. Rare. Acco Plain, Coast of Carmel, Sharon Plain; Jordan Valley.
Area: Native of N. and C. America.

Adventive. Not recorded in the Floras of Boissier or Post.

* Not illustrated.

89. TRICHOLAENA   Schrad. ex Schult. & Schult. fil.

Perennials, rarely annuals. Culms usually branched. Leaf-blades flat or rolled; ligule a ciliate rim. Inflorescence a lax or contracted panicle, with fairly long filiform branches and elongated pedicels. Spikelets small, laterally compressed, without a definite orientation towards axis, falling entire from pedicels at maturity, 2-flowered; lower floret staminate, the upper hermaphrodite. Glumes very unequal; the lower reduced to a minute membranous scale, veinless, or absent; upper glume and lemma of staminate floret (lower lemma) as long as spikelet and similar in shape and texture; lemma of hermaphrodite floret (upper lemma) much shorter, indurated, glabrous and glossy, clasping the margins of palea. Lodicules 2. Stamens 3. Caryopsis ovate-oblong to elliptic-oblong in outline; hilum punctiform.

About 12 species, mainly in Africa and S. W. Asia.

1. **Tricholaena teneriffae** (L. fil.) Link, Handb. 1:91 (1829); Stapf & C.E. Hubbard, in Fl. Trop. Afr. 9:913; Bor, Grasses Burma Ceylon India Pakist. 369; Bor, Fl. Iran. 70:507. *Saccharum teneriffae* L. fil., Suppl. 106 (1781). *Panicum teneriffae* (L. fil.) R. Br., Prodr. Fl. Nov. Holl. 189 (1810); Boiss., Fl. 5:434; Post, Fl. 2:695. [Plate 419]

Perennial, 30–60 cm, with a woody rhizome. Culms much branched at base; nodes pubescent. Leaves glaucous, rigid; mouth of sheaths hairy. Panicle 5–10(–13) × 3–4(–7) cm, ovate to oblong in outline; rachis hairy at nodes; branches and pedicels slender, flexuose; pedicels cupuliform at tip. Spikelets pedicellate, densely long-pilose, 2.5–3 mm (excl. hairs), ovate to oblong, suffused with purple. Lower glume visibly remote from the upper, minute, covered with long silky hairs; upper glume and lower lemma membranous, ovate-lanceolate, more or less mucronate, both covered with long silky hairs much exceeding spikelet; upper lemma 1.5–2 mm, ovate, acute, glabrous, glossy. Stigmas brown when dried. Fl. April–August.

Hab.: Rocks. Lower Galilee, Esdraelon Plain, Judean Desert; Upper and Lower Jordan Valley, Dead Sea area, Arava Valley; Golan, Moav, Edom.

Area: Saharo-Arabian and Sudanian.

Tricholaena teneriffae, as also *Pennisetum asperifolium* and, to a smaller extent, *P. ciliare*, has spread from rocky habitats to roadsides during the last decade following the use of herbicides (cf. A. Danin, Roadside vegetation in Israel, Phytoparasitica 4:2, August 1976). *T. teneriffae* is found along roadsides also in Esdraelon Plain.

90. ANTHEPHORA   Schreb.

Perennials, rarely annuals. Culms erect or geniculately ascending. Leaf-blades flat; ligule membranous or scarious. Inflorescence a cylindrical false spike with an angular rachis bearing subsessile deciduous clusters, each of 3–11 spikelets. Spikelets* lanceolate, ovate or oblong, dorsally compressed, obtuse

---

* Cf. Fig. 193 in Jacques-Félix, Les Graminées d'Afrique Tropicale 1:261 (1962).

or acute to acuminate-setaceous, 2-flowered; outer spikelets of cluster differing from inner ones which are partly sterile. Glumes very dissimilar; lower glumes of outer spikelets coriaceous, smooth, as long as spikelets, turned away from centre of cluster and forming a kind of involucre to cluster, usually with gaps between their bases; inner side of lower glumes distinctly several-veined; lower glumes of inner spikelets hyaline, more or less reduced or absent; upper glumes mostly shorter than lower, ovate at base and abruptly acuminate-setaceous above, 1-veined. Lower floret of spikelet neuter, reduced to empty hyaline or membranous lower lemma; the upper hermaphrodite or staminate, with a somewhat firmer 3–5-veined upper lemma and a 2-veined palea similar in texture. Lodicules absent. Stamens 3; filaments fleshy at base; anthers linear. Stigmas plumose. Caryopsis ellipsoid-oblong, flattened at back; embryo about ½ length of caryopsis; hilum punctiform. x=9.

About 16 species, mainly of Trop. and S. Africa; some extending to S. Iran and Trop. America.

1. **Anthephora laevis** Stapf & C. E. Hubbard, in Fl. Trop. Afr. 9:937 (1930).* [Plate 420]

Perennial, tufted, 30–40 cm. Culms slender, terete, simple, glabrous and smooth, 2–5-noded. Leaf-sheaths glabrous, the basal pale and papery; mouth of sheaths broadly hyaline-margined, bearded; ligule membranous, up to 2–2.5 mm, glabrous; blade up to 6–7 cm long and 4–5 mm broad, firm, glaucous, glabrous. False spikes 4–7 × 0.2–0.3 cm, cylindrical, rigid, glabrous, pale green to purplish-black at maturity; rachis slender, wavy, 3-quetrous; branches reduced to minute stumps; clusters of spikelets oblong, about 5 × 2–2.5 mm, comprising 4–5 spikelets (1–2 often reduced). Spikelets 5 mm. Lower glume 4.5–5 mm, elliptic-oblong to lanceolate-oblong, acute, free almost to base, smooth, coriaceous, purplish-black, 4–6-veined; the veins visible only on inside; upper glume setaceous-subulate from a short ovate base, hyaline, 2.5 mm, minutely hairy. Lower lemma 4 mm, oblong, flat with inflexed margins, minutely hairy all over, 3–6-veined; upper floret hermaphrodite or sometimes staminate; upper lemma 3.5–4 mm, linear-lanceolate, acute, 3-veined, chartaceous, glabrous; palea similar. Anthers about 3 mm. Fl. March–April.

Hab.: Rock-crevices of cliffs facing the Dead Sea, compact conglomerate. N. W. of Dead Sea.

Area: Sudanian.

Described from Eritrea. Not recorded in the Floras of Boissier or Post. Recently collected in Israel for the first time (by A. Shmida).

Tribe 20. ANDROPOGONEAE Dumort. (1823). Annuals or perennials. Leaf-blades flat or rolled; sheaths split nearly to base, without auricles; ligule

* Identified by T. A. Cope (Kew).

membranous, with or without a rim of cilia. Inflorescence consisting of spike-like racemes arranged in a panicle or in a digitate group, or rarely of a single terminal raceme. Spikelets usually in pairs (rarely single or in threes); one of each pair sessile, the other pedicellate, both 2-flowered, differing in sex in most genera. *Sessile spikelets* with lower floret staminate or reduced to an empty lemma (*lower lemma*), upper floret hermaphrodite or pistillate. *Pedicellate spikelets* with 1 or 2 staminate florets or with 1 or 2 empty lemmas, rarely upper floret hermaphrodite. Glumes 2, equal in length, both as long as spikelet, firmer in texture than lemmas; lemmas more delicate; upper lemma of sessile spikelet awned, sometimes reduced to a hyaline base of an awn; awn arising from tip or sinus of lemma, geniculate. Lodicules 2, cuneate-truncate. Stamens (1–2–)3. Ovary glabrous; stigmas plumose. Caryopsis varying from narrowly ellipsoid to nearly globose, not grooved and without an apical appendage, loosely enclosed between lemma and palea; embryo large; hilum punctiform. Chromosomes small; x=5, 9–15, 17, 19.

About 100 genera mainly in trop. and subtrop. regions.

Literature: A. Yacobi, Species of Andropogoneae in Israel, MSc. Thesis, Hebrew University, Jerusalem, 1962 (in Hebrew). H. Jacques-Félix, *Les Graminées (Poaceae) d'Afrique Tropicale* 1:267–314, I.R.A.T., Bull. Sci. 8, Paris, 1962.

91. HEMARTHRIA R. Br.

Perennials, tufted or rhizomatous. Leaf-blades flat; ligule a membranous ciliate rim. Inflorescence of spike-like spathe-subtended racemes. Racemes compressed, single or digitate in fascicles at end of culm or branch; rachis tough or disarticulating tardily. Spikelets seemingly opposite owing to fusion of rachis-internode with pedicel, each pair consisting of a sessile spikelet and the pedicellate companion of the sessile spikelet on the node just below; spikelets hermaphrodite, dorsally compressed, 2-flowered, awnless; lower floret reduced to a hyaline empty lemma; upper floret hermaphrodite. *Sessile spikelets* appressed to hollow of rachis; lower glume flat with very narrowly inflexed margins, coriaceous, as long as spikelet and closing up cavity formed by adjacent rachis-internode and adnate pedicel; upper glume subequal to the lower in length, more delicate, adhering to cavity of rachis except at apex; lemmas hyaline. Lodicules 2, cuneate. Stamens 3. Stigmas exserted. Caryopsis oblong; embryo about ⅔ length of caryopsis; hilum punctiform. *Pedicellate spikelets* with more elongated acuminate glumes; upper glume free.

About 10 species in Trop. Africa and Asia.

1. Hemarthria altissima (Poir.) Stapf & C.E. Hubbard, Kew Bull. 1934:109 (1934). *Manisuris altissima* (Poir.) Hitchcock, Jour. Wash. Acad. Sci. 24:292 (1934). *Rottboellia altissima* Poir., Voy. Barb. 2:105 (1789). *R. fasciculata* Lam., Tabl. Encycl. Méth. Bot. 1:201 (1792). *H. fasciculata* (Lam.) Kunth, Révis. Gram. 153 (1829); Boiss., Fl. 5:467; Post, Fl. 2:712. [Plate 421]

Perennial, 20–60(–80) cm. Culms long-decumbent and ascending, rooting at nodes, branched, glabrous. Leaves glabrous, keeled; sheaths compressed, open, sometimes pilose at mouth. Racemes 4–10 × 0.3–0.4 cm, pedunculate, straight or curved, solitary or 2–6 in fascicle in axils of spathe-like leaves. *Sessile spikelets* 5–8 mm, oblong, glabrous, with obtriangular, more or less conspicuous glabrous callus. Lower glume coriaceous, oblong, 7–9-veined, somewhat constricted below subobtuse apex; upper glume oblong-lanceolate, scarious, acute, adhering to inner face of cavity except at apex. *Pedicellate spikelets* similar in sex and shape to the sessile, 6 mm, but lower glume narrower, more acuminate, the upper free, abruptly mucronate-subulate, the mucro somewhat exceeding the lower glume. Fl. April–December.

Hab.: Wet places, marshes, ditches. Acco Plain, Sharon Plain, Philistean Plain; Upper Galilee, Esdraelon Plain; Hula Plain, Upper Jordan Valley; Golan.

Area: Mediterranean and Trop.

92. LASIURUS Boiss.

Perennials, branched and more or less woody at base. Leaf-blades firm, flat or rolled; ligule a fringe of hairs. Inflorescence a terminal silky-villose spike-like raceme; rachis villose, more or less fragile, disarticulating at maturity at right angles; internodes flattened and excavated on one side; pedicels long-villose. Spikelets awnless, 2 or generally 3 at each node; 1 (or 2) sessile and hermaphrodite, and 1 pedicellate, staminate or neuter; sessile spikelets falling at maturity with laterally adjoining rachis-internode and with pedicellate companion spikelet. *Sessile spikelets* 2-flowered, with ring-like narrow ciliate callus at base; lower floret staminate, the upper hermaphrodite. Glumes unequal, coriaceous; lower glume longer, flat and densely villose on back, acuminate, with inflexed margins; upper glume boat-shaped, keeled, ciliate along keel; lemma and palea hyaline. Lodicules 2, cuneate with a tooth on one side. Stamens 3. Stigmas linear. Caryopsis oblong, loosely enclosed between lemma and palea; embryo about ½ length of caryopsis; hilum large. *Pedicellate spikelets* mostly smaller, with a smaller and glabrous callus. x=7.

A monotypic genus.

Literature: H. Jacques-Félix, *Les Graminées (Poaceae) d'Afrique Tropicale*, 1:278–279, I.R.A.T., Bull. Sci. 8, Paris, 1962. T.A. Cope, Nomenclatural and taxonomic notes on *Lasiurus* (Gramineae), *Kew Bull.* 35:451–452 (1980).

1. Lasiurus scindicus Henrard, Blumea 4:514 (1941); Cope, Kew Bull. 35:451–452 (1980). *Saccharum hirsutum* Forssk., Fl. Aeg.-Arab. 16 (1775), non *Rottboellia hirsuta* Vahl, nec *L. hirsutus* (Vahl) Boiss.; Bor, Fl. Iraq 9:538, t.206. *Elionurus hirsutus* (Vahl) Munro ex Benth., Jour. Linn. Soc. London (Bot.) 19:68 (1881) *comb. illeg.*; Boiss., Fl. 5:466; Post, Fl. 2:711. [Plate 422]

Perennial, 30–60(–80) cm. Rhizome indurated, covered with imbricate cataphylls. Culms erect, branched; nodes shortly hairy. Leaf-blades firm, attenuate to filiform tip; ligule reduced to a fringe of hairs. Racemes 5–10 × 1.5 cm, silky-villose, often enveloped at base by uppermost abbreviated sheath. Sessile and pedicellate spikelets 9–10 mm (excl. tail of glume), or pedicellate spikelets shorter. Lower glume flat on back with incurved margins, densely long-hirsute, acuminate, with a 2-dentate tail; upper glume boat-shaped, ciliate on keel. Fl. January–November.

Hab.: Sands and wadi beds in deserts. C. Negev; Arava Valley; Transjordan (Al-Eisawi).

Area: Saharo-Sindian and Sudanian (S. Palestine, Egypt, Ethiopia, Somalia, Mali, Iraq, Iran, Afghanistan, Pakistan, N.W. India).

Fodder grass in deserts.

93. IMPERATA Cyr.

Perennials, rhizomatous. Culms erect, solid. Leaves tough, long, canaliculate; ligule membranous, ciliate. Inflorescence a spike-like cylindrical silky pani-cle; main rachis and branches not disarticulating. Spikelets muticous, in pairs, one of each pair sessile or short-pedicellate, the other long-pedicellate, all alike, hermaphrodite, 2-flowered, enveloped by long hairs of callus and glumes, each spikelet falling at maturity from its pedicel or from rachis; lower floret neuter and reduced to a lemma or rarely staminate, the upper hermaph-rodite. Glumes subequal, longer than florets, muticous, enveloped by very long silky hairs; lemma of lower floret (lower lemma) hyaline, longer than that of hermaphrodite floret (upper lemma) and enveloping it; lemma and palea of hermaphrodite floret hyaline, fimbriate at apex; palea very short. Lodicules absent. Stamens 1–2. Stigmas exserted. Caryopsis ellipsoid; embryo ½ length of caryopsis; hilum punctiform.

About 10 species in the Tropics and Subtropics.

Literature: C.E. Hubbard, Taxonomy of *Imperata*, *Imp. Agr. Bur. Joint. Publ.* no. 7 (1944).

1. Imperata cylindrica (L.) Raeuschel, Nomencl. Bot. ed. 3, 10 (1797); Clay-ton, in Fl. Eur. 5:265; Bor, Fl. Iraq 9:532, t.205; Boiss., Fl. 5:452; Post, Fl. 2:703. *Lagurus cylindricus* L., Syst. Nat. ed. 10, 2:878 (1759). [Plate 423]

Perennial, 40–80(–120) cm, tufted; rhizomes long. Culms upright, sur-rounded at base by sheaths of dried-up leaves; nodes of culms glabrous, rarely bearded. Radical leaves crowded; leaf-blades stiff, upright, long-acuminate, rolled, scabrous at margins, with a broad white midvein; cauline leaves few, short. Panicle 8–15 cm, erect, dense, cylindrical, silky-white; branches appressed, short. Spikelets 4–5 mm, surrounded by 10–11 mm hairs of callus. Glumes long-pilose, oblong, denticulate at tip, often purplish; lemmas and palea hyaline, glabrous; stigmas brown or purple, conspicuous. Fl. mainly March–July.

Hab.: Alluvial or marshy soil, often overlayed with sandy soil. Common and abundant. Coastal Galilee, Acco Plain, Coast of Carmel, Sharon Plain, Philistean Plain, W. Negev; Upper Galilee, Esdraelon Plain, Samaria, Judean Desert; Dan Valley, Hula Plain, Upper Jordan Valley, Beit Shean Valley, Lower Jordan Valley, Dead Sea area, Arava Valley; ?Golan, Gilead, Ammon, Moav.

Area: Mediterranean, Irano-Turanian, Saharo-Arabian, Trop.

Bor (Fl. Iran. 70:514, 1970) remarks that owing to its rhizome, even a small piece of which is known to develop into a plant, this plant is extremely valuable in soil conservation.

94. SACCHARUM L.

Tall perennials, mostly growing near water. Leaves tough; ligules membranous, hairy on back, short-ciliate at margin. Inflorescence a large, open, densely silky panicle of numerous long fragile racemes; rachis of racemes disarticulating at maturity. Spikelets in pairs, one of each pair sessile, the other pedicellate, all alike, hermaphrodite, 2-flowered, enveloped in and obscured by long weak hairs of callus and of glumes; at maturity pedicellate spikelets falling from their pedicels, the sessile spikelets breaking off together with pedicel and with laterally adjoining internode of raceme-rachis. Glumes subequal, membranous, longer than florets; lower floret neuter, consisting of an empty lemma (lower lemma), or sometimes lemma absent; palea absent; upper floret hermaphrodite; upper lemma hyaline, muticous or awned; palea hyaline, narrower than lemma. Lodicules 2. Stamens 2–3. Stigmas exserted. Caryopsis oblong, free; embryo about ½ length of caryopsis; hilum punctiform.

Five trop. and subtrop. species, including *S. officinarum* L., the Sugar Cane, which probably originated in the area of New Guinea and is cultivated in most trop. regions.

1. Spikelets awnless; rachis of panicle long-villose; hairs enveloping spikelet about twice as long as spikelet.   **1. S. spontaneum** var. **aegyptiacum**
- Spikelets awned; rachis of panicle glabrous; hairs enveloping spikelet at most slightly longer than spikelet   2
2. Hair-envelope of spikelet 5–7 mm, about as long as spikelet (excl. awn).
                                                                                                          2. S. ravennae
- Hair-envelope of spikelet ½–⅔ length of spikelet (excl. awns) and not exceeding 3–4 mm.   **3. S. strictum**

1. Saccharum spontaneum L., Mantissa Alt. 183 (1771) var. **aegyptiacum** (Willd.) Hackel in DC., Monogr. Phan. 6:115 (1889); Clayton, in Fl. West Trop. Afr. ed. 2, 3:466. *S. biflorum* Forssk., Fl. Aeg.-Arab. 16 (1775); Post, Fl. 2:704. *S. aegyptiacum* Willd., Enum. Pl. Horti Berol. 82 (1809); Boiss., Fl. 5:454. [Plate 424]

Perennial, 2–3 m. Culms terete, solid, 1 cm in diameter at base, pubescent below panicle, appressed-puberulent under nodes. Leaves stiff, glaucescent, long, 5–10 mm broad, canaliculate, long-attenuate, serrulate at margins, keeled, with a conspicuous white broad midrib; sheaths bearded at mouth; ligule short, brown, ciliate. Panicle 50–60 × 6–10 cm, plumose, erect; racemes long; rachis of panicle and of racemes long-silky-villose. Spikelets 4–6 mm, awnless, enveloped in long white silky hairs of callus, which are about twice as long as spikelets. Glumes pale purplish, oblong-lanceolate, acuminate, long-ciliate at margins; lemmas muticous; lemma and palea shorter than glumes, short-ciliate. Fl. September–January.

Hab.: Banks of rivers, marshes. Acco Plain, Sharon Plain, Philistean Plain; Upper Galilee, Mt. Carmel; Upper Jordan Valley, Dead Sea area; Transjordan (Al-Eisawi).

Area: Mediterranean, Irano-Turanian, Saharo-Arabian, Trop. of the Old World.

Clayton (*loc. cit.*): "Var. *spontaneum* differs in having a deltoid ligule, and the leaf-blade narrowed to the midrib at its insertion on the sheath".

2. Saccharum ravennae (L.) Murr., Syst. Veg. ed. 13, 88 (1774); Bor, Fl. Iraq 9:539, t.207. *Andropogon ravennae* L., Sp. Pl. ed. 2, 1481 (1763). *Erianthus ravennae* (L.) Beauv., Agrost. 14 (1812); Boiss., Fl. 5:454; Post, Fl. 2:704. [Plate 425]

Perennial, 2–3 m, in large crowded tufts. Culms terete, solid, 8–10 mm thick at base. Leaf-blades flat or canaliculate, very long, long-acuminate, keeled, with a broad white midrib, serrulate at margins; lower sheaths villose; ligule long-ciliate. Panicle 20–50 × 8–12 cm, erect, plumose, silky, whitish, greyish or purplish; rachis of panicle glabrous; racemes on more or less long branches; rachis of racemes long-silky-villose. Spikelets awned, 4–5.5 mm (excl. awn), enveloped by long silky hairs (5–7 mm) of callus and glumes; hair-envelope as long as to slightly longer than spikelet. Glumes lanceolate, acute to acuminate; lemma of hermaphrodite floret lanceolate, prolonged into a conspicuous exserted awn about as long as lemma proper (up to 7 mm). Fl. August–October.

Hab.: Banks of streams, marshes, sandy soil. Acco Plain, Sharon Plain, Philistean Plain; Hula Plain, Upper Jordan Valley, Beit Shean Valley, Dead Sea area; Transjordan (Al-Eisawi).

Area: Mediterranean and Irano-Turanian.

3. Saccharum strictum (Host) Spreng., Pugill. 2:16 (1815); Bor, Fl. Iraq 9:540, t.208; Bor, Fl. Iran. 70:519. *Andropogon strictus* Host, Gram. Austr. 2:2, t.3 (1802). *Erianthus strictus* (Host) Bluff & Fingerh., Comp. Fl. Germ. ed. 2, 1(1):105 (1836); Boiss., Fl. 5:455; Post, Fl. 2:705. *E. hostii* Griseb., Spicil. Fl. Rumel. 2:548 (1846); Aschers. & Graebn., Syn. Mitteleur. Fl. 2(1):35. [Plate 426]

Perennial, 1–2 m. Culms 4–5 mm thick at base, solid, scabridulous. All leaves cauline, long, pale-green, scabrous-margined, with a white midvein; sheaths glabrous; ligule of dense yellowish or reddish hairs. Panicle 15–30 × 3.5–6 cm, reddish to straw-coloured; racemes 2–4 cm; rachis and branches of panicle glabrous, erect. Spikelets awned, 3–4 mm (excl. awns), enveloped in silky hairs of callus and glumes (3–4 mm); hair-envelope somewhat shorter than spikelet. Glumes acute or acuminate; lemma of hermaphrodite floret lanceolate, with a terminal awn; awn somewhat longer than lemma. Fl. April–August.

Hab.: Banks of streams, springs. Rare. Dead Sea area; Golan.

Area: S.E. Europe (Austria, Hungary, Italy, Balkans), E. Mediterranean (Palestine, Lebanon, Syria, Turkey), Irano-Turanian (Iraq, Iran).

95. SORGHUM Moench

Tall annuals or perennials. Culms erect, woody in lower half. Leaf-blades flat; ligule membranous, usually densely hairy at back and margin. Inflorescence a large panicle with numerous branches. Spikelets in pairs on panicle-branches (terminal spikelets often in threes), one of the pair sessile and hermaphrodite, the other(s) pedicellate and staminate or neuter; each sessile spikelet falling at maturity with laterally adjoining rachis-internode and with pedicel(s) of its companion pedicellate spikelet(s) which is deciduous from its pedicel when mature. Rachis tough in cultivated plants. Florets 2; lower floret reduced to an empty lemma (lower lemma); the upper hermaphrodite in sessile spikelets, staminate or neuter in pedicellate spikelets. *Sessile spikelets* awned. Glumes subequal, muticous, the lower coriaceous, flattened, 2-keeled above, mucronulate, the upper shallowly boat-shaped, acute; lemmas hyaline, upper lemma emarginate or 2-lobed, with a geniculate and twisted awn arising from sinus; palea minute or absent. Lodicules 2. Stamens 3. Stigmas densely plumose. Caryopsis obovoid or ellipsoid; embryo ½ length of caryopsis; hilum circular or obovate. *Pedicellate spikelets* awnless, their glumes membranous and narrower than in sessile spikelets.

About 60 species in Trop. or Subtrop.

Literature: J.D. Snowden, *The Cultivated Races of Sorghum*, Alard, 1936; The wild fodder Sorghums of the section *Eu-Sorghum*, *Jour. Linn. Soc. London (Bot.)* 55:191–260 (1955). H. Doggett, *Sorghum*, London & Harlow, 1970. A. Danin, *Sorghum halepense* (L.) Pers. and *S. virgatum* (Hack.) Stapf in Israel, *Israel Jour. Bot.* 32:172–173 (1983).

1. Perennial with a long creeping rhizome. Panicle effuse, pyramidal. Sessile spikelets 5 mm. **1. S. halepense**
- Annual or sometimes perennant with a short rhizome. Panicle contracted, very narrow (up to 5 cm broad); branches more or less erect. Sessile spikelets 6.5–7 mm. **2. S. virgatum**

1. Sorghum halepense (L.) Pers., Syn. Pl. 1:101 (1805); Bor, Fl. Iraq 9:548, t.211; Boiss., Fl. 5:459; Post, Fl. 2:707. *Holcus halepensis* L., Sp. Pl. 1047 (1753). [Plate 427]

Perennial, 60–100(–150) cm, with a creeping rhizome. Culms tough, puberulent at nodes, often branched. Leaf-blades 6–15 mm broad, glabrous, long-acuminate, with a white midvein; ligule hairy, lacerate. Panicle lax, with erecto-patent verticillate branches and with tufts of hairs at base of branches. Spikelets tinged with purple. *Sessile spikelets* 4.5–5.5 mm, ovate-oblong; callus short, bearded. Glumes pubescent, coriaceous, glossy; the lower flattened, 5-veined in upper part, denticulate; the upper boat-shaped, 1-keeled in upper part; lower lemma as long as glumes; upper lemma short, 2-lobed, provided with an awn 2–3 times as long as spikelet, rarely muticous. *Pedicellate spikelets* staminate; glumes papery, narrow, veined, usually purple. 2n=40 (4x) (autotetraploid). Fl. May–December.

Hab.: Irrigated ground; weed in fields of summer crops. Common. Acco Plain, Coast of Carmel, Sharon Plain, Philistean Plain; Upper and Lower Galilee, Esdraelon Plain, Samaria, Judean Mts., Judean Desert, N. Negev; Hula Plain, Upper Jordan Valley, Beit Shean Valley, Lower Jordan Valley, Dead Sea area; Golan, Gilead, Ammon.

Area: Mediterranean, Irano-Turanian and Trop.

2. Sorghum virgatum (Hackel) Stapf, in Fl. Trop. Afr. 9:111 (1917); Täckholm & Drar, Fl. Eg. 1:520; Maire, Fl. Afr. Nord 1:267, f.154. *Andropogon sorghum* subsp. *halepensis* (L.) Hackel var. *virgatus* Hackel in DC., Monogr. Phan. 6:504 (1889).*

Annual, 1 m or more, or sometimes perennant with a short rhizome and rejuvenation buds at base of culms. Culms glabrous, usually with suberect branches. Leaf-blades up to 1.2(–2) cm broad. Panicle 15–60 × 1–5 cm, very narrow; rachis more or less hairy to villose at nodes; branches erect or suberect, the lowest up to 15 cm. Spikelets pale straw-coloured, greenish in upper half, rarely flushed with purplish-brown. *Sessile spikelets* 6–7 mm, narrowly lanceolate, finely awned; callus short, bearded. Glumes hairy to glabrescent; lower lemma about 5 mm, broadly oblong; the upper ovate, 2-lobed, awned; awn up to 2 cm, geniculate and twisted. *Pedicellate spikelets* linear-lanceolate, glabrous, usually staminate. 2n = 20 (Batia Pazy, unpubl.). Fl. summer.

Hab.: Irrigated ground, roadsides sprayed by herbicides. Sharon Plain, Philistean Plain, W. Negev; Lower Galilee, Esdraelon Plain, Shefela, Judean Mts.; Upper Jordan Valley, Arava Valley.

Area: Tropical Africa.

Adventive. Cultivated as a forage-grass.
New record for the Floras of Boissier and Post. Identified by A. Danin.

* Not illustrated.

96. DICHANTHIUM Willem.

Perennials, very rarely annuals. Leaf-blades flat; ligule membranous. Inflorescence subdigitate with 3-9 shortly pedunculate fragile spike-like racemes. Spikelets in pairs, 2-flowered; one of each spikelet-pair sessile and hermaphrodite, the other pedicellate and staminate or neuter; the lowermost 2-4 pairs in raceme neuter; each sessile spikelet falling at maturity with laterally adjoining rachis-internode and with pedicel of companion pedicellate spikelet which is deciduous from its pedicel at maturity. *Sessile spikelets* dorsally compressed, awned, with lower floret reduced to an empty lemma (lower lemma), upper floret hermaphrodite. Glumes equal in length; lower lemma hyaline, awnless; lemma of the hermaphrodite floret (upper lemma) reduced to a hyaline stipe passing into a slender twisted awn; palea minute or absent. Lodicules 2, minute. Stamens 3. Stigmas exserted. Caryopsis oblong, obtuse; embryo more than ½ length of caryopsis. *Pedicellate spikelets* awnless, with 1 or both florets staminate or both neuter. x=10.

About 8 species in trop. and subtrop. regions of the Old World.

Literature: C. Sauvage, *Bull. Soc. Sci. Maroc* 36:93-96 (1936).

1. Dichanthium annulatum (Forssk.) Stapf, in Fl. Trop. Afr. 9:178 (1917); Bor, Fl. Iraq 9:523, t.200. *Andropogon annulatus* Forssk., Fl. Aeg.-Arab. 173 (1775); Boiss., Fl. 5:463; Post, Fl. 2:710. [Plate 428]

Perennial, 30-60(-100) cm, tufted, with a woody rhizome. Culms glabrous, long-bearded at nodes. Leaf-blades flat, tapering to a fine point, usually pilose; ligule membranous, obtuse. Racemes 3-8, 3-6 × 0.2 cm (excl. awns), often purplish; rachis and pedicels long-ciliate. Spikelets numerous, overlapping, 3-4 mm (excl. awns), oblong; callus very short. *Sessile spikelets* awned; lower glume ciliate along keels and more or less hairy, hairs with tubercle at base; awn of upper lemma brown, geniculate and twisted, 4-5 times as long as spikelet; lower lemma much shorter than glumes. 2n=40, 60 (4×, 6×). Fl. nearly all year round.

Hab.: Irrigated places, ditches, banks of water-reservoirs. Common. Sharon Plain, Philistean Plain; Upper Galilee, Esdraelon Plain, Shefela, Judean Mts., Judean Desert; Hula Plain, Upper and Lower Jordan Valley, Dead Sea area; Golan, Gilead.

Area: Subtrop. and Trop.

D. annulatum has spread and established itself along roadsides since herbicides have come into use in the seventies of the current century.

97. EREMOPOGON Stapf

Perennials, rhizomatous. Culms slender, simple below, more or less branched above; branches often in fastigiate bundles. Inflorescence a solitary spike-like raceme at end of a culm or branch, subtended by a bladeless spathe-like sheath; rachis of raceme jointed, fragile; internodes of rachis and pedicels of spikelets filiform. Spikelets in pairs, 2-flowered; one of each spikelet-pair sessile and hermaphrodite, the other pedicellate and staminate or neuter; the lowermost 1–3 pairs in raceme homogamous; each sessile spikelet falling at maturity with laterally adjoining rachis-internode and with pedicel of its companion pedicellate spikelet, which is deciduous from its pedicel when mature. *Sessile spikelets* dorsally compressed, awned, with lower floret reduced to an empty lemma (lower lemma), upper floret hermaphrodite. Glumes equal in length, thinly chartaceous to membranous; lower glume 2-keeled with inflexed margins, the upper 3-veined, acutely keeled; lower lemma hyaline, awnless; lemma of the hermaphrodite floret (upper lemma) reduced to a linear stipe passing into a slender twisted awn; callus small, shortly bearded. Lodicules 2. Stamens 3. Stigmas exserted, longer than styles. *Pedicellate spikelets* awnless, with the lower floret staminate or neuter, the upper suppressed, or both suppressed. x=10.

Three or four species in trop. and subtrop. regions of Old World.

Literature: M.B. Raizada & S.K. Jain, The genus *Eremopogon* Stapf and its affinities, etc., *Jour. Bombay Nat. Hist. Soc.* 54:858–865 (1957).

1. Eremopogon foveolatus (Del.) Stapf, in Fl. Trop. Afr. 9:183 (1917); Bor, Fl. Iran. 70:537. *Andropogon foveolatus* Del., Fl. Eg. 16, t.8 f.2 (1813–1814); t.8 f.2 (1826); Boiss., Fl. 5:462; Post, Fl. 2:709. [Plate 429]

Perennial, 30–40 cm, rhizomatous and tufted. Rhizomes short, much branched. Culms very slender, geniculate at base, few-noded; nodes bearded. Leaf-blades folded or flat, very narrow, sparsely long-ciliate near base and at mouth of sheaths; sheaths shorter than internodes; ligule hyaline, lacerate, truncate, continued at sides into sheath-margins. Racemes 4–5 × 0.3 cm (excl. awns), dense; rachis joints and pedicels ciliate along margins. Sessile and pedicellate spikelets subequal, purplish. *Sessile spikelets* glossy, with a long-bearded short callus, awned. Glumes of sessile spikelets glabrous, the lower flattened at back, minutely 2-denticulate at tip, with a round pit above middle between 2 keels; lemma of upper (hermaphrodite) floret narrow, passing into an awn; awn glabrous, geniculate, 3–4 times as long as spikelet, divided into a slender brown twisted column and an equally long pale bristle. *Pedicellate spikelets* staminate or neuter, often devoid of lemma. 2n=40 (4x). Fl. January–April.

Hab.: Sandy soils in deserts. Rare. Lower Jordan Valley, Dead Sea area, Arava Valley; Edom.

Area: Saharo-Sindian and Sudano-Deccanian (S. Palestine, Egypt, N. and Trop. Africa, Canary Islands, Trop. Arabia to India).

98. ANDROPOGON L.

Mostly perennials. Leaf-blades coarse, often folded; sheaths somewhat inflated in upper leaves; ligule membranous, hairy at back, ciliate at margin. Inflorescence a single pair of spike-like racemes or digitate. Spikelets in pairs, 2-flowered; one spikelet of each pair sessile and hermaphrodite, the other pedicellate and staminate or neuter; each sessile spikelet falling at maturity with laterally adjoining rachis-internode and with the pedicel of its companion pedicellate spikelet, which is deciduous from its pedicel at maturity. *Sessile spikelets* awned, with lower floret reduced to an empty lemma (lower lemma), upper floret hermaphrodite. Glumes equal or subequal in length; lower glume dorsally flattened, 2-keeled with sharply inflexed margins; lower lemma hyaline, awnless; lemma of hermaphrodite floret (upper lemma) 2-fid or 2-dentate, awned; callus short, shortly bearded; palea short or absent. Lodicules 2. Stamens 3. Stigmas exserted. Caryopsis narrowly lanceolate or oblong; embryo about ½ length of caryopsis. *Pedicellate spikelets* staminate or neuter, differing from the sessile in size and shape, awnless or aristulate. x=9.

Over 100 species, mostly in Trop. and Subtrop. of both hemispheres.

1. **Andropogon distachyos** L., Sp. Pl. 1046 (1753); Stapf, in Fl. Trop. Afr. 9:218. *Pollinia distachya* (L.) Spreng., Syst. Veg. ed. 16, 1:288 (1824); Boiss., Fl. 5:456; Post, Fl. 2:705. [Plate 430]

Perennial, 30–80 cm, tufted, with a short rhizome. Culms erect, glabrous, ending in a pair of spike-like racemes. Leaf-blades flat, tapering to fine point; uppermost leaf-sheath often enclosing inflorescence at base; ligule hairy. Racemes 4–10 cm, rather thick, spreadingly awned, greenish often tinged with dull purple; raceme-rachis and pedicels ciliate along outer margin and often puberulent. *Sessile spikelets* heterogamous, 9–12 mm, broadly lanceolate, acuminate; callus short, shortly bearded. Glumes unequal in texture; the lower subcoriaceous, glabrous or shortly pubescent, lanceolate, with prominent veins, 2-keeled and 2-winged; keels prolonged above into 2 mucros or short bristles; upper glume shorter, membranous, boat-shaped, 1-keeled above, tip minutely 2-dentate with a 7–10 mm long slender awn in sinus; lower floret reduced to a ciliate hyaline lemma; upper floret hermaphrodite, its lemma hyaline, glabrous, 2-fid to beyond middle into lanceolate lobes and with a geniculate 2–2.5 cm awn twisted below middle. Stigmas brown. Caryopsis 4 mm, oblong, with a large embryo. *Pedicellate spikelets* shorter (about 8 mm) and narrower; lower glume ending in a slender awn; awn about 6–7 mm; upper glume much shorter, shortly aristulate. 2n=36, 54 (4×, 6×). Fl. April–June.

Hab.: Batha and maquis. Not common. Coastal Galilee, Coast of Carmel, Sharon Plain, Philistean Plain; Upper and Lower Galilee, Mt. Carmel, Mt. Gilboa, Samaria, Shefela, Judean Mts.; Gilead, Ammon.

Area: Mediterranean incl. Canary Islands, S. and Trop. Africa.

99. Hyparrhenia N. J. Andersson ex E. Fourn.

Perennials, rarely annuals. Leaves coarse; leaf-blades flat; ligule membranous, lacerate. Inflorescence a compound lax panicle subtended by a bladeless spathe and consisting of a few to many pairs of fragile spike-like racemes, each raceme-pair at end of a peduncle included in a *spatheole*. Spikelets in pairs, 2-flowered, one sessile and hermaphrodite, the other pedicellate and staminate or neuter; lowermost pair(s) of spikelets in raceme often homogamous; each sessile spikelet falling off at maturity with laterally adjoining rachis-internode and with pedicel of its companion pedicellate spikelet, which is deciduous from pedicel when mature. *Sessile spikelets* obtuse, awned, with lower floret reduced to an empty lemma (lower lemma), upper floret hermaphrodite. Glumes equal in length, more or less chartaceous; lower glume with a shallow longitudinal depression; lower lemma hyaline; lemma of hermaphrodite floret (upper lemma) stipe-like, shortly 2-dentate, ending in a long geniculate more or less hairy brown awn; callus short, bearded; palea short or absent. Lodicules 2. Stamens 3. Stigmas exserted. Caryopsis oblong; embryo about ½ length of caryopsis; hilum punctiform. *Pedicellate spikelets* often longer than sessile ones, acute, awnless. x=9.

About 50 species in Africa, Arabia, the Mediterranean area.

Literature: W. D. Clayton, *A Revision of the Genus Hyparrhenia*, Kew Bull. Add. Ser. 2, 196 pp., London, 1969.

1. Hyparrhenia hirta (L.) Stapf, in Fl. Trop. Afr. 9:315 (1919); Bor, Fl. Iraq 9:530, t.204. *Andropogon hirtus* L., Sp. Pl. 1046 (1753); Boiss., Fl. 5:464; Post, Fl. 2:710; incl. var. *glabriglumis* Opphr., Florula Transiord. 145, Bull. Soc. Bot. Genève ser. 2, 22:264 (1931). [Plate 431]

Perennial, 40–100 cm, densely tufted, coarse. Culms slender, wiry, glabrous, sometimes fastigiate, passing into a panicle subtended by a spathe. Leaf-blades narrow, with a whitish midvein and with a few scattered long tubercle-based hairs above ligule; cauline leaves usually rolled or folded, attenuate above to a fine point; ligule oblong, scarious or membranous, ciliate-denticulate. Panicle 10–40 cm; spatheoles turning pale reddish, acuminate, usually glabrous. Racemes 2–3 cm (excl. awns), slightly diverging, not deflexed, oblong, flattened, often purplish; one raceme in each pair sessile, the other on a 3–5 mm short-hairy peduncle; rachis and pedicels ciliate; spikelets of lowermost pair homogamous, staminate, of other pairs heterogamous. *Sessile spikelets* 6 mm, linear-oblong, reddish. Lower glume 9–11-veined, obtusely 2-keeled, more or less villose, rarely glabrescent (var. *glabriglumis* Opphr.); upper glume 3-veined, 1-keeled, ciliate; lower lemma about as long as glumes; upper lemma narrowly linear, hyaline, with 2 short lobes; awn 1.5–3.5 cm, slightly geniculate and twisted; column appressed-hairy. *Pedicellate spikelets* 7 mm. 2n=36, 45 (4x, 5x). Fl. February–June.

Hab.: Stony hillsides and hardened soils, open overgrazed ground. *Thymus capitatus–Hyparrhenia hirta* association, Ceratonieto–Pistacietum lentisci.

Very common and stand-forming. Coastal Galilee, Coast of Carmel, Sharon Plain, Philistean Plain; Upper and Lower Galilee, Mt. Carmel, Mt. Gilboa, Samaria, Shefela, Judean Mts., Judean Desert, W. and C. Negev; Hula Plain, Upper Jordan Valley, Beit Shean Valley, Lower Jordan Valley, Dead Sea area, Arava Valley; Golan, Gilead, Ammon, Moav, Edom.

Area: Mediterranean, Irano-Turanian and Saharo-Arabian extending into Trop. Africa and S. Africa.

In the Mediterranean parts of the country *H. hirta* has become the most prominent species along roadsides since herbicides have come into use in the seventies of the current century.

100. CYMBOPOGON Spreng.

Densely tufted perennials, often aromatic. Leaves coarse. Inflorescence a panicle subtended by a bladeless spathe-like leaf-sheath and consisting of fascicles of erect or pendulous raceme-pairs, each pair supported by a spatheole; one raceme of each pair sessile, the other pedunculate; rachis of racemes jointed, fragile. Spikelets in pairs, 2-flowered, one spikelet sessile and hermaphrodite, the other pedicellate and staminate or neuter; the lowest spikelet-pair of one or of both racemes homogamous, staminate or neuter; each sessile spikelet falling at maturity with the laterally adjoining rachis-internode and with the pedicel of its companion spikelet, which is deciduous from pedicel when mature. *Sessile spikelets* compressed, awned, with the lower floret reduced to an empty lemma (lower lemma), the upper hermaphrodite. Glumes equal or subequal, chartaceous, the lower flat or narrowly grooved, with margins sharply inflexed at least from middle upwards; upper glume boat-shaped, keeled, 1-veined; lower lemma hyaline, ciliate; lemma of hermaphrodite floret (upper lemma) 2-fid with an awn arising from sinus. Lodicules 2, cuneate, glabrous. Stamens 3. Caryopsis elliptic in outline; embryo more than ½ length of caryopsis; hilum obovate. *Pedicellate spikelets* awnless, their lemmas and paleas reduced or absent.

About 40 species, mostly in the Tropics of the Old World.

Many of the species yield essential oils used in perfumery, etc. Some species, like *C. citratus* (DC.) Stapf, are cultivated commercially for their aromatic oils.

1. Cymbopogon parkeri Stapf, Kew Bull. 1929:10 (1929); Bor, Grasses Burma Ceylon India Pakist. 122, 131; Bor, Fl. Iraq 9:518, t.198. [Plate 432]

Perennial, up to 1 m, densely tufted, with a lemon odour. Culms erect, glabrous; nodes of culms brown, mostly bearded on one side. Basal leaf-sheaths woolly or sparsely hairy; leaf-blades very narrow, nearly filiform, flexuose to curled; ligule membranous, triangular. Panicle 8–25(–40) cm, erect, narrow, of raceme-pairs in short remote fascicles. Raceme-pairs spreading, 2–2.5 cm, with long white silky hairs borne on raceme-rachis and pedicels; lowermost joint of sessile raceme adnate to lowermost pedicel of pedicellate spikelet, both swollen, hard. *Sessile spikelets* linear-lanceolate, acute, pale green below, purplish in upper part. Glumes equal in length; lower glume scabridulous between keels, more or less grooved and minutely villose or scurfy in groove; keels shortly ciliate, apex minutely 2-denticulate; upper glume with broad hyaline margins, ciliate in the upper half; awn of lemma slender, about twice as long as spikelet, geniculate and twisted below middle; stigmas exserted, rust-coloured. *Pedicellate spikelets* purplish, with a prominently veined upper glume. Fl. March–June.

Hab.: Granite rocks. Rare. S. Negev; Arava Valley.

Area: Saharo-Arabian (S. Palestine, ?Egypt, S. Iraq, Iran, Afghanistan, N.W. India).

"This is a species of hot dry places and has often been confused with *C. schoenanthus*" (Bor, Fl. Iraq, *loc. cit.*).

124. PALMAE

Trees, mainly monoecious or dioecious, also shrubs and vines, with thick fibrous roots, sometimes monocarpic. Stem mostly unbranched, more rarely dichotomously branched. Leaves large, persistent, sheathed at base and petiolate, simple and variously divided, more often pinnatisect or flabellate. Inflorescence usually very large, simple or racemosely branched, subtended by a large spathe, mostly an axillary spike or head-like spadix, or frequently branched into a panicle. Flowers numerous, mostly unisexual, mostly actinomorphic, small, bracteate, anemophilous or entomophilous. Perianth in 2 whorls, 3-merous, free or connate, varying in texture; rarely perianth absent. Staminate flower: stamens 6 in 2 whorls, or 3, rarely 9–∞, free or connate at base; anther-thecae 2. Pistillate flower: ovary superior, usually 3-carpellate, usually syncarpous, 3–1-locular; locules 1-ovulate; mostly only 1 ovule fertile; styles as many as locules, free or connate; stigma mostly 1, 3-fid or 3-lobed. Fruit a berry, drupe or nut, much varying in size. Seed large with copious endosperm.

Over 200 genera and 2,500 tropical and subtropical species, several of them of economic importance.

Literature: O. Drude, Palmae, in: Engler & Prantl, *Pflanzenfam.* II, 3:1–93, 1889. P.B. Tomlinson, Systematics and ecology of the Palmae, *Ann. Rev. Ecol. Syst.* 1979, 10:85–107 (1979). P.E. Boissier, *Flora orientalis* 5:45–47, 1882. G.E. Post, *Flora of Syria, Palestine and Sinai*, ed. 2, 2:556–557, 1933.

1. Leaves pinnatisect. Stem unbranched.   **2. Phoenix**
  – Leaves flabellate. Stem branched above.   **1. Hyphaene**

1. Hyphaene Gaertn.

Dioecious trees, mostly with a dichotomously branched stem. Leaves flabellate; petiole with a ligule at tip. Inflorescence a branched spadix; each branch partly enclosed in a spathe, cylindrical, with densely imbricate hairy bracts. Flowers solitary, small, subsessile. Staminate flower: outer whorl of perianth 3-lobed, the inner of 3 free perianth-segments; stamens 6 with free filaments. Pistillate flower: segments of both whorls free, rounded; stamens rudimentary; ovary 3-2-locular, stigmas 3-2, sessile. Fruit a drupe, 1-locular by abortion of 2 carpels, with a fibrous-spongy pericarp and a woody stone; pericarp sugary, tasting like gingerbread. Endosperm horny, hollow; embryo small, at tip of seed.

Thirty species in warm Africa, Madagascar, Arabia and W. India.

Literature: V. Täckholm & M. Drar, *Flora of Egypt* 2:273-296, Cairo, 1950. C.X. Furtado, Asian species of *Hyphaene, Gardens' Bull., Singapore* 25:299-309 (1970).

1. Hyphaene thebaica (L.) Mart., Hist. Palm. 3:225, t.131-133 (1838-1845); Dyer, Fl. Trop. Afr. 8:120 (1901); Boiss., Fl. 5:46; Post, Fl. 2:557. *Corypha thebaica* L., Sp. Pl. 1187 (1753). *Cucifera thebaica* Del., Fl. Eg. 145 (1813-1814); tt.1 & 2 (1826); ?*H. sinaitica* Furtado, Gardens' Bull., Singapore 25:306, f.3 (1970). [Plate 433]

Tree, 5-10(-20) m. Stem dichotomously branched, ringed. Leaves 20-30, crowded at tips of branches, flabellate, 1 m broad; lobes 20-25, linear-lanceolate, acute; petiole flattened, canaliculate, with numerous hooked prickles. Spadix emerging from among leaves, surrounded by several spathes. Flowers small, pale yellow. Fruit (5-)6-7(-8.5) × (4-)4.5-6(-7) cm, obliquely ovoid to ellipsoid, obscurely carinate, obtuse, irregularly nodulose and minutely pitted, dark brown, glossy. Fl. March.

Hab.: Sandy saline soil; 20-30 m a. s. l. Arava Valley, Gulf of Elat coast.

Area: Sudanian.

The species of *Hyphaene* from the coast of the Gulf of Elat is given here under *H. thebaica*, though doubts remain whether this taxon is indeed identical with *H. thebaica* of Upper Egypt and Nubia. Furtado (*loc. cit.*) described *H. sinaitica* from Wadi Taba on the coast of the Gulf of Elat and differentiated it from the Indian *H. dichotoma* (White) Furtado, by its smaller fruits and especially by the lined and cracked epidermis of the fruit. However, this last feature is an artifact. For his study Furtado had received fruits which had been lying on the ground for a long period. Fruits collected directly from the *Hyphaene* trees on the Gulf of Elat coast show neither lines nor cracks and in surface and size resemble the fruits of *H. thebaica*, as shown on the plate by Delile (Fl. Eg., t.2, 1826).

Furtado (*op. cit.* 306) remarks: "The fruit of *H. thebaica* is always conspicuously carinate, broadly truncate at the base and much narrowed towards the apex. Its epidermis is light brown or spadiceous and remains intact. In *H. sinaitica*, on the other hand, the fruit is obovate and obscurely carinate, with its epidermis dark brown and partly discoloured in lines so as to give it a net-like appearance". In his opinion (*op. cit.* 299): "the true *H. thebaica* is a species restricted in its distribution from the mountains of the Sudan... and not to the seashore of the Mediterranean or the Gulf of Suez or the Red Sea, where there are many other branched species which cannot be said to be in any way closely related to *H. thebaica*".

Groups of trees of *Hyphaene* growing on the Gulf of Elat coast are known from 5 locations: 1. Evrona, 2 km N. of Elat; 2. Elot, at the N. end of Elat Airport; 3. Wadi Taba; 4. Bir Sweir; 5. Nuweiba (A. Danin, pers. communication). Fruits collected in Evrona, Wadi Taba and Nuweiba show considerable variability in size. They are (5-)6-7(-8.5) cm high to (4-)4.5-6(-7) cm thick.

More well-documented taxonomic studies of *Hyphaene* species with branched stems are still needed for the delimitation of species in the area along the Red Sea coasts.

2. PHOENIX L.

Dioecious trees. Stem unbranched, columnar, covered with old leaf-bases. Leaves pinnatisect. Inflorescence a much branched spadix enveloped in a simple spathe before flowering. Flowers sessile. Staminate flower: outer whorl of perianth cup-shaped, 3-dentate, the inner of 3 free valvate segments; stamens 6 or 3, anthers linear, subsessile. Pistillate flower: perianth similar to that of staminate flower; ovaries 3, free, globose-ovate; stigmas sessile, hooked; only a single carpel ripens. Fruit a 1-seeded berry (date). Seed with a longitudinal furrow; endosperm of hard cellulose; embryo small, embedded in endosperm half-way along side opposite furrow.

Seventeen species in Africa, S. Asia, Crete.

Literature: V. Täckholm & M. Drar, *Flora of Egypt* 2:165–273, Cairo, 1950.

1. Phoenix dactylifera L., Sp. Pl. 1188 (1753); Boiss., Fl. 5:47; Post, Fl. 2:557.*

Tree 10–20 m, slender. Leaves 2–3 m, glaucescent, oblong-lanceolate in outline, with a thick semi-terete rachis; leaf-segments linear-lanceolate, acuminate, folded lengthwise. Branches of spadix long, flexuous. Date edible, sessile, ovoid to ellipsoid. Fl. April–May.

Hab.: Cultivated and naturalized, and probably indigenous in salty marshes in the deserts; associated with *Juncus arabicus* (Aschers. & Buchenau) Adamson. Lower Jordan Valley, Dead Sea area, Arava Valley; Edom.

Area: Saharo-Arabian, penetrating into the Sudanian.

In Israel the date palm is cultivated commercially mainly along the Rift Valley, from the Upper Jordan Valley down to the Arava Valley.

125. ARACEAE**

Perennial herbs with underground rhizomes or tubers or, in the Tropics, climbing shrubs, epiphytes, etc., mostly monoecious. Stem mostly sympodial; each joint of sympodium with 1 or more cataphylls and 1 foliage leaf. Foliage leaves alternate, entire or variously divided, usually petiolate. Inflorescence a

---

* Not illustrated.
** In co-operation with Mr J. Koach (Tel-Aviv University).

spadix surrounded or subtended by a *spathe*; axis of spadix often prolonged beyond flower-bearing zones as a fleshy appendage. Flowers ebracteate, mostly unisexual, small, mostly devoid of perianth, mostly protogynous; pistillate flowers usually borne at or near base of spadix; staminate flowers above them; sterile flowers, usually filiform or subulate, present in some genera between the zones of pistillate and staminate flowers, or in some taxa also above the staminate zone. Staminate flowers of 1–8 stamens, free or united into a synandrium; filaments short or absent. Pistillate flowers of 1 pistil; ovary mostly superior, 1–3-locular; stigma usually sessile or subsessile. Fruit a berry. Seeds mostly with endosperm.

One hundred and fifteen genera and about 2,000 trop. and subtrop., more rarely temperate species. Many species contain latex.

Colocasia esculenta (L.) Schott in Schott & Endl. (*C. antiquorum* Schott) is one of the most ancient and important cultivated plants of S. E. Asia and Trop. Africa, known as *taro*. Its large rhizomes lose the poisonous substances they contain after being boiled and are eaten as valuable food. Probably indigenous in E. India and still cultivated in countries of the Middle East. It has been found as an escapee from cultivation on banks of streams in En Gedi (Dead Sea area) and Wadi Qilt (Judean Desert).

Literature: A. Engler, Araceae-Aroideae, in: Engler, *Pflanzenreich* 73 (IV. 23F): 1–249, 1920. J. Galil, The inflorescence of *Arum, Mada* (Science, in Hebrew) 3:41–44 (1958). H. Riedl, Araceae, in: K. H. Rechinger (ed.), *Flora iranica* 1:1–8, 1963; Kritische Untersuchungen über die Gattung *Eminium*, etc., *Ann. Naturh. Mus. (Wien)* 73:103–121 (1969). An. A. Fedorov (ed.), *Flora partis europaeae URSS* 4:314–317, Leningrad, 1979. Tutin et al. (eds.), *Flora europaea* 5:268–272, Cambridge, 1980. J. Koach, *Bio-Ecological Studies of Flowering and Pollination in Israeli Araceae*, Ph. D. Thesis, Tel-Aviv University (in preparation). P. E. Boissier, *Flora orientalis* 5:30–45, 1882. G. E. Post, *Flora of Syria, Palestine and Sinai* ed. 2, 2:549–556, 1933.

1. Leaves appearing after flowering, rarely during flowering, not hastate nor sagittate at base.   **3. Biarum**
– Leaves appearing before flowers, hastate or sagittate at base   2
2. Lateral lobes of leaf-blade pedately divided into several secondary lobes which are spirally twisted. Spathe wrinkled and warty on upper face.   **2. Eminium**
– Leaves and spathe not as above   3
3. No sterile flowers present on spadix. Margins of spathe-tube connate for most of its length.   **4. Arisarum**
– Subulate or filiform sterile flowers present between zones of pistillate and staminate flowers and above staminate flowers. Margins of spathe-tube not connate.
                                                                         1. Arum

1. ARUM L.

Scapose perennial herbs with tuber. Tuber often depressed-globose. Leaves all basal, appearing before flowers, long-petiolate, sheathing at base; blades sagittate to hastate. Spathe marcescent, convolute at base into an ovoid or oblong tube, slightly constricted at throat; limb longer than tube, oblong-lanceolate or ovate-lanceolate. Spadix lengthened above into a naked long appendage; appendage clavate or cylindrical, tapering. Flowers unisexual, devoid of perianth; pistillate flowers at base of spadix; staminate flowers above them; sterile flowers subulate to filiform, thickened and verruculose at base, occupying interspace between zones of pistillate and staminate flowers and also above zone of staminate flowers. Staminate flowers: anthers 2–5 together, subsessile to sessile, each with 2 thecae dehiscing by a pore. Pistillate flowers: ovary oblong, 1-locular, many-ovulate; stigma sessile, subhemispherical. Berry few-seeded. Seeds globose.

Fifteen European, Mediterranean and Irano-Turanian species.

Literature: J. Hruby, Le genre *Arum*, *Bull. Soc. Bot. Genève* Ser. 2, 4:113–160, 330–371 (1912).

1. Spathe-tube not exceeding 2 cm in diameter   2
- Spathe-tube 2.5 cm or more in diameter   3
2. Spathe-limb light green on inner side and purple along margins (except in very rare albino mutants); its upper part bent forwards; spathe-tube partly purple inside.   **3. A. hygrophilum**
- Spathe-limb dark purple on inner side (or sometimes yellowish), not purple-margined; spathe-tube whitish inside.   **4. A. elongatum**
3(1). Peduncle of inflorescence less than ½ length of petioles; inflorescence sitting close to soil surface. Inner side of spathe-limb with large rounded dark purple spots on a yellowish-green or in some cases on purple ground; spathe-tube greenish inside. Inflorescence emitting the smell of dung and carrion during anthesis.
                                                                      1. A. dioscoridis
- Peduncle of inflorescence about a long as petioles. Inner side of spathe-limb dark purple; inner side of spathe-tube purple in its upper half. Inflorescence usually emitting the smell of fermenting fruit, in some populations the smell of dung and carrion.   **2. A. palaestinum**

1. **Arum dioscoridis** Sm. in Sibth. & Sm., Fl. Graec. Prodr. 2:245 (1816) & Fl. Graeca 10:37, t.947 (1840) emend. Engler in DC., Monogr. Phan. 2:583 (1879); Engler, Pflanzenreich 73 (IV. 23F):72 (1920); Mout., Nouv. Fl. Liban Syrie 1:184; Boiss., Fl. 5:35; Post, Fl. 2:552. [Plate 434]

Tuber discoid, 4–8 cm in diameter. Petiole 15–40 cm, longer than blade; leaf-blade hastate to sagittate at base; terminal lobe of blade triangular-ovate. Inflorescence emitting a strong smell of dung and carrion; pollination by beetles and flies. Peduncle of inflorescence less than ½ length of petioles. Spathe-limb 15–45 × 5.5–15 cm, lanceolate, acuminate, velvety and with large rounded dark purple spots on a yellowish-green or purple ground on inner side; outer side greenish; spathe-tube oblong-ovoid, 3–6 cm, 2.5–4 cm in diameter, pale green on out- and inside. Sterile flowers purplish to yellowish, longer and more numerous above staminate zone than between two zones of flowers. Appendage of spadix dark purple (rarely yellowish in albino mutants). Fl. March–May.

Hab.: Batha, maquis, forest of *Quercus ithaburensis*, rocky places, fallow fields. Coastal Galilee, Acco Plain, Coast of Carmel, Sharon Plain, Philistean

Plain; Upper and Lower Galilee, Mt. Carmel, Mt. Gilboa, Samaria, Judean Mts., Judean Desert, N. Negev; Upper Jordan Valley; N. Golan, Gilead, Ammon, Moav.

Area: E. Mediterranean extending into the W. Irano-Turanian (Palestine, Lebanon, Syria, Asia Minor, Cyprus, Chios, Rhodes, Mesopotamia).

Several varieties have been described, but they are unclear and hardly merit separate status, as already remarked by Boissier.

2. Arum palaestinum Boiss., Diagn. ser. 1, 13:6 (1854); Engler, *op. cit.* 76 (1920); Hook. fil., Bot. Mag. 91:t.5509 (1865); Mout., Nouv. Fl. Liban Syrie 1:185; Boiss., Fl. 5:37; Post, Fl. 2:553. [Plate 435]

Tuber discoid, 4–8 cm in diameter. Petiole 15–50 cm, longer than blade; leaf-blade hastate to sagittate at base; terminal lobe of blade triangular-ovate. Peduncle about as long as petioles. Inflorescence mostly with a strong smell of decaying fruit and fermentation; pollinated by *Drosophila*; in some populations with a smell of dung and carrion and pollinated by beetles and flies. Spathe 16–50 × (3–)6.5–13.5 cm; tube of spathe 3–6 cm, 2.5–4.5 cm in diameter, oblong-ovoid; outer side green, inner side purple and velvety in its upper half, greenish in the lower; limb lanceolate, 3–7 times as long as tube. Appendage of spadix dark purple, somewhat shorter than limb of spathe. Sterile flowers dark purple, subulate or filiform, thickened and verruculose at base, numerous in interspace between the two zones, fewer above staminate zone. Fl. March–April.

Hab.: Batha, maquis, alluvial soils, rocky places. Philistean Plain; Mt. Carmel, Mt. Gilboa, Samaria, Shefela, Judean Mts., Judean Desert, N. Negev; Transjordan (Al-Eisawi).

Area: E. Mediterranean (Palestine, Lebanon).

Two ecotypes can be distinguished within this species. They differ in the smell emitted by the inflorescence during anthesis and consequently in their pollinators, as also in their habitat. The inflorescence of the typical ecotype of *A. palaestinum* emits the smell of fermenting fruit and is pollinated by *Drosophila* flies; the plants are known from the hills of Mt. Carmel, Samaria, the Judean Mts. and Ammon, on terra rossa derived from calcareous rocks and dolomites. The second ecotype, found and studied by J. Koach, differs in emitting the smell of dung and carrion and in its pollinators which are beetles and flies other than *Drosophila*; it is found on alluvial soil in the southern part of Shefela, on Eocene beds of Mt. Gilboa, on basalt in S. Golan and on Mt. Hermon.

With regard to the smell of the inflorescence, Mouterde (*loc. cit.*) writes (translated from French): "Engler attributes a pleasant odour to *A. palaestinum*. No such odour is noted in populations of Lebanon, which are practically devoid of smell. A plant found in Damour, which seems to be intermediate between the two species (*A. palaestinum* and *A. dioscoridis*), emitted a weak *A. dioscoridis* smell". Post (*loc. cit.*) remarks that *A. palaestinum* is "bad smelling".

Some of the variability within both *A. palaestinum* and *A. dioscoridis* may be due to introgression between the two species.

3. Arum hygrophilum Boiss., Diagn. ser. 1, 13:8 (1854); Engler, *op. cit.* 77, f.12A, B (1920); Mout., Nouv. Fl. Liban Syrie 1:187; Boiss., Fl. 5:37; Post, Fl. 2:553. [Plate 436]

Tuber ovoid to discoid, 2.5--5 cm in diameter. Petiole 20-50 cm, much longer than blade, broadly sheathing to about middle; leaf-blade erect, hastate at base, with lateral lobes obtuse to subobtuse; terminal lobe oblong-lanceolate. Peduncle as long as or somewhat shorter than petioles. Inflorescence without perceptible smell, attracting sandflies (Psychodidae). Spathe 8-16 × 2-3.5 cm, light green and purple-margined, velvety on inside; limb oblong-lanceolate, acute, erect, with upper part bent forward; tube oblong, 2.5-3.5 cm, 1.2-1.7 cm in diameter, partly purple within. Appendage of spadix purple, slender, ⅔ length of spathe-limb; sterile flowers verruculose and whitish at base, filiform and often purple at tip; those above the staminate zone more numerous than below it. Fl. February–April.

Hab.: Near water and shade of orchards. Coastal Galilee, Acco Plain, Coast of Carmel, Sharon Plain, Philistean Plain; Upper and Lower Galilee, Mt. Carmel, Esdraelon Plain, Mt. Gilboa, Samaria, Judean Mts., Judean Desert; Dan Valley, Upper Jordan Valley; Golan, Gilead, Moav, Edom.

Area: E. Mediterranean (Palestine, Lebanon, Syria, S. Turkey, Cyprus).

4. Arum elongatum Stev., Bull. Soc. Nat. Mosc. 30(3):67 (1857); Mout., Nouv. Fl. Liban Syrie 1:188. *A. orientale* Bieb., Fl. Taur.-Cauc. 2:407 var. *elongatum* (Stev.) Boiss., Fl. 5:39 (1882). *A. orientale* Bieb. subsp. *elongatum* (Stev.) Engler p. p., Pflanzenreich 73 (IV. 23F):79 (1920).*

Tuber discoid, 3-5 cm in diameter. Petiole 15-40 cm; leaf-blade hastate-sagittate; terminal lobe oblong-lanceolate. Peduncle about as long as to longer than petioles. Inflorescence devoid of smell; pollination by Ceratopogonidae (Diptera). Spathe 11-23 cm, green to dark purple on outside; limb of spathe lanceolate, acuminate, 2-3 cm broad, dark purple (or in some cases yellowish) and velvety on inside; tube narrowly cylindrical, 2.5-4 cm, 0.8-2.0 cm in diameter, whitish and glossy inside. Appendage of spadix purple or yellowish, as long as or shorter than limb. Sterile flowers thickened and verruculose at base, filiform and purple at tip, more numerous below the staminate zone than above it. Fl. April (in Edom); May–July (on Mt. Hermon).

Hab.: Among rocks. Edom; Mt. Hermon.

Area: E. Mediterranean and W. Irano-Turanian.

2. Eminium (Blume) Schott
[*Helicophyllum* Schott (1853) non Brid. (1827)]

Scapose perennial herbs with tuber. Tuber often flattened from above. Leaves all basal, appearing before flowers, long-petiolate, sheathing at base; blade

*Not illustrated.

hastate or intorto-pedatisect or -pedatipartite. Spathe marcescent, convolute at base into an ovoid or oblong subventricose tube; limb oblong or ovate-oblong, erect. Spadix shorter than spathe, prolonged above into a clavate or cylindrical appendage. Flowers unisexual, devoid of perianth; pistillate flowers at base of spadix; staminate flowers above them; sterile flowers occupying long interspace between the 2 zones. Staminate flowers: anthers 2, each with 2 thecae dehiscing by a slit. Pistillate flowers: ovary 1-locular, 2-ovulate; stigma sessile or subsessile, hemispherical. Berry usually 1-seeded.

Five E. Mediterranean and C. Asiatic species.

1. Eminium spiculatum (Blume) Schott, Aroid. 16 (1855). *Arum spiculatum* Blume, Rumphia 1:121 (1836). *Helicophyllum crassipes* (Boiss.) Schott, Syn. Aroid. 22 (1856); Boiss., Fl. 5:42; Post, Fl. 2:555. *A. crassipes* Boiss., Diagn. ser. 1, 13:9 (1854). [Plate 437]

Tuber depressed, 3–10 cm. Leaves 3–6;; petiole 15–40 cm; blade pedatipartite; terminal lobe oblong-lanceolate, acute; lateral lobes divided into 4–9 secondary lanceolate to linear lobes, spirally twisted and alternately directed up- and downwards. Scape shorter than petioles, hidden in the ground, about as thick as spathe-tube. Inflorescence emitting a strong smell of dung and carrion, attracting beetles and flies. Spathe convolute at base into an oblong thickened and stiff tube, with margins free to its base, greenish or whitish on outside, with or without purple spots; spathe-limb fleshy, oblong-ovate, dark purple, wrinkled and densely warty on upper (inner) side. Appendage of spadix 6–7 cm, 5–7 mm thick, clavate, tuberculate-rugose, dark purple to greenish-yellow. Interspace between zones of pistillate and staminate flowers longer than both zones together; sterile flowers in interspace elongate-filiform, tapering and curved upwards. Pistillate flowers yellowish, tuberculate; stigma sessile. Staminate flowers sessile, yellow to orange. Berries usually 1-seeded, white, with a leathery pericarp reticulate on surface. Seeds remaining in the ground, not dispersed.

Subsp. **spiculatum.** Incl. var. *tigrinum* (forma *tigrinum* Engler in DC., Monogr. Phaner. 2:600, 1879) & var. *albo-virens* Engler, Pflanzenreich 73 (IV. 23F):131 (1920). Plants generally large. Flowering during day-hours; flowering season: March–May. Terminal lobe of leaf 2.5–6.5 cm broad; lateral lobes (4–)5–9 on either side, lanceolate. Petioles mostly purple-spotted on abaxial side in their lower part, more rarely devoid of spots. Spathe-limb 12–20 × 7–14 cm, recurved from its base; its margins turned upwards; epidermis on inner side devoid of papillae, limb glossy on inner side; outer side mostly with scattered purple spots. Spathe-tube 8–12 cm, 2.5–4 cm in diameter, often purple-spotted on outside, more rarely devoid of spots.

Hab.: Fallow fields, cultivated ground. Coastal Galilee, Acco Plain, Coast of Carmel, Sharon Plain, Philistean Plain; Upper and Lower Galilee, Mt. Carmel, Esdraelon Plain, Samaria; Judean Mts., Judean Desert; Golan, Ammon.

Subsp. **negevense** Koach & Feinbr.* Incl. var. *angustatum* (Schott) Engler, Pflanzenreich 73 (IV. 23F):132 (1920). *Helicophyllum angustatum* Schott, Syn. Aroid. 22 (1856). Plants generally smaller. Flowering at night; flowering season: February–March. Terminal lobe of leaf 1–3 cm broad, often oblong; lateral lobes 4–6 on either side, narrow, oblong-lanceolate, often oblong. Petioles not purple-spotted. Spathe-limb 5–11 × 4–7 cm, opening up in the afternoon; its upper half recurved, margins not turned upwards; epidermis on inner (adaxial) side papillose, thus inner side velvety, not glossy; outer (abaxial) side mostly devoid of purple spots, or with purple spots limited to margins. Spathe-tube 5–8 cm, 1.7–3 cm in diameter, very rarely purple-spotted on outside.

Hab.: Sandy soil. W., N. and C. Negev; Sinai.

Area of species: E. Mediterranean, extending into the W. Irano-Turanian (Palestine, N. Egypt, Lebanon, Syria, Iraq).

3. BIARUM Schott

Scapose perennial herbs with tuber. Tuber globose, ovoid or discoid. Membranous cataphylls surrounding and concealing base of foliage-leaves and of inflorescence. Leaves all basal, appearing after flowering, rarely during flowering, petiolate, ovate, elliptic, oblong-lanceolate or linear. Scape short. Spathe marcescent, convolute at base, forming a subventricose or cylindrical tube with margins more or less connate at base or higher up; limb longer than tube, lanceolate to linear. Spadix prolonged above into a naked appendage, cylindrical or fusiform, tapering towards tip. Flowers unisexual, devoid of perianth; pistillate flowers at base of spadix; staminate flowers above them; sterile flowers filiform, thickened at base, in interspace between zone of pistillate and of staminate flowers, sometimes also above zone of staminate flowers. Staminate flowers: 1–3 anthers, each with 2 thecae dehiscent at tip by a slit or pore. Pistillate flowers: ovary 1-locular, 1-ovulate; style present or stigma sessile. Berry 1-seeded. Seed obovoid; seeds remaining in the ground, not dispersed.

Fifteen species, Mediterranean, Irano-Turanian and Saharo-Arabian.

Literature: S. Talavera, Revisión de las especies españolas del genero *Biarum* Schott, *Lagascalia* 6:275–296 (1976).H. Riedl, Tentative key for the identification of species in *Biarum* and *Eminium*, with notes on some taxa included in *Biarum, Aroideana* 3:24–31 (1980).

1. Tube of spathe connate nearly up to mouth   2
- Tube of spathe connate only at base or at most to ⅔ of its length   3
2. Limb of spathe narrowly linear or subulate, narrower than tube. Plant of desert sands in the S. of country.   **4. B. olivieri**
- Limb of spathe lanceolate, much broader than tube. Plant of habitats differing from the above.   **3. B. angustatum**

*See Appendix, p. 397.

3(1). Sterile flowers few, more or less straight and horizontal, clearly decreasing in
length in the upward direction.   **1. B. pyrami**
- Sterile flowers numerous, curved down and then upwards, reaching the wall of
tube and ending nearly at the same level. Plants of Upper Galilee and northward.
2. B. bovei

1. Biarum pyrami (Schott) Engler in DC., Monogr. Phan. 2:576 (1879);
Engler, *op. cit.* 139, f.21F–J (1920); Boiss., Fl. 5:33; Post, Fl. 2:550. *Ischarum
pyrami* Schott, Prodr. 66 (1860); Hook., Bot. Mag. 88:t.5324 (1862). [Plate
438]

Tuber depressed-globose, 4.5–8 cm in diameter. Leaves 4–11; blades obovate-
elliptic or oblong-elliptic, gradually tapering to a petiole, obliquely veined.
Tube of spathe urceolate, convolute, with margins connate only near base;
limb of spathe 9–25 × 3–9 cm, lanceolate, acuminate, often revolute, dark
purple on inside. Appendage of spadix dark purple, fusiform, up to 1 cm
thick. Zone of staminate flowers much longer than zone of pistillate flowers;
sterile flowers, in interspace between the 2 zones, few, stiff, more or less
straight and horizontal, clearly decreasing in length in the upward direction.
Style as long as ovary.

Var. **pyrami.** Flowering October to November. Leaves appearing after flowering. Leaf-blades often dark-spotted. Limb of spathe usually marked with
dense brown-purple spots on outer (abaxial) side.

Hab.: Batha, *Quercus ithaburensis* forest, stony ground and fallow fields.
Acco Plain; Lower Galilee, Mt. Carmel, Esdraelon Plain, Mt. Gilboa, Samaria, Shefela, Judean Mts.; Hula Plain, Upper and Lower Jordan Valley.

Var. **serotinum** Koach & Feinbr.* Flowering November to early January.
Leaves devoid of spots, often appearing more of less simultaneously with
flowers. Limb of spathe usually not spotted on outer (abaxial) side.

Hab.: As above. Upper Galilee; Golan.

Area of species: E. Mediterranean and W. Irano-Turanian (Palestine,
Lebanon, ?Syria, S. Turkey).

2. Biarum bovei Decne., Ann. Sci. Nat. Bot. ser. 2, 4:346 (1835); Boiss., Fl.
5:33; Post, Fl. 2:550; incl. var. *karsaami* (Schott) Boiss., Fl. 5:34 (1882). *B.
bovei* Blume, Rumphia 1:114, t.29 (1836) var. *blumei* Engler in DC., Monogr.
Phan. 2:577 (1879). *B. bovei* subsp. *blumei* (Engler) Engler, Pflanzenreich 73
(IV. 23F):140 (1920). [Plate 439]

Tuber depressed-globose, 5.5–8 cm in diameter. Leaves 4–10, sometimes
appearing towards the end of flowering; blades mostly ovate, abruptly or
gradually contracted to a petiole. Tube of spathe with margins connate up to
$\frac{1}{3}$–$\frac{2}{3}$ its length; limb 10–20 × 3–7 cm, lanceolate, acuminate, dark purple on

*See Appendix, p. 398.

inside, often recurved. Appendage of spadix dark purple, fusiform, up to 1 cm thick. Zone of pistillate flowers hemispherical to flattened, zone of staminate flowers longer, cylindrical; sterile flowers, next to the pistillate zone, numerous and dense, filiform, more or less equal (9–16 mm), curved down and then upwards, reaching the wall of tube and ending nearly at the same level. Style about as long as ovary. Fl. November–December.

Hab.: Maquis areas. Upper Galilee; Golan.

Area: E. Mediterranean (Palestine, Lebanon, Turkey).

The diagnostic importance of sterile flowers in the taxonomy of *Biarum*, namely their number, shape, orientation and position in relation to pistillate and staminate flowers, was noted by Talavera (*op. cit.* 280). These characters have been more or less neglected, especially in the treatment of *B. bovei*. This species was treated by Engler in a broad sense as comprising three subspecies, two of which are W. Mediterranean. Only later were these two taxa separated from *B. bovei* as distinct species and named *B. dispar* (Schott) Talavera (1976) and *B. carratracense* (Haenseler) Font Quer (1926); Boissier (Fl. 5:34, 1882) had already remarked that they deserved the rank of species. Photographs by Talavera of the flowering spadices of the two latter species emphasize the diagnostic value of characters pertaining to sterile flowers.

The conspicuous difference in sterile flowers between *B. pyrami* and *B. bovei* is pointed out by J. Koach in his still unpublished study of Araceae in Israel. Previously, the distinction between these two E. Mediterranean species was unclear. A reexamination of our herbarium material indeed revealed that *B. bovei* clearly differs from *B. pyrami* in its sterile flowers. These are numerous and finely filiform, flexuose, concentrated in the area next to the pistillate flowers and ending at one level, whereas those of *B. pyrami* are few in number, stiff, more or less horizontal, becoming shorter in an upward direction.

The two species differ also in geographic distribution and ecology. *Biarum pyrami* has been described from Tiberias. In Israel it grows in batha and fallow fields in Upper and Lower Galilee, Mt. Carmel, Esdraelon Plain, Mt. Gilboa, Judean Mts., Hula Plain, Golan. Its area extends into Syria, S. Turkey, Iraq. *B. bovei* has been described from the Lebanon. In Israel it is mainly confined to maquis areas in Upper Galilee and Golan. The extent of the species' general geographic area is unclear. The distribution given by Engler (1920) for *B. bovei* subsp. *blumei* needs verification based on examination of sterile flowers in the relevant material.

3. Biarum angustatum (Hook. fil.) N. E. Brown, Jour. Linn. Soc. London (Bot.) 18:225 (1880); Engler, *op. cit.* 142 (1920); Boiss., Fl. 5:34; Post, Fl. 2:550. *Ischarum angustatum* Hook. fil., Bot. Mag. 104:t.6355 (1878). [Plate 440]

Tuber depressed-globose, 3.5–7 cm in diameter. Leaves 4–13, appearing after flowering; blade oblong-lanceolate, subacute; petiole narrowly sheathing. Tube of spathe cylindrical, whitish, much narrower than limb, with margins connate high up; limb of spathe 7–21 × 4–10 cm, oblong-lanceolate to lanceolate, acute to acuminate, often revolute, dark purple on inside. Appendage of spadix dark purple, fusiform, up to 8 mm thick, somewhat shorter than spathe. Zone of pistillate flowers shorter than zone of staminate flowers; sterile flowers not numerous, more or less horizontal, confined to lower ½ of interspace between the 2 zones. Fl. October.

Hab.: Calcareous sandstone, basalt, nari, sandy and marly soil. Sharon Plain, Philistean Plain; Upper and Lower Galilee, Esdraelon Plain, ?Samaria, ?Judean Mts., Judean Desert, N. and C. Negev; Hula Plain, Upper Jordan Valley; Golan, Ammon.

Area: ?Endemic. E. Mediterranean (Palestine, ?Lebanon, ?S. Syria).

4. Biarum olivieri Blume, Rumphia 1:115 (1836) p. p.; Engler, *op. cit.* 142 (1920). *B. alexandrinum* Boiss., Diagn. ser. 1, 13:6 (1854); Boiss., Fl. 5:34; Post, Fl. 2:551. [Plate 441]

Tuber 1-2(-3.5) cm in diameter, globose to ovoid. Leaves and inflorescences often more or less covered by sand. Leaves 4-9; blade 2-5(-10) mm broad, linear, linear-lanceolate or lanceolate, often undulate at margin. Tube of spathe 1.5-2 cm, ovoid, much broader than limb, whitish outside, purple inside, with margins connate high up; limb of spathe 3-7.5 cm, linear or subulate, green or brownish. Appendage of spadix slender, 2 mm thick, brown to purple with light spots, about as long as spathe. Zone of pistillate flowers ovate; zone of staminate flowers cylindrical; interspace between 2 zones without or with only a few rudimentary sterile flowers. Fl. November-December.

Hab.: Deserts, sandy soil. W., N. and C. Negev; Lower Jordan Valley; Edom.

Area: E. Saharo-Arabian (Palestine, Egypt).

4. ARISARUM Mill.

Scapose perennial herbs with an ovoid or cylindrical tuber or rhizome. Cataphylls few. Leaves basal, generally 1-2, appearing with flowers, ovate-cordate or sagittate, long-petiolate, shortly sheathing at base. Scape purple-spotted, mostly about as long as leaves. Spathe marcescent; tube oblong, about 1 cm in diameter, straight, closed; limb curved forwards, hood-like, concealing upper part of spadix. Spadix shorter than or equalling spathe and adnate at back to lower part of spathe, ending in a filiform or clavate appendage. Flowers unisexual, devoid of perianth. Pistillate flowers 3-5, at base on one side of spadix; staminate flowers just above in a lax spike; sterile flowers absent. Staminate flowers short-pedicellate; stamen 1; anthers horseshoe-shaped; thecae dehiscing by a continuous slit. Pistillate flowers: ovary 1-locular, many-ovulate; style cylindrical-conical; stigma subhemispherical. Berries aggregated. Seeds about 6, ovoid, acute.

Three Mediterranean species.

Literature: J. Galil, Morpho-ecological studies on *Arisarum vulgare* Targ.-Tozz., *Israel Jour. Bot.* 27:77-89 (1978).

1. Arisarum vulgare Targ.-Tozz., Ann. Mus. Firenze 2(2):67 (1810); Engler, *op.*

cit. 145 (1920); Boiss., Fl. 5:44; Post, Fl. 2:555; incl. var. *veslingii* (Schott) Engler in DC., Monogr. Phan. 2:563 (1879). [Plate 442]

Tuber ovoid to shortly cylindrical. Petiole subterete, 2–5 times as long as blade; blade sagittate-ovate, usually short-apiculate; petioles and scape purple-spotted. Spathe 6–7 cm, cylindrical, pale green with 10 alternating whitish and purplish longitudinal stripes; tube as long as or slightly longer than limb. Appendage of spadix slender, curved forward or downward, exserted from tube of spathe. Pistillate flowers 3–7; staminate flowers 25–40. Seeds 2–6. Fl. October–April.

Hab.: Fallow fields and batha, on a variety of soils. Common. Coastal Galilee, Acco Plain, Coast of Carmel, Sharon Plain, Philistean Plain; Upper and Lower Galilee, Mt. Carmel, Esdraelon Plain, Samaria, Shefela, Judean Mts.; Dan Valley, Upper Jordan Valley; Golan.

Area: Mediterranean.

126. LEMNACEAE

Small monoecious fresh-water plants, floating or submerged, consisting of undifferentiated green thalloid fronds. Fronds solitary or several connected. Roots unbranched, on lower side of fronds, 1 or several or absent. Leaves lacking. Reproduction chiefly by buds developing from a single median pouch or groove under edge of frond, or from 2 lateral pouches. Flowering infrequent in most species. Flowers borne on surface of water, unisexual, naked. Inflorescence minute, consisting of 1 stamen and 1 pistil in a single cavity on upper surface of frond, or of 2 stamens and 1 pistil enclosed in a membranous sheath in one of the 2 bud-pouches. Anthers with 1 or 2 thecae. Ovary 1-locular, 1–7-ovulate; style and stigma simple. Fruit a utricle. Seeds 1–4, large, smooth or ribbed, with a prominent operculum; endosperm scarce or absent.

Six genera and about 29 species in both hemispheres, except in the polar regions.

During the last three to four decades some of the water bodies, in which Lemnaceae were previously recorded, have disappeared, mainly because of increasing withdrawal of water for human consumption. On the other hand, artificial ponds have been established in which some Lemnaceae have reappeared.

Literature: A. Engler, Lemnaceae, in: Engler & Prantl, *Pflanzenfam.* 2(3):154–164, 1889. F. Hegelmaier, Systematische Uebersicht der Lemnaceen, *Bot. Jahrb.* 21:268–305 (1895). A. Eig, *Bull. Inst. Agr. Nat. Hist. Tel-Aviv* 6:50–52 (1927). E.H. Daubs, A monograph of Lemnaceae, *Illinois Biol. Monogr.* 34:1–118, 1965. C. den Hartog & F. van der Plas, A synopsis of the Lemnaceae, *Blumea* 18:355–368 (1970). Y. Waisel & N. Liphschitz, *Water Plants of Israel*, Tel-Aviv, 1971 (in Hebrew). T.G. Tutin et al. (eds.), *Flora europaea* 5:273, 1980. D. Porath, Y. Efrat & T. Arzee, Morphological pattern and heterogeneity in populations of duckweeds, *Aquatic Botany* 9:159–168

(1980). P. Uotila, in: *Flora of Turkey* 8:64–67, 1984. P.E. Boissier, *Flora orientalis* 5:28–30, 1882. G.E. Post, *Flora of Syria, Palestine and Sinai,* ed. 2, 2:547–549, 1933.

1. Fronds not more than 1.5 mm in diameter, without roots.   **3. Wolffia**
- Fronds more than 1.5 mm in diameter, with 1 or more roots   2
2. Each frond with 2 or more roots.   **1. Spirodela**
- Each frond with a single root.   **2. Lemna**

1. SPIRODELA Schleiden

Fronds floating, solitary or 2–5 connected by ventrally attached short hyaline stipes, flat to inflated, 3–15-veined. Roots 2–18 in a cluster on each frond; base of roots covered by a ventral scale. Bud-pouches 2, lateral. Inflorescence of 2 stamens and 1 ovary, enclosed in a sheath in 1 of the lateral pouches. Utricle more or less globose, slightly winged. Seeds ovoid, smooth or ribbed.

Five cosmopolitan species.

1. Spirodela polyrhiza (L.) Schleiden, Linnaea 13:392 (1839); Boiss., Fl. 5:30. *Lemna polyrhiza* L., Sp. Pl. 970 (1753); Post, Fl. 2:549. [Plate 443]

Roots on each frond 3–15; root-cap acute. Fronds 3–10 mm, usually 2–5 connected by hyaline stipes, orbicular to ovate; lower surface flat to slightly convex, often purplish.

Hab.: Pools. Sharon Plain (apparently disappeared);; Hula Plain, Upper Jordan Valley; Golan.

Area: Borealo-Trop.

2. LEMNA L.

Fronds floating or submerged, solitary or remaining attached, 1–3(–5)-veined. Root single, peltately attached, with a root-cap visible under low magnification. Bud-pouches 2, lateral. Inflorescence consisting of 2 stamens and 1 ovary enclosed in a sheath in one of the bud-pouches. Utricle more or less globose, 1–6-seeded. Seeds longitudinally ribbed with transverse striations; operculum prominent.

Nine temperate and trop. species.

Literature: E. Landoldt, Morphological differentiation and geographical distribution of the *Lemna gibba–Lemna minor* group, *Aquatic Botany* 1:345–363 (1975). A. Witztum, An ecological niche for *Lemna gibba* L. that depends on seed formation, *Israel Jour. Bot.* 26:36–38 (1977).

1. Fronds submerged, oblong-lanceolate, more or less acute, translucent, with persistent stalks, 5–15 mm.   **3. L. trisulca**
- Fronds floating, round or ovate, with fugacious stalks, 2–6 mm   2
2. Fronds bulging underneath, nearly hemispherical, spongy on lower surface and appearing whitish when dried, usually 4–5-veined; veins branched from base. Root-cap acute.   **1. L. gibba**

– Fronds flat on both surfaces, usually obscurely 3-veined; veins branched at some distance from base. Root-cap obtuse. **2. L. minor**

1. Lemna gibba L., Sp. Pl. 970 (1753); Boiss., Fl. 5:30; Post, Fl. 2:549. [Plate 444]

Plants floating. Fronds 3.5–6 × 2.5–5 mm, orbicular to obovate in outline, thick, flat to slightly convex on dorsal surface, convex and inflated-spongy on ventral; ventral surface with 40–50 meshes. Root-cap acute. Fl. May–June (Witztum, 1977).

Hab.: Lakes and pools, often small pools formed on rocks. Sharon Plain; Upper and Lower Galilee, Judean Mts., Judean Desert; Hula Plain; Golan, Edom.

Area: Borealo-Trop.

Witztum (*loc. cit.*) studied populations of *Lemna gibba* in the hills north of Hebron during two consecutive rainy seasons. The populations grew in small pools that form every winter on flat rocks, and he observed seed-germination at the beginning of the rainy season.

2. Lemna minor L., Sp. Pl. 970 (1753); Boiss., Fl. 5:29; Post, Fl. 2:548. [Plate 445]

Plants floating, commonly cohering in groups of 2–5. Fronds 3–6 × 1.5–4 mm, suborbicular or elliptic-obovate, somewhat asymmetrical, flat on both surfaces; ventral surface with 10–20 meshes. Root-cap obtuse. Fl. summer.

Hab.: Pools and lakes. Fairly common. Sharon Plain; Judean Mts., Judean Desert; Hula Plain, Upper Jordan Valley; Golan, Edom.

Area: Borealo-Trop. Widely distributed.

3. Lemna trisulca L., Sp. Pl. 970 (1753); Boiss., Fl. 5:29; Post, Fl. 2:548. [Plate 446]

Plants submerged (except at flowering time), remaining attached in large numbers. Fronds 5–15 × 2.5–5 mm, oblong-lanceolate, more or less acute and serrulate at apex, flat on both surfaces, thin, translucent, 3-veined, tapering at base into persistent stalk, remaining attached to each other and forming chains. Root-cap acute.

Hab.: Pools and lakes. Sharon Plain; Hula Plain.

Area: Borealo-Trop.

3. WOLFFIA Horkel ex Schleiden

Minute floating rootless plants, rarely exceeding 1 mm in diameter, mostly numerous. Fronds globose to ovoid. Flowers borne in median pouch with a circular opening on upper surface, without a spathe; staminate flower 1, with 1 stamen; pistillate flower 1, with globose, 1-ovulate ovary, short style and circular stigma. Fruit a globose utricle. Seed large in proportion to frond, globose, smooth, with a prominent operculum.

Eight trop. and temperate species.

Literature: Y. Lipkin, *Wolffia arrhiza* (L.) Horkel ex Wimmer on the Golan Plateau, *Israel Jour. Bot.* 22:175–177 (1973).

1. Wolffia arrhiza (L.) Horkel ex Wimmer, Fl. Schles. ed. 3, 140 (1857); Post, Fl. 2:548. *Lemna arrhiza* L., Mantissa Alt. 294 (1771). [Plate 447]

Fronds ellipsoid or subglobose, 0.5–1.2 × 0.4–1 mm; upper surface somewhat convex with a clearly delimited rounded border or shoulder; lower surface strongly bulging.

Hab.: Ponds. Sharon Plain (apparently disappeared); Golan.

Area: Borealo-Trop.

127. SPARGANIACEAE

A single genus.

1. SPARGANIUM L.

Monoecious perennial herbs of marshes, banks of rivers, lakes, etc., with a creeping rhizome. Stems leafy, simple or branched, erect or floating, usually projecting above water. Leaves alternate, 2-stichous, strap-shaped, sheathing at base, devoid of ligule, gradually passing into bracts. Inflorescence panicle-like, of globose unisexual, sessile or pedunculate heads; staminate heads above pistillate ones. Perianth scale-like, persistent, of usually 3–6 linear or spathulate segments. Staminate flowers with 3 or more stamens; filaments mostly free; anthers basifixed, with 2 thecae. Pistillate flowers each subtended by a bract, with 1 pistil, sessile or short-pedicellate; ovary superior, 1–2-locular, 1- or 2-carpellate, 1–2-ovulate; style simple or forked, persistent; stigma unilateral. Fruit drupe-like, beaked, with a spongy exocarp and hard endocarp. Seed white, with a mealy endosperm and straight embryo.

Nineteen species in the temperate regions of the N. hemisphere, and also in Australia and New Zealand.

Literature: H.G.A. Engler, Sparganiaceae, in: Engler & Prantl, *Pflanzenfam.* 2(1):192–193, 1889. K.O.P.P. Graebner, in: *Pflanzenreich* 2(IV. 10):1–26, 1900. C.D.K. Cook, *Sparganium* in Britain, *Watsonia* 5:1–10 (1961). H. Riedl, Sparganiaceae, in: K.H. Rechinger (ed.), *Flora iranica* 59:1–4, 1969. J.E. Alexeev, Sparganiaceae, in: An. A. Fedorov (ed.), *Flora partis europaeae URSS* 4:322–326, t.41, Leningrad, 1979. C.D.K. Cook, *Sparganium*, in: T.G. Tutin et al. (eds.), *Flora europaea* 5:274–275, 1980. C.C. Townsend, Sparganiaceae, in: *Flora of Iraq* 8:208–211 (1984). P. Uotila, *Sparganium*, in: *Flora of Turkey* 8:555–558, 1984. P.E. Boissier, *Flora orientalis* 5:48, 1882. G.E. Post, *Flora of Syria, Palestine and Sinai*, ed. 2, 2:557–558, 1933.

1. Sparganium erectum L., Sp. Pl. 971 (1753); Post, Fl. 2:558. *S. ramosum* Huds., Fl. Angl. ed. 2, 2:401 (1778); Boiss., Fl. 5:48. [Plates 448, 449]

Plant 0.5–1 m, immersed at base. Stem robust, erect. Leaves erect, subcoriaceous, triangular in transverse section, keeled at back. Inflorescence branched; staminate heads borne on lateral branches. Perianth-segments thick, dark brown to black at apex. Seeds with 6–10 longitudinal ridges. Fl. summer.

1. Fruit obpyramidal, (3–)4–6(–7) mm broad, truncate above, dark brown to black, with a distinct shoulder between shorter upper and much longer lower part, 3–5-angled in cross-section, abruptly beaked; beak at most 2 mm; ovary usually 2-locular. Perianth-segments of pistillate flowers brown at tip.   subsp. **erectum**
- Fruit ellipsoidal, tapering at both ends, 2–3.5 mm broad, projecting from the head by about ½, light brown, with shoulder between upper and lower parts indistinct, barely angular in cross-section; beak 2 mm or longer; ovary 1-locular. Perianth-segments of pistillate flowers pale at tip.   subsp. **neglectum**
- Fruit ovoid, obovoid to globose, light brown, 5–8 × 4–7 mm, upper part hemispherical; ovary usually 1-locular, only part of ovaries fertile; apparently a hybrid between subsp. *erectum* and subsp. *neglectum*.   subsp. **oocarpum**

Subsp. **erectum**; Hegi, Ill. Fl. Mitteleur. ed. 3, 2(1):288, f.142 (1977). Subsp. *polyedrum* (Aschers. & Graebn.) Schinz & Thell. in Schinz & Keller, Fl. Schweiz ed. 3, 2:14 (1914); Riedl, Fl. Iran. 59:3. *S. ramosum* subsp. *polyedrum* Aschers. & Graebn., Syn. Mitteleur. Fl. 1:283 (1897). *S. polyedrum* (Aschers. & Graebn.) Juzep., in Fl. URSS 1:219 (1934). [Plate 448]
   Hab.: Banks of streams and non-saline marshes. Rare. Hula Plain.

Subsp. **neglectum** (Beeby) K. Richter, Pl. Eur. 1:10 (1890); Schinz & Thell. in Schinz & Keller, Fl. Schweiz ed. 3, 2:14, t.258 (1914). *S. neglectum* Beeby, Jour. Bot. (London) 23:26 & 193 (1885); Juzep., in Fl. URSS 1:220. [Plate 449]
   Hab.: Banks of streams and non-saline marshes. ?Acco Plain, Sharon Plain, Philistean Plain; Esdraelon Plain, Dan Valley; Hula Plain; Golan.

Subsp. **oocarpum** (Čelak.) Domin, Preslia 13–15:53 (1935); Cook, in Fl. Eur. 5:274. *S. neglectum* var. *oocarpum* Čelak., Österr. Bot. Zeitschr. 46:425 (1896). *S. ramosum* var. *oocarpum* (Čelak.) Aschers. & Graebn., Syn. Mitteleur. Fl. 1:282 (1897). *S. erectum* subsp. *neglectum* var. *oocarpum* (Čelak.) Hayek, Prodr. Fl. Penins. Balcan. 3:426 (1933).
   Hab.: Hula Plain.
   Area of species: Euro-Siberian, Mediterranean and Irano-Turanian.

128. TYPHACEAE

A single genus.

1. Typha L.

Monoecious perennials, usually tall herbs, of marshes and banks of rivers, lakes, etc., with a sympodial creeping rhizome. Upper part of stem projecting above water. Leaves mostly basal, distichous, with a long open sheath; blade erect, linear. Inflorescence spike-like with 2 superposed dense unisexual spikes; the upper staminate, the lower pistillate, each with a caducous bract at base. Flowers anemophilous, protandrous, often bracteolate. Staminate flowers reduced to 1–5 stamens with filaments fused below; anthers with 2 thecae, oblong, with the connective produced at apex; pollen grains simple or in tetrads. Pistillate flowers reduced to 1 pistil borne on a filiform long-hairy gynophore; ovary 1-locular, 1-ovulate; style long; stigma linear or lanceolate; sterile flowers with elongate or clavate ovary (carpodium) interspersed between fertile flowers. Fruit dry, 1-seeded, dispersed with the aid of long hairs of gynophore. Seeds with endosperm; embryo straight.

Ten to twenty species in temperate and trop. regions.

Well-developed pistillate spikes are needed for identification of *Typha* species.

Literature: K. O. P. P. Graebner, Typhaceae, in: Engler, *Pflanzenreich* 2(IV. 8): 1–18, 1900. J.-B. Gèze, Le *Typha domingensis* Pers. (sensu amplo), *Bull. Soc. Bot. Fr.* 58:457–461 (1911). S. Crespo & R. L. Perez-Moreau, Revisión del genere *Typha* en la Argentina, *Darwiniana* 14:413–429 (1966–1968). H. Riedl, Typhaceae, in: K. H. Rechinger (ed.), *Flora iranica* 71:1–8, 1970. D. N. Napper, Typhaceae, in: E. Milne-Redhead & R. M. Polhill (eds.), *Flora of Tropical East Africa*, 1971. T. G. Leonova, Typhaceae, in: An. A. Fedorov (ed.), *Flora partis europaeae URSS* 4:326–330, 1979. K. P. Sharma & B. Gopal, A note on the identity of *Typha elephantina* Roxb., *Aquatic Botany* 9:381–387 (1980). P. E. Boissier, *Flora orientalis* 5:47–51, 1882. G. E. Post, *Flora of Syria, Palestine and Sinai*, ed. 2, 2:558–559, 1933.

1. Leaves keeled.   **3. T. elephantina**
- Leaves convex at back, not keeled   2
2. Pistillate flowers without bracteole; pistillate and staminate spikes contiguous; pollen grains in tetrads.   **1. T. latifolia**
- Pistillate flowers with a bracteole; pistillate spike remote from staminate spike; pollen grains shed single.   **2. T. domingensis**

1. Typha latifolia L., Sp. Pl. 971 (1753); Riedl, Fl. Iran. 71:2; Boiss., Fl. 5:49; Post, Fl. 2:559. [Plate 450]

Plant 1.5–2.5 m. Leaves 9–20 mm broad. Staminate and pistillate parts of spike usually contiguous. Stamens mostly 3; pollen in tetrads. Pistillate spike dark brown at end of flowering, 10–32 × 1–3.5 cm; pistillate flowers ebracteolate; stigma spathulate-lanceolate, acute, as long as or longer than hairs; carpodia rounded at apex, ending in a short point. Fl. summer.

Hab.: Banks of rivers, ponds, lakes. Rather rare. Acco Plain, Philistean Plain; Hula Plain, Upper Jordan Valley; Transjordan (Al-Eisawi).

Area: Borealo-Trop.

2. Typha domingensis (Pers.) Poir. ex Steud., Nomencl. Bot. 860 (1824); Breistr., Bull. Soc. Bot. Fr. 110 (Sess. Extr.):54 (1966); Cook, in Fl. Eur. 5:276 (1980). *T. domingensis* (Pers.) Poir. in Lam., Encycl. Méth. Bot. Suppl. 3:596

(1814) *pro sp. inquirend. T. latifolia* L. subsp. *domingensis* Pers., Syn. Pl. 2:532 (1807). *T. australis* Schum. & Thonn. in Schum., Beskr. Guin. Pl. 401 (1827) & Kong. Danske Vid. Selsk. Biol. Skr. ser. 4, 4:175 (1829); Greuter in Greuter & Rechinger, Boissiera 13:195 (1967); Riedl, Fl. Iran. 71:5. *T. angustata* Bory & Chaub. in Bory, Expéd. Sci. Morée 3(2):338 (1833); Boiss., Fl. 5:50; Post, Fl. 2:559. [Plate 451]

Plant 1.5–3 m. Leaves often longer than stem, 5–12 mm broad. Staminate spike separated by 0.5–10 cm from pistillate spike (with rare exceptions); stamens 3; anthers on short branchlets of common filament; pollen grains single. Pistillate spike 10–28 x 1 2.5 cm, light brown; pistillate flowers bracteolate; bracteoles spathulate; stigma linear, about as long as bracteole; bracteole and stigma distinctly longer than hairs; sterile flowers intermixed with fertile ones. Fl. summer.

Hab.: Banks of rivers, ponds, lakes. Common. Acco Plain, Coast of Carmel, Sharon Plain, Philistean Plain, W. Negev; Upper and Lower Galilee, Mt. Carmel, Esdraelon Plain, Samaria, Shefela, Judean Mts., Judean Desert; Hula Plain, Upper and Lower Jordan Valley, Dead Sea area, Arava Valley; Golan, Gilead, E. Ammon, Edom.

Area: Mediterranean, Irano-Turanian, E. Asia, Africa.

3. Typha elephantina Roxb. Fl. Ind. ed. 2, 3:566 (1832); Hook. fil., Fl. Brit. India 6:488–489 (1893); Riedl, Fl. Iran. 71:1, 4; Sharma & Gopal, *op. cit.**

Plant 2.5–4 m. Rhizome deeply seated, very robust, horizontal, giving off vertical shoots, each terminating in an aerial shoot. Leaves 2 cm broad, keeled. Staminate and pistillate spikes contiguous or somewhat remote. Stamen 1; pollen in tetrads. Pistillate spike brown or dark brown at maturity, 15–25 × 1.5–2 cm; pistillate flowers mixed with clavate carpodia; pistil longer than hairs of gynophore; stigma lanceolate. Fl. summer.

Hab.: Marshy soil covered with sand. Sharon Plain.

Area: Irano-Turanian, extending into Saharo-Sindian and Sudanian.

This species has been discovered recently in Israel for the first time, forming one stand only (near Michmoret).

129. CYPERACEAE

Grass-like herbs, mostly perennial, usually tufted or with a creeping rhizome, monoecious, rarely dioecious. Stems generally solid, often 3-gonous, mostly with nodes only at base. Leaves mostly 3-stichous, usually linear to setaceous, sheathing, sometimes reduced to sheaths only; sheaths mostly closed; sometimes a membranous ligule or a short-hairy rim present at junction of blade and sheath. Flowers small, inconspicuous, usually anemophilous, unisexual

* Not illustrated.

or hermaphrodite, each in axil of a bract designated as a *glume*, arranged in 1- to many-flowered *spikelets*. Spikelets generally numerous and variously grouped in inflorescences which are raceme-, spike-, head-, umbel- or panicle-like, often subtended by involucral bracts; sometimes inflorescence reduced to a single spikelet. Perianth absent, flowers sometimes with accompanying hypogynous bristles or, more rarely, scales. Stamens 3 or 2, rarely 1 or up to 6; anthers basifixed. Pistil 1; carpels 2–3; ovary superior, 1-locular and with 1 basal anatropous ovule; style 1; stigmas 2–3, rarely more. Fruit a 1-seeded nut, generally 3-gonous in flowers with 3 stigmas, flattened in flowers with 2 stigmas, free or enclosed in a bracteole connate at margins (*utricle*). Seeds with testa not adhering to pericarp, erect; endosperm abundant, mealy or fleshy; embryo small.

About 90 genera and over 3,500 species in all parts of globe but widespread mainly in temperate and cold zones of the N. hemisphere.

Literature: V. Täckholm & M. Drar, Cyperaceae, *Flora of Egypt* 2:3–163, 1950. J.H. Kern & T.J. Reichgelt, Cyperaceae, in: *Flora neerlandica* 1(3):7–133, 1954; 1(4):7–52, 1956. W. Schultze-Motel, Cyperaceae, in: Hegi, *Illustrierte Flora von Mitteleuropa* ed. 3, 2(1):2–274, 1966–1980. S.S. Hooper & D.M. Napper, Cyperaceae, in: F.N. Hepper (ed.), *Flora of West Tropical Africa* ed. 2, 3:278–349, 1972. J.H. Kern, Cyperaceae, in: *Flora malesiana* Ser. 1, 7(3):435–753, 1974. T.V. Egorova, Cyperaceae, in: An. A. Fedorov (ed.), *Flora partis europaeae URSS* 2:83–219, 1976. T.G. Tutin et al. (eds.), Cyperaceae, in: *Flora europaea* 5:276:323, Cambridge, 1980. S.S. Hooper, Cyperaceae, in: C.C. Townsend et al. (eds.), *Flora of Iraq* 8:331–406, 1984. P.E. Boissier, *Flora orientalis* 5:362–431, 1882. G.E Post, *Flora of Syria, Palestine and Sinai* ed. 2, 2:669–687, 1933.

1. Glumes distichous. Spikelets flattened   2
- Glumes spirally arranged. Spikelets round in cross-section   3
2. Inflorescence head-like; involucral bracts 1–2; the lower dilated at base and ending in a subulate point. Nut creamy-white.   **5. Schoenus**
- Inflorescence rarely head-like and, if so, involucral bracts 3 or more and nut not creamy-white.   **7. Cyperus**
3(1). Flowers unisexual; spikelets unisexual with staminate spikelets above pistillate ones in inflorescence, or bisexual with staminate flowers mostly in upper part of spikelet.   **8. Carex**
- Flowers hermaphrodite (uppermost flowers of spikelets sometimes staminate)   4
4. Inflorescence a single terminal spikelet, devoid of involucral bracts.   **2. Eleocharis**
- Inflorescence of several or numerous spikelets or, if spikelet single, involucral bracts present   5
5. Plants 1–3 m tall. Leaves serrate at margins and keel. Inflorescence paniculate, 20–60 cm.   **6. Cladium**
- Plants not as above   6
6. Style thickened at base, deciduous with its thickened base.   **3. Fimbristylis**
- Style not thickened at base or, if thickened, base persistent   7
7. Inflorescence, especially peduncles, pedicels and involucral bracts, conspicuously pubescent.   **4. Fuirena**
- Peduncles, pedicels and involucral bracts glabrous.   **1. Scirpus**

1. Scirpus L. *s. l.*

Perennial or rarely annual herbs. Stems terete to 3-gonous. Leaves with flat or cylindrical blades; sometimes leaves reduced to sheaths. Inflorescence umbel-, spike- or head-like (sometimes reduced to single spikelet), terminal or seemingly lateral (pseudolateral) owing to erect lowest involucral bract simulating prolongation of stem. Involucral bracts 1 to several, leaf-like or scale-like. Spikelets 1 to many, sessile or pedunculate, 3- to many-flowered. Glumes spirally arranged, imbricate, the lower ones mostly longer, all fertile or lowermost 1–3 often empty. Flowers hermaphrodite; upper flowers in spikelets often staminate. Hypogynous bristles 6 or fewer, often retrorsely scabrous, or absent. Stamens usually 3. Style almost always glabrous, not articulated with ovary, sometimes thickened at base, deciduous; stigmas 3 or 2. Nut 3-gonous or flattened, obovate or elliptic to oblong in outline, smooth or regularly rugulose transversely.

About 300 species in all parts of globe.

Literature: E. Palla, Zur Kenntnis der Gattung "*Scirpus*", *Bot. Jahrb.* 10:293–301 (1888). A.A. Beetle, Studies in the genus *Scirpus*. I–VIII, *Amer. Jour. Bot.* 27 to 31 (1940–1944) & 33 (1946). T. Koyama, Taxonomic study of the genus *Scirpus* Linné, *Jour. Fac. Sci. Tokyo Univ. (Bot.)* 7:271–366 (1958); The genus *Scirpus* Linn. Some North American aphylloid species, *Canad. Jour. Bot.* 40:913–937 (1962). T. Nordlindh, Notes on the variation and taxonomy in the *Scirpus maritimus* complex, *Bot. Not.* 125:397–405 (1972).

1. Annuals, 3–20 cm tall. Inflorescence not exceeding 1 cm, composed of 1–6 sessile spikelets   2
- Perennials taller than the above. Inflorescence larger than in above   3
2. Involucral bract 2.5 cm. or longer. Nut transversely rugose.   **2. S. supinus**
- Involucral bract not exceeding 1 cm. Nut not rugose.   **1. S. cernuus**
3(1). Spikelets 3–4 mm, aggregated into dense globose heads, 5–10(–15) mm in diameter.   **3. S. holoschoenus**
- Spikelets larger. Inflorescence not as in above   4
4. Inflorescence terminal; involucral bracts 2–4, flat, leaf-like; lowest bract much longer than inflorescence. Stems leafy; leaves flat.   **4. S. maritimus**
- Inflorescence pseudolateral; involucral bracts 1–2, subulate; lowest bract rigid, simulating prolongation of stem, generally not, or only slightly, exceeding inflorescence. Leaves usually reduced to basal sheaths   5
5. Stems terete. Hypogynous bristles tapering towards tip, retrorsely scabrous. Glumes narrowly membranous-margined, ciliate to fimbriate.
                                                 5. S. lacustris subsp. **tabernaemontani**
- Stems 3-quetrous. Hypogynous bristles dilated and plumose towards tip. Glumes broadly white-membranous, not as in above.   **6. S. litoralis**

1. Scirpus cernuus Vahl, Enum. Pl. 2:245 (1805–1806); Dyer, Fl. Cap. 7:219 (1898). *Isolepis cernua* (Vahl) Roem. & Schult., Syst. Veg. 2:106 (1817). *Schoenoplectus cernuus* (Vahl) Hayek, Prodr. Fl. Penins. Balc. 3:153 (1932). *Scirpus* *savii* Seb. & Mauri, Fl. Rom. 22 (1818); Boiss., Fl. 5:380. Post, Fl. 2:676 as *Scirpus filiformis* Savi. [Plate 452]

Annual, 3-20 cm. Stems erect, filiform, striate. Leaves with short subulate blades. Inflorescence pseudolateral, of 1(-2) spikelets. Involucral bract 1, erect, setaceous, not exceeding 1 cm, dilated at base. Spikelet(s) 3-4 mm, ovoid, 6-20-flowered. Glumes ovate, mucronulate, greenish at keel, veined and often greyish-brown at sides, longitudinally finely plicate. Hypogynous bristles absent. Stigmas 3. Nut 3-gonous, broadly obovoid, finely punctulate, about ½ length of glume. Fl. March-September.

Hab.: Marshes and wet sandy soil. Acco Plain, Sharon Plain, Philistean Plain; Shefela.

Area: Euro-Siberian, Mediterranean, S. and E. Africa, Australia, New Zealand, N. America.

2. Scirpus supinus L., Sp. Pl. 49 (1753); Boiss., Fl. 5:380; Post, Fl. 2:676. *Isolepis supina* (L.) R. Br., Prodr. Fl. Nov. Holl. 221 (1810). *Schoenoplectus supinus* (L.) Palla, Bot. Jahrb. 10:299 (1888). [Plate 453]

Annual, 10-20 cm. Stems ascending, terete, striate. Lower leaves reduced to sheaths; 1-2 upper leaves with subulate blades. Inflorescence pseudolateral, of 1-6 sessile spikelets. Involucral bracts 1-2, the lower many times longer than inflorescence, erect. Spikelets 3-5 mm, 6-9-flowered, ovoid to oblong. Glumes ovate, obtuse, mucronate, greenish at keel, brownish and veined at sides. Hypogynous bristles absent or rudimentary. Stigmas 3. Nut 3-gonous, broadly obovoid, obtuse, mucronulate, transversely rugose. Fl. April-September.

Hab.: Marshy places. Rare, sporadic. Sharon Plain, Philistean Plain.

Area: Borealo-Trop.

3. Scirpus holoschoenus L., Sp. Pl. 49 (1753); Boiss., Fl. 5:381 *("holoschaenus")*; Post, Fl. 2:676. *Holoschoenus vulgaris* Link, Hort. Berol. 1:293 (1827). *Isolepis holoschoenus* (L.) Roem. & Schult., Syst. Veg. 2:115 (1817). *Scirpoides holoschoenus* (L.) Soják., Cas. Nar. Muz. (Prague) 140 (3-4):127 (1972); Hooper, in Fl. Iraq 8 (MS.). [Plate 454]

Perennial, 50-100 cm, with a creeping rhizome. Stems tufted, erect, terete, striate. Leaves reduced to sheaths; upper sheaths ending in short linear semiterete-canaliculate blades, usually shorter than inflorescence. Inflorescence pseudolateral, umbellate, of several pedunculate and sessile heads of spikelets; heads globose, very dense, 5-10(-15) mm in diameter, with numerous spikelets. Involucral bracts (1-)2, long, semiterete; the lowest erect, mostly longer than inflorescence. Spikelets 3-4 × 2 mm, obovoid, obtuse. Glumes 2-3 mm, obovate to elliptic, truncate to emarginate, mucronulate or nearly 3-lobed, ciliolate, brown to rust-coloured, green at keel. Hypogynous bristles absent. Stamens 3. Stigmas 3. Nut 3-gonous, obovoid, minute, brownish-white, with persistent style-base at tip. Fl. May-August.

Hab.: Sands and damp habitats. Common. Acco Plain, Coast of Carmel, Sharon Plain, Philistean Plain, W. Negev; Upper and Lower Galilee, Esdraelon Plain, Samaria; Dan Valley, Hula Plain, Upper Jordan Valley, Beit Shean Valley, Lower Jordan Valley, Dead Sea area, Arava Valley; Golan, Gilead, Ammon, Edom.

Area: Mediterranean and Irano-Turanian, extending into the Euro-Siberian region.

4. Scirpus maritimus L., Sp. Pl. 51 (1753); Boiss., Fl. 5:384; Post, Fl. 2:678 incl. var. *biformis* Post, Fl. ed. 1, 678 (1883-1896). *Bolboschoenus maritimus* (L.) Palla in Koch, Syn. Deutsch. Schweiz. Fl. ed. 3, 3:2532 (1905); Hooper, in Fl. Iraq 8 (1984). *S. tuberosus* Desf., Fl. Atl. 1:50 (1798). *S. maritimus* L. var. *tuberosus* (Desf.) Roem. & Schult., Syst. Veg. 2:139 (1817). [Plate 455]

Perennial, 30–100 cm, glabrous (except for glumes), with creeping branched rhizome; rhizome often producing subglobose tubers 1–2 cm in diameter at ends of rhizome branches. Stems 3-quetrous, leafy, usually erect. Leaf-blades flat, keeled, long and up to 1 cm broad. Inflorescence terminal, varying from compact to lax, often umbel-like, rarely of a single spikelet; rays unequal, up to 5 cm. Involucral bracts 2–4, leaf-like, unequal, nearly all much exceeding inflorescence. Spikelets 10–20 mm, ovoid to oblong, acute, many-flowered. Glumes ovate to oblong, 1–veined, not keeled, brown, finely hairy, acutely 2-fid or irregularly 4-lobed at tip, with an awn in the sinus. Hypogynous bristles 6 or fewer, retrorsely scabrous. Stamens 3. Stigmas 3, rarely 2 (the number varying even in same spikelet). Nut 2(–3) mm, 3-gonous-obovoid or plano-convex, tapering at base, apiculate at apex, brown, glossy. Fl. March–October.

Hab.: Marshes and river banks, mostly on somewhat saline soil. Acco Plain, Coast of Carmel, Sharon Plain, Philistean Plain; Upper and Lower Galilee, Esdraelon Plain, Samaria, Shefela, Judean Mts.; Dan Valley, Hula Plain, Upper Jordan Valley, Beit Shean Valley, Lower Jordan Valley, Dead Sea area; Golan, Gilead, Ammon.

Area: In all parts of globe except polar regions.

A polymorphic species. "A modern revision of the variation range of *S. maritimus* on a worldwide basis is still lacking" [Schultze-Motel in Hegi, Ill. Fl. Mitteleur. ed. 3, 2(1):19, 1966].

5. Scirpus lacustris L., Sp. Pl. 48 (1753) subsp. **tabernaemontani** (C.C. Gmel.) Syme in Sowerby, Engl. Bot. ed. 3, 10:64 (1870); R.A. de Filipps, in Fl. Eur. 5:278 (1980). *S. glaucus* Sm., Engl. Bot. 33:t.2321 (1812) non Lam. (1791). *S. tabernaemontani* C.C. Gmel., Fl. Bad. 1:101 (1805). *Schoenoplectus tabernaemontani* (C.C. Gmel.) Palla, Verh. Zool.-Bot. Ges. Wien 38:49 (1888) & Palla, Bot. Jahrb. 10:299 (1889). Boiss., Fl. 5:383 as *Scirpus lacustris* var. *digynus* Godr., Fl. Lorr. 3:90 (1844). Post, Fl. 2:677 as *Scirpus lacustris* L. [Plate 456]

Perennial, 0.5–3 m, glaucous, with a creeping rhizome. Stems stout, terete. Leaves reduced to sheaths, the lower brown to purple, the uppermost green;

sometimes submerged leaves with long strap-shaped blades. Inflorescence pseudolateral, umbel-like, with unequal rays. Involucral bract 1 at base of inflorescence, erect, dilated below, cylindrical-subulate above, as long as inflorescence or shorter. Spikelets 8–10 mm, ovoid to ovoid-oblong, mostly in clusters. Glumes 3–4 mm, ovate to elliptic, emarginate, ciliate to fimbriate at margins, brown, mostly with minute red-brown tubercles, with a prominent midvein excurrent into a short mucro. Hypogynous bristles 4–6, retrorsely scabrous, as long as or shorter than nut. Stamens 3. Stigmas 2, rarely 3. Nut 2–2.5 mm, lenticular, more rarely obscurely 3-gonous, smooth, short-mucronate. Fl. April–October.

Hab.: Marshes, banks of rivers. Acco Plain, Coast of Carmel, Sharon Plain, Philistean Plain, W. Negev; Hula Plain; Transjordan (Al-Eisawi).

Area of subspecies: Euro-Siberian and Mediterranean.

6. Scirpus litoralis Schrad., Fl. Germ. 142, t.5 f.7(1806); Kern, in Fl. Males. 7(3):510; Boiss., Fl. 5:383; Post, Fl. 2:677. *Schoenoplectus litoralis* (Schrad.) Palla, Bot. Jahrb. 10:299 (1889); Serbanescu & Nyárády, in Fl. Rep. Soc. România 11:640, t.115 f.2 (1966). [Plate 457]

Perennial, 1–1.5 m, with a stoloniferous rhizome. Stems 3-quetrous. Leaves reduced to sheaths, or sometimes with blades up to 20 cm. Inflorescence pseudolateral, usually a compound umbel with unequal rays. Lowest involucral bract subulate, 3-quetrous, erect, shorter than to about as long as inflorescence, rarely longer. Spikelets 6–15 mm, oblong-ovoid, solitary or in small clusters. Glumes 3.5–4 mm, ovate, broadly white-membranous at margin, brown at centre, obtuse or retuse, mucronate. Hypogynous bristles 4–6, dilated and plumose towards tip, about as long as nut. Stamens (2–)3. Stigmas 2. Nut 3 mm, plano-convex, obovate in outline, smooth or reticulate, brown. Fl. May–October.

Hab.: Brackish swamps and banks of rivers. Acco Plain, Coast of Carmel, Sharon Plain, Philistean Plain, W. Negev; Hula Plain, Upper Jordan Valley, Dead Sea area; E. Ammon, Moav.

Area: Mediterranean and Irano-Turanian regions and Tropics of the Old World.

2. ELEOCHARIS R. Br.

Perennial or annual herbs. Stems terete, somewhat compressed or 4(−3)-gonous. Leaves reduced to tubular sheaths. Inflorescence a terminal single spikelet (rarely 2 spikelets), devoid of involucral bracts. Spikelets few- or many-flowered. Glumes spirally arranged, imbricate, 1–2 lowest glumes usually empty or differing in shape from others. Flowers hermaphrodite. Hypogynous bristles present, usually retrorsely scabrous. Stamens 2–3. Stigmas 2–3. Nut 3-gonous or 2-convex with thickened style-base at tip.

About 150 species all over the world.

Literature: S. M. Walters, *Eleocharis* R. Br., in: Biological Flora of the British Isles, *Jour. Ecol.* 37:192–206 (1949). S. O. Strandhede, A note on *Scirpus palustris* L., *Bot. Not.* 113:161–171 (1960). A. R. Clapham, T. G. Tutin & E. F. Warburg, *Flora of the British Isles* ed. 2, 1061–1064, 1962. W. Schultze-Motel, *Eleocharis*, in: Hegi, *Illustrierte Flora von Mitteleuropa* ed. 3, 2(1):52–67, 1966.

1. Eleocharis palustris (L.) Roem. & Schult., Syst. Veg. 2:151 (1817) subsp. **palustris.** *Scirpus palustris* L., Sp. Pl. 47 (1753). *E. palustris* subsp. *microcarpa* Walters, Jour. Ecol. 37:194 (1949). Boiss., Fl. 5:386 (*"Heleocharis"*) & Post, Fl. 2:679 as *E. palustris.* [Plate 458]

Perennial, 10–60 cm, with a creeping rhizome. Stems rigid, erect, terete, glaucous. Leaf-sheaths brown to dark red, glossy, truncate. Spikelet 1–3.5 cm, many-flowered. Glumes oblong-ovate, obtuse, hyaline-margined, pale brown on either side of greenish keel; middle glumes 2.7–3.5 mm. Hypogynous bristles 3–4. Stigmas 2(-3). Nut 1.2–1.4(-1.5) mm, 2-convex, nearly round in outline, smooth; style-base constricted below. Dispersal of nuts by water fowl observed. Fl. March–July.

Hab.: Marshes and wet soils. Coastal Galilee, Acco Plain, Sharon Plain, Philistean Plain; Upper Galilee, Mt. Carmel, Esdraelon Plain, Samaria, Judean Mts.; Hula Plain, Lower Jordan Valley; Golan, Gilead, Ammon.

Area of subsp.: Euro-Siberian, Mediterranean and Irano-Turanian, ?S. Africa.

Eleocharis palustris is currently subdivided (cf. Schultze-Motel, 1966) into two subspecies, subsp. *palustris* with 2n=16 and subsp. *vulgaris* Walters, *loc. cit.*, with 2n=(37-)38(-39, 40). Subsp. *palustris* is characterized by smaller nuts, shorter middle glumes and a more southern and eastern distribution area than the more northern and western subsp. *vulgaris* with 1.5–2 mm nuts and 3.5–4.5 mm dark brown middle glumes. Though in W. Europe the distinction between the two seems somewhat unclear, our material can be definitely classed with subsp. *palustris*.

3. Fimbristylis Vahl

Annual or perennial herbs similar to *Scirpus* in habit. Stems slender, leafy. Leaves near base of stem, mostly with well-developed blades. Inflorescence terminal, simple or compound, umbel- or head-like, rarely reduced to a single spikelet. Involucral bracts leaf-like. Spikelets usually many-flowered. Glumes spirally arranged, imbricate, acropetally deciduous, fertile or 1-3(-6) lowest glumes empty. Flowers hermaphrodite, the uppermost sometimes neuter. Hypogynous bristles generally absent. Stamens 1-3. Style compressed, often hairy; base thickened, articulated with ovary and deciduous at maturity together with style; stigmas 2-3. Nut 3-gonous or lenticular.

About 220 trop. and subtrop. species.

Literature: D. M. Napper, *Fimbristylis, Scleria* and *Diplacrum* (Cyperaceae) in Tropical West Africa, *Kew Bull.* 25:435–446 (1971); *Fimbristylis,* in: F. N. Hepper (ed.), *Flora of West Tropical Africa* ed. 2, 3(2):318–325, 1972. J.H. Kern, *Fimbristylis*, in: *Flora malesiana* Ser. 1, 7(3):540–592, 1974.

1. Spikelets 3–5 mm broad. Glumes densely grey-puberulent. Nut smooth to minutely dotted.   **1. F. ferruginea**
- Spikelets 1.2–2 mm broad. Glumes glabrous. Nut ribbed and transversely striate.
                                                                               2. F. bisumbellata

1. Fimbristylis ferruginea (L.) Vahl, Enum. Pl. 2:291 (1805–1806); Boiss., Fl. 5:388; Post, Fl. 2:680. *Scirpus ferrugineus* L., Sp. Pl. 50 (1753). [Plate 459]

Perennial, 20–80 cm, tufted. Stems erect, striate, somewhat compressed. Leaves linear, canaliculate, rigid, shorter than stems. Inflorescence umbel-like, simple or subcompound, with unequal rays. Involucral bracts 2–3, unequal, the longest longer to shorter than inflorescence. Spikelets ovoid to oblong-ovoid, 7–20 × 3–5 mm, greyish-rust-coloured. Glumes densely imbricate, 2.5–4 mm broad, broadly ovate, scarcely keeled, apiculate, densely grey-puberulent at least in apical part. Stamens 3. Style hairy; stigmas 2. Nut lenticular, smooth to minutely dotted. Fl. April–October.

Hab.: Marshy places. Acco Plain, Sharon Plain, Philistean Plain; Upper and Lower Galilee, Judean Mts., Samaria; Dan Valley, Hula Plain, Upper and Lower Jordan Valley, Dead Sea area, Arava Valley; Golan, Moav.

Area: Trop.

2. Fimbristylis bisumbellata (Forssk.) Bubani, Dodecanthea 30 (1850); Kern, in Fl. Males. 7(3):579. *Scirpus bisumbellatus* Forssk., Fl. Aeg.-Arab. 15 (1775). *S. dichotomus* Rottb., Descr. Icon. 57, t.13 f.1 (1786) non L. (1753). *S. umbellatus* Post, Fl. ed. 1, Add. 21 (1883–1896); Post, Fl. ed. 2, 2:678. [Plate 460]

Annual, 15–40(–70) cm, tufted. Stems erect, striate. Leaves narrowly linear, flat, pubescent. Inflorescence umbel-like, compound to decompound; primary rays unequal. Involucral bracts 2–3, the longest as long as or longer than inflorescence. Spikelets oblong to oblong-ovoid, acute, angular, 1.2–2 mm broad, reddish-brown. Glumes densely imbricate, ovate, acute, sharply keeled, mucronate, membranous, glabrous. Stamen 1(–2). Style hairy; stigmas 2. Nut lenticular, obovate, beautifully latticed with ribs and transverse striae. Fl. April–August.

Hab.: Humid habitats. Acco Plain, Sharon Plain, Philistean Plain; Upper and Lower Galilee, Judean Mts.; Dan Valley, Hula Plain, Upper Jordan Valley, Beit Shean Valley, Lower Jordan Valley, Dead Sea area; Golan, Gilead, Moav, Edom.

Area: Mediterranean, Irano-Turanian and Trop.; also in Australia.

4. FUIRENA Rottb.

Annual or perennial herbs. Stems with nodes and leaves throughout their length. Leaves with elongate blades, closed sheaths and membranous ligules. Inflorescence paniculate, consisting of a terminal partial inflorescence and

1 to several axillary ones. Spikelets terete, many-flowered. Glumes spirally arranged, imbricate, puberulent, aristate, acropetally deciduous, the 1–2 lowest empty. Flowers hermaphrodite, the uppermost staminate or neuter. Hypogynous bristles 0 or up to 6, or the 3 outer bristle-like and the 3 inner scale-like. Stamens (2–)3. Style glabrous, not thickened at base; stigmas 3, longer than style. Nut obovoid or ovoid, 3-gonous.

About 30 trop. and subtrop. species.

1. Fuirena pubescens (Poir.) Kunth, Enum. Pl. 2:182 (1837); Bouloumoy, Fl. Liban Syrie 351; Mout., Nouv. Fl. Liban Syrie 1:169, t.49 f.7; Post, Fl. 2:679. *Carex pubescens* Poir., Voy. Barb. 2:254 (1789). *Scirpus libanoticus* Post, Fl. ed. 1, 833 (1883–1896). [Plate 461]

Perennial, (20–)40–100 cm, with rhizome. Stems erect, 3-gonous, pubescent at nodes and under inflorescence. Leaves flat, keeled, more or less pubescent. Inflorescence terminal, pubescent, reduced to a few clusters of spikelets. Involucral bracts 2–3, the lower exceeding inflorescence. Spikelets 2–5 in a cluster, many-flowered, greenish-brown. Glumes ovate to elliptical, keeled, puberulent, veined, aristate. Hypogynous bristles 3, short. Nut obovoid, 3-gonous, white to light grey, smooth, about $\frac{2}{5}$ length of glume, with a short cone at top. Fl. July–September.

Hab.: Swampy ground or damp sandy soil. Sharon Plain, Dan Valley; Hula Plain.

Area: Mediterranean, Trop., S. Africa, India.

Not collected in Palestine until 1926; apparently sporadic.

5. SCHOENUS L.

Perennial herbs, rarely annuals. Stems rigid, terete or obtusely 3-gonous. Leaves all basal or basal and cauline, linear, sometimes reduced to mucronate sheaths. Inflorescence terminal, dense, head-like, consisting of short-pedicellate spikelets. Involucral bracts 1–2, the lower dilated at base, sheathing inflorescence and ending in a leaf-like subulate point. Spikelets 1–4-flowered, flattened. Glumes distichous, the lower ones empty, smaller than the fertile glumes. Flowers hermaphrodite, the upper ones often staminate or neuter. Hypogynous bristles plumose or ciliate, 0–6. Stamens 3, rarely fewer or 4–6. Style slightly swollen at base, usually deciduous; stigmas 3, very rarely 2. Nut obsoletely 3-gonous.

About 80 species, especially in S.E. Asia, Australia and New Zealand; only a few species in Europe, Asia and extratrop. S. America.

1. Schoenus nigricans L., Sp. Pl. 43 (1753); Boiss., Fl. 5:393; Post, Fl. 2:681. [Plate 462]

Perennial herb, 40–150 cm, glabrous, tufted. Stems erect, surrounded at base by sheaths of withered leaves; each stem ending in a head. Leaves about $\frac{1}{2}$–$\frac{2}{3}$

length of stem, with broad, blackish-brown sheaths and narrow involute-margined rigid blades; sheaths blackish at base, reddish-brown above. Head 1-1.5 cm, ovoid. Lowest bract much longer than head. Spikelets (2-)5-10, 10-12 mm, lanceolate, acute. Glumes lanceolate, keeled, blackish-brown. Hypogynous bristles 3-4, usually shorter than nut, or absent. Nut white, glossy, about ⅓ length of glume. Fl. April–May.

Hab.: Humid habitats. Acco Plain, Coast of Carmel, Sharon Plain; Upper and Lower Galilee, Mt. Carmel, ?Samaria, Judean Mts.; Hula Plain; Transjordan (Al-Eisawi).

Area: Euro-Siberian, Mediterranean, Irano-Turanian, also S. Africa; introduced in the New World.

6. CLADIUM Browne

Perennial herbs. Stems erect, hollow, leafy. Leaves 3-stichous, with flat blades and long sheaths; ligule lacking. Inflorescence compound, paniculate, consisting of 1 terminal and several remote lateral corymb-like pedunculate panicles, with unequal rays and involucral bracts at base. Spikelets 2(-3)-flowered. Glumes spirally arranged, imbricate, the lower 3-5 empty. Flowers hermaphrodite, or one of them (the lower or the upper one) staminate or functionally staminate. Hypogynous bristles absent. Stamens 2(-3). Base of style somewhat thickened; stigmas (2-)3(-4). Nut crustaceous, ovoid or conical, inserted on a white disc (persistent style-base).

Two species, one, *C. mariscus,* nearly cosmopolitan, the other, *C. mariscoides,* confined to N. America.

1. Cladium mariscus (L.) Pohl, Tent. Fl. Bohem. 1:32 (1809); R. Br., Prodr. Fl. Nov. Holl. 236 (1810); Boiss., Fl. 5:392; Post, Fl. 2:681. *Schoenus mariscus* L., Sp. Pl. 42 (1753). [Plate 463]

Perennial herb, 1-3 m, with a thick creeping rhizome and horizontal stolons. Stems stout, terete or obtusely 3-gonous; upper internodes grooved on alternate sides. Leaves about as long as stems, up to 1.5 cm broad, linear, rigid, glaucous, keeled, serrate-spinulose at margins and keel; sheaths yellow-brown, becoming black-brown. Inflorescence 20-60 cm, much-branched. Spikelets 3-15 in small dense clusters, 3-5 mm, ellipsoid, tapering at both ends, mostly 2-flowered. Glumes acutish, 1-veined, light brown; empty glumes ovate, 1.5-2.5 × 1.5 mm; fertile glumes ovate-oblong, larger (4 × 2 mm). Nut 2-4 mm, ovoid, acute or acuminate, dark brown, glossy, attached at base to a white disc. Fl. April–September.

Hab.: Swamps. Acco Plain, Coast of Carmel, Sharon Plain, Philistean Plain; Hula Plain, Upper Jordan Valley, Beit Shean Valley; E. Ammon.

Area: Borealo-Trop.

According to Kükenthal (Feddes Repert. 51:185-192, 1942), who subdivided *C. mariscus* into 3 subspecies, the plants from our region belong to subsp. *mariscus* var. *martii* (Roem. & Schult.) Kük.

7. CYPERUS L. *s. l.*

Perennial or rarely annual herbs. Stems erect, solid, 3-gonous or subterete, usually leafy at base. Leaves generally linear, sometimes reduced to sheaths; sheaths closed. Inflorescence a terminal umbel, compound or decompound, or capitate by suppression of rays. Involucral bracts at base of inflorescence leaf-like; in some species each ray surrounded at base by tubular *prophyll*. During anthesis prophylls swell on adaxial side and aid spreading out of rays. Spikelets more or less compressed, usually many-flowered; rachilla often rendered winged by decurrent bases of glumes. Glumes distichous or, more rarely, arranged spirally, mostly keeled, usually falling off acropetally; 2 basal glumes usually empty; glume production at top of rachilla continues in some species also after beginning of anthesis. Flowers hermaphrodite, the uppermost often staminate or aborted. Hypogynous bristles or scales absent. Stamens 3, 2 or 1. Style continuous with ovary, not thickened at base, deciduous; stigmas 2 or 3. Nut 3-gonous or flattened, biconvex, plano-convex or convex-concave.

About 600 species, mostly hygrophilous, the majority in the Tropics or Subtropics.

Literature: G. Kükenthal, Cyperaceae-Scirpoideae-Cypereae, in: Engler, *Pflanzenreich* 101 (IV. 20):1–671, 1935–1936. T. Koyama, The Cyperaceae Tribe Cypereae of Ceylon, *Gardens' Bull.* 30:123–164 (1977).

The chromosome numbers quoted here for species of *Cyperus* are based on Shlomith Lerman-Shaddar, MSc. Thesis, Hebrew Univ., Jerusalem, 1961 (unpublished).

1. Primary rays of inflorescence 30–100.   **1. C. papyrus**
   - Primary rays of inflorescence fewer or absent   2
2. Spikelets arranged on an elongated rachis in dense oblong spikes; spikes 5–12(–20) mm broad and at least twice as long   3
   - Spikelets in broader spikes or arranged otherwise   5
3. Glumes long-mucronate, remote from one another. Spikelets erect at maturity, greyish-brown.   **4. C. eleusinoides**
   - Glumes mucronulate or muticous, densely imbricate. Spikelets spreading at maturity, straw-coloured to golden-brown   4
4. Spikelets ovate-lanceolate, 2–3 mm broad. Glumes flattened at back, not keeled, with margins incurved at maturity exposing edge of plano-convex nut. Stigmas 2, rarely in some flowers 3.   **3. C. alopecuroides**
   - Spikelets oblong, 1–1.5 mm broad. Glumes boat-shaped, keeled, densely imbricate, margins not incurved as above. Stigmas 3.   **2. C. dives***
5(2). Lowest involucral bract much shorter than inflorescence   6
   - Lowest involucral bract as long as or longer than inflorescence   7
6. Stems with transverse partitions that are prominent in dry state. Involucral bracts of inflorescence scale-like, about 1 cm, sharp-pointed.   **5. C. articulatus**
   - Stems devoid of transverse partitions or at most obscurely septate. Involucral bracts leaf-like, unequal, the longest 4–10 cm.   **6. C. corymbosus**

* Hybrids occur between *C. dives* and *C. alopecuroides*.

7(5). Stigmas 3. Nut 3-gonous or 3-quetrous 8
- Stigmas 2. Nut more or less flattened, biconvex, plano-convex or concave-convex. Rarely single flowers in same spikelet with 3 stigmas 16
8. Leaves canaliculate, rigid; stems more or less terete or obtusely 3-gonous. Perennial herbs of maritime and desert sands or of springs around Dead Sea 9
- Leaves flat; stem 3-quetrous, more rarely subcompressed-3-quetrous. Annual or perennial herbs of swamps and various damp habitats 11
9. Spikelets few-flowered, obscurely distichous; inflorescence a compact globose head. Lower glumes empty, large, 6–10 × 4–5 mm. **13. C. capitatus**
- Spikelets many-flowered, distinctly distichous; inflorescence a head or a short-rayed umbel. Glumes more or less equal, smaller than in above 10
10. Spikelets 8–16-flowered, somewhat turgid, straw-coloured. Plants of maritime or desert sands. **14. C. conglomeratus**
- Spikelets 16–40-flowered, strongly compressed, brownish, with very densely imbricate glumes. Plants growing near springs around Dead Sea. **15. C. jeminicus**
11(8). Leaves 1–1.5 cm broad. Glumes oblong-obovate, truncate and broadly hyaline at apex. **7. C. latifolius**
- Leaves narrower. Glumes not as above 12
12. Perennial herbs with a rhizome or with slender stolons and tubers. Clusters of spikelets digitate or spike-like 13
- Annuals. Clusters of spikelets capitate or subcapitate 14
13. Plants with underground slender stolons thickened at ends into ovoid-oblong dark brown tubers. Spikelets reddish-brown, 2–3 mm broad. **9. C. rotundus**
- Plants with a rhizome, devoid of tubers. Spikelets greyish-brown, 1–1.5 mm broad. **8. C. longus**
14(12). Spikelets 2–3(–4) mm broad. Glumes 2–2.5 mm, white-margined, purplish to brown and 3–4-veined on either side of green keel. **10. C. glaber**
- Spikelets 1–1.5 mm broad. Glumes 0.5–1.25 mm 15
15. Glumes 0.5–0.75 mm, orbicular-reniform, turgid, prominently white-margined. Spikelets in very dense, regularly shaped globose clusters. Plants confined to northern districts of country. **12. C. difformis**
- Glumes 1–1.25 mm, ovate, not or hardly white-margined. Spikelet-clusters not as above. Fairly common throughout country. **11. C. fuscus**
16(7). Broad side of nut turned to rachilla 17
- Narrow side of nut turned to rachilla 18
17. Inflorescence a very dense globose terminal head; involucral bracts 2–7, at least 2 of them longer than inflorescence; glumes oblong-lanceolate, tapering to a narrow green tip ¼–⅓ length of glume itself. **16. C. pygmaeus**
- Inflorescence pseudolateral, not as above; involucral bracts (1–)2; the lower only exceeding inflorescence; glumes not ending in a tip as above. **17. C. laevigatus**
18(16). Inflorescence a lax umbel with 3–9 cm long rays. Spikelets 3–4 mm broad.
18. C. nitidus
- Inflorescence a head or an umbel with shorter rays. Spikelets 1–2.5(–3) mm broad
19
19. Spikelets 1–1.5 mm broad. **21. C. polystachyos**
- Spikelets 2–2.5(–3) mm broad 20
20. Nuts round in outline. Spikelets pale yellow at anthesis. **20. C. flavescens**
- Nuts elliptic-oblong in outline. Spikelets brownish at anthesis. **19. C. flavidus**

1. Cyperus papyrus L., Sp. Pl. 47 (1753); Kük. in Engler, Pflanzenreich 101:45; Hooper, in Fl. W. Trop. Afr. ed. 2, 3:284; Täckholm & Drar, Fl. Eg. 2:99; Boiss., Fl. 5:374; Post, Fl. 2:674. [Plate 464]

Perennial, 3–5 m, with a thick creeping rhizome. Stems very thick, 3-gonous above. Leaves reduced to coriaceous brown sheaths; sterile shoots with short blades. Inflorescence a large decompound umbel; primary rays 30–100, very long (10–30 cm), slender, with 3 cm long tubular brown prophylls at base; secondary rays 3–5, short, bearing 1–6 short oblong spikes with many spikelets. Bracts of involucre 4–10, broadly lanceolate, acuminate, much shorter than rays; bracts of involucel 3–5, 1 mm broad, longer to shorter than spike-cluster. Spikelets 5–7 × 1 mm, 6–16-flowered, spaced along rachis of spike, spreading, linear, acute, subterete; rachilla winged. Glumes 1.5–2 × 0.8–1 mm, oblong, obtuse, boat-shaped, rust-coloured, white-margined, green at keel. Stamens 3. Stigmas 3. Nut oblong, 3-gonous, ½–⅗ length of glume. 2n=~100. Fl. May–October.

Hab.: Swamps, in water. Sharon Plain; Hula Plain.

Area: E. Trop. Africa; introduced or spontaneous in Egypt, Sicily, Malta, Palestine.

Though the main area of distribution of *C. papyrus* is Trop. Africa, the species is known to have occurred further north, in Egypt, Palestine and in the Mediterranean islands of Sicily and Malta since ancient times (cf. Täckholm & Drar, 1950). This northern distribution has been ascribed mainly to introduction and cultivation. The disappearance of *C. papyrus* from the flora of Egypt has been recorded by Boissier ("Olim in Egypto ubi destructus nunc esse videtur" *op. cit.*, 1882) and by later authorities (Täckholm & Drar, 1950; Täckholm, 1956; and others). Recently Täckholm reported (in Students Flora ed. 2, 790, 1974) that the species was rediscovered a few years ago in a lake in Wadi Natroun. The restriction in the distribution of this decorative hydrophyte in Sicily and Palestine is no doubt connected with the general progress of swamp reclamation.

It should be pointed out, however, that the role of migratory birds in the occasional reappearance of the species beyond the limits of Trop. Africa cannot be excluded.

As to *C. papyrus* in Palestine, it is worthwhile following up the available records on its occurrence since Boissier (1882) who quoted it from "lacus Tiberiadis et Merom". The data by Post (Fl. ed. 1, 830, 1883–1896) are: "Marshes along coast, and about el-Hûleh and the Lake of Tiberias; el Ghor". Dinsmore, in the second edition of Post's Flora (2:674, 1933), reports: "Pal. Ras-ul-Ayn (D), Khan Minyah, Hulah (PH), Hedera (Eig), ul Ghawr (P)". Eig (1926) recorded several plants found by him in 1924 in Hadera (Sharon Plain) where A. Aaronsohn had collected *C. papyrus* in 1906 (Florula Cisiordanica, Bull. Soc. Bot. Genève ser. 2, 31:174, 1941).

The Herbarium of the Hebrew University, Jerusalem, contains the following collections of *C. papyrus*: Sharon Plain, betw. Wadi Zerqa near the Kabbara marshes, near Benjamina in the north and the Yarkon River in the south; the plants were collected in the years 1924, 1925, 1926, 1927, 1941, 1949, 1956; Hula Plain, especially N. of Lake Merom in the years 1924, 1926, 1933, 1941, 1949, 1953, 1954. In 1926 Eig wrote (Bull. Inst. Agr. Nat. Hist. Tel-Aviv 4:42): "We found it in the mentioned three localities, apart from Huleh, where it covers hundreds of dunams of marshes, and being the principal raw material for mat industry there, it serves as the economical basis of a number of Beduin villages".

In the early fifties Lake Merom was drained and the main areas of *C. papyrus* were laid dry. The Nature Reserves Authority established the Hula Nature Reserve which is the single area in Palestine where *C. papyrus* is preserved at present.

Chiovenda (1931) subdivided *C. papyrus* into several subspecies, the bulk of them in Trop. Africa, which differ from each other mainly in the length of the projecting tip of connective. In subsp. *antiquorum* he distinguished a variety, var. *palaestinus* Chiov. Examination of our specimens (HUJ) showed too much variability in this character to allow for clear identification. Thus, it seems that for the time being it is preferable to refrain from the subdivision of the species.

2. Cyperus dives Del., Fl. Eg. 149 (1813–1814); t.4 f.3 (1826); Del., Fl. Aeg. Ill. 50 (1813) *nom. nud.*; Kük. in Engler, Pflanzenreich 101:68; Täckholm & Drar, Fl. Eg. 2:86; Hooper, in Fl. W. Trop. Afr. ed. 2, 3:284; Täckholm, Stud. Fl. Eg. ed. 2, 789. [Plate 465]

Perennial, 100–150 cm, with a thick rhizome. Stems erect, stout, 3-gonous, leafy. Leaves 0.5–2 cm broad, flat, with a prominent midrib and 2 lateral veins, scabrous at margins and keel, folded along lateral veins below. Inflorescence a large decompound umbel; rays 5–10, unequal, up to 20 cm; spikelets arranged on an elongated rachis in dense oblong spikes; clusters of spikes borne on secondary rays almost from their base. Bracts of involucre much longer than inflorescence; bracts of involucel usually not exceeding spike-cluster. Spikes cylindrical, 8–12 mm broad; rachis hidden by densely crowded spikelets. Spikelets very numerous, 4–8 × 1–1.5 mm at maturity, spreading at maturity, many-flowered, oblong, yellow; rachilla narrowly winged. Glumes densely imbricate, ovate, ending in a short straight mucro, boat-shaped, keeled, straw-coloured to golden-brown, 3–5-veined, green at keel. Stamens 3. Style long; stigmas 3. Nut elliptic in outline, 3-gonous, apiculate, about ½ length of glume. Fl. April–September.

Hab.: Banks of rivers and ponds, ditches and marshes. Coastal Galilee, Coast of Carmel; Hula Plain.

Area: Paleotrop.

Plants of *C. dives* in Israel have broader leaves than those of *C. alopecuroides*.

Hybrids occur between *C. dives* and *C. alopecuroides* (see below under *C. alopecuroides*).

3. Cyperus alopecuroides Rottb., Descr. Pl. Rar. Programm. 20 (1772); Rottb., Descr. Icon. Rar. Pl. 38, t.8 f.2 (1773); Kük. in Engler, Pflanzenreich 101:71; Täckholm & Drar, Fl. Eg. 2:89; Hooper, in Fl. W. Trop. Afr. ed. 2, 3:285; Kern, in Fl. Males. 7(3):603; Boiss., Fl. 5:376 excl. syn.; Post, Fl. 2:672. *Juncellus alopecuroides* (Rottb.) C.B. Clarke in Hook. fil., Fl. Brit. Ind. 6:595 (1893). [Plate 466]

Perennial, 80–120 cm, tufted, with a very short thick rhizome. Stems erect, stout, 3-gonous, leafy at base. Leaves 6–16 mm broad, flat. Inflorescence a

360   CYPERACEAE

large decompound umbel; rays 5–10, unequal, up to 20 cm; spikelets arranged
on an elongated rachis in dense oblong spikes; clusters of spikes borne on
secondary rays, almost from their base. Bracts of involucre much longer than
inflorescence; bracts of involucel shorter than spike-cluster. Spikes cylindrical,
8–12(–20) mm broad; spikelets very numerous, 4–7(–8) × 2–3 mm at maturity,
spreading at maturity, many-flowered, ovate-lanceolate; rachilla not or narrowly winged. Glumes imbricate, ovate-elliptic, short-mucronate, flattened at
back, not keeled, straw-coloured, 3–5-veined, with incurved margins exposing
edge of nut at maturity. Stamens 2(–3). Stigmas 2, rarely 3. Nut elliptic to
obovate-elliptic in outline, plano-convex, apiculate, about ½ length of glume.
2n=~112–114. Fl. June–October.

   Hab.: Banks of rivers, pools, ditches and marshes. Sharon Plain; Dan Valley, Hula Plain, Upper Jordan Valley, Beit Shean Valley, Lower Jordan Valley, Dead Sea area; Golan, Moav.
   Area: Trop.

Täckholm & Drar (Flora of Egypt 2:87 & 91, 1950) reported hybrids between *C.
dives* and *C. alopecuroides*, with variable characters, from Egypt. Similar hybrids
occur also in Israel. A hybrid plant was collected at Bteiha (near the entrance of the
Jordan into Lake Kinneret), with spikelets similar to those of *C. dives*, but much
longer and lanceolate in outline and with 2 and 3 stigmas.
   Uses (mainly for mat making) and cultivation of *C. alopecuroides* in present-day
and ancient Egypt are summarized by Täckholm & Drar (*op. cit.* 91–96).

4. Cyperus eleusinoides Kunth, Enum. Pl. 2:39 (1837); Kük. in Engler, Pflanzenreich 101:144; Hooper, in Fl. W. Trop. Afr. ed. 2, 3:288; Boiss., Fl. 5:371;
Post, Fl. 2:673. *C. nutans* Vahl, Enum. Pl. 2:363 (1805–1806) var. *eleusinoides*
(Kunth) Haines, Bot. Bihar Orissa 5:898 (1924); Kern, in Fl. Males. 7(3):610.
[Plate 467]
   Perennial, 60–120 cm, tufted. Root fibrous. Stems acutely 3-gonous, leafy
below. Leaves up to 5 mm broad, flat; sheaths long, yellow. Inflorescence a
compound umbel; primary rays 6–10, unequal, up to 15 cm; secondary rays
bearing congested short spikes almost from their base. Bracts of involucre 5–6,
longer than inflorescence, scabrous; bracts of involucel 5–6, narrow, nearly as
long as or longer than spikes. Spikes cylindrical, 20–30 × 5–8 mm; spikelets
7–10 × 2–3 mm, 10–15-flowered, linear-oblong, more or less appressed to
rachis, erect at maturity; rachilla not winged. Glumes rather remote, spreading, ovate, long-mucronate, keeled, greyish-brown, hyaline-margined, 3–4-
veined on either side of keel. Stamens 3. Stigmas 3. Nut obovoid-oblong, 3-
gonous, apiculate, ⅔ length of glume. Fl. April–June.
   Hab.: Damp places. Dead Sea area (En-Gedi, Ghawr-es-Safiyah, Hammam-
uz-Zarka).
   Area: Trop.

5. Cyperus articulatus L., Sp. Pl. 44 (1753); Kük. in Engler, Pflanzenreich
101:77; Hooper, in Fl. W. Trop. Afr. ed. 2, 3:285; Boiss., Fl. 5:374. [Plate 468]

Perennial, 1-2 m, with a stoloniferous rhizome. Stems stout, more or less terete, transversely partitioned, appearing articulate on drying. Leaves generally bladeless; sheaths thin, upper sheaths usually ending in a short blade. Inflorescence a compound umbel; rays 4-12, unequal, up to 10(-12) cm, each ending in a cluster of spikelets. Involucral bracts at base of inflorescence very short (about 1 cm), nearly equal, scale-like, lanceolate, sharp-pointed. Spikelets 10-25 × 1-1.25 mm, many-flowered, narrowly linear; rachilla winged. Glumes imbricate, oblong-lanceolate, obtuse, boat-shaped, reddish-brown, hyaline-margined, green at keel, 3-5-veined. Stamens 3. Stigmas 3. Nut oblong-elliptic in outline, 3-gonous, apiculate, $\frac{2}{5}$ length of glume. 2n=~112-114. Fl. April-September.

Hab.: Swamps. Rare. Acco Plain, Philistean Plain; Upper Jordan Valley, Beit Shean Valley.

Area: Trop.

6. Cyperus corymbosus Rottb., Descr. Icon. Rar. Pl. 42, t.7 f.4 (1773); Kük. in Engler, Pflanzenreich 101:80, incl. var. *longispiculatus* (O.Kuntze) Kük., *op. cit.* 82. *C. diphyllus* Retz., Obs. Bot. 5:11 (1788); Boiss., Fl. 5:374. *C. tegetiformis* Roxb., Hort. Bengal. 6 (1814); Eig & Feinbr., Palest. Jour. Bot. Jerusalem ser., 2:101 (1940). [Plate 469]

Very close to *C. articulatus* from which it differs mainly in stems being non-septate or obscurely septate, 3-gonous above, and in leaf-like unequal involucral bracts which are 4-10 cm. long (shorter than inflorescence). 2n=~106. Fl. May-September.

Hab.: Banks of rivers, marshes. Rare. Sharon Plain (Yarkon River), Philistean Plain (Wadi Rubin); Hula Plain, Upper Jordan Valley.

Area: Trop.

S.S. Hooper (Kew) found that the rays of some of our plants which she examined were unusually long. She remarked that differences from typical plants are perhaps to be expected when one considers that Israeli populations on the banks of Yarkon River and in the Hula Plain are situated at the N. limit of the species.

7. Cyperus latifolius Poir. in Lam., Encycl. Méth. Bot. 7:268 (1806); Kük. in Engler, Pflanzenreich 101:87, t.11; Hooper, in Fl. W. Trop. Afr. ed. 2, 3:285; Eig & Feinbr., Palest. Jour. Bot. Jerusalem ser., 2:101 (1940). [Plate 470]

Perennial, 1-1.5 m, stoloniferous. Stems stout, 3-quetrous above. Leaves 1-1.5 cm broad, flat, coriaceous, long-acuminate, prominently 3-veined, shorter than stem. Inflorescence a compound or decompound, broadly pyramidal umbel with numerous spikes; primary rays 5-10, unequal, up to 20 cm, with truncate prophylls at base. Involucral bracts 3-4, broad, similar to leaves, the longest overtopping the inflorescence. Spikes broadly ovate, with 4-16 spikelets. Spikelets 10-20 × 2 mm, remote, spreading at maturity, 14-24-flowered, linear-lanceolate, acute; rachilla narrowly winged. Glumes loosely imbricate, oblong-obovate, concave, truncate, broadly hyaline at apex, reddish-brown, yellowish at back and at involute margins, veined. Stamens 3. Stigmas 3, shortly exserted. Nut obovoid-ellipsoid, 3-gonous, apiculate, $\frac{2}{5}-\frac{1}{2}$ length of glume. Fl. July–October.

Hab.: Swamps. Rare. Sharon Plain.

Area: Trop. Africa.

8. Cyperus longus L., Sp. Pl. 45 (1753); Kük. in Engler, Pflanzenreich 101:97; Boiss., Fl. 5:375; Post, Fl. 2:674. Incl. var. *pallidior* Kük., *op. cit.* 100 (1935), var. *heldreichianus* (Boiss.) Boiss., *loc. cit.* (1882) & var. *pallidus* Boiss., *loc. cit.* non Boeckeler, Linnaea 36:280 (1870). [Plate 471]

Perennial, 50–130 cm, with a creeping rhizome. Stems 3-quetrous, leafy. Leaf-blades up to 7 mm broad, flat, keeled, as long as or shorter than stems. Inflorescence a lax simple or compound umbel; primary rays 6–15, unequal, up to 35(–50) cm, with pale brown prophylls; secondary rays slender, unequal; primary or secondary rays ending in short digitate clusters of 3–12 spikelets. Involucral bracts 3–6, flat, unequal, the longest longer than inflorescence. Spikelets (5–)8–20(–25) × 1–1.5 mm, many-flowered, linear, acute; rachilla broadly winged. Glumes imbricate, ovate, obtuse, greyish-brown, white-margined, green at keel, 3–5-veined. Stamens 3. Stigmas 3. Nut oblong-obovoid, 3-gonous, apiculate, $\frac{1}{3}-\frac{3}{4}$ length of glume. 2n=~120. Fl. April–October.

Hab.: Ditches and other wet habitats. Common. Acco Plain, Coast of Carmel, Sharon Plain, Philistean Plain; Upper and Lower Galilee, Esdraelon Plain, Mt. Carmel, Samaria, Shefela, Judean Mts., Judean Desert; Dan Valley, Hula Plain, Upper Jordan Valley, Beit Shean Valley, Lower Jordan Valley, Dead Sea area; Golan, Gilead, Ammon, Moav.

Area: Mediterranean and Irano-Turanian, extending into Euro-Siberian region.

9. Cyperus rotundus L., Sp. Pl. 45 (1753); Kük. in Engler, Pflanzenreich 101:107, t.13; Hooper, in Fl. W. Trop. Afr. ed. 2, 3:285; Kern, in Fl. Males. 7(3):604; Boiss., Fl. 5:376; Post, Fl. 2:675; incl. var. *macrostachyus* Boiss., *op. cit.* 377 (1882). [Plate 472]

Perennial, 15–40(–80) cm, with a slender rhizome; rhizome with red-brown scales and emitting wiry stolons ending in a subglobose or ellipsoid dark-brown tuber. Stems slender, 3-quetrous, leafy at base. Leaves up to 6 mm broad, flat, acuminate, often shorter than stem. Inflorescence a simple or compound umbel; rays 3–9, unequal, up to 5(–10) cm, each ending in a short spike or cluster of spikelets. Involucral bracts 2–4, flat, mostly longer than inflorescence. Spikelets 10–35(–50) × 2–3 mm, many-flowered, linear, strongly compressed; rachilla winged. Glumes imbricate, ovate, obtuse or acutish, boat-shaped, reddish to deep brown, with white-membranous margins, green at keel, 5–7-veined. Stamens 3. Stigmas 3. Nuts oblong-obovoid or ellipsoid, 3-gonous, $\frac{2}{5}$ length of glume, glossy, rarely maturing (reproduction almost entirely by stolons). 2n=~200. Fl. February–December.

Hab.: Swamps, ditches, alluvial and irrigated ground. Common. Acco

Plain, Sharon Plain, Philistean Plain; Upper and Lower Galilee, Esdraelon
Plain, Samaria, Judean Mts., Coastal Negev; Dan Valley, Hula Plain, Upper
and Lower Jordan Valley, Dead Sea area; Golan, Ammon, Edom.

Area: Mediterranean, Irano-Turanian and Trop.

Young tubers edible and used in popular medicine.
The description of subsp. *rotundus* as given by Kern *(loc. cit.)* fits our plants.

10. Cyperus glaber L., Mantissa Alt. 179 (1771); Kük. in Engler, Pflanzenreich
101:159; Boiss., Fl. 5:371; Post, Fl. 2:673. [Plate 473]

Annual, 10–40 cm, tufted. Stems somewhat compressed-3-quetrous. Leaves
2–3 mm broad, flat, shorter than stems. Inflorescence a simple or sometimes
compound 3–8-rayed umbel; rays up to 4(–6) cm, each ending in a globose to
ovoid cluster comprising numerous subsessile spikelets; the central ray very
short. Involucral bracts 2–5, unequal, the longest several times longer than
inflorescence. Spikelets numerous, 8–12 × 2–3(–4) mm, many-flowered, linear-
lanceolate; rachilla narrowly winged. Glumes 2–2.5 mm, ovate-oblong, short-
mucronate, keeled, purplish-brown, white-margined, green at keel and 3–4-
veined on either side. Stamens (2–)3. Stigmas 3. Nut obovoid, 3-gonous, ½–⅗
length of glume. Fl. July–October.

Hab.: Wet ground. Sharon Plain; Upper and Lower Galilee, Samaria,
Judean Mts., Judean Desert; Dan Valley, Hula Plain, Upper and Lower Jordan Valley, Dead Sea area; Golan.

Area: Mediterranean and Irano-Turanian, extending into E. Europe.

11. Cyperus fuscus L., Sp. Pl. 46 (1753); Kük. in Engler, Pflanzenreich
101:235; Boiss., Fl. 5:370; Post, Fl. 2:673, incl. var. *virescens* (Hoffm.) Vahl,
Enum. Pl. 2:336 (1805–1806) & var. *minimus* F. Zimm., Feddes Repert. 14:372
(1916). [Plate 474]

Annual, 10–40 cm. Stems tufted, compressed-3-quetrous, leafy below. Leaves
somewhat shorter than stems, up to 4(–5) mm broad, flat, with a distinct midvein. Inflorescence a simple or compound umbel, sometimes congested into a
head; rays 3–7, unequal, up to 3 cm, each ending in a subcapitate cluster of
sessile or pedicellate spikelets. Involucral bracts 2–5, flat, unequal, the lowest
usually several times longer than inflorescence. Spikelets numerous, 4–7 ×
1–1.5 mm, many-flowered, lanceolate to oblong. Glumes imbricate, 1–1.25
mm, ovate, subacute, mucronulate, boat-shaped, keeled, dark brown to greenish, green along keel, sometimes greenish altogether. Stamens 2. Stigmas
3(–2). Nut obovoid-ellipsoid, 3-quetrous, apiculate, nearly as long as glume.
2n=36. Fl. May–December.

Hab.: Wet places, also on sandy soil. Fairly common. Acco Plain, Sharon
Plain, Philistean Plain; Upper and Lower Galilee, Mt. Carmel, Esdraelon
Plain, Samaria, Shefela, Judean Mts., Judean Desert; Dan Valley, Hula Plain,
Upper and Lower Jordan Valley, Dead Sea area, Arava Valley; Golan, Gilead,
Ammon, Edom.

Area: Euro-Siberian, Mediterranean and Irano-Turanian.

Plants vary in height and in colour of spikelets. Var. *virescens* (Hoffm.) Vahl, Enum. Pl. 2:336 (1805-1806) & Kük., *op. cit.* 236 has been collected in Hula Plain.

12. Cyperus difformis L., Cent. Pl. 2:6 (1756) & Amoen. Acad. 4:302 (1759); Kük. in Engler, Pflanzenreich 101:237, t.27 ff.F–H; Hooper, in Fl. W. Trop. Afr. ed. 2, 3:290; Kern, in Fl. Males. 7(3):629; Mout., Nouv. Fl. Liban Syrie 1:163; Boiss., Fl. 5:370. [Plate 475]

Annual, 20-50 cm. Stems compressed to 3-quetrous. Leaves few, 2-5 mm broad, flat, smooth, shorter than stems. Inflorescence a simple (rarely compound) umbel of dense heads; rays 3-8, unequal, up to 5 cm, the central ray nearly 0; heads of spikelets 6-12 mm in diameter. Involucral bracts 2-4, leaf-like, flat, unequal, the longer overtopping inflorescence. Spikelets numerous, 4-8 × 1 mm, 10-30-flowered, linear, pale. Rachilla not winged. Glumes imbricate, minute (0.5-0.75 mm), orbicular-reniform, turgid, obtuse or retuse, boat-shaped, maroon to green, white-margined, keeled, 1-veined on either side. Stamens 1(-2-3). Stigmas 3. Nut obovoid-ellipsoid, 3-gonous, nearly as long as glume. Fl. June–October.

Hab.: Very wet habitats; weed in rice fields. Rare. Dan Valley, Hula Plain.

Area: Trop. and Subtrop. (Mediterranean, Irano-Turanian, Trop. and Subtrop. E. Asia, Saharo-Arabian and Sudanian regions, Trop. and S. Africa, Australia).

13. Cyperus capitatus Vand., Fasc. Pl. 5 (1771); Kük. in Engler, Pflanzenreich 101:267; Hooper, Israel Jour. Bot. 26:98 (1977). *C. mucronatus* (L.) Mabille, Rech. Cors. 1:27 (1867) non Rottb. (1772) nec Sm. in Sibth. & Sm. (1806) nec Steud. (1854-1855); Post, Fl. 2:672. *Schoenus mucronatus* L., Sp. Pl. 42 (1753). *Galilea mucronata* (L.) Parl., Fl. Palerm. 1:299 (1845). *C. kalli* (Forssk.) Murb., Lunds Univ. Arsskr. 35:24 (1899). *Scirpus kalli* Forssk., Fl. Aeg.-Arab. 15 (1775). *C. aegyptiacus* Gloxin, Obs. Bot. 20, t.3 (1785). *C. schoenoides* Griseb., Spicil. Fl. Rumel. 2:421 (1846) *nom. superfl.*; Boiss., Fl. 5:368. [Plate 476]

Perennial, glaucescent, 15-40(-50) cm, with a creeping rhizome. Stems stout, erect or curved, more or less terete; base clothed with tawny sheaths and leaves. Leaves 1-6 mm broad, rigid, canaliculate, incurved, as long as or longer than stems. Inflorescence a dense, hemispherical to globose, sessile head, 2-3.5 cm in diameter. Involucral bracts 3-6, unequal, dilated at base, several times as long as inflorescence. Spikelets 8-20 × 3-4 mm, few-flowered, ovate-oblong, obscurely distichous. Glumes 5-8 × 2.5-4 mm, broadly ovate, mucronate, rust-brown to maroon, white-margined, striate, keeled; the 2 lowest glumes larger (6-10 × 4-5 mm), empty. Stamens 3. Stigmas 3. Nut obovoid, 3-gonous, ⅓ length of glume. 2n=82. Fl. February–June.

Hab.: Semistabilised maritime sands and wavy sandy fields. Common in the *Artemisia monosperma–Cyperus capitatus (mucronatus)* association (Eig, 1939). Coastal Galilee, Acco Plain, Coast of Carmel, Sharon Plain, Philistean Plain, W. Negev.

Area: Around the Mediterranean coast, Canary Islands and Atlantic coast of Africa.

14. Cyperus conglomeratus Rottb., Descr. Pl. Nov. Programm. 16 no. 29 (1772); Rottb., Descr. Icon. Rar. Pl. 21, t.15 f.7 (1773); Kük. in Engler, Pflanzenreich 101:272 p.p.; Hooper, in Fl. W. Trop. Afr. ed. 2, 3:292; Boiss., Fl. 5:369 p.p.; Post, Fl. 2:673. [Plate 477]

Perennial, glaucescent, tufted, 30–40 cm, with a creeping rhizome and woolly roots. Stems rigid, erect, nearly terete, sulcate, leafy at base. Leaves rigid, glaucous, canaliculate, as long as or shorter than stems. Inflorescence a globose head, 1–4 cm in diameter, or an umbel with 3–4 rays. Involucral bracts 1–3, unequal, the longest much overtopping inflorescence. Spikelets 8–16-flowered, distichous, oblong-lanceolate, subturgid. Glumes 2.5–3.5 mm, imbricate, ovate to ovate-oblong, acuminate, straw-coloured, striate, white-margined. Stamens 3. Stigmas 3. Nut obovate in outline, obtusely 3-gonous, ⅓ length of glume. 2n=74. Fl. February–June.

Hab.: Maritime and desert sand dunes. Common in the *Ammophila arenaria–Cyperus conglomeratus* association (Eig, 1939). Acco Plain, Sharon Plain, Philistean Plain; W., N. and C. Negev.

Area: E. Saharo-Arabian, extending into the Sudanian region and into the Mediterranean territories of Palestine; reaching also Madagascar, S. India and Ceylon (T. Koyama, *op. cit.* 144, 1977).

15. Cyperus jeminicus Rottb., Descr. Pl. Nov. Programm. 24 (1772); Rottb., Descr. Icon. Rar. Pl. 25, t.8 f.1 (1773); Hooper, in Fl. W. Trop. Afr. ed. 2, 3(2):292. *C. conglomeratus* Rottb. var. *multiculmis* (Boeckeler) Kük. in Engler, Pflanzenreich 101:274 (1936). *C. conglomeratus* Rottb. var. *effusus* Boiss., Fl. 5:369 (1882). [Plate 478]

Perennial, glaucescent, 20–50(–80) cm, tufted, usually with wiry roots (in sandy habitats roots often woolly). Stems rigid, erect, 3-gonous, sulcate, leafy at base. Leaves rigid, glaucous, canaliculate, as long as or shorter than stems; sheaths of previous years dark reddish-brown. Inflorescence an umbel with few rays, often contracted into a head. Involucral bracts 2–3, unequal, the longest erect, longer than inflorescence. Spikelets 16–40-flowered, distichous, oblong-lanceolate, strongly compressed, generally somewhat longer than in *C. conglomeratus* but with twice as many flowers. Glumes 2.5–3.5 mm, densely imbricate, lanceolate, acute, mucronulate, striate, brownish. Stamens 3. Stigmas 3. Nut obovate in outline, obtusely 3-gonous, ½ length of glume. Fl. June–October.

Hab.: Mostly near hot springs. Shores of the Dead Sea; Sinai.

Area: Sudanian, extending into the Saharo-Arabian.

16. Cyperus pygmaeus Rottb., Descr. Icon. Rar. Pl. 20, t.14, ff. 4 & 5 (1773); Kern, in Fl. Males. 7(3):634; Boiss., Fl. 5:368 excl. syn. *C. diffusus* Roxb.; Post, Fl. 2:672. *C. michelianus* (L.) Del. emend. Link subsp. *pygmaeus*

(Rottb.) Aschers. & Graebn., Syn. Mitteleur. Fl. 2(2):273 (1904); Kük. in Engler, Pflanzenreich 101:312. *Juncellus pygmaeus* (Rottb.) C.B. Clarke in Hook. fil., Fl. Brit. Ind. 6:596 (1893). [Plate 479]

Annual, 5–25 cm. Stems numerous, tufted, erect or diffuse, 3-quetrous. Leaves 1–3 mm broad, flat, acuminate, shorter or longer than stems. Inflorescence a dense sessile globose to ovoid head, 7–14 mm in diameter. Involucral bracts 3–7, leaf-like, flat, spreading, several times longer than inflorescence. Spikelets densely crowded, very numerous, 3–5 × 1.5–2 mm, 8–24-flowered, oblong-lanceolate, strongly compressed. Glumes distichous, 1.2–2.5 × 0.5–0.6 mm, densely imbricate, oblong-lanceolate, ending in a narrow green tip, boat-shaped, whitish-green, pellucid, green-keeled, veined. Stamens 1–3. Stigmas 2, sometimes flowers with 3 stigmas in same spikelet. Nut oblong, plano-convex or 3-gonous, smooth, ½ length of glume. 2n=~86. Fl. June–October.

Hab.: Damp places. Acco Plain, Sharon Plain; Upper and Lower Galilee; Hula Plain, Upper and Lower Jordan Valley; Golan, Transjordan (Al-Eisawi).

Area: E. Mediterranean and Irano-Turanian, extending into E. and Trop. Africa and through S. and E. Asia to Australia.

Kern *(loc. cit.)* remarks that there is no correlation between the number of stigmas and the shape of nuts in *C. pygmaeus.*

C. pygmaeus is often treated as part of *C. michelianus* (L.) Del. emend. Link (1827). Thus, in Flora Europaea *C. michelianus* is subdivided into: subsp. *michelianus*, with glumes arranged in 3 rows, and subsp. *pygmaeus* (Rottb.) Aschers. & Graebn., with distichous glumes. S.S. Hooper in Flora of Iraq (MS.) delimits subsp. *michelianus* by: spikelets terete; glumes spirally arranged, sides not or faintly veined; subsp. *pygmaeus* by: spikelets flattened; glumes distichous, sides clearly 2–3-veined. In our specimens from Palestine the glumes are distichous and clearly veined at the sides.

17. Cyperus laevigatus L., Mantissa Alt. 179 (1771); Kük. in Engler, Pflanzenreich 101:321; Boiss., Fl. 5:366; Post, Fl. 2:671. [Plates 480, 481]

Perennial, 15–40 cm, with a creeping rhizome. Stems erect, 3-gonous above. Leaves 2–3, the lower reduced to sheaths, the upper with a narrow, linear, mostly short blade. Inflorescence a pseudolateral cluster of sessile spikelets. Involucral bracts 2, the lower 2–6 cm, subulate, erect, simulating prolongation of stem and exceeding inflorescence; the upper short. Spikelets few to numerous, many-flowered, 5–20 × (2–)2.5 mm, oblong or oblong-lanceolate, acute, compressed to subturgid. Glumes densely imbricate, ovate to suborbicular, obtuse. Stamens 3. Stigmas 2. Nut lenticular, more or less plano-convex, obovate or elliptic in outline, obtuse, ½–⅔ length of glume, its broad side turned to rachilla. Fl. April–September.

Subsp. **laevigatus.** *Juncellus laevigatus* (L.) C.B. Clarke in Hook. fil., Fl. Brit. Ind. 6:596 (1893). [Plate 480]. Spikelets 5–8 × 2–2.5 mm, often numerous; glumes straw-coloured, obtuse. 2n=86.

Hab.: Marshy and somewhat saline and sandy soils. Acco Plain, Coast of

Carmel, Sharon Plain, Philistean Plain, W. Negev; Shefela; Upper Jordan Valley.

Area of subspecies: Subtrop.-Trop. (Mediterranean, Irano-Turanian, Saharo-Arabian and Sudanian regions, Trop. and S. Africa, California, C. and S. America, Australia).

Subsp. **distachyos** (All.) Maire & Weiller in Maire, Fl. Afr. Nord 4:35 (1957). *C. distachyos* All., Auct. Fl. Pedem. 48, t.2 f.5 (1789); Boiss., Fl. 5:367; Post, Fl. 2:671. *C. laevigatus* L. var. *distachyos* (All.) Coss. & Durieu, Explor. Sci. Algérie, Bot. II, 251 (1868); Kük. in Engler, Pflanzenreich 101:324. *Juncellus distachyos* (All.) Turrill, Kew Bull. 1926:375 (1926). [Plate 481]. Spikelets 10-20 × 2-2.5 mm, often curved, 2-5 in an inflorescence; glumes purplish-brown or blackish (at least partly), narrowly white-margined, apiculate. 2n=86.

Hab.: Saline marshy soils. Acco Plain, Coast of Carmel, Sharon Plain; Esdraelon Plain, Samaria, N. Negev; Upper Jordan Valley, Beit Shean Valley, Lower Jordan Valley, Dead Sea area, Arava Valley; E. Ammon.

Area of subsp.: Mediterranean, Irano-Turanian, Saharo-Arabian.

18. Cyperus nitidus Lam., Tabl. Encycl. Méth. Bot. 1:145 (1791). *Pycreus nitidus* (Lam.) Raynal in Hooper & Raynal, Kew Bull. 23:314 (1969); Hooper, in Fl. W. Trop. Afr. ed. 2, 3:300. *C. lanceus* Thunb., Prodr. Pl. Cap. 18 (1794); Kük. in Engler, Pflanzenreich 101:333 incl. var. *palaestinensis* Kük., *op. cit.* 336 (1936); Eig & Feinbr., Palest. Jour. Bot. Jerusalem ser., 2:101 (1940). *P. lanceus* (Thunb.) Turrill, Kew Bull. 1925:67 (1925). [Plate 482]

Perennial, 30-80 cm, robust, with long stolons. Stems solitary, 3-gonous. Leaves up to 5(-7) mm broad, flat, plicate at base, linear, long-acuminate, coriaceous, longer or shorter than stem. Inflorescence a lax umbel, 3-7-rayed; rays 3-9 cm, bearing subspicate clusters of 3-14 spikelets. Involucral bracts 3-6, unequal, spreading, the longest overtopping inflorescence. Spikelets many-flowered, 8-14 × 3-4 mm, compressed, elliptical to oblong-lanceolate. Glumes densely imbricate, oblong-ovate, obtuse, concave, brown to maroon, hyaline-margined, keeled, obsoletely 3-5-veined. Stamens 3. Stigmas 2. Nut biconvex, obovate in outline, apiculate, ¼-⅓ length of glume, its narrow side turned to rachilla. Fl. June-September.

Hab.: In or near water, river banks, swamps. Rare. Sharon Plain, Philistean Plain; Hula Plain.

Area: Mainly Trop. and S. Africa.

19. Cyperus flavidus Retz., Obs. Bot. 5:13 (1788); Kern, in Fl. Males. 7(3):648. *Pycreus flavidus* (Retz.) Koyama, Jour. Jap. Bot. 51:316 (1976).* *C. globosus* All., Auct. Fl. Pedem. 49 (1789) non Forssk. (1775); Kük. in Engler, Pflanzen-

---

* This combination was kindly communicated to me by S.S. Hooper (Kew).

reich 101:352; Boiss., Fl. 5:364; Post, Fl. 2:671. *P. globosus* (All.) Reichenb., Fl. Germ. Excurs. 140 (1830) *nom. illeg.* [Plate 483]

Perennial or annual, 20–60 cm, tufted. Stems rigid, very slender, erect, 3-gonous, leafy at base. Leaves rigid, flat or folded lengthwise, about 2 mm broad, shorter than to equalling stems. Inflorescence a head or umbel with few rays up to 3(–6) cm. Involucral bracts 2–4, unequal, spreading, the lowest much overtopping inflorescence. Spikelets numerous, sessile, many-flowered, 8–20 × 2–2.5 mm, linear, strongly compressed, brownish. Glumes imbricate, oblong-ovate, obtuse, green-keeled and 3-veined along keel, straw-coloured and variously red-streaked at sides. Stamens ? Stigmas 2. Nut flattened, elliptic-oblong, longer than broad, apiculate, finely punctulate, ⅓–½ length of glume, its narrow side turned to rachilla. Fl. July–September.

Hab.: Near water. Rare. Sharon Plain; Upper Galilee, Judean Mts.; Dan Valley, Hula Plain; Golan.

Area: Mediterranean, Irano-Turanian and Trop.

20. Cyperus flavescens L., Sp. Pl. 46 (1753); Kük. in Engler, Pflanzenreich 101:398; Boiss., Fl. 5:364; Post, Fl. 2:670. *Pycreus flavescens* (L.) Beauv. ex Reichenb. in Mössler, Handb. ed. 2, 1802 (1828); Hooper, in Fl. W. Trop. Afr. ed. 2, 3:302. [Plate 484]

Annual, 5–30(–50) cm, tufted. Stems slender, erect, 3-gonous, with 2–3 leaves at base. Leaves about 2 mm broad, flat, flaccid, shorter to longer than stem. Inflorescence a head or umbel with few rays up to 2(–4) cm. Involucral bracts generally 3, very unequal, spreading, the longest much longer than inflorescence. Spikelets sessile, 2 to several, 10–20 × 2–2.5(–3) mm, 10–18-flowered, oblong-lanceolate, acutish, densely imbricate, compressed, pale yellow, darkening with age. Glumes ovate, obtuse, often mucronulate. Stamens 3. Stigmas 2. Nut flattened, lenticular, round to obovate, apiculate, transversely rugulose, about ⅓–½ length of glume, its narrow side turned to rachilla. Fl. July–September.

Hab.: Swamps and wet places. Dan Valley, Hula Plain; ?Golan.

Area: Euro-Siberian, Mediterranean and Trop.

21. Cyperus polystachyos Rottb., Descr. Pl. Rar. Programm. 21 no. 54 (1772); Rottb., Descr. Icon. Rar. Pl. 39, t.11 f.1 (1773); Kük. in Engler, Pflanzenreich 101:367; Kern, in Fl. Males. 7(3):649; Boiss., Fl. 5:365. *Pycreus polystachyos* (Rottb.) Beauv., Fl. Oware 2:48, t.86 f.2 (1816); Koyama, Gardeners' Bull. 30:150 (1977). [Plate 485]

Annual or perennial, 20–50 cm, with a short rhizome. Stems 3-gonous, rigid, leafy below. Leaves narrow, 2–3(–4) mm broad, flat, often shorter than stem. Inflorescence a short-rayed umbel or compact head. Involucral bracts 3–5, unequal, up to 2.5 mm broad, the lower much exceeding inflorescence. Spikelets 8–16 × 1–1.5 mm, acute, lanceolate, many-flowered. Glumes oblong-lanceolate, acutish, straw-coloured to brownish; keel 3-veined. Stamens (1–)2.

Stigmas 2. Nut flattened, oblong, $\frac{1}{2}$–$\frac{2}{3}$ length of glume, its narrow side turned to rachilla. Fl. July–September.

Hab.: Near water. Found only once at Birkat-Battikh, Sharon Plain. Apparently casual.

Area: Trop. and subtrop. regions, rare in the Mediterranean (Ischia, S. Italy).

8. CAREX L.

Perennial herbs, monoecious, rarely dioecious. Stems mostly 3-gonous, leafy at base or up to the inflorescence. Leaves linear, flat or canaliculate, keeled; sheaths closed; ligule present at junction of blade and sheath. Inflorescence a head, spike or raceme of 1 or many spikelets, subtended by more or less developed involucral bracts. Spikelets bisexual or unisexual; the bisexual similar in appearance, with staminate flowers usually near top of spikelet; the unisexual dissimilar in appearance, the terminal ones staminate, the lateral wholly or mainly pistillate. Glumes spirally arranged. Flowers unisexual. Hypogynous bristles or scales absent. Staminate flower of (2-)3 stamens. Pistillate flower of an ovary enclosed in a modified bracteole connate at margins, forming a bladder-like body, the *utricle*; stigmas 2 or 3, exserted from utricle; utricle enlarged when mature, contracted at tip, often beaked, deciduous with nut. Nut 3-gonous or flattened.

About 1,500–2,000 species in all parts of globe, mainly in temperate regions; in Tropics restricted to mountains.

Literature: G. Kükenthal, Cyperaceae-Caricoideae, in: Engler, *Pflanzenreich* 38 (IV. 20):67–824, 1909. V.J. Kreczetowicz, *Carex* L., in: V.L. Komarov (ed.), *Flora URSS* 3:111–464, 1935. J.H. Kern & T.J. Reichgelt, *Carex* L., in: *Flora neerlandica* 1(3):7–133, 1954. W. Schultze-Motel, *Carex* L., in: Hegi, *Illustrierte Flora von Mitteleuropa* ed. 3, 2:96–274, 1968–1969.

1. Spikelets bisexual, 0.5–1.5 cm, with staminate flowers in upper part of spikelet and pistillate flowers below. Stigmas 2   2
- Spikelets unisexual, mostly exceeding 1.5 cm; staminate spikelets 1 or more in upper part of inflorescence; pistillate spikelets below staminate ones; uppermost pistillate spikelet sometimes with staminate flowers in its upper part. Stigmas 3
                                                                                                               5
2. Plants with a creeping rhizome. Inflorescence 0.8–3 cm, brown at least at maturity
                                                                                                               3
- Plants tufted, with a short rhizome. Inflorescence 3–8 cm, green to light brown   4
3. Low plants, 5–15(-20) cm in height, of arid habitats (deserts, steppes, batha). Inflorescence 0.8–1.5(-1.8) cm, usually less than twice as long as broad. Leaves 1–2 mm broad. Utricle not or obscurely veined.   **1. C. pachystylis**
- Taller plants, of damp ground on alluvial soils. Inflorescence (1-)1.5–3 cm, at least twice as long as broad. Leaves 2–2.5 mm broad. Utricle ribbed, with numerous elevated veins.   **2. C. divisa**
4(2). Plants of maquis and forest. Leaves 2–3(-4) mm broad. Utricle 4–5 mm,

broadly ovate, 2.5-3 mm broad, abruptly short-beaked. Inflorescence 0.7-1 cm broad.
                                                                        3. C. divulsa subsp. leersii
- Plants of marshes and heavy alluvial soil along water courses. Leaves 5-7 mm broad. Utricle 4.5-6 mm, ovate, 2-2.25 mm broad, gradually long-beaked. Inflorescence generally broader than in above.   4. C. otrubae
5(1). Pistillate spikelets 1-1.5 cm; at least 1-2 upper ones sessile or subsessile, crowded at base of staminate spikelet   6
- Pistillate spikelets longer, more or less remote from staminate spikelet (or spikelets)   7
6. Pistillate spikelets 2-5-flowered; in addition to spikelets in upper part of stem, 1 or more long-pedunculate pistillate spikelets often arise from base of plant.
                                                                        11. C. hallerana
- Pistillate spikelets many-flowered; no long-pedunculate spikelets arise from base of plant.   10. C. extensa
7(5). Utricle obtuse, abruptly ending in a very short beak; beak terete, truncate or indistinctly 2-dentate   8
- Utricle acuminate or acute, tapering to a flat, distinctly 2-dentate beak   9
8. Utricle obovate in outline, ciliolate or denticulate at upper margin. Glumes in pistillate spikelets oblong-lanceolate, ending in a flat cusp or awn.   9. C. hispida
- Utricle ovate to elliptic in outline, not as above at upper margin. Glumes in pistillate spikelets ovate to oblong-ovate, ending in a mostly short subulate mucro.   7. C. flacca subsp. serrulata
9(7). Pistillate spikelets pendulous on slender long peduncles. Glumes in pistillate spikelets pale green.   6. C. pseudocyperus
- Pistillate spikelets erect. Glumes in pistillate spikelets maroon   10
10. Glumes in pistillate spikelets ovate, abruptly mucronate, shorter than utricle. At least lowest bract with a long entire sheath. Leaves up to 5 mm broad.
                                                                        8. C. distans
- Glumes in pistillate spikelets generally exceeding utricle in length, oblong-lanceolate tapering into a long green cusp. Lowest bract not or shortly sheathing. Leaves often exceeding 5 mm in width.   5. C. acutiformis

Subgen. 1. VIGNEA (Beauv. ex Lestib.) Kük. in Engler, Pflanzenreich 38 (IV. 20):111 (1909). Monoecious; inflorescence mostly spike-like; all spikelets similar in appearance, mostly bisexual, with pistillate flowers at base and staminate flowers in upper part of spikelet. Stigmas usually 2.

1. Carex pachystylis J. Gay, Ann. Sci. Nat. Bot. ser. 2, 10:301 (1838); Kreczetowicz, in Fl. URSS 3:197, t.13 f.8. C. stenophylla Wahlenb. B. pachystylis (J. Gay) Aschers. & Graebn., Syn. Mitteleur. Fl. 2(2):25 (1902); Kük. in Engler, Pflanzenreich 38:121 as forma pachystylis (J. Gay) Aschers. & Graebn.; Post, Fl. 2:682 as var. pachystylis. C. stenophylla Wahlenb. var. planifolia Boiss., Fl. 5:400 (1882). ?C. eremitica Paine, Palest. Explor. Soc. Third Statement 126 (1875); Post, Fl. 2:684. [Plate 486]
Plant 5-15(-20) cm, with a creeping rhizome. Stems slender, obtusely 3-gonous, producing sterile shoots and flowering stems. Leaves 1-2 mm broad, firm, flat or rolled, shorter than or as long as stem. Inflorescence head-like,

0.8–1.5(–1.8) × 0.7–1 cm, dark brown, dense, subglobose to ovoid. Spikelets 4–6, bisexual; upper flowers in spikelet staminate. Glumes rust-brown to maroon, ovate, acute, nearly as long as utricle. Utricle 4.5–5 mm, short-stipitate, plano-convex, ovate to elliptic in outline, obscurely veined or not veined, rather abruptly ending in a 2-dentate beak. Stigmas 2. Fl. January–April.

Hab.: Deserts, steppes and batha. Esdraelon Plain, Judean Mts., Judean Desert, N. and C. Negev; Upper and Lower Jordan Valley; Gilead, Ammon, Moav, Edom.

Area: Irano-Turanian.

Common in steppe associations of the Judean Desert, the Negev, Transjordan and Syrian Desert, in particular in association with *Poa sinaica* (Eig, Palest. Jour. Bot. Jerusalem ser., 3:201–202, 1946).

2. Carex divisa Huds., Fl. Angl. 348 (1762); Kük. in Engler, Pflanzenreich 38:125; Boiss., Fl. 5:401; Post, Fl. 2:683. [Plate 487]

Plant (15–)20–30(–50) cm, glaucous, with a dark brown creeping rhizome. Stems erect, 3-gonous, slender. Leaves 2–2.5 mm broad, more or less flat, shorter or longer than stem. Inflorescence spike-like or head-like, (1–)1.5–3 cm, ovoid to oblong, at least twice as long as broad, sometimes interrupted, of several spikelets, light brown, becoming darker at maturity. Bracts subtending spikelets long-aristate, the lower bract sometimes as long as inflorescence. Spikelets 3–7, bisexual, their upper florets staminate. Glumes reddish to dark brown, pale at keel and margins, tapering to an awn-like point. Utricle 3.5–4 mm, plano-convex, ovate in outline, dark brown and glossy at maturity, with numerous elevated veins, ending in a short 2-dentate beak. Stigmas 2. Fl. February–May.

Hab.: Damp places. Fairly common. Coastal Galilee, Acco Plain, Coast of Carmel, Sharon Plain, Philistean Plain; Upper and Lower Galilee, Mt. Carmel, Esdraelon Plain, Samaria, Shefela, Judean Mts.; Hula Plain, Upper Jordan Valley, Dead Sea area; Golan, Gilead, E. Ammon.

Area: Mediterranean and Irano-Turanian, extending into the Euro-Siberian region.

3. Carex divulsa Stokes in With., Arr. Brit. Pl. ed. 2, 2:1035 (1787) subsp. **leersii** (Kneucker) Walo Koch, Mitt. Bad. Landesver. Naturk. Freiburg nov. ser., 11:259 (1923); David & Chater, Watsonia 11:253 (1977); Chater, Carex, in Fl. Eur. 5:298 (1980). *C. echinata* var. *leersii* (F.W. Schultz) Kük. in Engler, Pflanzenreich 38:161 (1909). *C. muricata* var. *leersii* Kneucker in Seubert-Klein, Excurs. Fl. Baden. 52 (1891). [Plate 488]

Plant 30–60 cm, tufted. Stems erect, 3-gonous. Leaves 2–3(–4) mm broad, light green, rather flaccid, flat or somewhat canaliculate. Inflorescence a markedly interrupted narrow spike, 3–8 × 0.7–1 cm, light green, of 5–12 spikelets; lowest 2 spikelets separated from each other by a gap as long as their length. Bract subtending lowest spikelet subulate from a broad base, shorter then spikelet or rarely exceeding it in length. Spikelets 5–7 mm, bisexual, few-flowered, with staminate flowers near apex and pistillate flowers at base. Glumes of pistillate flowers ovate, concave, hyaline, green on veined keel, ending in a 1 mm mucro. Utricle 4–5 × 2.5–3 mm, broadly ovate in outline, plano-convex, longer than glume, brown, veined near base, abruptly tapering to a short 2-dentate beak. Stigmas 2. Fl. March–May.

Hab.: Maquis. Upper Galilee; Dan Valley; Golan, Gilead.

Area: Euro-Siberian, Mediterranean and Irano-Turanian.

4. Carex otrubae Podp., Publ. Fac. Sci. Univ. Masaryk 12:15 (1922); Schultze-Motel in Hegi, ed. 3, 2:131; Chater, in Fl. Eur. 5:297; Mout., Nouv. Fl. Liban Syrie 1:175; Boiss., Fl. 5:402 & Post, Fl. 2:683 as *C. vulpina.* [Plate 489]

Plant 30–60(–80) cm, tufted. Stems rigid, erect, 3-gonous, scabrous at angles above, surrounded at base by sheaths of leaves. Leaves 5–7 mm broad, flat, scabrous at keel and margins. Inflorescence 3–8 × 1–1.5 cm, spike-like, mostly oblong, green to light brown, of 5–10 spikelets, the lower mostly remote. Bracts subtending spikelets usually well developed, scabrous, subulate from a triangular base, all, or at least lower bracts, as long as or generally longer than spikelet they subtend. Spikelets bisexual, many-flowered, ovoid; the lower flowers pistillate. Glumes of pistillate flowers ovate, mucronate, light brown, hyaline at margin, green at keel. Utricle 4.5–6 × 2–2.25 mm, ovate in outline, plano-convex, longer and broader than glume, brown, prominently veined on both faces, gradually tapering to a long 2-dentate beak; beak not slit along back. Stigmas 2. Fl. March–May.

Hab.: Marshy alluvial soil. Sharon Plain, Philistean Plain; Upper and Lower Galilee; Dan Valley, Hula Plain, Upper Jordan Valley.

Area: Mediterranean, Irano-Turanian and Euro-Siberian.

Subgen. 2. CAREX. Monoecious; inflorescence mostly a raceme; spikelets 2 or more, dissimilar in appearance, usually unisexual; at least terminal spikelet staminate; pistillate spikelets lower down; sometimes upper pistillate spikelet contains staminate flowers in its upper part; stigmas mostly 3.

5. Carex acutiformis Ehrh., Beitr. Naturk. 4:43 (1789); Kük. in Engler, Pflanzenreich 38:733; Post, Fl. 2:687. *C. paludosa* Good., Trans. Linn. Soc. London 2:202 (1794); Boiss., Fl. 5:430. [Plate 490]

Plant 60–100 cm, with a creeping rhizome. Stems 3-quetrous, scabrous at angles, leafy. Leaves 5–10(–20) mm broad, flat or plicate, keeled, tapering, scabrous, as long as or longer than stem; lower leaf-sheaths purplish-black, disintegrating into fibres. Bracts leaf-like, not or shortly sheathing, the lowest longer than inflorescence. Spikelets unisexual, many-flowered; staminate spikelets (1–)2–3; pistillate spikelets (2–)3–5, remote, the upper sessile, often with staminate flowers in their upper part, the lower short-pedunculate.

Glumes in pistillate spikelets oblong-lanceolate, narrower than utricles and about as long or longer, maroon with green 3-veined keel gradually excurrent into a long cusp. Utricle 3–3.5 mm (incl. beak), ovoid-oblong, flattened-3-gonous, many-veined, scabridulous in its upper part, tapering to a 2-dentate beak. Stigmas 3. Fl. April–June.

Hab.: Marshes and ditches. Dan Valley, Hula Plain.

Area: Euro-Siberian, Mediterranean and Irano-Turanian.

6. Carex pseudocyperus L., Sp. Pl. 978 (1753); Kük. in Engler, Pflanzenreich 38:695; Boiss., Fl. 5:428. [Plate 491]

Plant 30–100 cm, tufted. Stems erect, 3-quetrous, scabrous at angles. Leaves up to 10 mm broad, light green, flat, keeled, scabrous. Lower bract leaf-like, shortly sheathing, long-overtopping inflorescence. Spikelets unisexual, many-flowered; staminate spikelet 1, terminal; pistillate spikelets 3–5(–6), cylindrical, long-pedunculate, pendulous. Glumes in pistillate spikelets scabrous, long-subulate from lanceolate base, pale green at 3-veined keel, membranous at margins, shorter than utricle. Stigmas 3. Utricle 5–7 mm (incl. beak), spreading, ovoid-lanceolate, prominently veined, tapering to a long deeply 2-cuspidate beak. Stigmas 3. Fl. March–June.

Hab.: Humid wooded sites. Sharon Plain; Hula Plain.

Area: Euro-Siberian, Mediterranean and Irano-Turanian.

7. Carex flacca Schreb., Spicil. Fl. Lips. App. 178 no. 669 (1771) subsp. **serrulata** (Biv.) Greuter in Greuter & Rech. fil., Boissiera 13:167 (1967); Chater, in Fl. Eur. 5:306. *C. glauca* Scop.* subsp. *serrulata* (Biv.) Arcangeli, Comp. Fl. Ital. ed. 2, 92 (1894). *C. serrulata* Biv., Stirp. Rar. Sic. Descr. 4:9 (1818); Boiss., Fl. 5:417 *pro syn. C. glauca* Scop. subsp. *cuspidata* (Host) Suess. *C. glauca* subsp. *cuspidata* (Host) Vicioso, Bol. Inst. Forest. Invest. 79:100 (1959). *C. cuspidata* Host, Gram. Austr. 1:71, t.97 (1801); Boiss., Fl. 5:417 *pro syn. C. diversicolor* Crantz, Inst. Rei Herb. 1:405 (1766) (*nom. illeg.*) var. *cuspidata* (Host) Dinsmore in Post, Fl. 2:685. [Plate 492]

Plant 20–50 cm, with a stoloniferous rhizome. Stems 3-gonous, smooth. Leaves 2–4 mm broad, glaucous, flat, firm, scabridulous at margin and mostly strongly scabrous at apex. Bract subtending lowest spikelet leaf-like, mostly as long as inflorescence or longer. Spikelets unisexual, many-flowered; staminate spikelets terminal, 1–3(–4), erect, narrowly cylindrical; pistillate spikelets (1–)2–3(–4), remote, more or less pedunculate, cylindrical to ovoid. Glumes in pistillate spikelets ovate to oblong-ovate, mucronate, red-brown to light brown and green along midline. Utricle up to 4 mm, ovate to elliptic in outline, obtuse, flattened, at maturity subinflated-3-gonous, papillose-scabrous, with 2 marginal veins, abruptly ending in a very short obtuse beak. Stigmas 3. Fl. March–May.

* Scop. (1772); non *C. glauca* Murr. (1770) *nom. nud.*

Hab.: Maquis and batha. Upper Galilee, Mt. Carmel; Golan.

Area of subspecies: Mediterranean.

Area of species: Euro-Siberian, Mediterranean and Irano-Turanian.

8. Carex distans L., Syst. Nat. ed. 10, 2:1263 (1759); Kük. in Engler, Pflanzenreich 38:663; Boiss., Fl. 5:425; Post, Fl. 2:686 incl. var. *minor* Post, Fl. ed. 1, 838 (1883–1896) & forma *atripes* Nábělek & forma *longibracteata* Nábělek, Publ. Fac. Sci. Univ. Masaryk 105:47 (1929). [Plate 493]

Plant 20–80(–100) cm, tufted. Stems obtusely 3-gonous, smooth, leafy in their lower part. Leaves 3–5 mm broad, firm, flat, tapering, glaucescent, mostly shorter than stem; old leaf-sheaths brown, more or less split into fibres. Bract of lowest spikelet much longer than spikelet it subtends (but not exceeding whole inflorescence), leaf-like, with an entire cylindrical sheath. Spikelets unisexual, many-flowered; staminate spikelets 1(–2), terminal; pistillate spikelets 2–3(–4), lateral, 2–3(–4) cm, sometimes pedunculate, remote from one another and from staminate spikelet. Glumes in pistillate spikelets ovate, abruptly mucronate, copper-brown, green at 3-veined keel, shorter than utricles. Utricle 4–5 mm (incl. beak), ovoid to ellipsoid, obtusely 3-gonous, veined, tapering to a 2-dentate, rather long beak. Stigmas 3. Fl. March–June.

Hab.: Marshy soil and shady places. Fairly common. Acco Plain, Coast of Carmel, Sharon Plain, Philistean Plain; Upper Galilee, Mt. Carmel, Esdraelon Plain, Samaria, Shefela, Judean Mts.; Dan Valley, Hula Plain, Upper Jordan Valley; Golan, Gilead, Ammon.

Area: Euro-Siberian, Mediterranean and W. Irano-Turanian.

Pistillate spikelets vary in size within *C. distans*. According to the size of pistillate spikelets, our plants could be referred to forma *sinaica* (Nees) Boeckeler, Linnaea 41:269, 1877, which Kükenthal (*op. cit.* 664) recorded from the Lebanon.

9. Carex hispida Willd. in Schkuhr, Beschr. Abbild. Riedgr. 1:63, t.S f.64 (1801); Willd., Sp. Pl. 4:302 (1805); Kük. in Engler, Pflanzenreich 38:420; Post, Fl. 2:685. *C. echinata* Desf., Fl. Atl. 1:338 (1798) non Murr. (1770); Boiss., Fl. 5:417. *C. mediterranea* C.B. Clarke in Post, Fl. ed. 1, 837 (1883–1896); Post, Fl. 2:685; Kük., *op. cit.* 423. Cf. Mout., Nouv. Fl. Liban Syrie 1:179 in obs. (1966). [Plate 494]

Plant 40–150 cm, with a thick creeping rhizome. Stems stout, obtusely 3-gonous, leafy, with broad brown sheaths at base. Leaves 4–8 mm broad, flat, rigid, glaucous, acutely keeled, often scabrous. Bract subtending lowest spikelet leaf-like, about as long as inflorescence. Spikelets 5–8(–10), unisexual, many-flowered; the upper 3–6 staminate, more or less crowded together, subsessile; the lower 2–3 pistillate, thicker, very dense, often pedunculate. Glumes in pistillate spikelets oblong-lanceolate, narrower than utricles, green with tawny margins, ending in a long flat scabrous cusp or awn. Utricles 4.5–6 mm, densely imbricate, obtuse, obovate in outline, plano-convex, marginate, ciliolate or denticulate at upper margin, broader than glume, abruptly ending in a very short beak. Stigmas 3. Fl. March–July.

Hab.: Marshy ground, ditches. Fairly common. Acco Plain, Coast of Carmel, Sharon Plain, Philistean Plain; Upper Galilee, Esdraelon Plain; Beit Shean Valley; Gilead.

Area: Mediterranean.

10. Carex extensa Good., Trans. Linn. Soc. London 2:175, t.21 f.7 (1794); Kük. in Engler, Pflanzenreich 38:666; Boiss., Fl. 5:424; Post, Fl. 2:686. [Plate 495]

Plant (25-)50-70 cm, tufted. Stems obscurely 3-gonous, nearly smooth. Leaves 2-3 mm broad, rigid, scabrous, partly canaliculate. Inflorescence usually short. Bract subtending lowest spikelet long-overtopping inflorescence. Spikelets unisexual, many-flowered; staminate spikelet 1(-3), terminal; pistillate spikelets 2-3(-4), 1-2 cm, ovoid to oblong-ovoid; the upper 1-2 at base of staminate spikelet, sessile; the lower somewhat remote, short-pedunculate. Glumes in pistillate spikelets ovate, mucronate, tawny, green-veined, ½-¾ length of utricle. Utricle 3-4 mm (incl. beak), ovoid, prominently veined, glabrous, tapering to a short 2-dentate beak. Stigmas 3. Fl. April–June.

Hab.: Marshes. Acco Plain, Sharon Plain; Mt. Carmel, Samaria, Judean Mts.; Upper Jordan Valley.

Area: Mediterranean, extending to Caspian Sea and along Atlantic coast; also in S. Africa; naturalized in N. and S. America.

11. Carex hallerana Asso, Syn. Stirp. Arag. 133, t.9 f.2 (1779; *"halleriana"*); Kük. in Engler, Pflanzenreich 38:487; Post, Fl. 2:684. *C. gynobasis* Vill., Hist. Pl. Dauph. 2:206 (1787) *nom. superfl.*; Boiss., Fl. 5:409. [Plate 496]

Perennial, 10-35 cm, tufted, with a thick rhizome. Stems more or less erect, 3-gonous. Leaves 1-2 mm broad, firm, narrow, revolute-margined, tapering, often shorter than inflorescence; old leaf-sheaths brown, finely fibrous-reticulate. Bract at base of pistillate spikelets subulate, dilated at base, shorter or longer than inflorescence. Spikelets unisexual; staminate spikelet 1, terminal, many-flowered; pistillate spikelets 1-3, 2-5-flowered, short; the upper 1-2 sessile or subsessile, at base of staminate spikelet; in addition 1-3 long-pedunculate pistillate spikelets often arise from leaf-axils at base of plant. Glumes in pistillate spikelets ovate-lanceolate, acute to acuminate, tawny, green-keeled and hyaline-margined, about as long as utricle. Utricle 4-5 mm (incl. beak), obovoid, 3-gonous, veined, short-hairy above, tapering at base to a stalk, abruptly ending above in a short beak. Stigmas 3. Fl. April.

Hab.: Batha. Judean Mts. (collected only once, at Motza by A. Cohen).

Area: Mediterranean and W. Irano-Turanian.

130. ORCHIDACEAE

Perennial herbs, mostly terrestrial, with root-tubers, shoot-tubers or a rhizome, sometimes saprophytic or (in the Tropics) epiphytic, generally with mycrorrhiza. Stems leafy or leafless. Leaves entire, spirally arranged or 2-stichous, rarely subopposite, often sheathing at base; in saprophytes reduced to sheaths or scales. Inflorescence a spike, raceme or panicle, rarely flowers solitary. Flowers generally hermaphrodite, zygomorphic, epigynous, 3-merous, often resupinate, often entomophilous, with most variable and elaborate adaptations to pollination. Perianth-segments 6, in 2 whorls, free or partly connate, usually petaloid; the posterior (median) inner segment (*labellum*) differing from others in shape and larger size; labellum generally polymorphic and heterochromous, becoming abaxial by resupination, serves as a landing platform for insects, in several genera labellum forms a sac or a nectariferous spur; sometimes labellum divided by constriction into basal *hypochile* and apical *epichile*. Mostly 1 fertile stamen (median outer stamen), rarely 2; sometimes 1–3 staminodes or glands also present; anther(s) sessile or subsessile, with 2 thecae and mostly containing 2 *pollinia*; each pollinium consisting of a mass of aggregated pollen-grains and often borne on a sterile stalk or *caudicle*. Ovary inferior, long and resembling a stalk, often twisted, 1-locular, with parietal placentation, rarely 3-locular; ovules numerous; stigmas 3, fertile, or more often 2 lateral fertile, the third sterile and transformed into a beak-like structure, the *rostellum*, situated between anther and fertile stigmas; rostellum often forming 1 or 2 viscid bodies, *viscidia*, to which pollinia are attached by caudicles; viscidia often enclosed in 1 or 2 pocket-like outgrowths of rostellum, the *bursicles*. Anther(s) and stigmas connate and borne on a *column* or *gynostemium* formed by fused filaments and style or as an outgrowth of flower-axis. Fruit a capsule, dehiscing by 3 or 6 longitudinal slits; seeds very numerous, minute, appearing like sawdust and dispersed by wind; embryo undifferentiated; endosperm absent.

About 750 genera and 20,000 species, abundant in the Tropics.

Literature: H. G. Reichenbach, in: *Icones Florae germanicae et helveticae* 13 & 14, 1850–1851. E. Pfitzer, Orchidaceae, in: Engler & Prantl, *Pflanzenfam.* 2(6):52–220, 1888–1889. P. F. A. Ascherson & K. O. P. P. Graebner, *Synopsis der Mitteleuropäischen Flora* 3:613–925, Leipzig, 1907. E. G. Camus, P. Bergon & A. Camus, *Monographie des Orchidées de l'Europe*, etc., Paris, 1908. E. G. Camus & A. Camus, *Iconographie des Orchidées d'Europe et du Bassin Méditerranéen*, Paris, 1921; Texte, Paris, 1929. G. Keller, R. Schlechter & R. von Soó, *Monographie und Iconographie der Orchideen Europas und des Mittelmeergebietes* 1–5, Berlin, 1925–1944. S. Nevski, Orchidaceae, in: V. L. Komarov et al. (eds.), *Flora URSS* 4:589–730, Leningrad, 1935. N. Feinbrun, Materials for a revised Flora of Palestine I, *Proc. Linn. Soc. London*, Session 157, 1944–45, Pt. 1:48–53 (Dec. 1945). E. & O. Danesch, *Orchideen Europas Südeuropa*, Bern & Stuttgart, 1969. M. Sundermann, *Europäische und mediterrane Orchideen*, Hildesheim, 1975. J. Landwehr, *Wilde Orchideeën van Europa*, 1–2, s'Graveland, 1977. J. Renz, Orchidaceae, in: K. H. Rechinger (ed.), *Flora iranica* 126:1–148, 1978. J. G. & A. E. Williams & N. Arlott, *A Field Guide to the Orchids of Britain & Europe with N. Africa & the Middle East*, Collins, London, 1978. A. Dafni, *Orchids in Israel* (in Hebrew), Tel-Aviv, 1981. P. E. Boissier, *Flora orientalis* 5:51–94, 1882. G. E. Post, *Flora of Syria, Palestine and Sinai* ed. 2, 2:559–582, 1933.

1. Flowers spurred   2
- Flowers without spur   7
2. Plants devoid of green leaves; stem with violet-suffused sheaths. Saprophyte growing on decaying roots of pines and oaks.   **3. Limodorum**
- Plants with green leaves   3
3. Labellum undivided, strap-shaped.   **4. Platanthera**
- Labellum 3-lobed, or if entire, not strap-shaped   4
4. Middle lobe of labellum linear, 4–5 cm, twisted, several times as long as lateral lobes.   **7. Himantoglossum**
- Middle lobe of labellum not as above   5
5. Base of labellum with 2 parallel lamellae. Caudicles of the 2 pollinia connected by a single viscid gland.   **8. Anacamptis**
- Base of labellum without parallel lamellae. Each caudicle with a gland of its own   6
6. Outer perianth-segments 3–4 mm. Spur 1–2 mm.   **5. Neotinea**
- Perianth-segments and spur longer than in above.   **6. Orchis**
7(1). Flowers white; tip of labellum yellow.   **2. Cephalanthera**
- Flowers not white   8
8. Labellum 2.5–4 cm, divided into a rounded basal hypochile and a long acute distal epichile, the latter abruptly refracted. Flowers maroon or purple.   **9. Serapias**
- Labellum smaller, not as above   9
9. Plants with rhizome. Flowers pedicellate. Labellum divided into an incurved hypochile and a recurved cordate or subcordate epichile.   **1. Epipactis**
- Plants with globose-ovoid root-tubers. Flowers sessile. Labellum not as above.   **10. Ophrys**

Subfam. ORCHIDOIDEAE (*Monandrae*). Anterior stamen of outer whorl fertile, the others absent or staminodial; filaments fused with style to form column.

1. EPIPACTIS Zinn

Rhizomatous perennial herbs. Rhizome short or creeping. Stem leafy, often flexuose, glabrous or puberulent. Leaves many-veined; cauline leaves not sheathing; upper leaves decreasing in size, forming a transition to bracts. Inflorescence a lax raceme, several- to many-flowered, often one-sided. Bracts herbaceous, the lower exceeding flower. Flowers short-pedicellate, spreading, or nodding, nearly orbicular in outline. Perianth-segments spreading, rarely connivent; the outer segments more or less similar, ovate or lanceolate, the inner somewhat shorter; labellum without a spur, pointing forward, constricted at about middle and divided into a basal *hypochile* and an apical *epichile*, sometimes with an elastic hinge between them; hypochile concave or boat-shaped, nectar-secreting; epichile broader, more or less cordate or triangular, directed downwards. Column short. Anther erect or slightly curved forward, more or less mobile; pollinia 2, devoid of caudicle and attached to a single gland; pollen powdery-granulate; rostellum short or lacking. Ovary clavate, glabrous or puberulent, slightly twisted.

About 30 species, mainly in temperate Asia as far as China and Japan, in Europe, Mediterranean, C. Africa, C. and N. America.

Literature: Y. Ivri & A. Dafni, Pollination ecology of *Epipactis consimilis* Don (Orchidaceae) in Israel, *New Phytol.* 79:173–177 (1977).

1. Perianth-segments yellowish, purple-margined. Intermediate cauline leaves lanceolate, acuminate, at least 3 times as long as broad. Plants growing near springs and water courses.   **2. E. veratrifolia**
– Perianth-segments greenish, sometimes tinged with purple. Intermediate cauline leaves ovate, elliptic or suborbicular, acute or obtuse, at most 2½ times as long as broad. Plants growing in shade of maquis.   **1. E. helleborine**

1. Epipactis helleborine (L.) Crantz, Stirp. Austr. ed. 2, 467 (1769); Renz, Fl. Iran. 126:37, t.12 f.2 & t.57. *Serapias helleborine* L. & var. *latifolia* L., Sp. Pl. 949 (1753); L., Syst. Nat. ed. 10, 2:1245 (1759). *E. latifolia* (L.) All., Fl. Pedem. 2:152 (1785); Boiss., Fl. 5:87; Post, Fl. 2:581. *Helleborine latifolia* (L.) Moench, Meth. Suppl. 311 (1802). [Plate 497]

Plant 25–110 cm. Rhizome very short. Stem glabrous below, puberulent above. Leaves 5–9, spreading, scabrous at margin; intermediate leaves ovate, elliptic or suborbicular, acute or obtuse. Inflorescence many-flowered, puberulent, sometimes one-sided. Bracts lanceolate. Flowers broadly campanulate, greenish, tinged with purple. Outer perianth-segments up to 12 mm, ovate-lanceolate, green or olive-green, sometimes brownish-veined; the inner slightly shorter, ovate, pale green, tinged with pink; labellum shorter than outer segments; hypochile 4–6 mm, semi-globose, concave; epichile 5 mm, cordate with a reflexed acute tip, undulate margins and 2 bosses at base. Column short. Ovary glabrous or puberulent. Ripe capsule much thickened, subglobose, pedicellate, more or less spreading. Fl. April–June.

Hab.: Maquis, mainly in the shade, with preference for chalky and basalt soils. Rare. Upper Galilee, Mt. Carmel, Samaria; Golan.

Area: Euro-Siberian, Mediterranean and Irano-Turanian.

Protected by law.

2. Epipactis veratrifolia Boiss. & Hohen. in Boiss., Diagn. ser. 1, 13:11 (1854); Renz, Fl. Iran. 126:41, tt.13,14,16 f.1; Boiss., Fl. 5:87; Post, Fl. 2:580. *E. consimilis* Wall. ex Hook. fil., Fl. Brit. Ind. 6:126 (1890) *nom. illeg.* non D. Don (1825); P.F. Hunt, Kew Bull. 22:490 (1968). *Helleborine veratrifolia* (Boiss. & Hohen.) Bornm., Beih. Bot. Centralbl. 33 Abt. II:205 (1915). [Plates 498, 499]

Plant 25–120 cm, erect or pendulous. Rhizome branched, bearing several flowering stems. Stems glabrous below, puberulent above. Leaves 3–9, long, spreading, somewhat scabrous at margin, much varying in size, ovate-lanceolate to lanceolate, acuminate. Inflorescence many-flowered, lax, puberulent. Bracts ovate to lanceolate, decreasing in size upwards. Flowers large, spreading. Perianth-segments spreading, ovate-lanceolate, yellowish, purple-margined, puberulent; outer segments up to 20 mm; the inner slightly shorter; labellum curved, not exceeding outer segments; hypochile up to 10–12 mm, narrowly boat-shaped; epichile 9–11 mm, tongue-shaped, truncate-subcordate at base, oblong, abruptly narrowed towards middle, sub-3-lobed; middle lobe lanceolate, acute, maroon near base, with a white blotch at apex and a small orange protuberance above apex. Column somewhat elongate. Ovary and pedicels puberulent. Ripe capsule thickened, ellipsoid, pedicellate, pendulous. Fl. December–March (Dead Sea area); April–May (Dan Valley).

Hab.: Near springs and water courses, usually on wet travertine soils with *Phragmites* and *Typha*. Rare. Judean Desert; Dan Valley (Nature Reserve in the Tal Wood), Lower Jordan Valley, Dead Sea area, Arava Valley; Golan, Transjordan (Al-Eisawi).

Area: E. Mediterranean, Irano-Turanian, including also S. Asia (Nepal, India) and Trop. E. Africa (Somalia).

According to Ivri & Dafni (*op. cit.*), plants of *E. veratrifolia* growing in the Orchid Nature Reserve in the Dan Valley differ from those growing at En Gedi (Dead Sea area) in several features: lower stature, smaller flowers, flowering time, leaves drying in the winter season while evergreen at En Gedi. Pollination is by hover flies (Syrphidae).

Protected by law.

2. Cephalanthera L.C.M. Richard

Rhizomatose perennial herbs. Rhizome horizontal or vertical, with numerous slender roots. Stem erect, leafy or leafless. Leaves, when present, well developed, flat or plicate. Inflorescence a lax many-flowered spike. Bracts usually herbaceous. Flowers pink, snow-white or pale cream, showy. Outer perianth-segments subequal, more or less connivent, clasping labellum; the inner similar to outer, usually smaller; labellum mostly without spur, deeply constricted at middle and divided by a joint into a basal *hypochile* and an apical *epichile*; hypochile 2-lobed, concave or rarely with a sac-like spur, with sides erect, clasping column at base; epichile ovate, acute or acuminate, recurved, with longitudinal crests at base. Column long, erect; anther hinged to tip of column; pollinia 2-lobed, devoid of caudicles; pollen in powdery masses.

About 12 species in Europe, N. and Middle Asia.

1. Cephalanthera longifolia (L.) Fritsch, Österr. Bot. Zeitschr. 38:81 (1888). *Serapias helleborine* L. var. *longifolia* L., Sp. Pl. 950 (1753). *S. longifolia* (L.) Huds., Fl. Angl. ed. 2, 341 (1778) p.p. *C. ensifolia* (Murr.) L.C.M. Richard, Mém. Mus. Hist. Nat. Paris 4:60 (1818); Boiss., Fl. 5:85; Post, Fl. 2:579. [Plate 500]

Plant 25–50 cm. Rhizome abbreviated. Stems leafy. Leaves in 2 rows, linear-lanceolate, acute or acuminate. Bracts shorter than ovary. Flowers pure white, turning brown during anthesis, glabrous. Outer perianth-segments 14–18 mm, elliptic-lanceolate, acute, somewhat concave; median segment 4–5 mm broad, the lateral 5–6 mm broad, oblique; inner segments ovate, obtuse, rounded above, minutely papillose; labellum without a spur, 7–10 mm, shorter than sepals, erect; hypochile concave or shortly saccate at back, clasping column by its lateral obtuse lobes; epichile ovate, broader than long, keeled, with 4–6 yellow parallel crests along veins, tip yellow, densely papillose. Column 7–8 mm, slender. Ovary glabrous. Fl. March–May.

Hab.: Shade of maquis and pine forest, on terra rossa and basalt soil. Upper Galilee, Mt. Carmel, Samaria, Judean Mts.; Golan, Gilead, Ammon.

Area: Euro-Siberian, Mediterranean and Irano-Turanian.

Protected by law

3. LIMODORUM Boehmer

Perennial saprophytic herbs, with a short rhizome and dense fleshy roots. Stem robust, somewhat fleshy, glabrous, clothed with greenish-violet leaf-sheaths, devoid of green leaves. Inflorescence a spike-like raceme, mostly many-flowered. Bracts greenish-violet. Flowers large, pedicellate. Outer perianth-segments broader than the inner; labellum spurred, nearly equal in length to outer segments, divided into a deltoid hypochile and an ovate and obtuse epichile; spur pointing downwards. Column elongate, slightly dilated towards apex; rostellum short; pollinia powdery-granulate, attached to a single gland. Ovary shortly pedicellate, not twisted.

Two species, Mediterranean and Irano-Turanian.

1. Limodorum abortivum (L.) Swartz, Nova Acta Reg. Soc. Sci. Upsal. 6:80 (1799); Renz, Fl. Iran. 126:21, t.5 f.2; Boiss., Fl. 5:89; Post, Fl. 2:581. *Orchis abortiva* L., Sp. Pl. 943 (1753). [Plate 501]

Plant 30–80(–150) cm. Flowers violet, nearly erect. Outer perianth-segments somewhat spreading, the median 20–25 mm, obovate-lanceolate, incurved-concave; the lateral ones narrower, lanceolate; inner perianth-segments 16–19 mm, oblong, acute, violet towards apex, whitish towards base; labellum about 17 × 8–9 mm; hypochile ascending along column; epichile slightly undulate at margin, violet-veined, with 3 thicker median veins; spur about 15 mm, pale violet, narrowly cylindrical, straight. Fl. April–May.

Hab.: Pine forest, maquis and garigue. Coast of Carmel; Upper Galilee, Mt. Carmel, Samaria, Judean Mts.; Golan; Transjordan (Al-Eisawi).

Area: Mediterranean, extending into the Euro-Siberian and Irano-Turanian.

Protected by law.

4. PLATANTHERA L.C.M. Richard

Tuberous perennial herbs. Tubers 2, undivided, turnip-shaped or oblong. Leaves all cauline or 1–2(–3) basal. Inflorescence a many-flowered cylindrical spike. Bracts herbaceous. Flowers white, yellowish or greenish. Perianth-segments nearly equal in length, all connivent, or the outer lateral segments spreading; labellum spurred, more or less strap-shaped, entire or 3-lobed, somewhat exceeding outer lateral segments in length; spur filiform or narrowly cylindrical, often clavate at apex, usually longer than ovary. Column short, truncate; anther-thecae parallel or divergent; staminodes distinct; glands naked, attached to rostellum. Ovary sessile, twisted, glabrous.

About 80–100 species, mainly in E. Asia, also in cold and temperate regions of N. and S. America, in Europe and the Mediterranean.

1. Platanthera holmboei Lindb. fil., Årsbok Soc. Sci. Fenn. 20 B:No. 7 (1942); Iter Cyprium, Acta Soc. Sci. Fenn. nov. ser. B, 2, 7:11, f.7 (1946); Feinbr., Bull. Res. Counc. Israel 8D:172 (1960). [Plate 502]

Plant 30–40 cm. Basal leaves 2(-3), oblanceolate-elliptic, subobtuse, tapering towards base, 2.5–5 cm broad; cauline leaves much smaller. Inflorescence lax. Bracts acuminate. Flowers pale yellowish-green. Lateral outer perianthsegments 10 × 4 mm, spreading, obliquely ovate, the median 6–7 × 5–6 mm, deltoid, obtuse; inner segments 5–7 × 1–2 mm, narrowly lanceolate; labellum 7–10 × 1.5–2 mm, obtuse, green at tip; spur 15–23 mm, more or less horizontal, slender, cylindrical, not clavate, greenish towards tip. Anther-thecae divergent. Fl. April–May.

Hab.: Shade of *Quercus calliprinos–Q. boissieri* maquis. Rare. Upper Galilee; Golan.

Area: E. Mediterranean (N. Palestine, Cyprus).

Protected by law.

5. NEOTINEA Reichenb. fil.

Tuberous perennial herbs. Inflorescence a short, many-flowered, dense spike. Bracts acute, ½ length of ovary or nearly so. Flowers small. Outer perianthsegments connivent, forming a hood, connate below, free above; the lateral ones subsaccate at base; inner segments almost equalling median outer segment; labellum spurred, flat, deeply 3-lobed; spur very short. Column very short; anther-thecae nearly parallel; tip of connective short; pollinia with very short caudicles; glands 2, distinct, included in 2-celled bursicle. Ovary linear, hardly twisted.

A monotypic genus.

Literature: W.T. Stearn, Multum pro parvo: the nomenclatural history and synonymy of *Neotinea maculata* (Orchidaceae), *Ann. Mus. Goulandris* 2:69–81 (1974).

1. Neotinea maculata (Desf.) Stearn, Ann. Mus. Goulandris 2:79 (1975). *Satyrium maculatum* Desf., Fl. Atl. 2:319 (June–July 1799). *Orchis intacta* Link in Schrad., Jour. für die Bot. 1799, 2:322 (April 1800). *N. intacta* (Link) Reichenb. fil., Pollin. Orchid. 29 (1852); Post, Fl. 2:563. *Tinaea intacta* (Link) Boiss., Fl. 5:58 (1882). [Plate 503]

Plant 10–30 cm. Basal leaves 2–4, ovate-oblong, usually purplish-brownspotted; cauline leaves smaller. Inflorescence 2–7 cm, oblong. Bracts membranous, acute, 1-veined, dentate at apex. Perianth-segments pinkish, flushed with purple, each with a darker midvein, sometimes yellowish (forma *luteola*); outer segments 3–4 mm, ovate-lanceolate, acute; the inner linear, very narrow; labellum 3–5 mm, purplish or yellowish, pale at centre and purple-dotted, its base hidden in hood; lateral lobes of labellum linear; middle lobe broader and somewhat longer, varying from retuse to 2-lobulate, sometimes with a tooth between lobules; spur 1–2 mm, several times shorter than ovary. Fl. April–May.

Hab.: Batha. Rare. Upper Galilee, Samaria, Judean Mts.

Forma **luteola** (Renz) Landwehr, Wilde Orchideeën 11 (1977). *N. intacta* forma *luteola* Renz, Feddes Repert. 27:210 (1929). Flowers greenish-yellowish. Leaves unspotted.

Hab.: Together with the typical form. Rare.

Area of species: Mediterranean.

Protected by law.

6. ORCHIS L.

Tuberous perennial herbs. Tubers undivided, globose or ovoid. Stem erect. Leaves mainly basal; cauline leaves mostly sheath-like. Inflorescence usually many-flowered, bracteate. Flowers brightly coloured, pink, purple or yellow, rarely white; perianth-segments spreading or connivent forming a hood; labellum spurred, entire to variously 3-lobed; middle lobe often 2-lobulate. Column short, erect; stamen 1; anther-thecae parallel; rostellum 3-partite; pollinia club-shaped; caudicles short; glands separate, included in a single 2-locular bursicle. Stigma slightly concave. Ovary slightly twisted, glabrous.

About 35 Euro-Siberian, Mediterranean (incl. Canary Isles and Madeira) and Irano-Turanian species.

All *Orchis* species protected by law in Israel.

1. Labellum distinctly 3-lobed; median lobe larger than lateral ones, sometimes median lobe further divided into secondary lobes or lobules   2
 - Labellum entire, sometimes crenate or denticulate, or if lobed, median lobe much smaller than lateral ones   9
2. Median lobe of labellum divided into 2 lobules, with a tooth interjected in sinus between them   3
 - Median lobe of labellum entire or denticulate   6
3. Lobules of median lobe at least twice as long as broad   4
 - Lobules of median lobe broader than long   5
4. Labellum 15–20 mm; lobules acute; tooth between lobules of median lobe 3 mm or longer. Leaves undulate at margin.   **6. O. italica**
 - Labellum not exceeding 10 mm; lobules obtuse; tooth between lobules of median lobe about 1 mm. Leaves not undulate at margin.   **7. O. galilaea**
5(3). Flowers pale purple to purplish-pink, rarely white. Outer perianth-segments and bracts long-acuminate or subulate at apex. Inflorescence conical at beginning of anthesis.   **5. O. tridentata**

- Flowers yellowish-green with lobes of labellum brownish at apices. Outer perianth-segments and bracts obtuse. Inflorescence cylindrical.   **8. O. punctulata**

6(2). Median lobe of labellum oblong-ligulate, about twice as long as broad. Outer perianth-segments apiculate or acuminate   7

- Median lobe of labellum as long as broad, or broader than long. Outer perianth-segments obtuse to acute   8

7. Median lobe of labellum about 1 cm; lateral lobes serrate. Spur strongly curved forward.   **4. O. sancta**

- Median lobe of labellum 4 mm; lateral lobes crenate. Spur slightly curved.   **3. O. coriophora**

8(6). Labellum broader than long, pale purple and marked with 4–6 purple dots near base; median lobe broader than lateral lobes. Spur horizontal or pointing downwards. Flowers opening from top of inflorescence downwards. Leaves not spotted.   **2. O. israëlitica**

- Labellum not broader than long, purple (rarely white), marked with many purple dots along middle; median lobe not broader than lateral lobes. Spur pointing upwards. Inflorescence flowering from base upwards. Leaves usually dark-spotted.   **10. O. anatolica**

9(1). Spur sack-like, at most 3 times as long as broad, shorter than labellum.   **9. O. saccata**

- Spur cylindrical, gradually tapering towards apex, at least 5 times as long as broad, as long as or longer than labellum   10

10. Perianth-segments conspicuously veined. Labellum longer than broad, flat. Plants of batha and garigue.   **1. O. papilionacea**

- Perianth-segments not or obscurely veined. Labellum broader than long, deflexed along midline on both sides, obscurely lobed. Plants of marshes and other damp habitats.   **11. O. laxiflora**

1. Orchis papilionacea L., Syst. Nat. ed. 10, 2:1242 (1759); Boiss., Fl. 5:60; Post, Fl. 2:565. [Plate 504]

Plant 12–30 cm. Stem surrounded at base by basal leaves and clothed nearly up to inflorescence with cauline leaf-sheaths. Leaves linear-lanceolate, blackish in herbarium. Inflorescence cylindrical, many-flowered. Bracts lanceolate, purplish, 3–5-veined, longer than ovary. Flowers purplish-red. Perianth-segments dark-veined, obtuse; outer segments oblong-lanceolate, oblique, forming a hood; the inner oblong, oblique, shorter and narrower than the outer; labellum 10–12 × 7–8(–9) mm, longer than broad, paler at middle, usually purple-dotted, more or less flat, obovate or obovate-rhombic, cuneate at base, denticulate or crenate at margin; spur 8–10 mm, paler, cylindrical-subulate, descending, slightly curved, somewhat shorter than ovary. Fl. February–April.

Hab.: Batha and garigue. Common, with preference for terra rossa and rendzina. Acco Plain, Coastal Galilee, Coast of Carmel, Sharon Plain, Philistean Plain; Upper and Lower Galilee, Mt. Carmel, Esdraelon Plain, Mt. Gilboa, Samaria, Shefela, Judean Mts.; Dan Valley, Upper Jordan Valley; Golan, Gilead, Ammon, Moav.

Area: Mediterranean.

Boissier (Fl. 5:60) referred plants of *O. papilionacea* from the S. Mediterranean to var. *rubra* (Jacq.) Reichenb., 1834 (cited in 1943 by Renz as var. *rubra* Lindl., Gen. & Sp. 268,1835 = var. *parviflora* Willk., 1870). Together with specimens from Greece and Asia Minor he cited two specimens from Palestine and one from littoral Lebanon under var. *rubra*. Baumann & Dafni (AHO Mitteilungsblatt 11, 4:264–266, 1979) discuss the taxonomy of the Palestinian *O. papilionacea*: "The position of the taxon in Israel is not yet clarified. It has especially small flowers, but otherwise shows the characteristic traits of the species. Within its omnimediterranean distribution area *O. papilionacea* s.l. displays considerable polymorphism which led to numerous combinations... The small-flowered strains of this aggregate were so far described from the area between the Black and Caspian Seas". The authors prefer the binomial *O. caspia* Trautv. (1873) and mention var. *bruhnsiana* Gruner (Bull. Soc. Imp. Nat. Moscow 40, 2:453, 1867) as the earliest designation in the rank of variety (cf. Soó, Ann. Univ. Sci. Budapest, Sect. Biol. 11:62, 1969).

In *O. papilionacea* a cline apparently exists in the west-easterly direction, with a gradual decrease in flower size not correlated with noticeable changes in other morphological traits. Data on size of labellum are: 20–25 mm in Spain (var. *grandiflora* Boiss.); 12–16 mm in S. Europe; 10–12 mm in Palestine; 11 mm in Transcaucasia (*O. caspia* Trautv.). In our view, it is preferable to maintain *O. papilionacea* L. as a unit at the species rank.

Recently a hybrid between *O. papilionacea* and *O. saccata* has been found by W. Schmidt near Matsuba in W. Galilee.

A hybrid between *O. papilionacea* and *O. israëlitica* has been described by Baumann & Dafni as *O. × feinbruniae* (see p. 385).

2. Orchis israëlitica Baumann & Dafni, AHO Mitteilungsblatt 11, 4:261, ff.1–5 (1979).

Feinbr., Proc. Linn. Soc. London Session 157:51 (1945) as *O. picta* Lois. *O. syriaca* Boiss. & Bl. in Boiss., Diagn. ser. 2, 4:91 (1859) *nom. nud.* *O. morio* L. var. *albiflora* Boiss., Fl. 5:60 (1882); Post, Fl. 2:265. [Plate 505]

Plant 10–30 cm. Basal leaves oblong, cauline leaves sheathing. Inflorescence several- to many-flowered, fairly lax, with flowers opening from top downwards. Bracts oblong-ovate, acute or irregularly dentate at apex, often suffused with pale purple, shorter than ovary. Flowers very pale purple. Perianth-segments connivent, forming a hood; the outer 7 × 4–5 mm, green-veined, obliquely ovate, the inner 4–5 × 2.5 mm, green-veined, ovate; labellum 7–10 mm broad, broader than long, paler than outer segments, minutely puberulent, marked with 4–6 purple dots near base, deflexed on sides along midline, 3-lobed; median lobe broader than lateral ones, more or less entire; spur about 1 cm, cylindrical, slender, tapering towards tip, horizontal, about as long as ovary and nearly appressed to it; anther-thecae purple. Fl. March–April.

Hab.: Batha, garigue and maquis, mostly on terra rossa, rarely on marls. Rare. Upper and Lower Galilee.

Area: E. Mediterranean. Endemic in N. Palestine, Lebanon and S. W. Anatolia.

Described from Upper Galilee.

In order to ascertain whether *O. israëlitica* is endemic in Galilee only (cf. Baumann & Dafni), we examined specimens, kindly lent to us by the Conservatoire Botanique, Genève, of *O. morio* L. var. *albiflora* Boiss. cited in Boissier's Flora orientalis 5:60, 1882 (syn. *O. syriaca* Boiss. & Bl., MS.). The specimens collected by Blanche and by Gaillardot in Lebanon and by Heldreich in Lycia (S. W. Anatolia) proved to be *O. israëlitica*. The area of the species can thus be designated as E. Mediterranean; it stretches from Galilee through W. Lebanon to S. Turkey.

Renz (Feddes Repert. 27:209, 1929) described *Orchis picta* subsp. *libani* Renz from the Lebanon, Mt. Amanus (S. E. Turkey) and Cyprus, growing in forests and garigue, in Lebanon with *Orchis anatolica*. It is described as differing from *O. picta* in a laxer spike and a labellum devoid of spots. Mouterde (Nouv. Flore Liban Syrie 1:337) records subsp. *libani* from various localities in Lebanon and Syria, and gives as its area S. Turkey, Syria, Lebanon and Palestine. There is little doubt that *O. picta* subsp. *libani* Renz is a strain of *O. israëlitica* devoid of spots on the labellum.

O. israëlitica is very similar to *O. boryi* Reichenb. fil., which is endemic in Crete and S. Greece. The inflorescence of *O. israëlitica* opens basipetally as does that of *O. boryi*. Its flowers are considerably paler in colour than those of *O. boryi* and the perianth-segments somewhat smaller in size; the median lobe of the labellum is broader than the lateral lobes. Similar to *O. boryi*, *O. israëlitica* hybridizes with *O. papilionacea* L. The hybrid has been described as *O.* × *feinbruniae* Baumann & Dafni (*op. cit.* ff.7–9, 1979). A large hybrid swarm has been found in Lower Galilee. Hybridization points to affinity between *O. israëlitica* and *O. papilionacea* despite the differences in perianth characters.

It should be noted that Renz (Feddes Repert. 28:241, t.12, 1930) described from Crete *O.* × *lasithica*, which is a hybrid between *O. papilionacea* and *O. boryi*.

Similarities with *O. quadripunctata* Cyr. in shape of perianth have also been pointed out. The labellum of *O. israëlitica* and of *O. boryi* is not unlike that of *O. quadripunctata* which is marked with 4 purple dots near its base. However, *O. quadripunctata* differs in shape of its outer perianth-segments and in the acropetal opening of the inflorescence.

3. Orchis coriophora L., Sp. Pl. 940 (1753); Renz, Fl. Iran. 126:108, t.68; Boiss., Fl. 5:61; Post, Fl. 2:565. *O. fragrans* Pollini, Elem. 2:155 (1811). [Plate 506]

Plant 20–40 cm. Stem rather stout, leafy up to inflorescence. Basal leaves 4–10, often marcescent at anthesis. Inflorescence cylindrical, many-flowered, long, dense. Bracts lanceolate, acuminate, 1-veined, longer than ovary. Flowers rather small, from dusky-purple to greenish-pink, mostly fragrant. Outer segments connate up to middle, with free tips, forming a beaked hood, the lateral ones 7–10 × 3–4 mm, ovate-lanceolate, acuminate-apiculate, 1-veined, the median somewhat smaller; inner segments smaller than outer, 1–2 mm broad, narrowly lanceolate, acute or acuminate, 1-veined; labellum 5–7(–11) mm, deflexed, with dark purple dots and minute papillae at base and centre, 3-lobed; lateral lobes rhombic, crenate at margins; median lobe 4 mm, entire, oblong, subacute, longer and narrower than lateral lobes; spur conical, slightly curved, shorter than ovary, nectariferous. Fl. March–May.

Hab.: Damp ground on alluvial soils and travertine, and batha on terra rossa. Acco Plain, Sharon Plain; Upper Galilee, Esdraelon Plain; Dan Valley.

Area: Mediterranean and Irano-Turanian, extending into the Euro-Siberian.

About ten plants regarded as hybrids between *O. coriophora* L. and *O. saccata* Ten. (?*O. collina* Banks & Sol.) were found in the Horshat Tal Orchid Nature Reserve (Dan Valley). Apparently, cross-pollination was effected by honey bees (Dafni & Ivri, New Phyt. 83:181–187, 1979).

4. Orchis sancta L., Syst. Nat. ed. 10, 2:1242 (1759); L., Fl. Palaest. 22 no. 512 (1756) *nom. nud.*; Boiss., Fl. 5:62; Post, Fl. 2:566. [Plate 507]

Plant 30–50 cm. Stem rather stout, leafy up to inflorescence. Basal leaves often marcescent at anthesis; cauline leaves sheath-like, membranous, acuminate. Inflorescence cylindrical, many-flowered, dense. Bracts lanceolate, acuminate, membranous, the lower 5-veined, longer than ovary. Flowers dusky-purple. Outer segments 10–13 mm, lanceolate, acuminate, connate to beyond middle, forming a beaked hood; inner segments linear-lanceolate, acuminate; labellum 12–15 mm, cuneate, 3-lobed; lateral lobes rhombic, acutely serrate, dark-veined; median lobe nearly twice as long as broad, oblong, acute, entire or denticulate; spur conical, strongly curved forward, $\frac{1}{3}$–$\frac{1}{2}$ length of ovary. Fl. April–May.

Hab.: Batha and garigue, mainly on calcareous sandstone (kurkar), also on terra rossa and rendzina. Coastal Galilee, Acco Plain, Sharon Plain, Philistean Plain; Upper and Lower Galilee, Mt. Carmel, Mt. Tabor, Samaria, Judean Mts.; Gilead.

Area: E. Mediterranean (Palestine, Lebanon, Syria, S. W. Turkey, Cyprus, Rhodes, Chios, Cos).

5. Orchis tridentata Scop., Fl. Carn. ed. 2, 2:190 (1772) subsp. **commutata** (Tod.) Nyman, Consp. 691 (1882); Soó, Orchis, in Fl. Eur. 5:339. *O. tridentata* Scop. var. *commutata* (Tod.) Reichenb. fil. in Reichenb., Icon. Fl. Germ. 13–14:24 (1851); Renz, Fl. Iran. 126:110, incl. var. *albiflora* Post, Fl. ed. 1, 758 (1883–1896). *O. commutata* Tod., Orch. Sic. 24 (1842); Boiss., Fl. 5:62 *pro syn.*; Post, Fl. 2:566. [Plate 508]

Plant 15–40 cm. Stem slender. Leaves oblong-lingulate to lanceolate, often apiculate. Inflorescence many-flowered, usually dense, conical to subglobose at beginning of anthesis. Bracts lanceolate, long-acuminate to subulate, membranous, 1-veined, about as long as ovary. Flowers pale purple to purplish-pink, rarely white (var. *albiflora* Post). Perianth-segments purple-veined, connivent, forming a hood; the outer 8–10 mm, oblong-lanceolate, long-acuminate; the inner linear-lanceolate, acute, $\frac{2}{3}$ length of the outer; labellum 8–10 mm, variegated with purple spots, deeply 3-lobed; lateral lobes divaricate, oblong or rhombic, truncate and sometimes denticulate; median lobe 2-lobulate, with a little tooth at sinus between lobules; lobules diverging, denticulate; spur cylindrical, $\frac{1}{2}$–$\frac{2}{3}$ length of ovary. Fl. March–May.

Hab.: Batha, garigue, and sometimes maquis, mainly on terra rossa and rendzina. Common. Sharon Plain; Upper and Lower Galilee, Mt. Carmel, Samaria, Judean Mts.; Gilead, Ammon.

Area of subsp.: Mediterranean, slightly extending into the W. Irano-Turanian.

Area of species: Mediterranean, extending into the Euro-Siberian and W. Irano-Turanian.

6. Orchis italica Poir. in Lam., Encycl. Méth. Bot. 4:600 (1798). *O. longicruris* Link in Schrad., Jour. für die Bot. 1799 (2):323 (1800); Boiss., Fl. 5:65; Post, Fl. 2:568. [Plate 509]

Plant 30–40 cm. Tubers oblong, entire, rarely 2-fid. Leaves oblong-lanceolate to lanceolate, wavy-margined, sometimes brown-dotted. Inflorescence many-flowered, dense, conical to oblong. Bracts very short, scale-like, membranous, 1-veined, ovate-lanceolate, acuminate. Flowers pink. Perianth-segments purple-veined, connivent, forming a hood; the outer lanceolate, acuminate, 4-veined; the inner much smaller, linear, 1-veined; labellum 15–20 mm, pink with purple dots, rarely white (forma *albiflora*), elongate, glabrous, deeply 3-lobed; lobes long and narrow, acute, the lateral linear, somewhat falcate; median lobe longer and broader, divided into 2 long linear acute lobules, with an awn-like tooth 3 mm or longer in sinus; spur cylindrical, obtuse, often slightly 2-fid at apex, about ½ length of ovary, slightly curved. Fl. April.

Hab.: Batha and maquis, on terra rossa and rendzina. Rare. Upper and Lower Galilee.

Area: Mediterranean.

7. Orchis galilaea (Bornm. & M. Schulze) Schlechter, Feddes Repert. 19:47 (1923); Post, Fl. 2:568. *O. punctulata* Stev. var. *galilaea* Bornm. & M. Schulze in Bornm., Verh. Zool.-Bot. Ges. Wien 48:634 (1898). *O. punctulata* Stev. subsp. *galilaea* (Bornm. & M. Schulze) Soó, Feddes Repert. 24:28 (1927). [Plate 510]

Plant 20–40 cm. Leaves 3–5, oblong to ovate-oblong. Inflorescence cylindrical, many-flowered, long, dense, often basipetal. Bracts very short, ovate-lanceolate, membranous. Flowers varying in colour; ground colour white, greenish-yellow or purplish-pink. Perianth-segments purple-veined, forming a hood; outer segments oblong-ovate, 3–4-veined; the inner linear, 1-veined; labellum 7–9(–10) mm, purplish-pink or yellowish, pink- or purple-dotted at base, deeply 3-lobed; lateral lobes 5 mm, linear, divergent; median lobe 4 mm, 2-lobulate, with a toothlet about 1 mm in sinus between lobules; spur cylindrical, short (about 3 mm), ⅓ length of ovary, slightly curved. Fl. February–April.

Hab.: Batha, garigue and maquis, on terra rossa and rendzina. Common. Coastal Carmel, Acco Plain; Upper and Lower Galilee, Mt. Carmel, Samaria, Judean Mts.; Gilead, Ammon.

Area: E. Mediterranean (Palestine, Lebanon).

8. Orchis punctulata Stev. ex Lindl., Gen. Sp. Orchid. 273 (1835); Renz, Fl. Iran. 113, t.41; Boiss., Fl. 5:64; Post, Fl. 2:568. *O. sepulchralis* auct. non Boiss. & Heldr. (1854). [Plate 511]

388 ORCHIDACEAE

Plant robust, up to 80 cm. Stem stout. Leaves oblong to oblong-lanceolate. Inflorescence cylindrical, many-flowered, dense. Bracts short, ovate or lanceolate, membranous. Flowers fragrant. Perianth-segments yellowish-green, veined, connivent, forming a hood; the outer up to 13 mm, oblong-ovate or oblong, obtuse, with purplish-brown dots along veins; the inner up to 9 mm, linear, obtuse; labellum about 11 mm, deeply 3-lobed, yellowish-green near base, brown or purplish-brown towards apices of lobes; lateral lobes linear-falcate, divergent, 1 mm broad; median lobe fan-shaped, 2-lobulate; lobules truncate with a small tooth between them; spur pale green, cylindrical, ½–⅓ length of ovary. Fl. March April.

Hab.: Batha and maquis, mainly on marly soil. Fairly rare. Upper Galilee, W. Galilee, Mt. Carmel; Dan Valley (known in one large population on travertine), Hula Plain; Gilead, Ammon.

Area: E. Mediterranean and W. Irano-Turanian (Palestine, Lebanon, Syria, Cyprus, W. Turkey, Transcaucasia, Crimea, N. and W. Iran).

9. **Orchis saccata** Ten., Fl. Nap. 1, Prodr.: 53 (1811–1815); Boiss., Fl. 5:67; Post, Fl. 2:569. ?*O. collina* Banks & Sol. in Russ., Nat. Hist. Aleppo ed. 2, 2:264 (1794) *descr. insuffic.*; Renz, Fl. Iran. 126:98, t.35. [Plate 512]

Plant 20–30 cm. Stem stout. Leaves elliptic to ovate-oblong, up to 4 cm broad. Inflorescence cylindrical, few- to many-flowered. Bracts ovate-oblong, longer than ovary, often purplish. Outer perianth-segments up to 12 mm, deep purple to olive-green, oblong, obtuse; the lateral spreading, oblique, the median slightly concave, connivent to form a hood with inner segments; inner segments up to 10 mm, oblong-lanceolate, obtuse; labellum up to 12 × 11 mm, purple-red to purplish-pink, ovate or nearly orbicular, undivided, crenulate at margin; spur whitish, short and thick, cylindrical, saccate, slightly curved, obtuse, ½ length of ovary, not nectariferous. Fl. February–April.

Hab.: Batha on calcareous soils, terra rossa. Acco Plain, Sharon Plain, Philistean Plain; Upper and Lower Galilee, Mt. Carmel, Mt. Tabor, Mt. Gilboa, Esdraelon Plain, Samaria, Shefela, Judean Mts., N. Negev; Dan Valley; Golan, Ammon.

Area: Mediterranean, extending into the W. Irano-Turanian.

10. **Orchis anatolica** Boiss., Diagn. ser. 1, 5:56 (1844); Renz, Fl. Iran. 126:96, t.65; Boiss., Fl. 5:70; Post, Fl. 2:570. [Plate 513]

Plant 10–40 cm. Leaves oblong or lanceolate, usually dark-spotted. Inflorescence lax, few- to many-flowered; rachis purplish. Bracts lanceolate, purplish, shorter than ovary. Flowers reddish-purple, very rarely white (forma *leucantha* Renz). Lateral outer segments 10 × 4 mm or smaller, spreading, oblong, the median slightly broader, oblong, curved forward at apex; inner segments smaller, 8 × 3 mm, obliquely ovate forming a lax hood; labellum up to 14 mm, cuneate-obovate in outline, 3-lobed, reddish-purple and generally purple-dotted along pale centre and base; lateral lobes rounded; median lobe retuse or truncate; spur 1.5–2.5 cm, slender, tapering towards tip, horizontal or pointing upwards, longer than ovary. Fl. March–April.

Hab.: Common in batha and garigue, less common in maquis, mostly on terra rossa. Upper and Lower Galilee, Mt. Carmel, Samaria, Judean Mts.; Golan, Gilead, Ammon.

Forma **leucantha** Renz, Feddes Repert. 30:100 (1932). Flowers white. Leaves not spotted.

Hab.: Rare. Upper and Lower Galilee, Mt. Carmel, Samaria, Judean Mts.; Golan.

Area of species: E. Mediterranean, extending into the W. Irano-Turanian (Palestine, Lebanon, Syria, Cyprus, Aegean Islands, W. Turkey, Iraq, W. Iran).

11. **Orchis laxiflora** Lam., Fl. Fr. 3:504 (1779); Renz, Fl. Iran. 126:101; Evenari in Opphr. & Evenari, Florula Cisiord., Bull. Soc. Bot. Genève ser. 2, 31:203 (1941) as subsp. *ensifolia* (Vill.) Aschers. & Graebn.; Boiss., Fl. 5:71; Post, Fl. 2:570. Incl. var. *dinsmorei* Schlechter in Soó, Feddes Repert. 24:29 (1927) & Schlechter, in Monogr. Icon. Orch. Eur. 1:191 & forma *albiflora* Guss., Fl. Sic. Syn. 2:535 (1844). *O. dinsmorei* (Schlechter) Baumann & Dafni, AHO Mitteilungsblatt 13 (3):311–336 (1981). [Plate 514]

Plant 30–60(–70) cm. Stem stout, mostly leafy up to inflorescence. Leaves oblong to linear-lanceolate; lower leaves long. Inflorescence many-flowered; rachis purple. Bracts linear-lanceolate, purplish, as long as or longer than ovary. Flowers deep purple, rarely white (forma *albiflora* Guss.). Outer perianth-segments 8 mm, oblong-ovate, obtuse; the 2 lateral segments reflexed; inner segments nearly as long as outer; labellum 8–10 × 10–12 mm, cuneate, white and purple-spotted near base, broadly triangular-obcordate, broader than long, reflexed on both sides along midline, entire and emarginate, or obscurely 3-lobed with lateral lobes broad, rounded; spur cylindrical, slightly curved upwards, shorter than ovary; column white. Fl. February–April.

Hab.: Damp places, marshes, banks of streams. Acco Plain, Coast of Carmel, Sharon Plain, Philistean Plain; Upper and Lower Galilee, Esdraelon Plain, Samaria, Shefela; Dan Valley, Hula Plain, Beit Shean Valley; Golan, Gilead, E. Ammon.

Area: Mediterranean and W. Irano-Turanian, extending into the Euro-Siberian.

Contrary to Soó (1926, 1927) and Post *(loc. cit.)*, Renz (Fl. Iran. 126:99) considers subsp. *dielsiana* Soó as belonging to *O. palustris* Jacq. (not to *O. laxiflora* Lam.) and hardly deserving a separate status. Var. *dinsmorei* Schlechter in Soó (1927) is recorded from Palestine, but it, too, cannot be accorded the status of a separate taxon.

7. HIMANTOGLOSSUM Koch

Robust tuberous perennial herbs. Leaves oblong-lingulate. Inflorescence a many-flowered spike. Flowers large, greenish. Perianth-segments connivent, forming a hood; labellum spurred, 3-lobed; lateral lobes undulate, obliquely linear or triangular; median lobe much longer, narrow, linear, 2-fid at apex. Column short; tip of connective short, obtuse; staminodes small, verrucose; pollinia long-caudiculate, with a single gland in bursicle. Ovary cylindrical, twisted.

Four Euro-Siberian, Mediterranean and Irano-Turanian species.

1. **Himantoglossum affine** (Boiss.) Schlechter, Feddes Repert. 15:287 (1918); Renz, Fl. Iran. 126:89, tt.31, 32. *Aceras affinis* Boiss., Fl. 5:56 (1882). *Loroglossum affine* (Boiss.) Camus & Bergon, Orch. Eur. 83 (1908); Post, Fl. 2:562. *Himantoglossum bolleanum* (Siehe) Schlechter, Feddes Repert. 15:287 (1918). *Aceras bolleana* Siehe, Gard. Chron. ser. 3, 23:365 (1898). [Plate 515]

Plant robust, 40–80 cm. Tubers large. Leaves oblong-lanceolate or obovate. Inflorescence cylindrical, long. Bracts lanceolate, longer than ovary. Flowers spreading, greenish, partly suffused with purplish-brown. Outer segments connivent forming an obtuse hood, green, brownish at margin, ovate, obtuse; the inner smaller, lanceolate, acute; labellum up to 6 cm, whitish at centre; lateral lobes 3–6 mm, subfalcate; median lobe 4–5 × 0.25–0.35 cm, green to purplish-brown, spirally twisted in bud, later strap-shaped and stretched outwards, 2-fid, with lobules 6–8 mm, acute; spur 3.5–5 mm, broadly conical, oblique, incurved. Column about 5 mm. Fl. May–June (August).

Hab.: Maquis. Very rare. Upper and Lower Galilee; Golan, Transjordan (Al-Eisawi).

Area: E. Mediterranean (N. Palestine, Lebanon, Syria, W. and S. Turkey, N. Iraq, Iran).

Protected by law.

8. ANACAMPTIS L.C.M. Richard

Tuberous perennial herbs. Tubers undivided, globose. Stem erect. Leaves linear to narrow-lanceolate, the upper sheath-like. Inflorescence a many-flowered spike. Flowers pink or purple, very rarely white. Perianth-segments more or less equal in length; lateral outer segments spreading; median outer and inner segments connivent; labellum spurred, 3-lobed, with 2 parallel lamellae at base. Column erect; tip of connective obtuse; pollinia connected to a single gland which is enclosed in a bursicle.

A monotypic genus.

1. **Anacamptis pyramidalis** (L.) L.C.M. Richard, Mém. Mus. Hist. Nat. Paris 4:55 (1818); Boiss., Fl. 5:57; Post, Fl. 2:563. *Orchis pyramidalis* L., Sp. Pl. 940 (1753). [Plate 516]

Plant 25–50 cm. Leaves long. Inflorescence conical at first, later oblong to cylindrical. Bracts linear-lanceolate, long-acuminate, nearly as long as ovary. Flowers pink to pink-purple. Median outer segment ovate-lanceolate, as long as inner segments; lateral outer segments somewhat longer, spreading; labellum pointing forward, flabellate-cuneate, 3-lobed up to its middle into nearly equal, oblong, obtuse or truncate lobes; spur filiform, as long as or longer than ovary. Ovary glabrous, twisted. Fl. April–May.

Hab.: Batha, fairly common, more rare in garigue and maquis. Acco Plain; Upper Galilee, Mt. Carmel, Mt. Gilboa, Samaria, Judean Mts.; Gilead, Ammon.

Area: Mediterranean, extending into the Irano-Turanian and Euro-Siberian.

Protected by law.

9. SERAPIAS L.

Tuberous perennial herbs. Stem more or less leafy. Inflorescence a several- to many-flowered spike. Flowers maroon, violet-brown or purple. Outer perianth-segments erect, lanceolate, acute, veined, more or less connate, forming an acute hood; the inner hidden by hood, free, dilated at base and long-cuspidate; labellum without a spur, 2-gibbous at base, 3-lobed; lateral lobes rounded, veined, erect, together with base of labellum forming the hypochile; median lobe, the epichile, much longer, lanceolate to cordate, acute, reflexed, long-hairy at its base. Apex of column acuminate, petaloid; anther vertical, entirely adnate; thecae parallel; pollinia borne on separate caudicles; gland single, included in a bursicle. Ovary not twisted.

Ten species, mainly in the Mediterranean.

1. Serapias vomeracea (Burm. fil.) Briq., Prodr. Fl. Corse 1:378 (1910); Post, Fl. 2:561. *Orchis vomeracea* Burm. fil., Nova Acta Acad. Leop.-Carol. 4, App. 237 (1770). *S. pseudocordigera* Moric., Fl. Venet. 374 (1820); Boiss., Fl. 5:54 [Plate 517]

Plant 30–40 cm. Stem stout, erect, purplish above. Leaves linear-lanceolate, acute. Inflorescence lax, 4–8-flowered. Bracts violet-brown, lanceolate, acuminate, many-veined, exceeding perianth-hood. Flowers maroon; perianth-segments veined; the outer ovate-lanceolate, acute; the inner shorter, broadly ovate and undulate at base, long-acuminate at tip; labellum 2.5–4 cm; epichile strongly refracted, broadly-lanceolate, covered with long hairs at its base; hypochile ½ length of epichile; lateral lobes of hypochile erect, oblique, rounded, usually partly covered by hood. Ovary cylindrical. Fl. March–April.

Hab.: Batha and damp habitats, on various soils. Fairly common. Coastal Galilee, Acco Plain, Coast of Carmel, Sharon Plain, Philistean Plain; Upper and Lower Galilee, Mt. Carmel, Mt. Gilboa, Samaria, Judean Mts.; Dan Valley, Beit Shean Valley.

Area: Mediterranean, extending into S. Russia.

Protected by law.

10. Ophrys L.

Tuberous perennial herbs. Tubers undivided, globose, ovoid or ellipsoid, sometimes stipitate. Stem provided at base with 1–2 sheath-like cataphylls. Leaves basal and sometimes also cauline, oblong-lingulate or oblong-ovate. Inflorescence lax, few- or several-flowered. Bracts mostly longer than ovary. Flowers showy. Outer perianth-segments nearly equal, mostly erecto-patent; inner segments smaller; labellum without a spur, spreading, flat or convex, entire or more or less distinctly 3-lobed, with lateral lobes often 2-gibbous near base, velvety, and with a conspicuous usually glabrous central area (*speculum*). Column short, erect; anther-thecae parallel, connective ending in an often beak-like tip; each pollinium with a gland inserted into bursicle of its own. Ovary glabrous, somewhat twisted.

About 30 Euro-Siberian, Mediterranean and Irano-Turanian species.

Literature: E. Nelson & H. Fischer, *Monographie und Ikonographie der Gattung Ophrys*, Chernex-Montreux, 1962. R. Soó, Species and subspecies of the genus *Ophrys*, *Acta Bot. Acad. Sci. Hung.* 16:373–392 (1970); *ibid.* 18:379–384 (1973); *Ophrys*, in: Tutin et al. (eds.), *Flora europaea* 5:344–349, 1980.

All *Ophrys* species protected by law in Israel.

1. Inner perianth-segments linear or narrowly lanceolate, 6–10 mm   2
- Inner perianth-segments triangular, up to 4 mm long   5
2. Labellum strongly convex, entire or shallowly lobed at sides, with a short acute appendage at apex; speculum with two parallel bluish-silvery lines.
                                                                                      4. O. transhyrcana
- Labellum not as above   3
3. Labellum with a broad glabrous yellow zone at margin surrounding its brown or brownish-purple middle part.   **1. O. lutea** subsp. **galilaea**
- Labellum without a broad glabrous yellow zone, densely villose up to margin, especially on distal part   4
4. Labellum 15–22 mm; base of labellum with a V-shaped notch; speculum brilliantly metallic-blue.   **2. O. iricolor**
- Labellum not exceeding 15 mm; base of labellum not notched; speculum not as above.   **3. O. fleischmannii**
5(1). Labellum deeply 3-lobed, with a broad sinus between median and each lateral lobe   6
- Labellum not or shallowly 3-lobed   7
6. Appendage at apex of labellum longer than broad. Outer perianth-segments pink or purple, 12–16(–20) mm.   **7. O. apifera**
- Appendage at apex of labellum broader than long. Outer segments green or white, up to 10 mm.   **8. O. carmeli** subsp. **carmeli**
7(5). Labellum 7–9 mm. Outer segments pale green to whitish and green-veined; inner segments 2–3 mm.   **5. O. bornmülleri**
- Labellum 11–15 mm. Outer segments purplish-pink to light purple; inner segments 4 mm.   **6. O. fuciflora**

1. Ophrys lutea Cav., Icon. Descr. 2:46 (1793) subsp. **galilaea** (Fleischm. & Bornm.) Soó, Notizbl. Bot. Gart. Berlin 9:906 (1926); Mout., Nouv. Fl. Liban Syrie 1:330. *O. galilaea* Fleischm. & Bornm., Ann. Naturh. Mus. (Wien) 36:12 (1923); Post, Fl. 2:573. Boiss., Fl. 5:75 & Post, Fl. 2:573 as *O. lutea* Cav. [Plate 518]

Plant 10–30 cm. Inflorescence few-flowered. Outer perianth-segments 8–9 mm, yellowish-green, broadly ovate, revolute-margined, median segment arched forward; inner segments 7 mm, brownish- or yellowish-green, linear, obtuse, glabrous, often undulate-margined; labellum flat, with a broad yellow margin, otherwise brown or brownish-purple, hairy, very shortly 3-lobed in distal part; median lobe retuse; speculum grey-violet, extensive. Fl. December–March.

Hab.: Batha on calcareous soils. Sharon Plain; Upper Galilee, Mt. Carmel, Samaria, Shefela, Judean Mts.

Area: Mediterranean.

2. Ophrys iricolor Desf., Ann. Mus. Hist. Nat. (Paris) 10:224, t.13 (1807); Post, Fl. 2:572. *O. fusca* Link var. *iricolor* (Desf.) Reichenb. fil. in Reichenb., Icon. Fl. Germ. 13–14:73 & 75, t.444 f.4, t.445 (1851); Boiss., Fl. 5:75. *O. fusca* subsp. *iricolor* (Desf.) O. Schwarz, Feddes Repert. 36:77 (1934). [Plate 519]

Plant 10–35 cm. Inflorescence 1–5-flowered. Outer perianth-segments about 15 mm, light green and green-veined, ovate-oblong, spreading, median segment arched forward; inner segments about ⅔ length of the outer, linear, rounded or truncate at apex, brownish, often slightly undulate; labellum 15–22 mm, more or less flat, dark brown or blackish-purple, densely villose, broadened and 3-lobed in distal half, narrowed towards base; base with a V-shaped notch, white-hairy; speculum brilliantly metallic-blue, with dark lines, turning dirty brown later. Fl. February–April.

Hab.: Maquis and batha, mostly on chalky and marly soil. Fairly rare. Sharon Plain; Upper Galilee, Mt. Carmel, Samaria, Shefela, Judean Mts; Transjordan (Al-Eisawi).

Area: C. and E. Mediterranean.

3. Ophrys fleischmannii Hayek, Feddes Repert. 22:388 (1926); Prodr. Fl. Penins. Balcan. 3:375 (1933); Baumann & Dafni, Beih. Veröff. Naturschutz Landsch. Pflege Bad.-Württ. 19:129–153 (1981). *O. heldreichii* Fleischm., Österr. Bot. Zeitschr. 74:182, t.2 f.6 (1925)* non Schlechter (1924). *O. fusca* Link subsp. *fleischmannii* (Hayek) Soó, Notizbl. Bot. Gart. Berlin 9:905 (1926); Feddes Repert. 24:26 (1927). Boiss., Fl. 5:75 & Post, Fl. 2:572 as *O. fusca*. [Plate 520]

Plant 15–30 cm. Inflorescence 1–4-flowered. Outer perianth-segments about 1 cm, light green, ovate-oblong, obtuse, glabrous, spreading, the median outer

---

* The designation of drawings for plate 2 on p. 194 of Fleischmann's paper is erroneous.

and inner segments arched forward; inner segments 7–8 mm, brownish-green, linear, truncate; labellum up to 15 mm, without a V-shaped notch at base, dark purple-brown or greenish-brown, densely velvety-villose, 3-lobed in distal ⅓; lateral lobes rounded-triangular, non-gibbous, recurved; median lobe much longer and broader, recurved, deeply emarginate; speculum 2-partite, with white or bluish lines or spots and surrounded by a white or yellowish broad margin; basal end of labellum white-hairy. Fl. December–March.

Hab.: Mainly batha on terra rossa. Common. Upper and Lower Galilee, Mt. Carmel, Samaria, Judean Mts.; Gilead.

Area: E. Mediterranean (Palestine, Lebanon, ?S. Turkey, Cyprus, Crete, Aegean Islands).

O. *fleischmannii* is part of the Mediterranean O. *fusca* aggregate comprising several species. Two of these species, O. *fusca* Link s.s. and O. *iricolor* Desf., are noted by a V-shaped notch at the base of the labellum; the others, O. *fleischmannii* Hayek, O. *omegaifera* Fleischm., O. *atlantica* Munby and O. *dyris* Maire, are without a notch. The species differ also in geographical distribution. O. *iricolor* is spread over most of the Mediterranean and is usually not common. O. *fusca* s.s. occurs mainly in the C. Mediterranean. O. *atlantica* and O. *dyris* are confined to the W. Mediterranean. The area of O. *fleischmannii* comprises Palestine, Lebanon, ?S. Turkey, Cyprus, Crete and S. Aegean. O. *omegaifera* is restricted to Crete, Karpathos and Rhodes (Baumann & Dafni, Differenzierung und Arealform des *Ophrys omegaifera*-Komplexes im Mediterrangebiet, Beih. Veröff. Naturschutz Bad.-Württ. 19:129–153, 1981).

4. Ophrys transhyrcana Czerniak., Not. Syst. (Leningrad) 4:1 (15 January 1923). O. *sphegodes* Mill., Gard. Dict. ed. 8, no. 8 (1768) subsp. *transhyrcana* (Czerniak.) Soó, Acta Bot. Acad. Sci. Hung. 5:444 (1959); Renz, Fl. Iran. 126:70, t.22 f.1. O. *sintenisii* Fleischm. & Bornm., Ann. Naturh. Mus. (Wien) 36:10 (February 1923); Post, Fl. 2:575 *pro syn.* O. *sphegodes* subsp. *sintenisii* (Fleischm. & Bornm.) Nelson, Monogr. Icon. Ophrys 181 (1962). [Plate 521]

Plant 20–50 cm. Inflorescence few-flowered. Outer perianth-segments 12–17 mm, greenish, lanceolate; inner segments shorter (up to 10 mm), 2–3 mm broad, linear, often undulate-margined, puberulent; labellum up to 15 mm, dark purple or brown-purple, velvety, broadly obovate, strongly convex, entire or shallowly lobed, usually not gibbous near base, with a short acute appendage at apex; centre velvety except on two parallel bluish-silvery lines of speculum. Tip of connective long. Fl. February–April.

Hab.: Batha, garigue and maquis. Acco Plain, Coast of Carmel, Sharon Plain, Philistean Plain; Upper and Lower Galilee, Mt. Carmel, Mt. Gilboa, Samaria, Judean Mts.; Dan Valley; Golan, Gilead, Ammon.

Area: E. Mediterranean and Irano-Turanian (S. Anatolia, Lebanon, Syria, Palestine, Iraq, W. and N. Iran, Turkmenia).

5. Ophrys bornmülleri M. Schulze, Mitt. Thür. Bot. Ver. 13:127 (1899); Renz, Fl. Iran. 126:73; Post, Fl. 2:574; incl. forma *grandiflora* Fleischm. & Soó in Soó, Feddes Repert. 24:26 (1927). M. Schulze in Bornm., Verh. Zool.-Bot. Ges. Wien 48:635 (1898) as *O. tenthredinifera* Willd. var. *sine nomine cum descriptione* M. Schulze. [Plate 522]

Plant 10–30 cm. Inflorescence 2–9-flowered. Outer perianth-segments 7–9 × 4–6 mm, pale green or whitish, with a dark green midvein and 2 greenish lateral veins, ovate-oblong, obtuse; lateral outer segments reflexed or spreading, the median arched forward; inner segments 2–3 mm, triangular, pale green or whitish, velvety; labellum 7–9 mm, brown to purplish-brown, velvety, flat to somewhat convex, entire, emarginate, with a subquadrate glabrous yellowish-green appendage in sinus, with two small conical protuberances near base and a yellowish glabrous marking resembling the letter H. Tip of connective short, apiculate. Fl. February–March from Mt. Carmel southwards, April–May northwards.

Hab.: Maquis and batha, mostly on terra rossa. Rather rare except on Mt. Carmel. Acco Plain; Upper and Lower Galilee, Mt. Carmel, Samaria, Shefela, Judean Mts. (Post); Golan.

Area: E. Mediterranean, extending into the W. Irano-Turanian (Palestine, Lebanon, Syria, Cyprus, S. Anatolia, Iraq).

6. Ophrys fuciflora (F. W. Schmidt) Moench, Meth. Suppl. 311 (1802); Wirth, Acta Bot. Acad. Sci. Hung. 23:285–293 (1977); Post, Fl. 2:574. *Arachnites fuciflora* F.W. Schmidt, Fl. Boëmica 1:76 (1793). *Orchis holosericea* Burm. fil., Nova Acta Acad. Leop.-Carol. 4, App. 237 (1770) *nom. confus.* ?*Ophrys holosericea* (Burm. fil.) Greuter, Boissiera 13:185–186 (1967). [Plate 523]

Closely related to *O. bornmülleri* and differs mainly in the following: flower somewhat larger; outer perianth-segments purplish-pink to light purple; the inner 4 mm, pinkish to brownish-pink; labellum 11–15 mm, broader than in *O. bornmülleri* and subquadrate. Fl. end of April–May.

Hab.: Maquis and batha. Less common than *O. bornmülleri*. Upper and Lower Galilee, Mt. Carmel, Samaria, Judean Mts.

Area: E. Mediterranean (Palestine, Lebanon, Syria), extending into S. and Middle Europe.

Transition forms between *O. bornmülleri* and *O. fuciflora* (referred to as *O. holosericea*) and occasional hybrids were studied on Mt. Carmel and reported by G. Neeman (MSc. Thesis, Tel-Aviv University, 1972, in Hebrew).

7. Ophrys apifera Huds., Fl. Angl. 340 (1762); Renz, Fl. Iran. 126:74, t.23 f.2; Boiss., Fl. 5:79; Post, Fl. 2:578. [Plate 524]

Plant 20–40 cm. Inflorescence few-flowered. Flowers rather large. Outer perianth-segments 12–16(–20) × 9 mm, pink or purple, green-veined, ovate to ovate-oblong, obtuse; inner segments 3–4 mm, green or brownish, triangular, densely whitish-velvety; labellum 11 mm, dark red-brown, deeply 3-lobed near base; speculum collar-shaped, encircling a red-brown basal zone; lateral lobes reflexed, conical, densely hairy; median lobe ovate, strongly convex, ending in a reflexed glabrous yellowish-green acute appendage. Tip of connective long, flexuous, beak-like. Ovary sessile, twisted. Fl. March–April.

Hab.: Maquis and batha. Fairly rare. Coastal Galilee, Acco Plain; Upper Galilee, Mt. Carmel, Mt. Gilboa, Samaria, Judean Mts.; Dan Valley, Hula Plain; Gilead.

Area: Euro-Siberian and Mediterranean, reaching N. Iran.

Known to be self pollinated, though cross-pollination also occurs.

8. **Ophrys carmeli** Fleischm. & Bornm., Ann. Naturh. Mus. (Wien) 36:7 (February 1923) subsp. **carmeli**; Renz, Fl. Iran. 126:78, t.27 f.1; Post, Fl. 2:577. *O. dinsmorei* Schlechter, Feddes Repert. 19:46 (1 March, 1923); Post, Fl. 2:576. [Plate 525]

Plant 20–40 cm. Inflorescence 5–9(–12)-flowered, lax. Outer perianth-segments 8–12 mm, pale green, rarely white, ovate-oblong, obtuse, 3-veined, the median arched forward; inner segments small, 2.5–3 mm, olive-green, triangular-ovate, villose; labellum 7–10 mm, dark brown (becoming paler during anthesis), deeply 3-lobed; lateral lobes small, deflexed, velvety, forming hollow conical protuberances near base of labellum; median lobe strongly convex with revolute margins, ovate, puberulent except on yellowish markings; sinuses between median and lateral lobes broad; apical appendage small, upcurved, glabrous, yellowish-green. Tip of connective short. Fl. January–April.

Hab.: Batha and maquis. Various soils. Fairly common. Coastal Galilee, Coast of Carmel, Acco Plain, Sharon Plain, Philistean Plain; Upper and Lower Galilee, Mt. Carmel, Esdraelon Plain, Mt. Gilboa, Samaria, Judean Mts., N. Negev; Dan Valley; Golan.

Area of subspecies: E. Mediterranean and W. Irano-Turanian (Palestine, Lebanon, Syria, Iraq, S. and N. Iran).

APPENDICES

DIAGNOSES PLANTARUM NOVARUM

Leopoldia longipes (Boiss.) Losinsk. subsp. **negevensis** Feinbr. & Danin subsp. nov. [Tab. 101]

Perennis, 20–40 cm. Bulbus 3–6 cm in diam. Folia 3–4(-5), late lanceolata, canaliculato-plicata, plerumque falcata, 10–20 cm longa, 2–4 cm lata. Differt a subsp. *longipedi* foliis minus numerosis, brevioribus et latioribus, bulbo majore, habitatione in saxosis.

Holotype: Central Negev, Mt. Sagi, rocks and hill slopes, 19.5.1967, *A. Danin* 967 (HUJ).

Allium curtum Boiss. & Gaill. in Boiss., Diagn. ser. 2, 4:116 (1856) subsp. **palaestinum** Feinbr., Palest. Jour. Bot. Jerusalem ser., 3:14, ff.6 & 25 (1943) var. **negevense** Kollmann var. nov.

Perigonium 4–5 mm longum; segmenta pallide viridia, margine albida, nervo medio lato, purpureo, viridi-purpureo vel viridi. Filamenta parte superiore purpurea, cuspis centralis cuspidibus lateralibus aequilonga; antherae purpureae. Stylus purpureus. Floret Februario–Aprili.

Holotype: Palestine, Central Negev, Mitzpeh Ramon, opposite Radar station, with *Chiliadenus (Varthemia) iphionoides* and *Origanum ramonense*, 19.4.1967, *F. Kollmann* 383 (HUJ).

Lolium rigidum Gaudin subsp. **negevense** Feinbr. subsp. nov. [Tab. 297]

Culmus erectus, 30–50 cm. Spica recta, (5-)10–20 cm; rachidis internodia spiculis ²⁄₃ vel plus longiora; rachis laevis vel scabra. Spiculae 14–25 mm, 6–8(-10)-florae, pallide virides; flores conspicue remoti; rachillae internodia 1.5–2 mm. Glumae 5–9(-11) mm, spiculis dimidio vel plus breviores, marginibus et apicibus membranaceae vel hyalinae; lemma 5–7.5 mm longa, parte superiore hyalina. Floret Martio–Maio.

Holotype: N. Negev, about 10 km S. of Beersheva, sand, 15.4.1928, *A. Eig, M. Zohary, N. Feinbrun* 928 (HUJ).

Eminium spiculatum (Blume) Schott subsp. **negevense** Koach & Feinbr. subsp. nov.

Planta quam typo humilior. Floret nocte, Februario–Martio. Folii lobus terminalis 1–3 cm latus; laciniae laterales utrinque 4–6, oblongae vel oblongo-lanceolatae. Petiolus non purpureo-maculatus. Spathae tubus 5–8 cm, 1.7–3 cm diametro, extus rarissime purpureo-maculatus; spathae lamina 5–11 × 4–7 cm, apertio post meridiem; laminae dimidia pars superior recurva, margine non involuta; laminae epidermis superne papillosa, velutina.

Hab.: W., N. and C. Negev, Sinai.

Holotype: Central Negev, Rotem Plain, sandy soil, 10.12.1984, *J. Koach* (photo) 84 (HUJ).

Biarum pyrami (Schott) Engler var. **serotinum** Koach & Feinbr. var. nov.

Floret Novembri–Januario. Folia non maculata, saepe plus minusve synanthia. Spathae lamina extus plerumque non maculata.

Holotype: Golan, Katsrin, hills, basalt rocks and soil, 22.12.1979, *Revital Heiman* 79 (HUJ).

ADDENDA AND CORRIGENDA TO PART THREE

p. 25, line 9 from top, instead of 1. **Solenostemma oleifolium** (Nect.) Bullock & Bruce in Bullock, Kew Bull., read:

1. **Solenostemma oleifolium** (Nect.) Bullock & Bruce ex Maire, Mém. Soc. Hist. Nat. Afr. Nord 3(2):170 (1933); Bullock & Bruce in Bullock, Kew Bull., etc.

p. 28, lines 10-12 from top, instead of 1. **Pentatropis spiralis** (Forssk.) Decne., etc., read:

1. **Pentatropis nivalis** (J.F. Gmel.) D.V. Field & J.R.I. Wood, Kew Bull. 38:215 (1983). *Asclepias nivalis* J.F. Gmel., Syst. Nat. ed. 13, 2:444 (1791), based on *A. nivea* Forssk., Fl. Aeg.-Arab. (1775) *nom. illeg.* non L. (1753). *P. spiralis* sensu Decne. (1838) & auctt. non *A. spiralis* Forssk., Fl. Aeg.-Arab. 49 (1775). [Plate 40]

p. 38, lines 9-12 from bottom, instead of 9. **Convolvulus glomeratus** Choisy, etc., read:

9. **Convolvulus auricomus** (A. Richard) Bhandari, Bull. Bot. Surv. Ind. 6:327 (1964) & Fl. Ind. Desert 244 (1978). *Ipomoea auricoma* A. Richard, Tent. Fl. Abyss. 2:67 (1851). *C. glomeratus* Choisy in DC., Prodr. 9:401 (1845) non Thunb. (1792); Sa'ad, *op. cit.* 182; Boiss., Fl. 4:102; Post, Fl. 2:206, incl. var. *sericeus* Dinsmore in Post, *loc. cit.* [Plate 58]

p. 44, before 1. **Calystegia sepium**, insert:

1. Leaves reniform. Stems not or slightly twining.	**2. C. soldanella**
- Leaves not reniform. Stems twining.	**1. C. sepium**

p. 44, before Subfam. 2. CUSCUTOIDEAE, insert:

2. **Calystegia soldanella** (L.) R. Br., Prodr. Fl. Nov. Holl. 484 (1810); Brummitt, in Fl. Eur. 3:78 (1972); Boiss., Fl. 4:112; Post, Fl. 2: 211. *Convolvulus soldanella* L., Sp. Pl. 159 (1753).

Perennial herb, glabrous, procumbent, not or slightly twining. Leaves 2-3 cm, somewhat fleshy, reniform, cordate, as broad as long or broader; apex obtuse or emarginate. Bracteoles large, ovate to suborbicular. Sepals shorter than bracteoles. Corolla 35-50 mm, pink, 4 times as long as calyx. Fl. May-July.

Hab.: Maritime sands. Coastal Galilee (S. of Rosh Hanikra).

Area: Mediterranean, extending along the coasts of Europe, California, S. America and Australia.

New to the Flora of Palestine. Cited by Post (*loc. cit.*) from Sidon and Tripoli on the coast of Lebanon.

p. 45, line 6 from bottom, add: Arava Valley.

p. 58, lines 17–18 from top, after *H. undulatum* Vahl, Symb. Bot. 1:13 (1790), add: *nom. illeg.*

p. 59, line 8 from bottom, instead of 4. MEGASTOMA (Benth. & Hook. fil.) Bonnet & Barratte, read:

4. OGASTEMMA Brummitt
Megastoma (Benth. & Hook. fil.) Bonnet & Barratte (1895) non Grassi (1881).

p. 60, lines 7–9 from top, instead of **1. Megastoma pusillum,** etc., read:
1. Ugastemma pusillum (Coss. & Durieu ex Bonnet & Barratte) Brummitt, Kew Bull. 36:679–680 (1982). *Megastoma pusillum* Coss. & Durieu ex Bonnet & Barratte, Ill. Phan. Tunisie t.11 ff.4–11 (1895) *icon.*; Barratte in Bonnet & Barratte, Cat. Raison. Pl. Vasc. Tunisie 301 (1896) *descr.* [Plate 96]

p. 78, line 22 from bottom, instead of **2. Alkanna tinctoria,** read:
2. Alkanna tuberculata (Forssk.) Meikle, Kew Bull. 34:823 (1980). *Anchusa tuberculata* Forssk., Fl. Aeg.-Arab. 41 (1775). *Alkanna tinctoria* Tausch, Flora (Regensb.) 7:234 (1824) *nom. illeg.*; Boiss., etc.

p. 81, line 17 from bottom, before Judean Desert, add: E. Samaria,

p. 86, lines 10–11 from bottom, instead of **1. Symphytum palaestinum** Boiss., etc., read:
1. Symphytum brachycalyx Boiss., Diagn. ser. 1, 4:43 (1844); Boiss., Fl. 4:173; Wickens, in Fl. Turkey 6:384 (1978). *S. palaestinum* Boiss., Diagn. ser. 1, 11:94 (1849); Boiss., Fl. 4:173; Post, Fl. 2:229. [Plate 141]

p. 87, line 3 from top, instead of Var. **palaestinum.** Corolla white, read:
Var. **brachycalyx.** Corolla white.

p. 87, line 6 from top, read:
Var. **violaceum** (Feinbr.) Feinbr. comb. nov. *S. palaestinum* Boiss. var. *violaceum* Feinbr., Israel Jour. Bot. 25:79 (1976). Corolla pale violet.

p. 111, line 14 from top, after Fl. 2:367., read: *Tapeinanthus persicus* Boiss. ex Benth. in DC., Prodr. 12:436 (1848) non Herbert (1837);

p. 111, line 20 from bottom, before Edom, add: E. Gilead, Moav,

p. 119, lines 24–26 from top, read:

5(2). Bracteoles ovate to oblanceolate, flat, 3–10 mm broad. Desert plants.
                                                                    4. P. platystegia
– Bracteoles lanceolate-subulate, usually canaliculate, incurved, 1.25–2(–3) mm broad. Non-desert plants   6
6. Calyx-teeth subulate-acuminate from a broad base, unequal, the longer ones 4–7 mm.
                                                                    5. P. viscosa
– Calyx-teeth broad and very short, equal, each ending in a short prickly mucro.
                                                                    6. P. chrysophylla

p. 119, line 6 from bottom, before Gilead, add: Golan,

p. 121, before **5. Phlomis viscosa**, etc., insert:
4a. Phlomis chrysophylla Boiss., Diagn. ser. 1, 12:89 (1853); Boiss. Fl. 4:788; Post, Fl. 2:397.

Shrub, 1 m or more, appressed-woolly, golden. Leaves 1.5–4.5 cm, petiolate, ovate to oblong-ovate, reticulately veined, crenulate, cordate, rounded at apex, yellowish-green, devoid of smell; floral leaves as long as or longer than flowers. Verticillaster 1, terminal, or 2 to few, remote; flowers pedicellate; bracteoles linear-subulate, spinescent, somewhat shorter than calyx. Calyx 10–12 mm, golden-pannous, ribbed-grooved, truncate; teeth short, broader than long; primary ribs ending in a minute prickly mucro. Corolla yellow, twice as long as calyx. Filaments appendiculate at base. Fl. April.
  Hab.: Rock crevices; S. slope. Upper Galilee, Judean Mts. (Post).
  Area: E. Mediterranean (N. Palestine, Lebanon, Syria, S. Turkey).

p. 123, line 19 from bottom, add: Arava Valley (Elat).

p.164, before 4. SOLANUM L., insert:

3a. PHYSALIS L.

Annual or perennial herbs. Flowers solitary, axillary, nodding. Calyx campanulate, 5-dentate, accrescent, inflated in fruit. Corolla broadly campanulate, 5-lobed. Stamens inserted near base of corolla. Fruit a globose berry, usually enclosed by the inflated calyx.
  About one hundred species, especially in America.

1. Physalis angulata L., Sp. Pl. 183 (1753); Hawkes, in Fl. Eur. 3:196.
  Annual, 40–60 cm, subglabrous. Leaves lanceolate to ovate, cuneate at base, entire or sinuate-dentate. Pedicels filiform. Corolla yellow. Anthers 1.5–2 mm. Fruiting calyx 3–5 cm. Berry 10–12 mm, yellowish-green. Fl. March.
  Hab.: Casual weed in a garden. Esdraelon Plain.
  Area: Trop. America.

p.165, line 10 from top, add:

A. Gilli, Bestimmungsschlüssel der Subgenera und Sektionen der Gattung *Solanum*, *Feddes Repert.* 81:429–435 (1970). J. M. Edmonds, Taxonomic studies on *Solanum* section *Solanum* (*Maurella*), *Jour. Linn. Soc. London* (*Bot.*) 75:141–178 (1977).

p. 165, before **1. Solanum luteum**, insert:
1. Solanum dulcamara L., Sp. Pl. 185 (1753); Boiss., Fl. 4:285; Post, Fl. 2:258.
  Perennial herb or shrub, up to 2 m. Stems woody below. Leaves petiolate, ovate to ovate-lanceolate, acute to acuminate, cordate or truncate at base, subglabrous or pubescent, entire or with 1–2 pairs of small lobes at base of blade. Calyx 2–3 mm. Corolla 12–18 mm in diameter, stellate, mostly violet; lobes

lanceolate-triangular. Anthers 5 mm, yellow, opening by 2 pores at tip. Berry 7–10 mm, red, glossy, ovoid-globose. Fl. summer.

    Hab.: Near water. Rare. Golan (Banias), Gilead.

    Area: Euro-Siberian and Mediterranean.

p. 165, line 16 from bottom, instead of **1. Solanum luteum** Mill., Gard. Dict. ed. 8, no. 3 (1768), read:

1a. Solanum villosum Mill., Gard. Dict. ed. 8, no. 2 (1768). *S. luteum* Mill., Gard. Dict. ed. 8, no. 3 (1768). Cf. Edmonds, Jour. Linn. Soc. London (Bot.) 78:214 (1979).

    According to Edmonds (*op. cit.*), *S. villosum* can be subdivided, following Kirschleger (Flore d'Alsace, etc., 1852), into 2 subspecies as follows:

Subsp. **villosum.** Berries yellow.

    Hab.: Waste places. Common. As given on p. 165 for *S. luteum.*

Subsp. **puniceum** (Kirschleger) Edmonds, Jour. Linn. Soc. London (Bot.) 78:215 (1979). *S. nigrum* L. subsp. *puniceum* Kirschleger, Fl. Alsace 1:532 (1852). *S. puniceum* C.C. Gmel., Fl. Bad. 4:176 (1826) *nom. superfl. S. miniatum* Bernh. ex Willd., Enum. Pl. Horti Berol. 236 (1809); Boiss., Fl. 4:284. *S. luteum* Mill. subsp. *alatum* (Moench) Dostál, Květena ČSR 1270 (1949); Hawkes & Edmonds, in Fl. Eur. 3:198 (1972). *S. alatum* Moench, Meth. 474 (1794); Post, Fl. 2:257. Berries red.

    Hab.: Less common. Philistean Plain; Judean Mts.

p. 166, line 1 from bottom, add: Arava Valley (Elat).

p. 167, line 15 from top, add: Coast of Carmel; Judean Mts.; Beit Shean Valley.

p. 168, line 1 from bottom, add: Hula Plain.

p. 170, line 21 from top, instead of **6. Chaenorhinum**, read:
6. Chaenorrhinum

p. 175, line 2 from bottom, add to first footnote: Sincere thanks are expressed to Dr. A. Huber-Morath for his advice in connection with the hybrid *Verbascum fruticulosum* Post × *sinuatum* L.

p. 193, line 7 from bottom, read: 6. CHAENORRHINUM (DC. ex Duly) Reichenb.

p. 194, line 1 from top, instead of **1. Chaenorhinum rytidospermum,** read:
1. Chaenorrhinum calycinum (Banks & Sol.) P.H. Davis, Notes Roy. Bot. Gard. Edinb. 36:4 (1978). *Antirrhinum calycinum* Banks & Sol. in Russ., Nat. Hist. Aleppo ed. 2, 2:256 (1794). *C. rytidospermum* (Fisch. & Mey.) Kuprian., etc.

p. 201, line 16 from bottom, instead of *Gratiola monnieri* L., Cent. Pl. 2:no. 120 (1756), read: *Lysimachia monnieri* L., Cent. Pl. 2:9 (1756). *Gratiola monnieri* (L.) L., Syst. Nat. ed. 10, 2:851 (1759); Philcox, Kew Bull. 33:679 (1979).

p. 203, line 9 from bottom, after Lower Jordan Valley, add: ; Moav.

p. 210, line 5 from top, after Gilead, add: , Edom.

p. 211, line 15 from bottom, before Gilead, add: Golan,

p. 213, before **5. Orobanche cernua** Loefl., insert:
4a. Orobanche coerulescens Stephan ex Willd., Sp. Pl. 3:349 (1800); G. Beck, Orobanchaceae, in Engler, Pflanzenreich 96 (IV. 261):118 (1930); Gilli in Hegi, Ill. Fl. Mitteleur. ed. 2, 6(1):481, f.230 (1974).
Parasitic herb, 15–30 cm, more or less arachnoid-woolly. Stems numerous. Spikes dense, abundantly white-woolly. Bracteoles absent; bract ovate-lanceolate, woolly, shorter than flower. Calyx about half as long as corolla, woolly, of 2 ovate-lanceolate segments, each deeply 2–3-dentate. Corolla 15–23 mm, amethyst-violet, tubular, constricted above the insertion of filaments, broadened above and curved forwards. Stamens inserted at about the lower $\frac{1}{3}$ of corolla-tube; filaments glabrous except in their lower part; anthers mostly glabrous. Fl. March.
Hab.: Fields of tomatoes. Together with *O. cernua*. Arava Valley (En Yahav).
Area: E. and C. Asia; C. and E. Europe.

Not known here until recently. Apparently a new introduction. Recorded in the literature as parasitic on species of *Artemisia*.

p. 216, lines 17–19 from bottom, instead of **2. Utricularia gibba**, etc., read:
2. Utricularia exoleta R. Br., Prodr. Fl. Nov. Holl. 430 (1810). *U. gibba* L., Sp. Pl. 18 (1753) subsp. *exoleta* (R. Br.) Taylor, Mitt. Bot. Staatssamm. (München) 4:101 (1961); Taylor, *op. cit.* 197 (1964); Taylor, in Fl. Eur. 3:296. [Plate 366]

p. 216, line 10 from bottom, add: Recently rediscovered in Gonen, Hula Plain.

p. 263, Dipsacus laciniatus L. collected again in Gan-Shemuel (Sharon).

p. 286, in left column of *Synopsis,* instead of 17. Asteropterus, read: 17. Leysera

p. 286, in left column of *Synopsis,* instead of 19. Varthemia, read: 19. Chiliadenus

p. 287, in left column of *Synopsis,* delete: 66. Rhaponticum

p. 289, line 18 from top, instead of **66. Rhaponticum**, read: **65. Serratula**

p. 290, line 14 from bottom, instead of **19. Varthemia**, read: **19. Chiliadenus**

p. 291, line 9 from top, instead of **17. Asteropterus**, read: **17. Leysera**

p. 291, lines 17–18 from top, read:

62. Florets yellow or pink. Cauline leaves, if present, not linear-filiform. Perennial
herbs.   **65. Serratula**

p. 301, before Tribe 3. INULEAE, insert:

5a. HETEROTHECA Cass.

Coarse erect herbs. Leaves alternate. Heads heterogamous, radiate, disposed in
terminal corymbose panicles. Involucral bracts narrow, imbricate in several
rows. Receptacle naked, alveolate. All florets yellow; ray florets 1-seriate, pistillate, fertile; achenes thick, 3-angular-compressed; pappus none or deciduous;
disc-florets tubular, hermaphrodite and fertile or some of them staminate;
achenes flattened with 2-seriate pappus, the outer short, setose, the inner
longer, capillary.
Thirty species in N. America.

1. **Heterotheca subaxillaris** (Lam.) Britton & Rusby, Trans. N. Y. Acad. Sci.
7:10 (1887); Britton & A. Brown, Ill. Fl. U.S. Canad. ed. 2, 3:372, f.4194
(1913). *Inula subaxillaris* Lam., Encycl. Méth. Bot. 3:259 (1789).
Annual or biennial, 50–100 cm, hispid, glandular above. Stem stout,
branched above. Leaves ovate to lanceolate, subentire; the lower petiolate, the
upper subcordate-clasping. Heads numerous. Involucre 6–8 mm, canescent.
Florets yellow; ray-florets 20–28; disc-florets 40–60. Receptacle naked. Achenes
glabrous. Pappus reddish. Fl. summer.
Hab.: Sandy soil, roadsides. Acco Plain, Sharon Plain.
Introduced from the U.S.A. as sand-binder and escaped.

p. 312, line 7 from top, before N. Negev, insert: Mt. Carmel, Judean Mts.; Dan
Valley.

p. 313, line 3 from top, instead of 17. ASTEROPTERUS, etc., read: 17. LEYSE-
RA L.

p. 313, lines 13–17 from top, instead of 1. **Asteropterus leyseroides**, etc., read:
1. **Leysera leyseroides** (Desf.) Maire, Bull. Soc. Hist. Nat. Afr. Nord 20:186
(1929); Bremer, Bot. Not. 131:379 (1978). *Gnaphalium leyseroides* Desf., Fl.
Atl. 2:267 (1799). *Asteropterus leyseroides* (Desf.) Rothm., Feddes Repert.
53:5 (1944). *L. capillifolia* (Willd.) Spreng., Nov. Prov. 25 (1819) *nom.
illeg.*; Boiss., Fl. 3:240; Post, Fl. 2:36. [Plate 521]

p. 314, line 16 from top, add: Arava Valley.

p. 315, instead of lines 9–17 from top, read:

19. CHILIADENUS Cass.
Varthemia apud Boiss., Fl. Or. 3:212 p.p. non DC.

Chamaephytes, woody at base, with a tap-root penetrating the rock. Stems
branched from base, unarmed. Leaves entire. Heads homogamous, discoid.

Involucre imbricate; outer involucral bracts herbaceous, the inner scarious. Receptacle naked, alveolate. Florets hermaphrodite, fertile, tubular, 5-dentate, purple-tipped. Achenes oblong, compressed, striate, hairy all over, glandular near apex. Pappus in 2 rows; the outer of very short free bristles; the inner of numerous long scabrous hairs.

Eight species in the Mediterranean and C. Sahara.

Literature: V. Botschantsev, Additamenta ad floram Aegypti, *Nov. Syst. Pl. Vasc. (Leningrad)* 1964:364–365 (1964). S. Brullo, Taxonomic and nomenclatural notes on the genera *Jasonia* Cass. and *Chiliadenus* Cass. (Compositae), *Webbia* 34:289–308 (1979–1980).

p. 315, lines 18–19 from top, read:

1. **Chiliadenus iphionoides** (Boiss. & Bl.) Brullo, Webbia 34:301 (1979–1980). *Jasonia iphionoides* (Boiss. & Bl.) Botsch., Nov. Syst. Pl. Vasc. (Leningrad) 1964:365 (1964). *Varthemia iphionoides* Boiss. & Bl. in Boiss., Diagn. ser. 2, 3:9 (1856); Boiss., Fl. 3:212; Post, Fl. 2:29. [Plate 525]

p. 315, lines 7–9 from bottom, read:

2. **Chiliadenus montanus** (Vahl) Brullo, Webbia 34:301 (1979–1980). *Jasonia montana* (Vahl) Botsch., Nov. Syst. Pl. Vasc. (Leningrad) 1964:365 (1964). *Chrysocoma montana* Vahl, Symb. Bot. 1:70 (1790). *Varthemia montana* (Vahl) Boiss., Fl. 3:212 (1875); Post, Fl. 2:29. *V. conyzoides* (DC.) Boiss., Diagn. ser. 2, 3:10 (1856). [Plate 526]

p. 318, instead of **3. Pulicaria desertorum** DC. in Decne., read:

3. **Pulicaria incisa** (Lam.) DC., Prodr. 5:479 (1836); Jeffrey & Lack, Taxon 29:695 (1980). *Inula incisa* Lam., Encycl. Méth. Bot. 3:256 (1789). *P. desertorum* DC. in Decne., etc.

p. 324, line 4 from top, add: Judean Mts.; Dan Valley.

p. 324, before 27. ECLIPTA L., insert:

26a. PARTHENIUM L.

Perennial or annual herbs, mostly pubescent or canescent. Leaves alternate. Heads heterogamous, radiate, in corymbose or paniculate inflorescences. Involucre broadly campanulate or hemispheric, of imbricate obtuse nearly equal bracts in 2 or 3 rows. Receptacle with membranous bracts subtending the disc-florets. Florets white or yellow; ray-florets about 5, pistillate, fertile, with short and broad 2-dentate or obcordate ligules; disc-florets tubular, with 5-dentate corolla, sterile. Achenes compressed, margined, broadly obovoid, black. Pappus of 2–3 pales or awns.

About 12 species in America.

1. **Parthenium hysterophorus** L., Sp. Pl. 988 (1753); Britton & A. Brown, Ill. Fl. U.S. Canad. ed. 2, 3:465, f.4432 (1913).

Annual aromatic herb, up to 120 cm, diffusely branched. Leaves ovate to oblong in outline, 1–2-pinnatipartite into linear or lanceolate, dentate or pinnatifid, obtuse segments. Heads about 5 mm broad, in lax panicles on slender peduncles. Involucre saucer-shaped, of 2 mm ovate bracts. Florets white. Achenes 2 mm, broadly obovoid, black, bearing persistent ray-florets on the summit. Fl. summer.

Hab.: Near fish ponds. Beit Shean Valley.

Area: America; introduced into the Old World.

26b. VERBESINA L.

Annual or perennial herbs. Leaves simple, alternate or opposite. Heads heterogamous, radiate. Involucral bracts in 3–4 rows. Receptacle with scales. Rayflorets yellow or white, pistillate, fertile or sterile; disc-florets hermaphrodite, fertile. Achenes flattened laterally and more or less winged. Pappus usually of 2 awns.

Over 100 species of warm-temperate to trop. N. and S. America.

1. **Verbesina encelioides** (Cav.) Benth. & Hook. fil. ex A. Gray in Brewer, S. Watson & Gray, Bot. Calif. 1:350 (1876). *Ximenesia encelioides* Cav., Icon. Descr. 2:60, t.178 (1793); Britton & A. Brown, Ill. Fl. U.S. Canad. ed. 2, 3:489, f.4491 (1913).

Annual, 20–120 cm, much branched, pubescent. Stems erect. Leaves opposite below, alternate above, petiolate, lanceolate to ovate-deltoid, coarsely serrate, green on upper face, canescent on the lower; petioles broadly auriculate. Heads terminal, solitary. Involucral bracts lanceolate, acute, subequal. Rayflorets 10–15, orange-yellow, 3–lobed at apex. Achenes of ray-florets rugose, often wingless; of disc-florets obovate, winged, pubescent, with pappus of 2 subulate awns. Receptacular scales as long as achenes. Fl. summer.

Hab.: Sandy fields and roadsides. W. Negev, Sharon Plain.

Area: N. America.

Adventitious.

p. 325, line 1 from top, add: Lower Jordan Valley.

p. 326, line 16 from top, after N. Negev, add: Mt. Carmel; Dan Valley, Jordan Valley.

p. 339, instead of 30. ORMENIS Cass., read: 30. ORMENIS (Cass.) Cass.

p. 353, line 3 from bottom, before Gilead, add: Upper Jordan Valley;

p. 355, line 8 from top, after Arava Valley, add: Transjordan (Alexander, Notes Roy. Bot. Gard. Edinb. 37:420, 1979).

p. 355, line 17 from bottom, after Post, Fl. 2:69., insert: *S. leucanthemifolius* Poir. var. *vernalis* (Waldst. & Kit.) Alexander, Notes Roy. Bot. Gard. Edinb. 37:403 (1979).

p. 356, line 7 from top, after (1932), insert: *S. glaucus* L. subsp. *glaucus* sensu Alexander, Notes Roy. Bot. Gard. Edinb. 37:412 (1979).

p. 377, instead of 56. NOTOBASIS Cass., read: 56. NOTOBASIS (Cass.) Cass.

p. 388, after line 15 from bottom, insert:

1. Outer involucral bracts with a deflexed appendage ending in a straight spine. Head large, sessile in centre of the leaf-rosette. Florets pink.   **2. S. pusilla**
- Outer involucral bracts acute, appressed. Heads not as above. Florets yellow.
                                                                        1. S. cerinthifolia

p. 389, delete lines 4–11 from top.

p. 389, line 12 from top, instead of **1. Rhaponticum pusillum**, read:
2. Serratula pusilla (Labill.) Dittrich, Candollea 36: 350 (1981). *Rhaponticum pusillum*, etc.

p. 391, before 69. CENTAUREA L., insert:

68a. MANTISALCA Cass.

Annual or perennial herbs close to *Centaurea*, but differing by pappus. Involucral bracts in many rows, imbricate, coriaceous, entire. Receptacle covered with firm thread-like fibres. Pappus of numerous persistent scales around a ring which on one side runs into a single scale.

Ten Mediterranean species.

1. **Mantisalca salmantica** (L.) Briq. & Cavillier, Arch. Sci. Phys. Nat. (Genève) ser. 5, 12:111 (1930). *Centaurea salmantica* L., Sp. Pl. 918 (1753). *Microlonchus duriaei* Spach, Ann. Sci. Nat. Bot. ser. 3, 4:166 (1845); Täckholm, Stud. Fl. 49 (1956); Boiss., Fl. 3:700.

Annual, glabrous except in lower part. Stems virgate. Lower leaves pinnatifid or lyrate, the upper narrowly linear, nearly entire. Heads long-pedunculate, medium-sized, ovate, conical. Involucral bracts wax-yellow, glossy, acute, often reddish-brown at tip. Florets purple, not radiating. Achenes striate, minutely rugose.

Hab.: Sandy wadis. Very rare. N. Negev.

Area: S. Mediterranean (N. Africa, Spain, S. France, Corsica, Sardinia, Sicily, N. Egypt).

Found only once. Recorded as very rare also from N. Egypt.

p. 398, before **14. Centaurea lanulata** Eig, insert:
13a. Centaurea postii Boiss., Fl. 3:688 (1875); Post, Fl. 2:114.

Perennial herb, 30–60 cm, tomentose-canescent. Stems prostrate, leafy. Leaves oblong in outline, pinnatipartite into oblong prickly-pointed lobes. Heads at forks of branches and terminal, sessile or short-pedunculate, subtended by upper leaves. Involucre ovoid, 1 cm, woolly; involucral bracts entire with a terminal spine; spine broad at base, mostly provided with short spin-ules at base, the lower as long as bracts, the upper 2–3 times as long as bracts. Florets pink. Fl. April–June.

    Hab.: Desert; alluvial clay flat. E. Ammon (Azraq).

    Area: W. Irano-Turanian (Transjordan, Syria).

p. 408, line 19 from bottom, after Post, Fl. 2:124, insert: *C. endivia* L. subsp. *pumilum* (Jacq.) Jeffrey, Candollea 34:309 (1979).

p. 415, line 2 from bottom, after (Nazareth), add: , Samaria; S. Golan.

p. 418, line 4 from top, add:

H.W. Lack, *Die Gattung Picris L. s. l. im ostmediterran-westasiatischen Raum*, Ph.D. Thesis, Wien, 1974.

p. 418, lines 20–24 from top, instead of **1. Picris sprengeriana**, etc., read:

1. Picris altissima Del., Fl. Eg. 260 (1813–1814), t.41 f.2 (1826). *P. sprengeri-ana* var. *altissima* (Del.) Aschers. & Schweinf., Mém. Inst. Eg. 2:98 (1887). Boiss., Fl. 3:738 & Post, Fl. 2:131 as *P. sprengeriana*. [Plate 705]

p. 418, after line 8 from bottom, add:

According to Lack (pers. communication), *Hieracium sprengerianum* L., Sp. Pl. 804 (1753), the basionym of *Picris sprengeriana* (L.) Chaix, proved to be a *Crepis*, with pappus of soft non-feathery hairs.

p. 419, line 5 from top, instead of **1. Picris radicata**, read:

1. Picris asplenioides L., Sp. Pl. 793 (1753). *P. radicata*, etc.

p. 419, line 19 from top, after Area: Saharo-Arabian, add: (Palestine, Sinai, Egypt, Libya, Tunisia).

p. 419, lines 20–21 from top, instead of **4. Picris damascena**, etc., read:

4. Picris longirostris Sch. Bip., Mus. Senck. 3:60 (1839); Lack, *op. cit.* 144. *P. damascena* Boiss. & Gaill. in Boiss., Fl. 3:740 (1875); Post, Fl. 2:132; Eig, *op. cit.* 72–75, incl. *P. blancheana* Boiss., Fl. 3:741 (1875); Post, Fl. 2:132. [Plate 708]

    P. longirostris was described from a plant collected by Rüppell in Arabia.

p. 419, lines 8 and 10 from bottom, instead of var. **damascena**, read: var. **longirostris**

p. 420, line 5 from bottom, add: 2n=10.

p. 421, line 12 from top, read: (*Helminthia* Juss. 1789)

p. 426, line 19 from bottom, after Moav, add: , Edom. Edible.

p. 429, line 2 from top, after 7:36 (1946), insert: *T. transjordanicum* van Soest, Acta Bot. Neerl. 19:27, f.4 (1970).

p. 443, before **5. Crepis syriaca**, insert:

4a. Crepis pulchra L., Sp. Pl. 806 (1753); Babc., Crepis 2:661, f.204; Boiss., Fl.
   3:846; Post, Fl. 2:154.

Annual, 50–100 cm. Stem erect, glandular-pubescent, branched above.
Leaves rosulate, oblanceolate, tapering to a petiole, dentate to pinnately
lobed; cauline leaves sessile. Heads medium-sized. Involucre calyculate, cylindrical, 8–11 mm, glabrous; inner involucral bracts narrowly lanceolate, acute,
membranous-margined, spongy-thickened in fruit, ultimately reflexed. Receptacle glabrous. Florets light yellow, hardly longer than involucre; corolla-tube
densely pubescent with fine tortuous hairs. Achenes narrowed at base, beak-
less; marginal achenes 5–6 mm, usually without pappus; inner achenes 4–5
mm, terete, bearing copious pappus. Pappus very fine, white, not deciduous.
2n=8. Fl. May.

Hab.: Volcanic tuff, N. slope, near maquis of *Quercus boissieri*. Golan (Tel
Abu Nida).

Area: Mediterranean, Irano-Turanian and Euro-Siberian.

New to the flora of Palestine. Collected and identified by A. Liston.

4b. Crepis pterothecoides Boiss., Fl. 3:850 (1875); Babc., Crepis 2:671, f.210;
   Post, Fl. 2:155.

Annual, 10–40 cm, glandular-hairy. Stem erect. Leaves rosulate, oblanceolate, tapering to a petiole, acute, dentate to pinnately lobed, with acute lobes;
cauline leaves few, small. Heads medium-sized. Involucre calyculate,
cylindrical-campanulate, 8–14 mm, glandular-pubescent; outer bracts
unequal, lanceolate, acute to acuminate; inner bracts 8–10, lanceolate, acute
to acuminate. Receptacle areolate, glabrous. Florets yellow, longer than
involucre. Achenes all beaked; beak about ⅓ length of to as long as achene
proper. Pappus white, deciduous. 2n=8. Fl. March–May.

Hab.: Rocks or stony places, about 900 m. C. Negev; N. Golan.

Area: W. Irano-Turanian (Palestine, Antilebanon, Djebel Druze).

New to the flora of Palestine. Identified by A. Liston.

p. 451, line 13 from bottom, add: Lower Jordan Valley, Arava Valley.

p. 469, left column, after line 1 from top, insert: (Lactuca) aculeata Boiss. &
   Ky. ex Boiss. 454

LIST OF SPECIES WHICH WERE NOT RECORDED
FROM JORDAN IN PART THREE

The following species are cited in the recently published "List of Jordan vascular plants" by D. M. Al-Eisawi (Mitt. Bot. Staatssamm. München 18:79–182, 15 Dec. 1982).

p.* 8 Plumbago europaea L.
p. 10 Limonium meyeri (Boiss.) O. Kuntze
p. 19 Centaurium spicatum (L.) Fritsch
p. 32 Cressa cretica L.
p. 33 Ipomoea cairica (L.) Sweet
p. 37 Convolvulus lanatus Vahl
p. 37 C. spicatus Peter
p. 38 C. coelesyriacus Boiss.
p. 38 C. auricomus (A. Richard) Bhandari (C. glomeratus Choisy)
p. 42 C. fatmensis Kunze
p. 42 C. palaestinus Boiss.
p. 54 Heliotropium supinum L.
p. 56 H. suaveolens Bieb.
p. 56 H. hirsutissimum Grauer
p. 57 H. maris-mortui Zohary
p. 58 H. arbainense Fresen.
p. 60 Ogastemma pusillum (Coss. & Durieu ex Bonnet & Barratte) Brummitt (Megastoma pusillum Coss. & Durieu ex Bonnet & Barratte)
p. 65 Paracaryum intermedium (Fresen.) Lipsky
p. 69 Arnebia tinctoria Forssk.
p. 69 A. hispidissima (Lehm.) DC.
p. 71 Onosma frutescens Lam.
p. 77 Echium rauwolfii Del.
p. 79 Alkanna galilaea Boiss.
p. 85 Anchusa milleri Willd.
p. 90 Hormuzakia aggregata (Lehm.) Guşul.
p. 91 Myosotis ramosissima Rochel
p. 93 Verbena officinalis L.
p. 104 Teucrium scordioides Schreb.
p. 104 T. spinosum L.
p. 105 T. divaricatum Sieb. ex Heldr.
p. 105 T. leucocladum Boiss.
p. 112 Marrubium alysson L.
p. 125 Moluccella spinosa L.
p. 130 Stachys viticina Boiss.
p. 130 S. longispicata Boiss. & Ky.

* Page on which the name of the species appears in Part 3 of *Flora Palaestina*. To add Jordan to Hab. of respective species.

p. 302 Pluchea dioscoridis (L.) DC.
p. 306 Filago eriocephala Guss.
p. 311 Gnaphalium luteo-album L.
p. 314 Inula crithmoides L.
p. 314 I. graveolens (L.) Desf.
p. 317 Iphiona scabra DC.
p. 317 Pulicaria dysenterica (L.) Bernh.
p. 318 P. inuloides (Poir.) DC.
p. 318 P. incisa (Lam.) DC. (P. desertorum DC.)
p. 322 Ambrosia maritima L.
p. 323 Xanthium strumarium L.
p. 324 X. spinosum L.
p. 324 Eclipta alba (L.) Hassk.
p. 329 Anthemis hyalina DC.
p. 330 A. cornucopiae Boiss.
p. 333 A. zoharyana Eig
p. 335 A. chia L.
p. 337 A. cotula L.
p. 339 Ormenis mixta (L.) Dumort.
p. 348 Chlamydophora tridentata (Del.) Ehrenb. ex Less.
p. 354 Senecio flavus (Decne.) Sch. Bip.
p. 360 Calendula pachysperma Zohary
p. 368 Carlina curetum Heldr. ex Halácsy
p. 371 Atractylis prolifera Boiss.
p. 378 Cirsium phyllocephalum Boiss. & Bl.
p. 379 C. gaillardotii Boiss.
p. 386 Onopordum alexandrinum Boiss.
p. 386 O. jordanicolum Eig
p. 387 Zoegea purpurea Fresen.
p. 390 Amberboa crupinoides (Desf.) DC.
p. 397 Centaurea verutum L.
p. 398 C. aegyptiaca L.
p. 400 C. procurrens Sieb. ex Spreng.
p. 405 Cnicus benedictus L.
p. 406 Scolymus maculatus L.
p. 411 Hyoseris scabra L.
p. 419 Picris asplenioides L. [Picris radicata (Forssk.) Less.]
p. 421 Helminthotheca echioides (L.) Holub
p. 425 Scorzonera syriaca Boiss. & Bl.
p. 428 Chondrilla juncea L.
p. 432 Launaea capitata (Spreng.) Dandy
p. 432 L. spinosa (Forssk.) Sch. Bip. ex O. Kuntze
p. 436 Lactuca serriola L.
p. 445 Crepis micrantha Czerep.
p. 446 C. aculeata (DC.) Boiss.
p. 447 C. senecioides Del.

ADDENDA AND CORRIGENDA TO PART FOUR

p. 28, after line 7 from bottom, insert:

Boulos, Jallad & Laham (Bot. Not. 128:369, 1975) reported *Colchicum croci-folium* Boiss. from Ras-en-Naqb in S. Edom (coll. in fruit). Since the area of *C. crocifolium* Boiss. (1844) lies further to the northeast, one may presume that the plant in question was *C. tuviae* Feinbr. (1953) which resembles *C. crocifolium* in leaves. However, identification of *Colchicum* of this group is impossible without flowers (note the densely fringed lamellae at base of perianth-lobes in *C. tuviae*).

p. 31, insert at bottom of page:

The binomial *Colchicum jordanicolum* (Regel) Stefanoff (*op. cit.* 46, 1926) has been inadvertently omitted from the account of *Colchicum* (pp. 26–31). The following passage from the paper on *Colchicum* by Feinbrun (1953) concerns this binomial.

"The flowers of one specimen collected in Jerusalem by Z. Bumstein had leaves of perigonium free down to their base and lacking stamens altogether (Plate III). The specimen is especially interesting in connection with the puzzling *C. jordanicolum* (Regel) Stefanoff which has never been collected since Regel. Neither Stefanoff nor myself were able to examine Regel's type specimen. According to the description, *C. jordanicolum* is synanthous and possesses free perigonium leaves, being therefore assigned to *Merendera* by Regel. The existence of this species is doubtful, and its description may have been made from an anomalous specimen like our specimen of *C. hierosolymitanum*, but also another species (*C. Ritchii?*) might be involved in this case" (Feinbr., Palest. Jour. Bot. Jerusalem ser., 6:86 & 95, 1953).

p. 32, before 1. **Androcymbium palaestinum** insert:

Literature: W. Greuter, Contributiones floristicae austro-aegaeae 10–12, *Candollea* 22:242–252 (1967).

p. 87, line 4 from top, instead of: Subsp. **decaisnei** (C. Presl) Kollmann comb. nov. *A. decaisnei* C. Presl, Bot., read:

13a. **Allium decaisnei** C. Presl, Bot.

p. 87, before 14. **Allium albotunicatum,** insert:

On page 87 we treated *A. decaisnei* C. Presl (1844) as a subspecies of *A. stamineum* Boiss. (1859). During the printing of the text it was pointed out to us by Dr Irene Gruenberg-Fertig that the epithet *A. decaisnei* (1844) has priority over the epithet *A. stamineum* (1859). It seems, however, undesirable to sink the species *A. stamineum*, an E. Mediterranean taxon well known and well distinguished in Palestine, in *A. decaisnei* as a subspecies, thus obscuring its image. We prefer therefore to use the epithet *A. decaisnei* C. Presl as a name for the vicarious Irano-Turanian (Sinai and the Negev) taxon at species rank. Though being a weak species and very close to *A. stamineum*, it differs from it in several morphological features and in having a distinct distribution area.

p. 91, before **17. Allium scorodoprasum**, insert:

16a. Allium trachycoleum Wendelbo (Sect. *Allium*), Bot. Not. 122:35, f.1H (1969); Wendelbo, Fl. Iran. 76:t.5 f.78 (1971); Kollmann, Israel Jour. Bot. 26:133–134, f.2 (1977).

Bulb about 1.5–2 cm in diameter; outer tunics blackish, breaking up into fibres; bulblets 0.7–1 cm, yellowish-brown, shiny. Stem (50–)60–100 cm. Leaves 3–5, 3–7 mm broad, flat, canaliculate, with scabridulous veins, sheathing the lower ⅓ of stem; sheaths scabridulous Spathe caducous. Umbel spherical, 3.5–5(–6) cm in diameter, dense. Flowering pedicels unequal, 1–3 cm, with laciniate bracteoles at base. Perianth campanulate; segments 4–5(–6) mm, white or whitish with a green midvein; outer segments more or less narrowly elliptic-ovate, usually more or less scabridulous-papillose on back and margins, obtuse; the inner ovate, glabrous, subtruncate at apex. Filaments somewhat longer than segments; inner filaments 3-cuspidate; median cusp of inner filaments shorter than the slightly contorted lateral cusps and about ½ length of basal lamina; anthers brown. Style exserted. Capsule 3 mm, globose. 2n=32. Fl. July.

Hab.: Basalt slopes, clearings in maquis. Golan.

Area: Irano-Turanian (Mt. Hermon, S. and S.E. Anatolia, Iraqi Kurdistan). Close to *Allium qaradaghense* Feinbr.

p. 176, before **11. Aegilops geniculata**, insert:

The name *Aegilops biuncialis* Vis. (1842) is validly published according to Article 44 of International Code of Botanical Nomenclature 1961, since figure 2 of table 1 by Visiani clearly shows the essential characters of the species. (See also P.A. Gandilian, Determiner to the wheats, Aegilopses, ryes and barleys,p. 100, Acad. Sci. Armenian SSR Erevan, 1980.)

p. 178, add the following list of literature on *Triticum dicoccoides*:

A. Aaronsohn & G.A. Schweinfurth, Die Auffindung des wilden Emmers (*Triticum dicoccum*) in Nordpalästina, *Altneuland* 3:216 (1906). G.A. Schweinfurth, Die Entdeckung des wilden Urweizens in Palästina, *Ann. Serv. Antiqu. Egypte* 5:193 (1907); Über die von Aaronsohn ausgeführten Nachforschungen nach dem wilden Emmer (*T. dicoccoides*), *Ber. Deutsch. Bot. Ges.* 26:310 (1908). A. Aaronsohn, Über die in Palästina und Syrien wildwachsend aufgefundenen Getreidearten, *Verh. Zool.-Bot. Ges. Wien* 59:485 (1909); Agricultural and botanical explorations in Palestine, *U.S. Dept. Agric. Bull.* 180 (1910). A. Schulz, Über eine neue spontane *Eutriticum*-Form: *T. dicoccoides* Koern. forma *Straussiana*, *Ber. Deutsch. Bot. Ges.* 31:226 (1913); Über eine Emmerform aus Persien und einige andere Emmerformen, *Ber. Deutsch. Bot. Ges.* 33:233 (1915). C.A. Flaksberger, Contribution to the study of wild *monococcum* and *dicoccum* and their phylogenetic connections, *Bull. Appl. Bot. Pl.-Breed. (Leningrad)* 16 (3):201 (1926). A. Eig, *Bull. Inst. Agr. Nat. Hist. Tel-Aviv* 6:70 (1927). M.M. Jakubziner, A contribution to the knowledge of wild wheat in Transcaucasia, *Bull.*

Appl. Bot. Pl.-Breed. (*Leningrad*) Ser. 5, 1:147 (1932); The wheats of Syria, Palestine and Transjordania, cultivated and wild, *Bull. Appl. Bot. Pl.-Breed.* (*Leningrad*) Suppl. 53 (1932). M.G. Tumanyan, Die wildwachsenden Verwandten der kultivierten Weizen in Armenien, *Zeitschr. Zücht.* A 20:352 (1935).

p. 193, line 22 from top, after (1939)., add: *B. madritensis* L. var. *delilei* Boiss., Fl. 5:649 (1884).

p. 217, before 29. GASTRIDIUM Beauv., insert:

28a. AGROSTIS L.

Perennials or annuals. Leaves flat or setaceous; ligule membranous. Inflorescence a panicle. Spikelets 1-flowered, hermaphrodite. Rachilla disarticulating above the glumes. Glumes persistent, longer than floret; lemma $\frac{2}{3}$–$\frac{3}{4}$ length of glumes, truncate; palea shorter than lemma. Stamens 3.

Differs from *Polypogon* in persistent glumes.

1. **Agrostis stolonifera** L., Sp. Pl. 62 (1753). *A. alba* auct. non L.

Stoloniferous perennial, 80–100 cm. Panicle 15–30 cm, lax, narrow, contracted, branches spreading only at anthesis. Spikelets 2–3 mm, greenish to purplish. Glumes lanceolate, acute, nearly equal; lemma truncate, awnless. Anthers 1–1.5 mm. Fl. June.

Hab.: Wadis streaming in winter. Upper Galilee; Golan.

Area: Eurosiberian-Boreoamerican and Mediterranean.

p. 248, line 17 from top, after (1906)., add: *P. sylvicola* Guss., Enum. Pl. Inar. 371 (1854).

p. 248, before **6. Poa annua** L., insert:

5a. Poa moabitica Bor, Notes Roy. Bot. Gard. Edinb. 31:396 (1972).

Perennial, rhizomatous, up to 110 cm. Leaf-blades flat; ligule 1 mm, truncate, covered on back with a very short furry indumentum. Panicle up to 35 cm × 5 cm; branches 4–5 at nodes. Spikelets 4–4.5 mm, 3–4-flowered, elliptic in outline; glumes nearly equal, shorter than spikelet, 1–3-veined; lemma 2.5 mm, glabrous, 5-veined, with intermediate lateral vein not particularly marked, scabridulous on veins, particularly on keel; callus with a few short strands of wool; anthers 1.5 mm.

Hab.: Moav, ?on the plains (W.A. Hayne).

Area: E. Mediterranean. Endemic.

Species conspicuous in its truncate ligule, densely and shortly woolly on back.

p. 371, line 7 from bottom, before *C. echinata* var. *leersii*, insert:

C. leersii F.W. Schultz, Flora (Regensb.) 53:459 (1870) & 54:25 t.2 (1871) p. p. non Willd., Prodr. Fl. Berol. 28 (1787).

LIST OF BIBLICAL NAMES OF PLANTS
CITED IN FLORA PALAESTINA*

The Part of *Flora Palaestina* in which the name appears is indicated in bold-faced type, followed by the page number.

Acacia Mill. **2**:26 (Isaiah xli:19) — *shitta* שִׁטָּה; (Exodus xxvi:15 and elsewhere) — *shittim* שִׁטִּים

Amygdalus communis L. **2**:22 (Genesis xxx:37) — *luz* לוּז; (Ecclesiastes xii:5 and elsewhere) — *shaqed* שָׁקֵד

Atriplex L. **1**:145 (Job xxx:4) — *malluaḥ* מַלּוּחַ

Balanites aegyptiaca (L.) Del. **2**:258 (Genesis xliii:11 and elsewhere) — *ẓori* צְרִי

Capparis spinosa L. **1**:243 (Ecclesiastes xii:5) — *aviyona* אֲבִיּוֹנָה; (Talmud) — *ẓalaf* צָלָף

Ceratonia siliqua L. **2**:32 (Talmud) — *ḥaruv* חָרוּב

Conium maculatum L. **2**:406 (Deuteronomy xxxii:32 and elsewhere) — *rosh* רֹאשׁ, רוֹשׁ

Coriandrum sativum L. **2**:402 (Exodus xvi:31; Numbers xi:7) — *gad* גַּד

Crocus sativus L. **4**:130 (Song of Solomon iv:14) — *karkom* כַּרְכֹּם

Cupressus sempervirens L. **1**:19 (1 Kings ix:11) — *berosh* בְּרוֹשׁ; (Isaiah xli:19 and elsewhere) — *te'ashur* תְּאַשּׁוּר

Cyperus papyrus L. **4**:358 (Exodus ii:3; Job viii:11; Isaiah xviii:2, xxxv:7) — *gome* גֹּמֶא

Eruca sativa Mill. **1**:314 (2 Kings iv:39) — *orot* אֹרֹת

Ficus carica L. **1**:38 (Genesis iii:7; Song of Solomon ii:13 and elsewhere) — *te'ena* תְּאֵנָה

Ficus sycomorus L. **1**:38 (1 Kings x:27; Isaiah ix:9 and elsewhere) — *shiqma* שִׁקְמָה

Helichrysum sanguineum (L.) Kostel. **3**:312. (Popular name) — Blood of the Maccabees דַּם הַמַּכַּבִּים

Hordeum L. **4**:179 (Deuteronomy viii:8 and elsewhere) — *se'ora* שְׂעֹרָה

Juniperus phoenicea L. **1**:20 (Jeremiah xvii:6) — *ar'ar* עַרְעָר

Lycium L. **3**:159 (Judges ix:14-15; Psalms lviii:10) — *atad* אָטָד

Majorana syriaca (L.) Rafin. **3**:153 (Exodus xii:22; 1 Kings v:13; Psalms li:9 and elsewhere) — *ezov* אֵזוֹב

Malva L. **2**:317 (Job vi:6) — *hallamut* חַלָּמוּת

Mandragora autumnalis Bertol. **3**:167 (Genesis xxx:14-16; Song of Solomon vii:14) — *dudaim* דּוּדָא (דּוּדָאִים)

Mesembryanthemum forsskalii Hochst. ex Boiss. **1**:77 (Numbers xxiv:6; Proverbs vii:17) — *ahalim* אָהָל (אֲהָלִים)

Myrtus communis L. **2**:372 (Isaiah xli:19 and elsewhere) — *hadas* הֲדַס

Nigella ciliaris DC. **1**:195 (Isaiah xxviii:25, 27) — *qeẓaḥ* קֶצַח

* Hebrew names of plants treated in Part Four appear only in this list.

Olea europaea L. 3:14 (Judges ix:8-9; Jeremiah xi:16; Amos iv:9; Psalms cxxviii:3) — *zayit* זַיִת

Phoenix dactylifera L. 4:330 (Psalms xcii:13 and elsewhere) — *tamar* תָּמָר

Phragmites australis (Cav.) Trin. 4:270 (1 Kings xiv:15 and elsewhere) — *kane* קָנֶה

Pinus halepensis Mill. 1:18 (Isaiah xli:19 and elsewhere) — *ez shemen* עֵץ שֶׁמֶן; (Isaiah xliv:14) — *oren* אֹרֶן

Pistacia L. 2:298 (Genesis xxxv:4; Ezekiel vi:13 and elsewhere) — *ela* אֵלָה

Platanus orientalis L. 2:1 (Genesis xxx:37 and elsewhere) — *armon* עַרְמוֹן

Populus euphratica Oliv. 1:29 (Ezekiel xvii:5) — *zafzafa* צַפְצָפָה

Quercus L. 1:33 (Genesis xxxv:8; Hosea iv:13; Amos ii:9) — *allon* אַלּוֹן; *elon* אֵלוֹן

Retama raetam (Forssk.) Webb 2:48 (1 Kings xix:4-5; Job xxx:4 and elsewhere) — *rotem* רֹתֶם

Ricinus communis L. 2:269 (Jonah iv:6-7, 9-10) — *qiqayon* קִיקָיוֹן

Salix acmophylla Boiss. 1:26 (Isaiah xliv:4 and elsewhere) — *aravim* עֲרָבָה (עֲרָבִים)

Sarcopoterium spinosum (L.) Spach 2:15 (Isaiah xxxiv:13 and elsewhere) — *sirim* סִירָה (סִירִים)

Scirpus L. 4:348 (Isaiah ix:13; lviii: 5 and elsewhere) — *agmon* אַגְמוֹן

Solanum incanum L. 3:166 (Micah vii:4; Proverbs xv:19) — *hedeq* חֵדֶק

Styrax officinalis L. 3:13 (Genesis xxx:37; Hosea iv:13) — *livne* לִבְנֶה

Tamarix L. 2:351 (Genesis xxi:33 and elsewhere) — *eshel* אֵשֶׁל

Triticum L. 4:177 (Deuteronomy viii:8 and elsewhere) — *hitta* חִטָּה

Typha L. 4:344 (Exodus ii:3; Isaiah xix:6; Jonah ii:6) — *suf* סוּף

Urtica L. 1:39 (Isaiah lv:13; Ezekiel ii:6; Zephania ii:9) — *sirpad* סִרְפָּד

INDEX TO PART FOUR

Accepted names are in roman type, synonyms in italics; also italicized are names of taxa regarded unworthy of independent status and included within an accepted species or subspecies. The main page reference for each entry is given first, followed by subsidiary references. Subspecies are listed alphabetically under their species, and varietal names follow. Names of plants not belonging to the flora of Palestine are marked by an asterisk.

432

INDEX TO FAMILIES AND GENERA
OF PARTS ONE TO FOUR

The Part of *Flora Palaestina* in which each name appears is indicated here in bold-face type, followed by the page number. Family names are in capitals, family and generic synonyms in italics. Names of genera not belonging to the flora of Palestine are marked by an asterisk.

ERRATA

p. 135, line 15, instead of *C. pallasi* (Baker) Boiss., read: *C. pallasii* Goldb.

p. 195, line 13 from bottom, instead of 13. BOISSIERA Hochst. ex Steud., read: 13. BOISSIERA Hochst. ex Ledeb.*

p. 196, line 4, instead of Steud., Syn. Pl. Glum. 1:200 (1854), read: Ledeb., Fl. Ross. 4:405 (1852).

p. 402, line 8 from bottom, instead of Duly, read Duby

p. 416, line 2 from bottom, instead of Nigella ciliaris DC., read: Nigella sativa L.

p. 417, line 4, after Trin., add: ex Steud.

p. 421, right column, after line 19 add: subsp. *arenaria 214*

p. 427, left column, line 12 from bottom, after Maw, add: ex Boiss.

p. 434, left column, line 18 from bottom, read: *Ixia bulbocodium* (L.) L. *128*

p. 438, left column, line 7 from bottom, read: *dinsmorei* (Schlechter) Baumann

p. 440, right column, above line 15 from bottom, add: var. ciliatum (Boiss.) Halácsy 224

* Our thanks to Dr G. Zijlstra of the University of Utrecht for informing us of the correct citation.

MAPS

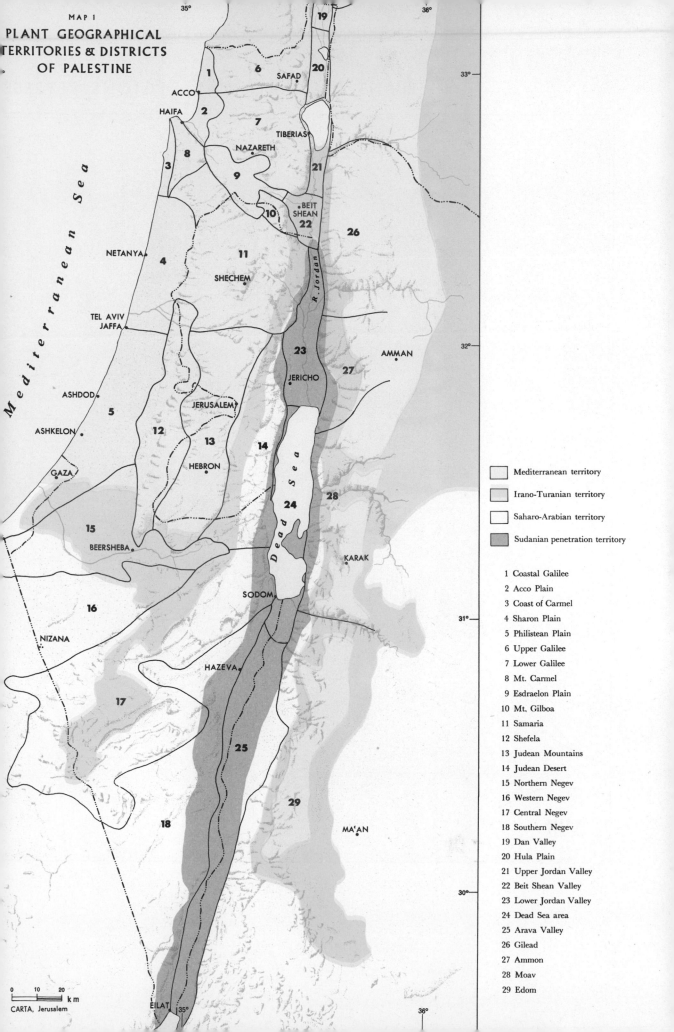

MAP I

PLANT GEOGRAPHICAL
TERRITORIES & DISTRICTS
OF PALESTINE

Mediterranean Sea

Dead Sea

R. Jordan

35°   36°
33°
32°
31°
30°

Cities/places labeled:
SAFAD, ACCO, HAIFA, TIBERIAS, NAZARETH, BEIT SHEAN, NETANYA, SHECHEM, TEL AVIV JAFFA, AMMAN, ASHDOD, JERUSALEM, ASHKELON, JERICHO, GAZA, HEBRON, BEERSHEBA, SODOM, KARAK, NIZANA, HAZEVA, MA'AN, EILAT

Legend:

▢	Mediterranean territory
▨	Irano-Turanian territory
▢	Saharo-Arabian territory
▨	Sudanian penetration territory

1 Coastal Galilee
2 Acco Plain
3 Coast of Carmel
4 Sharon Plain
5 Philistean Plain
6 Upper Galilee
7 Lower Galilee
8 Mt. Carmel
9 Esdraelon Plain
10 Mt. Gilboa
11 Samaria
12 Shefela
13 Judean Mountains
14 Judean Desert
15 Northern Negev
16 Western Negev
17 Central Negev
18 Southern Negev
19 Dan Valley
20 Hula Plain
21 Upper Jordan Valley
22 Beit Shean Valley
23 Lower Jordan Valley
24 Dead Sea area
25 Arava Valley
26 Gilead
27 Ammon
28 Moav
29 Edom

0   10   20
k m

CARTA, Jerusalem

MAP 2
PLANT GEOGRAPHICAL REGIONS REPRESE

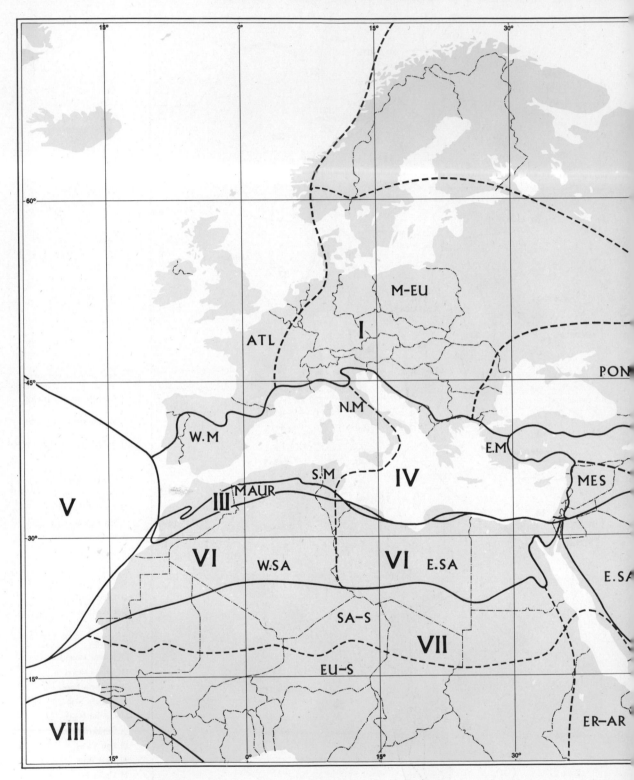

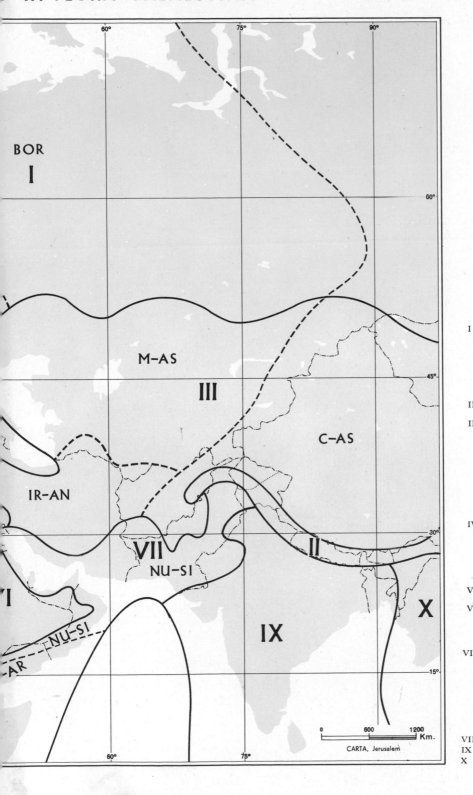

I   Euro-Siberian region

West Euro-Siberian subregion
ATL   *Atlantic province*
BOR   *Boreal province*
M-EU   *Medio-European province*
PON   *Pontic province*

II   Sino-Japanese region

III   Irano-Turanian region

West Irano-Turanian subregion
MAUR   *Mauritanian steppes province*
MES   *Mesopotamian province*
IR-AN   *Irano-Anatolian province*
M-AS   *Medio-Asiatic province*

East Irano-Turanian subregion
C-AS   *Centro-Asiatic province*

IV   Mediterranean region

W.M   *West Mediterranean subregion*
N.M   *North Mediterranean part*
S.M   *South Mediterranean part*
E.M   *East Mediterranean subregion*

V   Macaronesian region

VI   Saharo-Arabian region

W.SA   *West Saharo-Arabian subregion*
E.SA   *East Saharo-Arabian subregion*

VII   Sudanian region

West Sudanian subregion
SA-S   *Sahelo-Sudanian province*
EU-S   *Eu-Sudanian province*
E.S   *East Sudanian subregion*
NU-SI   *Nubo-Sindian province*
ER-AR   *Eritreo-Arabian province*

VIII   Guineo-Congolese region
IX   Indian region
X   Malaysian region

CARTA, Jerusalem

0   600   1200 Km.